12th Hungarian Conference on Materials Science (HMSC12)

AA007627

IOP Conference Series: Materials Science and Engineering
Volume 903

Balatonkenese, Hungary
13 - 15 October 2019

Editors:

K. Majlinger
P.J. Szabo

ISBN: 978-1-7138-3186-0
ISSN: 1757-8981

Printed from e-media with permission by:

Curran Associates, Inc.
57 Morehouse Lane
Red Hook, NY 12571

Some format issues inherent in the e-media version may also appear in this print version.

Institute of Physics
Dirac House, Temple Back
Bristol BS1 6BE UK

Phone: 44 1 17 929 7481
Fax: 44 1 17 920 0979

techtracking@iop.org

Additional copies of this publication are available from:

Curran Associates, Inc.
57 Morehouse Lane
Red Hook, NY 12571 USA
Phone: 845-758-0400
Fax: 845-758-2633
Email: curran@proceedings.com
Web: www.proceedings.com

TABLE OF CONTENTS

NOVEL CONCEPTS FOR ESTABLISHING EXPERT SUPPORT SYSTEMS TO INVESTIGATE THE DEFECT OCCURRING IN METALLURGICAL PHASES IN THE TECHNOLOGY OF ISD DUNAFERR ZRT. 1
K Wizner, A Kovári

INVESTIGATION OF THE MELTING POINT DEPRESSION OF COPPER NANOPARTICLES 7
L Somlyai-Sipos, D Janovszky, A Sycheva, P Baumli

EFFECT OF THE SAMPLING METHOD ON THE RESULTS OF MELT QUALITY ASSESSMENT OF ALUMINUM ALLOYS WITH COMPUTED TOMOGRAPHY 14
G Gyarmati, Gy Fegyverneki, M Tokár, T Mende

DEVELOPMENT OF NEW CLADDING TYPES FOR NUCLEAR FUEL 22
Z Hózer, T Novotny, E Perez-Feró, M Horváth, A Pintér Csordás, P Szabó, L Illés, M Schyns, R Delville, D Kim

THE SIMULATION OF THE STEADY-STATE CONDITIONS OF ALUMINIUM DIRECT-CHILL CASTING 30
Daniel Molnar, Viktor Karpati, Imre Budavari

THE EFFECT OF PRIMARY PHASE GRAIN REFINEMENT OF AL-SI ALLOYS ON HEAT CRACKING 38
M. Bubenko, D. Molnar, T. Monika, Gy. Fegyverneki

INVESTIGATION OF THE FORMABILITY OF POLYMERIC ELEMENTS IN DIFFERENT SOFTWARE ENVIRONMENTS FOR THE AUTOMOTIVE INDUSTRY 46
Péter Zoltán Kovács, Bence Tóka

INFLUENCE OF 3D PRINTING PARAMETERS 53
P Ficzere, N L Lukács

THE EFFECT OF TITANIUM DIOXIDE ON THE MOISTURE ABSORPTION OF POLYAMIDE 6 PREPARED BY T-RTM 59
Orsolya Viktória Semperger, András Suplicz

MECHANICAL PROPERTY EVALUATION OF CLOSED AND OPEN-CELL FOAM STRUCTURES WITH FINITE ELEMENT METHOD 67
K Kulcsár, J Kónya, I Zsoldos

DETECTION OF NEUTRON IRRADIATON INDUCED DEGRADATION OF REACTOR STEEL BY MAGNETIC METHOD 75
G Vértesy, A Gasparics, I Uytdenhouwen, I Szenthe, F Gillemot, R Chaouadi

SEMI-CONTINUOUS CASTING AND MICROSTRUCTURE INVESTIGATION OF THE ALSI12 ALLOY 81
Agota Kazup, Viktor Karpati, Balazs Hegedus, Zoltan Gacsi

FRACTURE MECHANICAL ANALYSES OF HIGH STRENGTH STEELS APPLYING EXPERIMENTS AND SIMULATION 89
Zs Koncsik, Zs Lukács

MEASUREMENT POSSIBILITIES OF WATER CONTENT IN POLYAMIDE 97
D V Terjék, E Kókai

INVESTIGATION OF THE TRIBOLOGICAL PROPERTIES OF NANO-SCALED ZRO2 AND CUO ADDITIVE IN AUTOMOTIVE LUBRICANTS 102
Á. D. Tóth, J. Knaup

EXAMINATION OF WELDED JOINT OF TITANIUM ALLOY USED IN ORAL SURGERY 110
J Kónya, K Kulcsár, H Hargitai

SEMI-CONTINUOUS CASTING OF ALUMINIUM ALLOYS 116
Viktor Karpati, Balazs Hegedus, Agota Kazup, Tibor Ferenczi, Daniel Molnar, Zoltan Gacsi

CHARACTERIZATION OF CHEMICALLY FOAMED POLY(LACTIC ACID) 122
K Litauszki, Á Kmetty

DEFORMABILITY TESTS OF PURE NIOBIUM 129
V Mertinger, I Aviles Santillana, M Benke, A Gallifa Terricabras, A Hlavács, D Janovszky, V Kárpáti, Gy Krállics, T Mikó, A B Palotás

ENHANCING THE FIREWORK ALGORITHM ECOSYSTEM FOR THE RECONSTRUCTION OF THE HTC FUNCTION 137
Z Fried, I Felde, S Szénási

THE EFFECT OF MICROSTRUCTURE ON THE DYNAMIC MECHANICAL PROPERTIES OF CARBON FIBER AND CARBON NANOTUBE REINFORCED MULTI-SCALE COMPOSITES WITH A POLYAMIDE 6 MATRIX 145
R Petrény, L Mészáros

MODELLING OF SEGREGATION DURING INJECTION MOULDING 153
Dorottya Réka Fris, Ferenc Szabó

CHARACTERISATION OF INHOMOGENEOUS PLASTIC DEFORMATION OF ALMG SHEET METALS DURING TENSILE TESTS 161
Sz Szalai, D Harangozó, I Czinege

MODELING THE DEFORMATION BEHAVIOR OF POLYMER SANDWICH STRUCTURES WITH INHOMOGENEOUS CORE 169
Laszlo Takacs, Ferenc Szabó

DEVELOPMENT OF HIGH POWER FEMTOSECOND LASER MICROSTRUCTURES ON AUTOMOTIVE STAINLESS STEEL 177
Ferenc Tajti, Miklós Berczeli

OPTIMIZATION OF MATERIAL REMOVAL PARAMETERS BY FEMTOSECOND LASER PULSES 185
Miklós Berczeli, Ferenc Tajti

PRODUCING COMPRESSIVE STRESSES IN THE SURFACE LAYER OF SPHEROIDAL GRAPHITE CAST IRON BY LASER HARDENING 193
Anna Maloveczky, Ádám Filep, Gábor Buza, Ádám Vida

APPLICATION OF RIETVELD REFINEMENT IN ORIENTATED MATERIAL STRUCTURES 199
E Nagy, F Kristály, V Mertinger

THE EFFECTS OF THE STEEL'S SURFACE QUALITY ON THE PROPERTIES OF ANTI-CORROSION COATINGS 207
B. Herbath, K. Kovacs

MICROSTRUCTURE EXAMINATION OF ADDITIVE MANUFACTURED 316L STEEL 215
Kemény Dávid Miklós, Pammer Dávid, Kovács Dorina

CHARACTERIZATION OF SHORT FIBER-REINFORCED POLYLACTIC ACID COMPOSITES PRODUCED WITH FUSED FILAMENT FABRICATION (FFF)...................................... 221
Tóth Csenge, Krisztián Norbert Kovács

SIMULATION OF HOT ROLLING BY CELLULAR AUTOMATA... 229
T Hegyes, P Barkóczy

DETERMINING THE OPTIMUM HEATING TIME OF SMALL SIZED TEST SPECIMEN MADE FROM WELDABLE MILD STEEL 235
F Tancsics, T Ibriksz

THE EFFECT OF DIFFERENT GRAIN SIZES AND HEAT INPUT ON THE GAS PRESSURE INSIDE ARTIFICIAL RESIN-BONDED SAND CORES... 243
L Mádi, I Budavári, L Varga

WETTING OF NON-REACTIVE AND REACTIVE METALLIC SUBSTRATES BY BRAZING LIQUIDS.. 251
Z Weltsch

LIGHTWEIGHT MANUFACTURING OF AUTOMOTIVE PARTS.. 258
T Wadas, M Tisza

CONSERVATION PURPOSE MATERIAL TESTING OF CORROSION PRODUCTS ON OUTDOOR BRONZE STATUES 269
Szilvia Szilvia, Melinda Nagy, Laura Juhász, Tamás Bubonyi, Barkóczy Péter, György Forgács, Eszter Szatmáriné Bakonyi

COMPARISON OF CT AND METALLOGRAPHIC METHOD FOR EVALUATION OF MICROPOROSITIES OF DYE CAST ALUMINUM PARTS 277
Tamás Bubonyi, Péter Barkóczy, Zoltán Gácsi

EFFECT OF THE HOMOGENIZATION AND COLD DEFORMATION ON THE MECHANICAL PERFORMANCE OF AL8006 ALUMINIUM ALLOY.. 283
J Pázmán, K Fehér, V Gonda, B Verö

MAGNETO-ACOUSTIC INVESTIGATION ON STEEL SAMPLES.. 291
O. Szabados, R. Baki, Zs, Csincsi, J. Molnar, Á. Pámer, P. Szabó, G. Por

DEFORMATION-SOFTENING IN ULTRAFINE-GRAINED MATERIALS 298
Jenö Gubicza, Nguyen Quang Chinh

PREPARATION OF SOFT MAGNETIC COMPOSITE FROM FE-6.9WT%SI BY DIFFERENT HEAT TREATMENT STRATEGIES. 306
Bence Kocsis, Lajos Károly Varga, Ibolya Zsoldos

ROOT CAUSE ANALYSIS OF METALLURGICAL DEFECTS IN CONTINUOUS CAST STEEL SLABS AT ISD DUNAFERR ZRT. 312
K Wizner, A Kovári

ON THE NECESSITY OF DOMESTIC RESEARCH ON NUCLEAR MATERIALS SCIENCE 320
Peter Trampus

MECHANICAL PROPERTIES OF ALSI10MNMG MATRIX SYNTACTIC FOAMS FILLED WITH LIGHTWEIGHT EXPANDED CLAY PARTICLES ... 328
A Szlancsik, D Kincses, I N Orbulov

JOINING ALUMINIUM AND POLY(LACTIC ACID) SPECIMENS BY LASER BEAM – A FEASIBILITY STUDY ... 334
T Temesi, T Czigany

INFLUENCE OF PACK CEMENTATION TIME ON THE MICROSTRUCTURE OF CU NANOFOAMS PROCESSED BY DEALLOYING ... 340
Jenei Péter, Gigap Han, Pham Tran Hung, Heeman Choe, Jeno Gubicza

FLAME RETARDANCY OF RECYCLED PET FOAM ... 347
V A Szabó, G Dogossy

INVESTIGATION OF THE INTERFACIAL ADHESION OF GLASS BEAD-FILLED MULTICOMPONENT INJECTION MOULDED COMPOSITES ... 355
A Suplicz, R Boros, O V Semperger

FATIGUE TESTS OF ZINC ALUMINIUM MATRIX SYNTACTIC FOAMS FILLED WITH EXPANDED PERLITE ... 362
A Kemény, B Katona, N Movahedi, T Fiedler

INVESTIGATION OF VVER-1200 REACTOR PRESSURE VESSEL'S MATERIAL ... 368
Dorina Kovács, Dávid Kemény

NEUTRON IRRADIATION OF ZIRCONIUM ALLOYS IN THE BUDAPEST RESEARCH REACTOR ... 373
I Szenthe, F Gillemot, M Horvath, Z Hózer

MANUFACTURING AND INVESTIGATION OF ALUMINIUM MATRIX BIMODAL METAL FOAMS ... 380
B Leveles, A Kemény, I N Orbulov

EFFECT OF BERACHA-I TYPE COLOR ETCHANT ON THE FERRITE PHASE IN DIFFERENT TYPE FE-C ALLOYS ... 387
József Bálint Renkó, Attila Bonyár, Péter János Szabó

NUMERICAL MODELLING OF THE THINNING BEHAVIOUR OF SHEET METAL PARTS ... 393
V Gál, Zs Lukács

A COMPARATIVE ANALYSIS OF CHARACTERISTICS OF CUTTING FORCES AT MICRO-MILLING OF HARDENED STEELS ... 399
B Z Balázs, M Takács

PREPARATION AND CHARACTERIZATION OF POLY(LACTIC ACID)/BOEHMITE ALUMINA COMPOSITES FOR ADDITIVE MANUFACTURING ... 407
L Lendvai, I Fekete

THE INFLUENCE OF THE BASE MATERIAL - FILLER METAL PAIRING ON THE FATIGUE CRACK PROPAGATION OF 700 MPA STRENGTH CATEGORY WELDED JOINTS ... 414
J Lukács, Hfh Mobark

INVESTIGATION OF THE ENERGY ABSORPTION PROPERTIES OF CROSS-LINKED POLYETHYLENE FOAMS ... 422
M Tomin, Á Kmetty

EXAMINATION OF CORRELATION BETWEEN THE GRANULOMETRIC PROPERTIES OF MOLDING AND CORE SAND MIXTURES AND THEIR PRODUCTION PARAMETERS..................... 430
Henrietta Hudák, Dr. László Varga

Author Index

IOP Conf. Series: Materials Science and Engineering **903** (2020) 011001 doi:10.1088/1757-899X/903/1/011001

Preface

The series "Hungarian Conference and Exhibition on Materials Science, Testing and Informatics" (later simply "Hungarian Conference on Materials Science") was intended to provide a forum where Hungarian and foreign scientists and research groups interested in metals and alloys, silicates, polymers and composites get an opportunity to exchange and publish ideas and establish new and integrated partnerships. The 12th Hungarian Conference on Materials Science was held at the Balaton lakeshore in Balatonkenese, October 13-15, 2019.

The winner of the photo competition of the 12th Hungarian Conference on Materials Science was Dr. Márton Benke, associate professor, Institute of Physical Metallurgy, Metal Forming and Nanotechnology, Faculty of Materials Science and Engineering, University of Miskolc

There can be seen the thermoelastic martensitic microstructure of a Cu-Al-Ni-Mn shape memory alloy

The organisers present this volume as a peer-reviewed selection of papers submitted to publication by participants

The organisers express their gratitude to all sponsors. Thanks are due to organising societies, as the Hungarian Materials Science Society and to other Hungarian scientific and engineering societies for their support.

Budapest, July, 2020.

Peter J. SZABÓ
Chairman of Organising Committee

József GYULAI and András ROÓSZ
Co-chairmen of Scientific Committee

IOP Conf. Series: Materials Science and Engineering **903** (2020) 011002 doi:10.1088/1757-899X/903/1/011002

12th Hungarian Conference on Materials Science

Edited by

K. Májlinger and P. J. Szabó
Budapest University of Technology and Economics, Hungary

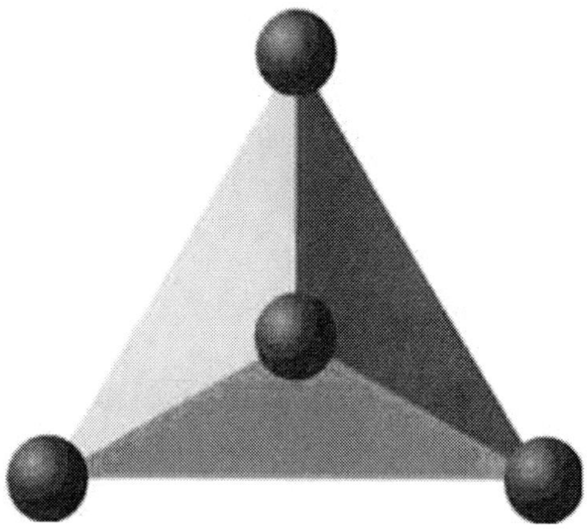

IOP Conf. Series: Materials Science and Engineering **903** (2020) 011002 doi:10.1088/1757-899X/903/1/011002

Scientific Committee

József Gyulai – co-chairman

András Roósz – co-chairman

István Bársony	Zoltán Kónya
Dezső Beke	István László
Károly Belina	János Lukács
Imre Bertóti	Béla Pécz
Tibor Czigány	Mihály Réger
Zoltán Gácsi	Tamás Réti
János Ginsztler	János Szépvölgyi
Jenő Gubicza	Miklós Tisza
György Kaptay	Péter Trampus

Ibolya Zsoldos

Organising Committee

Péter J. Szabó – chairman

Dénes Zsámbók honorary chairman

Tamás Bárány	György A. Gémes
Jánosné Fehér	Balázs Verő
Valéria Mertinger	Zoltán Weltsch

Conference Secretariat

Ildikó Kónya – secretary

e-mail: oatk@oatk.hu

IOP Conf. Series: Materials Science and Engineering **903** (2020) 011002 doi:10.1088/1757-899X/903/1/011002

Sponsors of the Conference

Gold sponsor

Association of Hungarian Iron and Steel Industry

Silver sponsor

CORWELD Ltd. Hungary

Other sponsors

Carl Zeiss Hungary Ltd.

Hungarian Academy of Science

Firstname	Surname	Institution	E-mail address
Krisztián	Wizner	ISD Dunaferr Zrt.	wizni006@gmail.com
László	Somlyai-Sipos	Department of Nanotechnology, University of Miskolc	femsomlyai@uni-miskolc.hu
Gábor	Gyarmati	University of Miskolc	gygabor007@gmail.com
Zoltán	Hózer	Centre for Energy Research, Budapest	hozer.zoltan@energia.mta.hu
Dániel	Molnár	University of Miskolc, Institute of Foundry	daniel.molnar@uni-miskolc.hu
Marianna	Bubenkó	University of Miskolc, Institute of Foundry	ontbm@uni-miskolc.hu
Péter Zoltán	Kovács	University of Miskolc, Institute of Material Science and Technology	metkpz@uni-miskolc.hu
Péter	Ficzere	Budapest University of Technology and Economics, Faculty of Transportation Engineering and Vehicle Engineering, Department of Vehicle Elements and Vehicle-Structure Analysis	ficzere@kge.bme.hu
Orsolya Viktória	Semperger	HD Composite ZRt.	orsolya.semperger@hd-composite.com
Klaudia	Kulcsár	Dent-Art-Technik Ltd.	kulcsar.klaudia@dentarttechnik.hu
Gábor	Vértesy	Hungarian Academy of Sciences, Centre for Energy Research	vertesy.gabor@energia.mta.hu
Ágota	Kazup	Institute of Physical Metallurgy, Metalforming and Nanotechnology	femagota@uni-miskolc.hu
Zsuzsanna	Koncsik	University of Miskolc, Institute of Material Science and Technology	zsuzsanna.koncsik@uni-miskolc.hu
Eszter	Kókai	John von Neumann University, GAMF Faculty of Engineering and Computer Science Department of Materials Technology	kokai.eszter@gamf.uni-neumann.hu
Álmos Dávid	Tóth	Széchenyi István University, Department of Internal Combustion Engines and Propulsion Technology	toth.almos@sze.hu
János	Kónya	Dent-Art-Technik Ltd.	labor@dentarttechnik.hu
Viktor	Kárpáti	Institute of Physical Metallurgy, Metalforming and Nanotechnology	femkv@uni-miskolc.hu
Katalin	Litauszki	Department of Polymer Engineering, Faculty of Mechanical Engineering, Budapest University of Technology and Economics	litauszkik@pt.bme.hu
Valéria	Mertinger	Institute of Physical Metallurgy, Metalforming and Nanotechnology	femvali@uni-miskolc.hu
Zoltán	Fried	John von Neumann Faculty of Informatics Óbuda University Budapest	fried.zoltan@stud.uni-obuda.hu
Roland	Petrény	Department of Polymer Engineering, Faculty of Mechanical Engineering, Budapest University of Technology and Economics	petrenyr@pt.bme.hu
Ferenc	Szabó	Department of Polymer Engineering, Faculty of Mechanical Engineering, Budapest University of Technology and Economics	szabof@pt.bme.hu
Dóra	Harangozó	Department of Materials Science and Engineering, Széchenyi István University	harangozo.dora@sze.hu
László	Takács	eCon Engineering Kft	laszlo.takacs@econengineering.com
Ferenc	Tajti	Department of Materials Technology, GAMF Faculty of Engineering and Computer Science, John Von Neumann University	cbfferi@gmail.com
Miklós	Berczeli	Department of Materials Technology, GAMF Faculty of Engineering and Computer Science, John Von Neumann University	berczeli.miklos@gamf.uni-neumann.hu
Anna	Maloveczky	Bay Zoltán Alkalmazott Kutatási Közhasznú Nonprofit Kft Ipari Lézertechnológiai Csoport	anna.maloveczky@bayzoltan.hu
Erzsébet	Nagy	Institute of Physical Metallurgy, Metalforming and Nanotechnology	femzsofi@uni-miskolc.hu
Beáta	Herbáth	BPW-Hungária Kft.	herbathbea@gmail.com
Dávid Miklós	Kemény	Paks II Zrt.	kemenydavid@gmail.com
Norbert	Kovács	Department of Polymer Engineering, Faculty of Mechanical Engineering, Budapest University of Technology and Economics	kovacsn@pt.bme.hu
Tibor	Hegyes	Arconic Köfém Kft	Tibor.Hegyes@arconic.com
Tamás	Ibriksz	Department of Materials Science and Engineering, Széchenyi István University	ibriksz@sze.hu
Laura	Mádi	University of Miskolc, Institute of Foundry	madi.laura.johanna@gmail.com
Zoltán	Weltsch	Department of Materials Technology, GAMF Faculty of Engineering and Computer Science, John Von Neumann University	weltsch.zoltan@gamf.uni-neumann.hu
Tatiane	Wadas	University of Miskolc, Institute of Material Science and Technology	tatiane_wadas@hotmail.com
Péter	Barkóczy	Institute of Physical Metallurgy, Metalforming and Nanotechnology	fembarki@uni-miskolc.hu
Tamás	Bubonyi	Institute of Physical Metallurgy, Metalforming and Nanotechnology	fembubo@uni-miskolc.hu
Judit	Pázmán	University of Dunaújváros	pazman@uniduna.hu
Gábor	Pór	University of Dunaujvaros	porg@uniduna.hu
Jenő	Gubicza	Department of Materials Physics, Eötvös Loránd University	jeno.gubicza@ttk.elte.hu
Bence	Kocsis	Wigner Research Center for Physics Institution for Solid State Physics and Optics	kocsis.bence@sze.hu
Peter	Trampus	University of Dunaújváros	trampus@uniduna.hu
Attila	Szlancsik	Budapest University of Technology and Economics, Faculty of Mechanical Engineering, Department of Materials Science and Technology	szlancsik@eik.bme.hu
Tibor	Czigány	Department of Polymer Engineering, Faculty of Mechanical Engineering, Budapest University of Technology and Economics	czigany@eik.bme.hu
Péter	Jenei	Department of Materials Physics, Eötvös Loránd University	jenei@metal.elte.hu
Veronika Anna	Szabó	Department of Materials Science and Engineering, Széchenyi István University	szabo.veronika.anna@ga.sze.hu
András	Suplicz	Department of Polymer Engineering, Faculty of Mechanical Engineering, Budapest University of Technology and Economics	suplicz@pt.bme.hu
Alexandra	Kemény	Budapest University of Technology and Economics, Faculty of Mechanical Engineering, Department of Materials Science and Technology	alexa@eik.bme.hu
Dorina	Kovács	Budapest University of Technology and Economics, Faculty of Mechanical Engineering, Department of Materials Science and Technology	dorina@eik.bme.hu
Ildikó	Szenthe	Centre for Energy Research, Budapest	szenthe.ildiko@energia.mta.hu
Borbála	Leveles	Budapest University of Technology and Economics, Faculty of Mechanical Engineering, Department of Materials Science and Technology	borbala.leveles@edu.bme.hu
József	Renkó	Budapest University of Technology and Economics, Faculty of Mechanical Engineering, Department of Materials Science and Technology	renko.jozsef@edu.bme.hu
Viktor	Gál	University of Miskolc	metgv@uni-miskolc.hu
Barnabás Zoltán	Balázs	Budapest University of Technology and Economics, Faculty of Mechanical Engineering, Department of Manufacturing Science and Engineering	balazs@manuf.bme.hu
László	Lendvai	Széchenyi István University, Audi Hungaria Faculty of Vehicle Engineering	lendvai.laszlo@sze.hu
János	Lukács	University of Miskolc,	janos.lukacs@uni-miskolc.hu
Ákos	Kmetty	Department of Polymer Engineering, Faculty of Mechanical Engineering, Budapest University of Technology and Economics	kmetty@pt.bme.hu
Henrietta	Hudák	University of Miskolc, Institute of Foundry	ontheni@uni-miskolc.hu

IOP Conf. Series: Materials Science and Engineering **903** (2020) 011005 doi:10.1088/1757-899X/903/1/011005

Peer review declaration

All papers published in this volume of IOP Conference Series: Materials Science and Engineering have been peer reviewed through processes administered by the Editors. Reviews were conducted by expert referees to the professional and scientific standards expected of a proceedings journal published by IOP Publishing.

- **Type of peer review:** *Single-blind*
- **Conference submission management system:** *EasyChair*
- **Number of submissions received:** *62*
- **Number of submissions sent for review:** *62*
- **Number of submissions accepted:** *60*
- **Acceptance Rate (Number of Submissions Accepted / Number of Submissions Received X 100):** *97 %*
- **Average number of reviews per paper:** *3,05 with editorial reviews*
- **Total number of reviewers involved:** *75*
- **Any additional info on review process:**
  Manuscript with track changes and detailed responses to the reviewer's comments was a requirement for every manuscript submitted after revisions.
- **Contact person for queries:** *Dr. Kornél MÁJLINGER email: vmakornel@gmail.com*

Novel concepts for establishing expert support systems to investigate the defect occurring in metallurgical phases in the technology of ISD DUNAFERR Zrt.

K Wizner[1] and A Kővári[2]

[1] Technology department senior associate, ISD Dunaferr Zrt. Dunaújváros, Hungary
[2] University of Dunaújváros, Dunaújváros, Hungary

E-mail: wizner.krisztian@isd-dunaferr.hu

Abstract. Competition among steel companies is intense in order to retain customers and get/gain new ones. Economic operators are much more critical of their suppliers and of the quality of their products due to increasing quality requirements. In addition to quality improvement measures to meet quality demands, cost optimization measures pose a major challenge for manufacturers.
Complaints always require an overview of the manufacturing process of the particular product, as it is essential for product development and can highlight technological issues that can eliminate similar types of complaints (improve product quality).
The investigation of the complaints requires extensive professional experience. Deep theoretical knowledge of the technological process, practical experience and knowledge of the local circumstances are essential. This amount of knowledge is available to the experts in varying degrees, and therefore, in addition to examining the main parameters, the aspects of the investigation may vary considerably. By creating or expanding a unified knowledge base, certain parts of the investigation can be analyzed and automated in higher detail. With this solution, the burden on experts is reduced, uniform investigation becomes possible, and examination aspects can be classified. Further statistical analysis may even make it possible to filter out technological parameters that are critical for quality.
The system described below is primarily designed from a metallurgical point of view, but with the expansion of its knowledge base it is also possible to study additional technological phases.

1. Introduction

During industrial processes, like steelmaking process, many parameters influence the characteristics and quality of the product released. Such parameters are e.g. the mass, temperature, composition, for which the knowledge of technology, operational practice and conditions is essential [1, 2, 3, 4]. It is important to have the required knowledge of the operating principles of the measuring devices / samplers, the criteria of their usability, the difficulties associated with them due to the operating conditions, and the specificity arising from the place of sampling. The temperature and composition homogeneity of the sampled material can also significantly influence the relevance of the measurements.

In order to judge the main characteristics of the process, it is not enough to check whether certain parameters of the process are within certain range, but even complex analysis of several parameters is necessary.

IOP Conf. Series: Materials Science and Engineering 903 (2020) 012001 doi:10.1088/1757-899X/903/1/012001

During the Dunaferr steelmaking process, first and foremost, chemical compliance and castability are the main viewpoints. Unfortunately, the pursuit of safety (e.g. warmer production, higher overheating) often increases the chances of downgrading / complaints of rolled products in later stages. For example, based on previous studies [5], the overheating [6] of the steel transferred from the converter to the casting machine was significantly related to the degree of downgrading of the rolled product.

In this article, using the experience gained so far, the most important factors of steel production / casting, which are of greater importance for quality and especially for subsequent complaints, are reviewed and summarized. It is very difficult to evaluate the technological parameters and the underlying processes, as well as the characteristics that influence the quality of the product by studying the raw data series. Therefore, in recent years, we have aimed to develop data analysis methods that provide a clear view of certain details of the manufacturing process, making it easier to detect possible manufacturing errors. The mechanism of the effect of each parameter on product quality is also analyzed.

2. Examining data

When analyzing the production parameters, their nature and their influencing factors must be given special attention, so that they can be used properly in the further analysis and the conclusions drawn in this way are also correct. The following is a summary of the principles of evaluating technological data that are necessary to reach valid conclusions.

2.1. Validity of the data

The first step in investigation is to verify the correctness of the data [7]. Due to human-, measurement–, data transmission (IT) errors, data may be available which is not justified by the manufacturing process. It requires special attention because it may indicate a technology error or it might even be correct data, if other related data are incorrect. It requires great expertise to judge correctly.

It is also important to know the methods of measuring instruments [8]. For example, a „lollypop sampler" used to take steel samples use vacuum to suck up the sample. The sampler used after blowing in the high oxygen content steel bath, an aluminium wire deoxidizes the steel to prevent gas formation, which generates a high aluminium content of about 4% during analysis. This aluminium value must be ignored because it cannot be a real value thermodynamically.

From an IT point of view, the text in the comment field of the database is a serious problem because it cannot be handled properly, but may contain important information. E.g. in case of bottom argon stirring, the operator indicates that the argon gas has not entered the ladle and therefore the measured data are not accurate.

2.2. Range checking

The second step is to check the data within the required range. Out of range parameters do not necessarily mean error. Often, they only occur because of unexpected parameters resulting from operating conditions. It also needs a great professional experience to judge the justification of differences and their impact.

Certain treatments are often applied based on several considerations, which must be taken into account when assessing their justification. For example, the duration and intensity (or amount) of argon stirring (bottom and top) primarily affects the homogeneity of the steel bath, but also plays a role in temperature control and inclusion removal. At high intensities, a free steel surface may be formed, which may lead to additional oxygen / nitrogen uptake and inclusion formation, requiring additional deoxidizing agent.

2.3. Time course of the data

The third step is to evaluate the time course of the data, e.g. temperature, composition (steel, slag), dosing and order of ladle metallurgy treatments. The time course is very important for thermodynamic processes and because Dunaferr technology does not have heating possibilities outside of the converter. Incorrectly selected order may result in e.g. a critical steel temperature drop, inability of casting or reduction in the period (in critical cases skipping) of ladle treatments, which affect the purity of steel, due to the forced process acceleration.

2.3.1. Composition and temperature changes

The compositions should reflect technological steps as well as the downstream nature of the temperatures (no heating possible). The added deoxidizing agent should reduce the active oxygen level in the steel and the FeO content in the slag. If this is not the case, and no other technological intervention can justify it (e.g. due to strong argon stirring, the unprotected steel surface absorbs oxygen from the environment) then the quality of the samples can be questioned.

2.3.2. Technological steps of steelmaking

The sequential nature of the technological steps can determine, among other things, the efficiency of the alloying process, explain the reasons for the addition of a greater amount of deoxidizing agent, and, for example, unpredicted technological steps before blowing because of unexpected scrap condition / chemical composition, or unexpected ladle condition.

The large drop in temperature between measurements (due to the lack of heat transfer outside the converter) poses a technological risk. A ladle that is red-hot inside, but not hot in its total cross-section, can be deceptive. Experience has shown that the strongest cooling occurs under intense mixing (argon stirring) and, therefore, due to the potential of high cooling, its duration can be significantly reduced, which is detrimental to homogeneity and inclusion content.

A program depicting the technological steps of the steel production phases, which can be run from a browser, has been created to facilitate the investigation of complaints. The program displays the required diagrams by entering the heat number (Figure 1).

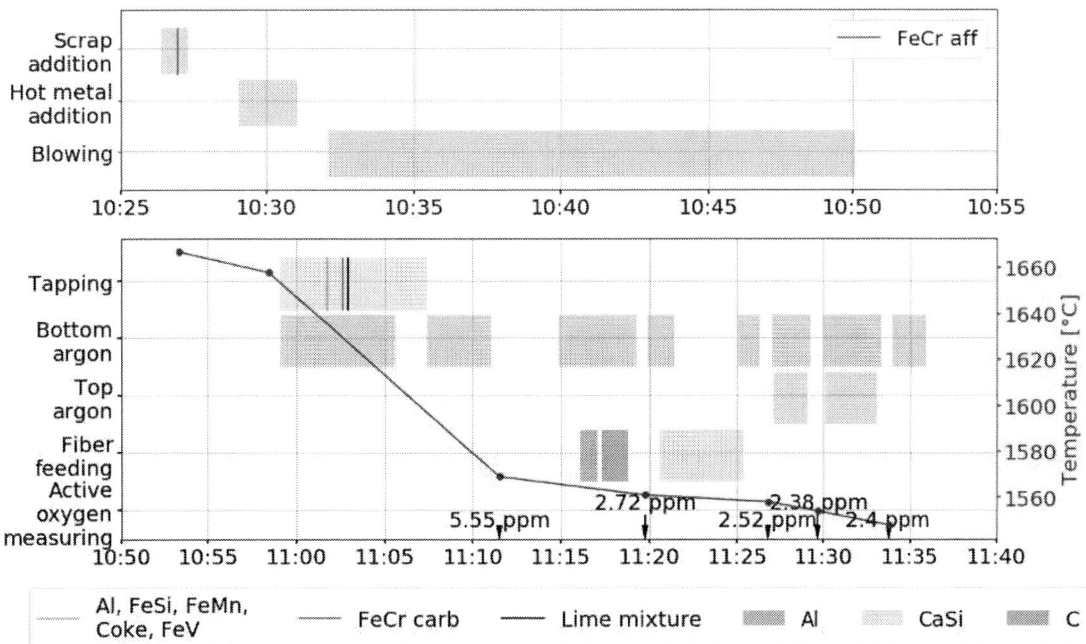

Figure 1. Time course of the technological process

The diagrams show the technological progress over time. It is also evident that the most intense cooling is caused by the addition of alloys and upper argon stirring. The charts also help to understand the followings:

- The deoxidizing agent used has been deoxidizing the steel or the slag.
    - In the case of the deoxidizing agent added to the slag, a large amount of primary slag flow was likely to be compensated in order not to oxidize the deoxidized steel.
- Order of alloys adding based on thermodynamic and technological aspects.
- Compensation of the sulphur or phosphorus content of the scrap by after blowing and slagging.

2.3.3. Casting data inspection

After investigating the steelmaking, the analysis of the casting is the next step [9]. During casting, several heats are sequentially cast. Due to the peculiarity of the process, it is also necessary to consider the environment of the heat / slab in question, which further increases the complexity of the task. The program for analyzing the casting process greatly facilitates the identification of problematic strand sections. Visualization of the strand sections burdened with non-steady state, the tendency of a given slab to surface or internal defects can be more easily determine..

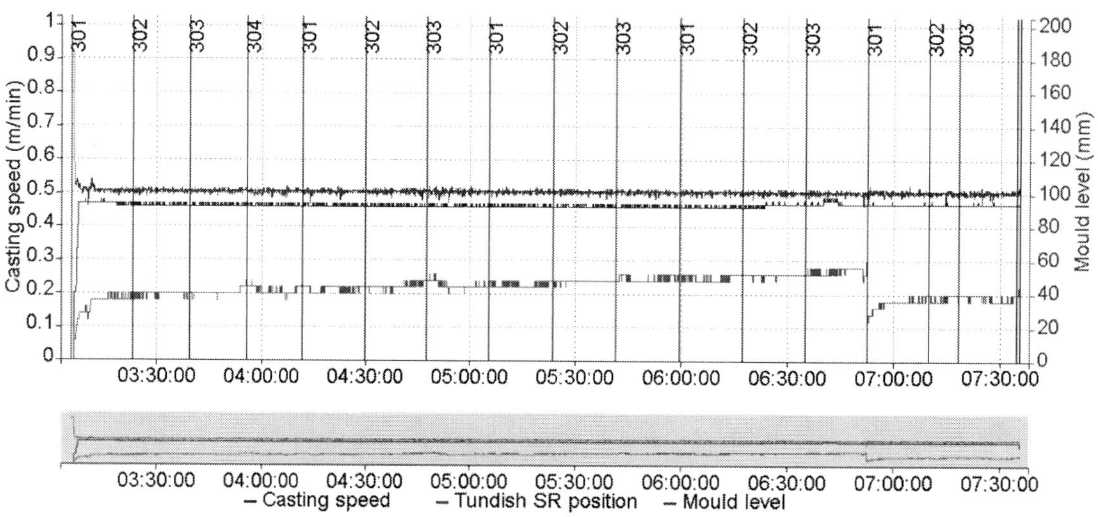

Figure 2. Casting curves

The curves in Figure 2 shows the events with the greatest impact on the strand:

- In the beginning of the casting, the temperature of the casting machine is not considered to be constant, and therefore the starter slab is often subject to surface defect formation based on operational experience.
- During the casting of the ending slab, the steel level in the tundish is lowered, causing contaminants / particles that are concentrated near the surface to enter into the mould due to the vortex effect.
- In the strand sections affected by deceleration, the stability of the crystallization process changes significantly, which, though to varying degrees, affects the strand in the entire casting machine (~ 10 m)
- Sudden high casting speed change or fluctuation of mould steel level can cause serious surface defects, especially at the liquid steel meniscus.

- Changes in the position of the stopper rod of the steel flow regulator in the tundish can be inferreded from inclusion cluster adhesion, which affect the flow conditions, and inclusion cluster detachment, which cause the inclusion cluster to move into the mould.
- Formation of a transitional strand segment (so-called "mixed casting") resulting from the joining of heats with different compositional requirements (due to technological constraints) [10, 11].
- The strand sections affected by interventions are cut out if their nature requires it [12].

As mentioned above, the strand sections affected by major problems (e.g. casting speed change, level fluctuation, interventions) are cut / removed according to technology, thus causing no problem in the finished product sold, which has been confirmed by previous examinations. It is up to the skilled expert to judge other disorders affecting slabs.

Operating parameters often differ from the recommended manufacturing parameters. This difference is justified in most cases and is compensated by another technological parameter (which is also different, hence from the recommended value). In the fourth step, the system of compensation of these parameters should be examined.

- High steel temperature at the end of blowing or slower than expected cooling rate:
  - Longer argon stirring in the converter
  - Adding cooling scrap
  - Longer then prescribed bottom argon stirring in ladle
  - Upper argon stirring
- High level of active oxygen in the raw steel due to post-blowing
  - Longer argon stirring in the converter
  - Addition of deoxidizing agent in excess of prescribed
- Existence / absence of slag closure
  - The length of the tap time refers to the age of the tap stone. No new slag closure is required for new stone, as tapping stone formation and wear do not yet require
- High overheating
  - Reduction of/Reduced casting speed

3. Tendency factors

Based on experience, none of the steelmaking / casting differences / problems mentioned in the previous chapter cause problem in the rolled product by all means, they can only be considered as tendency factors. It can only be assumed that the combined effect of several factors may have caused the complained defect. The nature of the defect also depends to a large extent on whether the particular manufacturing conditions may have caused the defect.

Based on experience, the following circumstances generally justify the occurrence of some of the complained defects:

- High degree of overheating, above 40°C and starter slab
- Low quantity argon stirring combined with rapid cooling

4. Conclusions

It follows from the above that the investigation of the steelmaking and casting process of each complained product requires great expertise and attention. Most of the investigation aspects can be summarized in a constantly expanding knowledge base that can be programmed and thus automated. Expanded time diagram of the steelmaking process as well as the data displayed on the basis of the analysis of the casting curves are of great help to experts. Situations due to special circumstances (such as malfunctions due to human factors) and the continual expansion of the knowledge base will require the involvement of a specialist in the investigation process, but will greatly facilitate his / her work and thus allow deeper understanding and further analysis of the processes.

By exploring the relationships, it is possible to identify the conditions that influence quality in the technological process. Suspected defective strand sections can already be removed on the casting machine, or by using appropriate surface treatment (grinding, flame scarfing) to prevent surface defects on the rolled product. The measures taken on the basis of the above will reduce the amount of metallurgical defects in rolled products as well as internal problems which are not visible by surface inspection and which are detected only during processing at the customer. For these reasons, the system is beneficial from the points of view of both customer satisfaction and internal outcome.

Acknowledgment
The study is funded by „EFOP-3.6.1-16-2016-00003 K+F+I folyamatok hosszú távú megerősítése a Dunaújvárosi Egyetemen" project.

References

[1] Tanabe H and Nakada M 2003 Steelmaking technologies contributing to steel industries *NKK technical review* **88** pp 18-27
[2] Weigel M et al 2016 Multicriteria analysis of primary steelmaking technologies *Journal of cleaner production* **112** pp 1064-1076
[3] Iosif A et al 2008 Process integrated modelling for steelmaking life cycle inventory analysis *Environmental Impact Assessment Review* **28**(7) pp 429-438
[4] Mazumdar D and Evans J W 2009 Modeling of steelmaking processes *CRC Press* ISBN 978-1-4200-6243-4
[5] Wizner K and Kovari A 2020 Effect of steel production / casting parameters on the degree of downgrading of the hot rolled product *Acta Materialia Transylvanica* in press
[6] Huang X et al 1992 Modeling superheat removal during continuous casting of steel slabs *Metallurgical and Materials Transactions* **B 23**(3) pp 339-356
[7] Nakamura S et al 1990 A New Energy Control System Using Real Time Expert Systems *IFAC Proceedings Volumes* **23**(3) pp 381-386
[8] Ericsson O 2009 An experimental study of liquid steel sampling Doctoral dissertation, KTH Stockholm, Sweden
[9] Thomas B G 2001 Continuous casting of steel Chapter 15 pp 499-540
[10] Huang X and Thomas B G 1993 Modeling of steel grade transition in continuous slab casting processes *Metallurgical Transactions* **B 24**(2) pp 379-393
[11] Thomas B G 1997 Modeling study of intermixing in tundish and strand during a continuous-casting grade transition *Iron and Steelmaker* **24**(11) pp 83-96
[12] Offerman C 1981 The affect on the Quality and Internal Structure in Continuously Cast Slabs by the Metal Flow in the Mould *Scandinavian J. Metallurgy* **10** pp 25-28

IOP Conf. Series: Materials Science and Engineering **903** (2020) 012002 doi:10.1088/1757-899X/903/1/012002

Investigation of the Melting Point Depression of Copper Nanoparticles

L Somlyai-Sipos[1], D Janovszky[2], A Sycheva[2] and P Baumli[1]

[1] Department of Nanotechnology, University of Miskolc, Miskolc-Egyetemváros, H-3515, Hungary
[2] MTA-ME Materials Science Research Group, Miskolc-Egyetemváros, H-3515, Hungary

E-mail: femsomlyai@uni-miskolc.hu

Abstract. In this paper copper nanoparticles (Cu NPs) have been produced by solvothermal method. One of the objectives of the present research was to create Cu NPs using aqueous medium technology. The analysis of the produced particles was performed by X-ray diffraction (XRD) and Scanning Electron Microscopy (SEM). In the course of the experiments, copper core and organic shell nanoparticles were created successfully. It has been found that the average diameter of the created Cu NPs was 50 nm. The thermal behaviour of these particles was investigated by Differential Scanning Calorimetry (DSC). The melting temperature depression of the one-pot synthesized Cu NPs was 633 °C.

1. Introduction

The melting point is one of the most important thermodynamic properties of all metal materials. The melting point characterizes the bulk materials independently on the size in the microscopic range. It is well known that the large surface-to-volume ratio in nano-sized particle (NPs) systems has significant influence on their thermodynamic properties and phase relations. NPs have high surface to volume ratio, and as a result the melting temperature will be size-dependent in the nano-range. Pawlow [1] in 1909 postulated that the melting point depression exists due to the large influence of the surface energy on the energy and structure of small particles. Allen et al. [2] have investigated the effect of reduction in size of lead, tin, indium and bismuth. Oxidation can also be significant in nano-size range, so a decrease in the melting point is measurable only in an oxide-free environment. A number of similar researches have been carried out on alumina [3], SAC (Sn3.5Ag0.5Cu) [4], gold [5][6], lead [7], indium [7][8], tin [9], bismuth [9][10]. Similarly, the melting temperature depression of Sn-based nanoalloys synthesized via various techniques has also been regarded as a promising property for applications in electronics [11]. The melting point depression of nanoparticles can vary from a few to several hundred degrees.

This size reduction naturally opens up new ways for the industry to use copper. The properties of copper nanoparticles have been investigated for the following systems, such as electronic and optical nano-devices [12], catalysts [13], lubricants [14], thermal conductive nanoliquids [15], antibacterial agents [16], biosensors [17], as well as solar collectors [18].

In line with modern challenges, there is a growing demand for the development of technologies that allow joining of heat-sensitive materials. The main requirements for brazing are to be reliable, low-temperature processes that allow the widespread use of heat-sensitive materials in industrial application. Nanoparticles can be the solution to such difficulties, as nano-effect, such as melting point depression, or wetting properties can open up new opportunities for technological challenges [19].

Copper is very important in brazing technologies of the steel structures in the industry. Its importance is even greater when we consider microelectronic application. Increasing cost of production of raw materials, as well as the reduction in size mainly in the electronics industry, have led researchers to set smaller targets for stable copper nanoparticles. There have been relatively few studies on the melting point depression of Cu nanoparticles. A sol-gel method was used to form a SiO_2 matrix composite reinforced with Cu/Cu_xO particles by Yeshchenko et al. [20]. They separated two melting regimes based on the temperature dependence of the surface-plasmon resonance energy and its width. The aim of the current research work is to create copper nanoparticles (Cu NPs) by solvothermal method in order to develop braze material with melting point depression. Another goal is to analyse the thermal properties and the microstructure of the obtained nanoparticles.

2. Experimental

2.1. Materials

All of the chemicals were of analytical grade and used as purchased without further purification. Copper(II)chloride hexahydrate ($CuCl_2 \cdot 6H_2O$) salt was dissolved in distilled water. Polyethylene glycol (PEG 400) was used as the capping agent. Sodium borohydrate ($NaBH_4$) and ascorbic acid ($C_6H_8O_6$) were used as reducing agents. Sodium hydroxide (NaOH) was also used to adjust the pH and accelerate the reduction reaction in water (Table 1.).

2.2. Synthesis of copper nanoparticles

The first step of the preparation of copper nanoparticles starts with dissolving copper(II)chloride hexahydrate salt ($CuCl_2 \cdot 6H_2O$) in distilled water to obtain a greenish solution. The solution was in a beaker under stirring constantly in a water bath at 60 °C. Dissolved polyethylene glycol 400 was added to the copper(II)-chloride solution. Sodium borohydrate ($NaBH_4$), ascorbic acid ($C_6H_8O_6$) and sodium hydroxide (NaOH) were measured and dissolved in distilled water and added to the solution dropwise. The colour of the solution changed from greenish via yellowish to black/red. This colour change confirmed that reduction reaction had been started and finished. The mixture was further stirred rapidly (500 rpm) for 1 hour in closed system to complete the reaction.

Table 1. The concentration of the reagents and the parameters of the experiment

Component	Concentration	Others
Copper(II)-chloride hexahydrate	2.4 g/L	
Polyethylene glycol (PEG 400)	35.6 g/L	
Sodium hydroxide	4.0 g/L	
Ascorbic acid	3.5 g/L	
Sodium borohydrate	3.8 g/L	
Temperature		60 °C
Reaction time		1 hour
Water bath, stirring constantly, closed system		

2.3. Characterization

The synthesized particles were separated from the solution with an Eppendorf Centrifuge 5804 R centrifuge, using 5000 rpm. For the separation distilled water and ethanol were used. The temperature was 20 °C during the centrifugation.

The synthesized samples were studied applying Scanning Electron Microscopy (Hitachi S-4800 SEM) equipped with a BRUKER AXS type energy-dispersive X-ray spectrometer (EDS) and X-Ray Diffraction (Bruker D8 Discovery). Differential Scanning Calorimetry (Netzsch STA 449 F3 Jupiter TG-DSC) was used for the determination of the melting point of the copper nanoparticles. The tests were carried out in vacuum (10^{-4} mbar) to minimize the oxidation; the heating rate was 10 °C/min. The equipment measures the change of the mass of the sample continuously.

3. Results and discussion

It can be seen on the SEM micrograph that the particles obtained are connected to each other and have a diameter of 60-70 nm (Figure 1a.). The EDS analysis shows that the significant part of the connected particles is Cu (Figure 1b) which means that the synthesis method is pure.

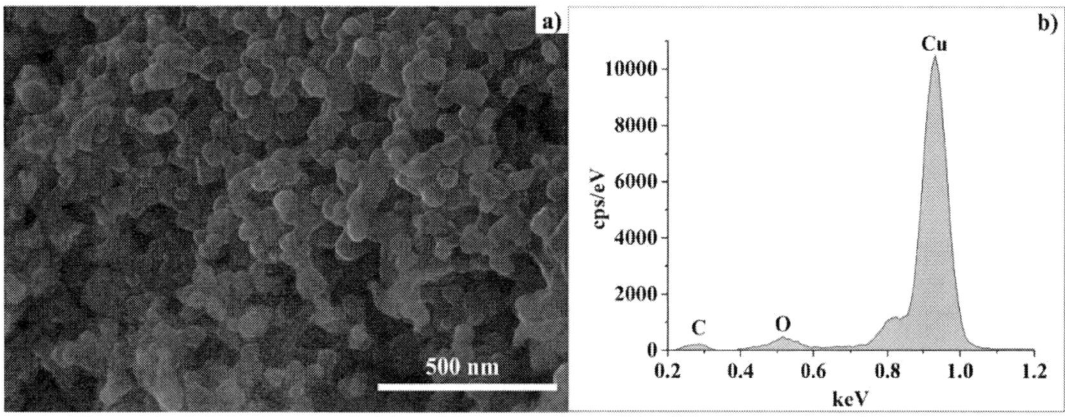

Figure 1. SEM micrograph (a) and EDS spectrum (b) of the copper-containing samples

It can be observed in the larger magnification image (Figure 2.) that these particles are stuck together due to the PEG 400 addition to the system. The nanoparticle core and the PEG organic shell surrounding the particle are clearly visible in Figure 2b featuring SEM image obtained using transmission electron (TE) detector. It proves that an inorganic core/organic shell structure was obtained. The PEG layer is also responsible for separating the particles, i.e. preventing their aggregation. PEG 400 is a low-molecular-weight grade of polyethylene glycol which is strongly hydrophilic. PEG 400 changes the available surface energy of the particles so that the surface tension decreases. The Kelvin barrier moves, allowing more particles to avoid the aggregation process and as a result lowering the mean particle size.

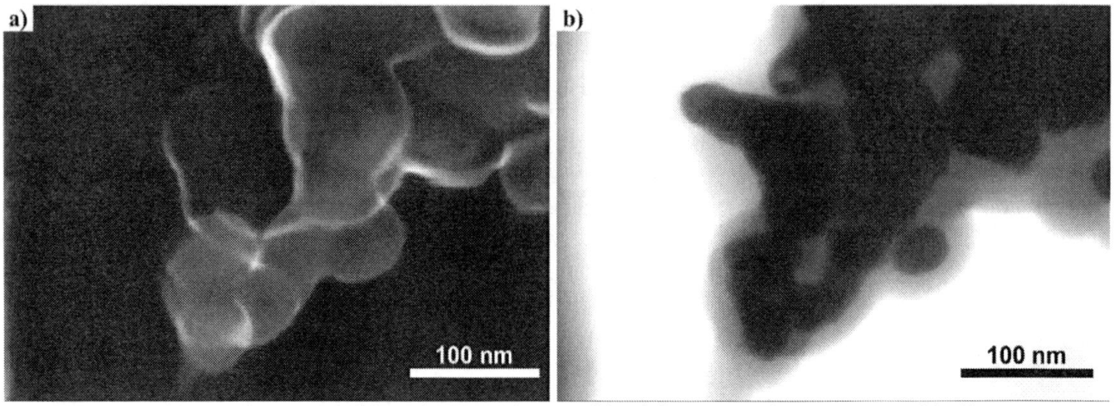

Figure 2. SEM micrograph (a) and STEM image (b) of the copper/PEG400 core/shell structure

Using STEM mode, the real particle size of the copper particles can be estimated being of 50 nm in average (Figure 3). It means that the particles, which can be seen in the Figure 1. do not agglomerate but they are generally connected by PEG.

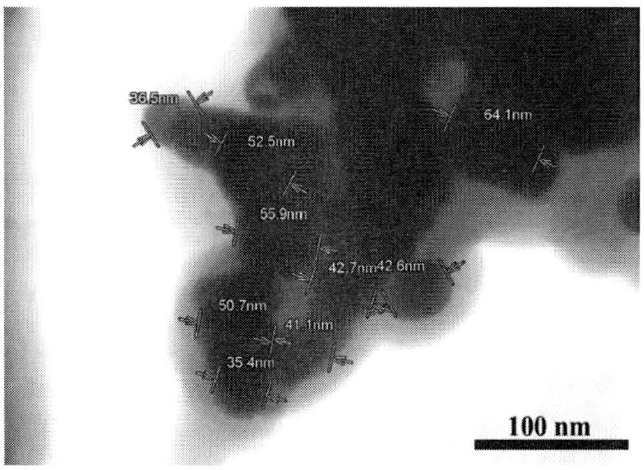

Figure 3. The particle size of the copper particles (STEM image)

The phase identification was performed by X-ray diffraction (Figure 4.) which shows that some of the produced particles are pure copper, however a significant amount of Cu_xO was also formed. A large amount of Cu NPs has an X-ray amorphous structure based on the amorphous halo in the beginning of the diffractogram (Figure 4).

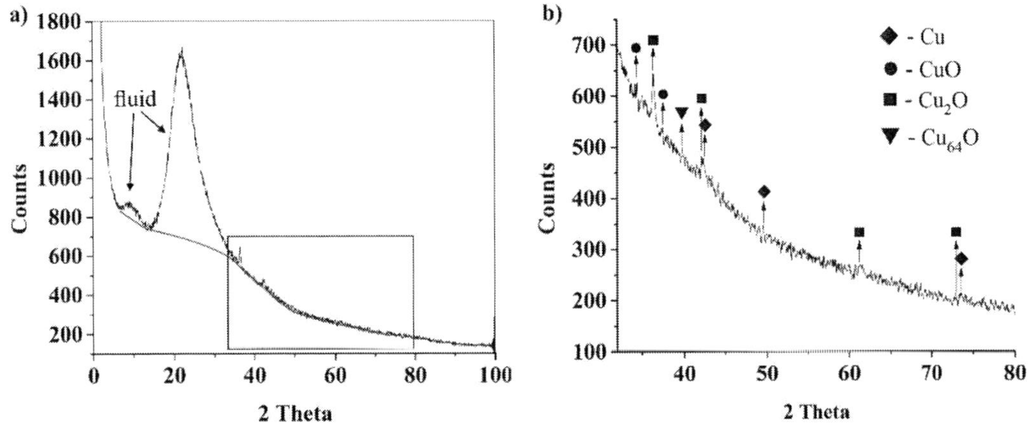

Figure 4. Results of X-ray diffraction analysis (a) and an enlarged detail of the area marked with a rectangle (b) of the copper-containing sample

In the current research work, the melting point of the copper-containing particles has been measured by Differential Scanning Calorimetry (DSC). The results showed (Figure 5.) that the endothermic peak at 473 °C is followed by an exothermic peak, from which one can deduce a melting and a crystallization. This may mean that the pure copper nanoparticles melted at 447 °C and then crystallized after the fusion of the particles again with a peak temperature of 563 °C. After the melting process of Cu particles an intensive oxidation starts, which proceeds up to 803 °C.

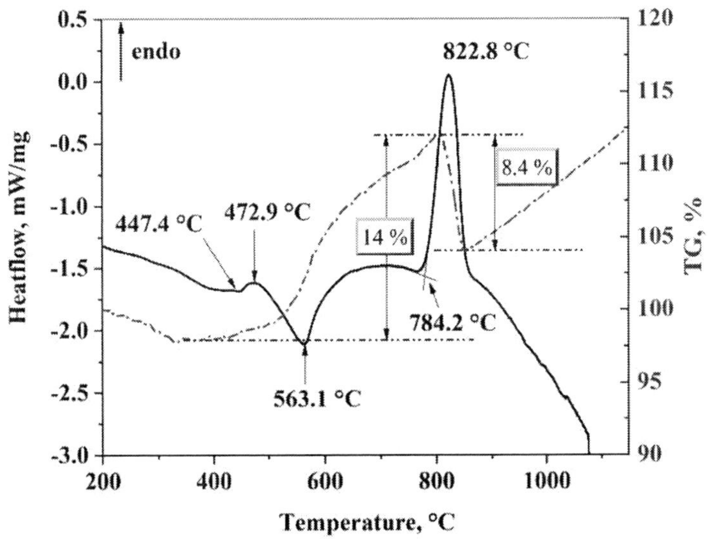

Figure 5. Investigation of the melting point and mass change of copper-containing particles by TG-DSC

The melting point measured is in agreement with experiment of Yeshchenko et al. [20]. The difference between the data reported and the current measurement is that the authors used SiO_2 matrix to determine the melting point of the Cu NPs. Here, the Cu NPs were separated only by the organic shell, for which the boiling point is less than 250 °C. The volume melting point of 440 °C determined in [20] is equivalent to the melting point measured for 50 nm diameter particles. Fig5. shows an endothermic peak at 820 °C. Observing the TG line a mass loss step of 8 wt.% was detected under this endothermic process. Heinemann et al. [21] dealt with stability of various copper oxides. It was found that the reduction of Cu_2O to pure copper occurred at a pressure of 10^{-4} torr and at around 820 °C.

Based on the literature and the data obtained in the current research, it can be concluded that the pure copper nanoparticles produced have melted at 450 °C in the DSC and then recrystallized as they collapsed. Further increase in temperature leads to decomposition of Cu_2O started at 785 °C and indicated by high endothermic peak in Figure 5. The Cu NPs formed during the decomposition of Cu_2O were oxidized again during the further temperature increase according to TG measurement. Due to oxygen reduction the pressure changed in the chamber; the oxidization can start again after decomposition of Cu_2O.

4. Conclusions

In this paper, copper nanoparticles were successfully synthesized by a chemical reduction method in water. The main conclusions, which can be drawn are as follows:

1. The size of the Cu NPs obtained was around 50 nm. Due to the small size and high surface area NPs easily oxidized.
2. Based on SEM results, it can be stated that a metal core and an organic shell structure has been created. This type of structure is beneficial for determination of the melting point of metal NPs since they are separated from each other.
3. The reduction of the melting point due to the nano-size range was detected using differential scanning calorimetry method (DSC).
4. The copper nanoparticles melted at 450 °C, which means that the depression of the melting point of bulk copper was equal to 633 °C.

Acknowledgements

The described article was carried out as part of the GINOP-2.3.2-15-2016-00027 "Sustainable operation of the workshop of excellence for the research and development of crystalline and amorphous nanostructured materials" project implemented in the framework of the Szechenyi 2020 program. The realization of this project is supported by the European Union. Supported by the ÚNKP-19-3 New National Excellence Program of the Ministry for Innovation and Technology.

Courtesy to Innovacios Laboratorium Ltd., (Miskolc, Hungary) for support of Field Emission Scanning Electron Microscope (SEM) Hitachi S-4800 equipped with Bruker AXS Energy-dispersive X-ray Spectrometer (EDS) system. Special thanks for Dr. Ferenc Kristály for X-ray investigations.

References

[1] Pawlow P 1909 Ober die Abhängigkeit des Schmelzpunktes von der Oberflächenenergie eines festen Körpers (Zusatz.) *Zeitschrift für Phys. Chemie* **65U**

[2] Allen G L, Bayles R A, Gile W W and Jesser W A 1986 Small particle melting of pure metals *Thin Solid Films* **144** 297–308

[3] Joshi N, Mathur N, Mane T and Sundaram D 2018 Size effect on melting temperatures of alumina nanocrystals: Molecular dynamics simulations and thermodynamic modeling *Comput. Mater. Sci.* **145** 140–53

[4] Kim H J, Jang S P, Lee J-H and Park Y-J 2016 Transformation of SAC (Sn3.0Ag0.5Cu) nanoparticles into bulk material during melting process with large melting-point depression *Micro Nano Lett.* **11** 840–3

[5] Sambles J R 1971 An electron microscope study of evaporating gold particles: the Kelvin equation for liquid gold and the lowering of the melting point of solid gold particles *Proc. R.*

Soc. London. A. Math. Phys. Sci. **324** 339–51
[6] Buffat P and Borel J-P 1976 Size effect on the melting temperature of gold particles *Phys. Rev. A* **13** 2287–98
[7] Coombes C J 1972 The melting of small particles of lead and indium *J. Phys. F Met. Phys.* **2** 441–9
[8] Zhang M, Efremov M Y, Schiettekatte F, Olson E A, Kwan A T, Lai S L, Wisleder T, Greene J E and Allen L H 2000 Size-dependent melting point depression of nanostructures: Nanocalorimetric measurements *Phys. Rev. B* **62** 10548–57
[9] Frongia F, Pilloni M, Scano A, Ardu A, Cannas C, Musinu A, Borzone G, Delsante S, Novakovic R and Ennas G 2015 Synthesis and melting behaviour of Bi, Sn and Sn–Bi nanostructured alloy *J. Alloys Compd.* **623** 7–14
[10] Kellermann G and Craievich A F 2002 Structure and melting of Bi nanocrystals embedded in a B2O3−Na2O glass *Phys. Rev. B* **65** 134204
[11] Jiang H, Moon K and Wong C P 2013 Recent advances of nanolead-free solder material for low processing temperature interconnect applications *Microelectron. Reliab.* **53** 1968–78
[12] Wang Y H, Peng S J, Lu J D, Wang R W, Mao Y L and Cheng Y G 2008 Optical properties of Cu and Ag nanoparticles synthesized in glass by ion implantation *Vacuum* **83** 408–11
[13] Judai K, Numao S, Nishijo J and Nishi N 2011 In situ preparation and catalytic activation of copper nanoparticles from acetylide molecules *J. Mol. Catal. A Chem.* **347** 28–33
[14] Zhang B-S, Xu B-S, Xu Y, Gao F, Shi P-J and Wu Y-X 2011 Cu nanoparticles effect on the tribological properties of hydrosilicate powders as lubricant additive for steel–steel contacts *Tribol. Int.* **44** 878–86
[15] Bönnemann H, Botha S S, Bladergroen B and Linkov V M 2005 Monodisperse copper- and silver-nanocolloids suitable for heat-conductive fluids *Appl. Organomet. Chem.* **19** 768–73
[16] Stelzig S H, Menneking C, Hoffmann M S, Eisele K, Barcikowski S, Klapper M and Müllen K 2011 Compatibilization of laser generated antibacterial Ag- and Cu-nanoparticles for perfluorinated implant materials *Eur. Polym. J.* **47** 662–7
[17] Lu L-M, Zhang X-B, Shen G-L and Yu R-Q 2012 Seed-mediated synthesis of copper nanoparticles on carbon nanotubes and their application in nonenzymatic glucose biosensors *Anal. Chim. Acta* **715** 99–104
[18] Musa A ., Akomolafe T and Carter M . 1998 Production of cuprous oxide, a solar cell material, by thermal oxidation and a study of its physical and electrical properties *Sol. Energy Mater. Sol. Cells* **51** 305–16
[19] Janczak-Rusch J, Kaptay G and Jeurgens L P H 2014 Interfacial Design for Joining Technologies: An Historical Perspective *J. Mater. Eng. Perform.* **23** 1608–13
[20] Yeshchenko O A, Dmitruk I M, Alexeenko A A and Dmytruk A M 2007 Size-dependent melting of spherical copper nanoparticles embedded in a silica matrix *Phys. Rev. B* **75** 085434
[21] Heinemann M, Eifert B and Heiliger C 2013 Band structure and phase stability of the copper oxides Cu2O, CuO, and Cu4O3 *Phys. Rev. B* **87** 115111

Effect of the sampling method on the results of melt quality assessment of aluminum alloys with computed tomography

G Gyarmati[1], Gy Fegyverneki[1], M Tokár[1] and T Mende[1]

[1]University of Miskolc, Miskolc-Egyetemváros, Hungary

E-mail: ontgabor@uni-miskolc.hu

Abstract. Liquid metal quality is one of the key factors which determine the soundness of cast parts. This is the reason why the assessment of the melt quality is of critical importance prior to casting. The most common and most deleterious defects of liquid aluminum alloys are the so-called bifilms whose quantity can be characterized by the computed tomographic porosity analysis of reduced pressure test pieces. During the sampling of these specimens, however, generally pouring is involved, which is known to damage melt quality and introduce bifilms into the liquid metal. For this reason, a new sampling method was tested and compared with conventional pouring. It was found, that by using the new sampling method, the pore volume fraction of the test pieces can be lowered, however, regarding the pore number density, no clear difference could be observed. The results also suggest that fluxes have a remarkable effect on the structure of double oxide films, and in this way, on the susceptibility to pore formation.

1. Introduction

Regardless of the type of the casting process used for making cast components, in order to ensure decent mechanical properties and avoid casting defects like porosity, inclusions, and leakage, ensuring proper melt quality is essential [1-4]. The most frequently occurring, and in fact, the most harmful inclusions of aluminum casting alloys are the double oxide films (also known as bifilms), which have a remarkable effect on the microstructure and thus, the properties of cast metals [5-7]. Bifilms can easily form during melt handling and processing, as the smallest amount of disturbance of the surface oxide layer of the melt can cause its incorporation into the bulk liquid [8]. During this so-called entrainment process, the surface oxide layer folds over itself, and a doubled-over film is created, which has a central unbounded interface, thus bifilms constitute cracks in the microstructure of the cast metals [9]. During the solidification of the alloy, bifilms can initiate pore formation, as the hydrogen precipitation between their layers, and the local pressure drop in the interdendritic regions (caused by the solidification shrinkage) leads to the unfurling and inflation of the double oxide films [5, 10, 11]. This phenomenon is schematically illustrated in Figure 1. In fact, it was proposed by Campbell [5], that in the case of aluminum alloys, microporosity can only form if bifilms are present in the liquid metal. This theory is supported by extensive theoretical calculations [11-15] and experimental findings [16-23].

The quantification of the bifilm content of liquid alloys is particularly difficult, as double oxide films are usually small-sized and the thickness of their layers is often only a few nanometers [5, 6]. However, the ability of bifilms to initiate pore formation can be utilized for the characterization of bifilm quantity. For this purpose, the so-called reduced pressure test (RPT) can be a valuable tool, as it was revealed by Fox and Campbell [24, 25].

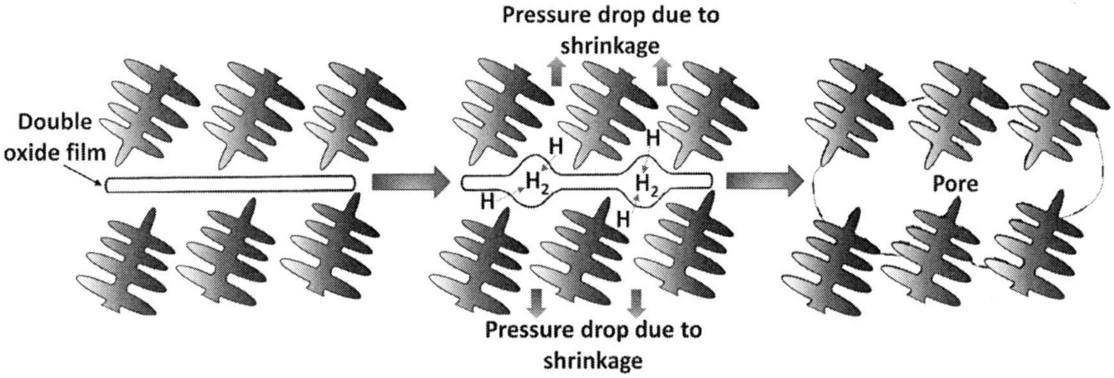

Figure 1. Bifilm initiated pore formation during the solidification of the alloy

Under reduced pressure, pore formation is enhanced during the solidification of the alloy, as the solubility of hydrogen is lowered and the entrapped air between the layers of the double oxide films is being expanded due to the lower pressure [10, 26]. In this way, small-sized bifilms in the liquid alloy can be also detected, as they expand into relatively large-sized pores during the solidification of the metal. The detection of the pores inside the RPT samples can be realized with the quantitative image analysis of the cross-section of the specimens or with non-destructive techniques such as computed tomography (CT) [23, 27, 28].

It was revealed by the authors [23], that volumetric pore number density, which can be determined with the CT-aided porosity analysis of RPT samples, can be used for the characterization of bifilm quantity. However, during the casting of the samples, the selection of pouring height could have a significant effect on the bifilm content of the specimens [17]. According to Campbell [5, 29], a fall of only 12.7 mm is enough to reach the critical melt velocity of aluminum alloys, which is 0.5 m/s. At velocities higher than this, surface turbulence occurs, and the entrainment of the surface oxide layer is unavoidable. On the other hand, the height of typical cups used for the casting of RPT samples is about 40 mm. This means that bifilm formation should be expected during the sampling process, which can significantly alter the results of the melt quality assessment. For this reason, the aim of this study was to investigate the effect of the sampling method on the results of CT-analysis of RPT specimens and to develop a casting technique, which minimizes the chance of bifilm formation during the sampling process.

2. Experimental

To perform the experiments, 2968 g of an AlSi7MgCu alloy was melted in a clay-graphite crucible in an electric resistance-heated furnace. The exact chemical composition of the alloy was evaluated with optical emission spectroscopic analysis. The results are shown in Table 1.

Table 1. Chemical composition of the alloy used for the experiments (wt. %)

Si	Fe	Cu	Mn	Mg	Ti	Sr
6.760	0.105	0.522	0.064	0.390	0.124	0.014

For the casting of RPT specimens, two sampling methods were compared:

- in the first case, conventional gravity casting conditions were used, in which the melt was poured into coated steel cups with the aid of a ladle (Figure 2a). In order to minimize the falling height, the pouring was realized directly from the top of the steel cups.
- in the second case, the steel cups were immersed into the liquid alloy and were immediately taken into the vacuum chamber of the RPT apparatus (Figure 2b). In this way, no pouring was necessary for the sampling of the test pieces.

IOP Conf. Series: Materials Science and Engineering **903** (2020) 012003 doi:10.1088/1757-899X/903/1/012003

Before the casting of the RPT specimens, the steel cups used for the sampling process were preheated to 200 °C. The samples stayed in the vacuum chamber of the RPT machine at 80 mbar pressure for 6 minutes.

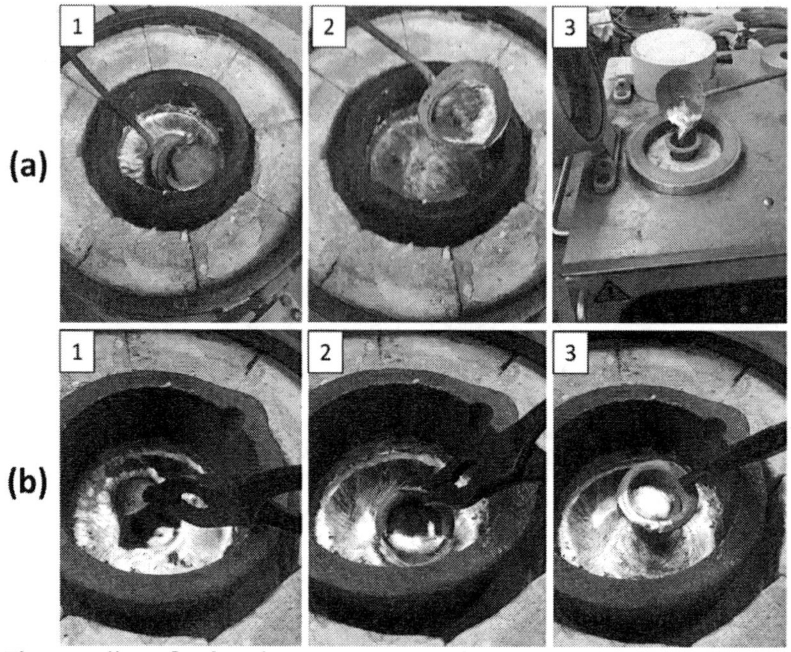

Figure 2. The sampling of reduced pressure test specimens by (a) pouring with a ladle and (b) immersing the cups into the melt

Following the melting of the charge material, 3 RPT samples were taken with both sampling methods. The temperature of the melt was 750 ± 15 °C, which was checked in each sampling step with a K-type thermocouple. After the casting of the samples, 3 g of a commercially available granular cleaning flux was manually stirred into the alloy. The chemical composition of the flux was previously characterized by X-ray powder diffraction measurements. The identified components of the flux are KCl, NaCl, CaF_2, and Na_2SO_4. Following the application of the flux, and a 10 minutes long resting period, the dross layer was removed from the top of the melt, and another series of RPT samples were taken.

Using the Archimedes principle the densities of the RPT specimens were evaluated. For the characterization of the porosity inside the RPT samples, computed tomography (CT) aided porosity analysis was conducted on the test pieces. The CT scanning of the specimens was executed with a GE Seifert X-Cube Compact 225kV apparatus with a tube current of 0.8 mA and an acceleration voltage of 135 kV. 900 projections were acquired during each rotation of the test pieces. For the image reconstruction and processing, VGSTUDIO MAX 3.2 software was used. The segmentation of the voids inside the samples was performed with the VGDefX algorithm, which is included in the porosity analysis module of the software. Based on the local contrast in the grey level of the voxels, a probability value was calculated by the software for each pore. Pores with a volume smaller than 0.05 mm^3 and those which had a probability value lower than 0.9 were excluded. Following the CT analysis, the RPT specimens were cut in half and were ground. For this purpose, SiC grinding papers were used with a grit size up to 1200. The inner surface of the pores found in the cross-section of the RPT samples was studied with a Zeiss EVO MA 10 scanning electron microscope (SEM) equipped with an energy-dispersive X-ray spectroscopic (EDS) detector.

3. Results and discussion

The results of the density evaluation of RPT samples are shown in Figure 3. As can be seen, the pieces which were taken by the immersion of the sampling cup into the melt had a slightly higher density. This can be attributed to the quantity of air being entrapped into the melt during the sampling process. As during pouring, the liquid metal has a falling height of approximately 4 cm, significantly more air can be entrained into the metal, than in the case of immersion of the sampling cup. This suggests that the melt flow is more laminar when no pouring is involved during the sampling process. Based on the results, it is also evident, that due to the fluxing treatment, the average densities of the specimens are remarkably higher, and the deviation of the density values are notably lower. During the experiments, no degassing was conducted, and no bubble formation could be observed during the fluxing treatment which indicates that the flux did not have any degassing effect on the melt. Therefore, the reason for the increment of the density values is either the effect of the flux on the bifilm quantity or on the ability of the double oxide films to grow into pores during the solidification of the alloy.

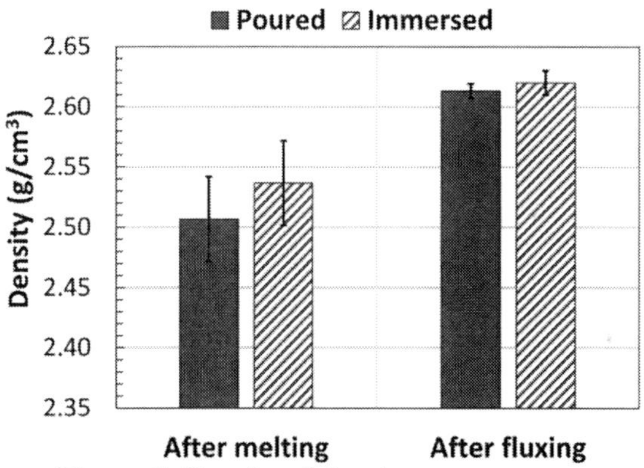

Figure 3. Results of density measurement

Figure 4. shows the pore number density and volume fraction results evaluated by CT-aided porosity analysis of reduced pressure test pieces. Regarding pore number density, no clear relationship could be observed between the method of sampling and the results, as after melting, the specimens taken by immersion had higher pore number density, while after the fluxing treatment, the values of the poured samples were significantly higher. According to the results of Tiryakioğlu et al. [17], the pore number density of the RPT samples increases with the falling height of the liquid metal. This is reasonable because increased melt velocity causes more turbulent flow, which in turn, results in the entrainment of more oxide films into the melt. However, this phenomenon could not be evinced during the experiments of this study. After fluxing, in the case of poured samples, the pore number density values had substantially increased, which indicates higher bifilm content. In practice, fluxing treatments should reduce the number of double oxide films present in the melt, however, this could not be achieved in this case. This can be attributed to the method of flux addition, as manual stirring was applied, which could lead to the entrainment of numerous oxide films into the melt. On the other hand, the results of the pieces taken by immersion were nearly the same as after melting.

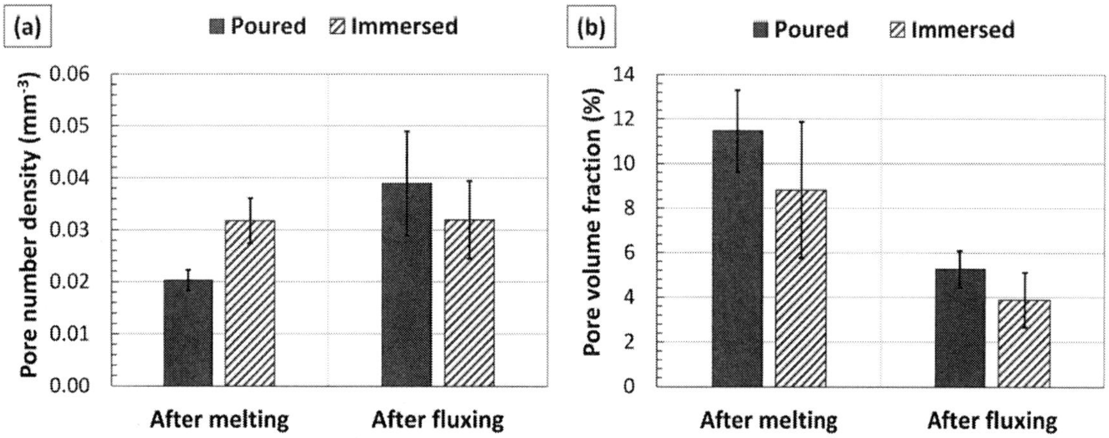

Figure 4. Average (a) pore number density and (b) pore volume fraction results

IOP Conf. Series: Materials Science and Engineering **903** (2020) 012003 doi:10.1088/1757-899X/903/1/012003

Based on the results shown in Figure 4b, the average pore volume fraction was remarkably higher when the samples were taken by pouring. This is in agreement with the density results and can be attributed to the air being entrapped into the liquid metal during the pouring process. After fluxing, significantly lower pore volume fraction values were determined, and as the average pore number density had not been decreased, it is evident, that the fluxing treatment resulted in smaller sized pores. This is clearly visible on the three-dimensional reconstructed CT images of the pores found in the samples (Figure 5).

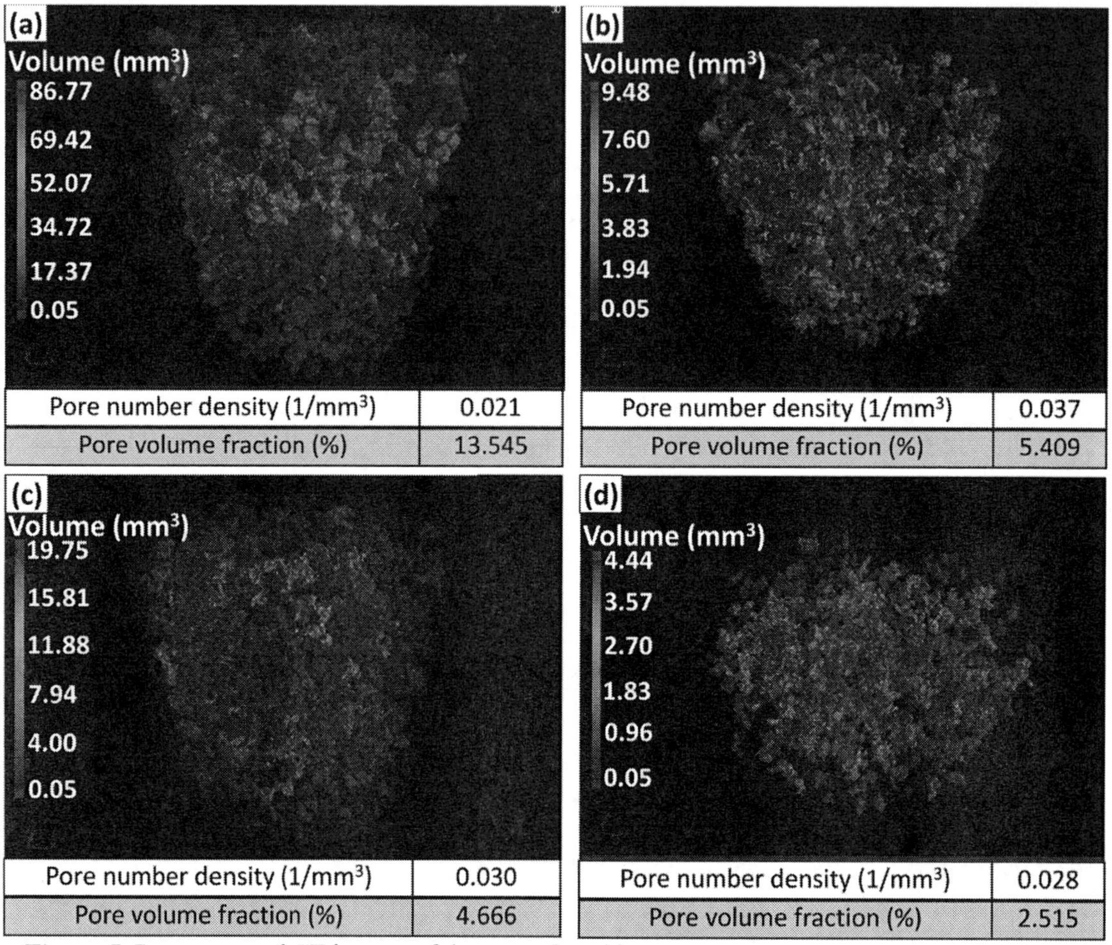

Figure 5. Reconstructed CT images of the pores found in RPT samples: after melting: (a),(b); after fluxing: (c),(d); taken by pouring: (a),(c); and by the immersion of the sampling cup: (b),(d)

During the investigations of the pores found in the RPT samples with SEM, in the case of specimens taken after melting, thin, wrinkled films were found which were covering the inner surface of the pores (as it is demonstrated in Figure 6). These films could be found in each pore, which supports the pore formation mechanism presented in Figure 1. In some cases, they could be easily detected (Figure 6a), while in other cases, only vanishingly thin creases indicated their presence (Figure 6b). However, these films were so thin, that no significant amount of oxygen could be detected during their EDS analysis. On the other hand, after fluxing, the morphology and the size, as well as the chemical composition of the surface of the pores had changed. Figure 7 shows the SEM image of the microstructure of a sample cast following the fluxing treatment. As can be seen, the pores mostly have interdendritic or tear-like morphology, and darker grey regions could be found linking them together. As the contrast of the BSD SEM images is dependent on the atomic number of the sample, these darker areas have a different chemical composition as the alloy.

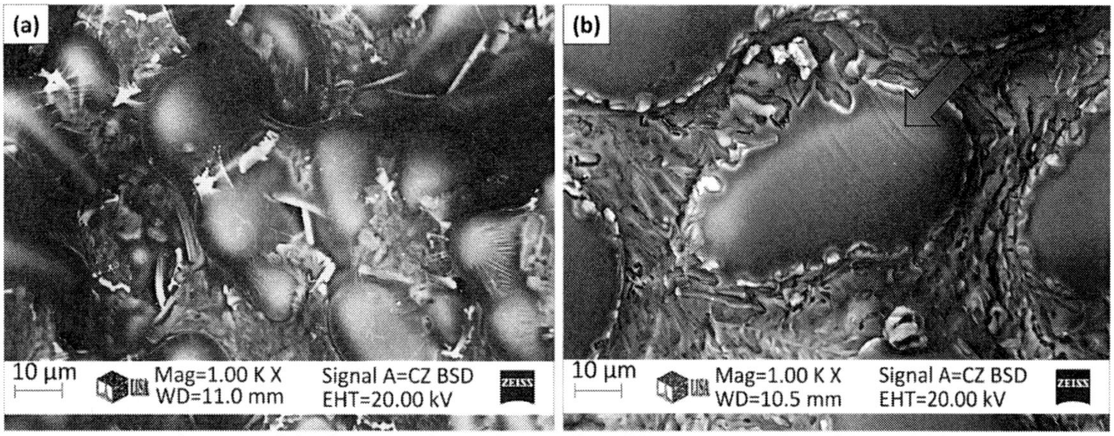

Figure 6. SEM images of the inner surface of pores found in RPT samples with vanishingly thin oxide films covering the α-Al dendritic and eutectic phases

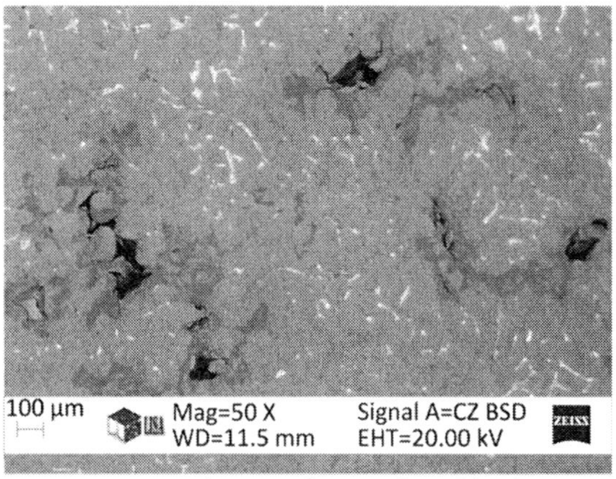

Figure 7. SEM image of pores found in the RPT samples after the fluxing treatment

Figure 8 presents the results of the EDS analysis of a dark grey area found in the vicinity of a small-sized pore. Based on the relatively high oxygen-concentration, the investigated area is most likely consists of Al_2O_3. This indicates that the difference in the size and morphology of the pores is the result of the altered structure of the double oxide films present in the liquid alloy. Based on the morphology of the pores, it can be assumed, that during the transformation of the thin oxide films (Figure 6) into thick oxide inclusions (Figure 7 and 8), some parts of the layers of the bifilm defects could bond together causing single layers of thick and apparently porous oxide inclusions. This phenomenon was also observed by Aryafar et al. [30], who found that in the case of AlSi7Mg0.4 (A356) alloys the two layers of double oxide films can bond together to some extent after a resting period of approximately 13 minutes at a holding temperature of 750 °C.

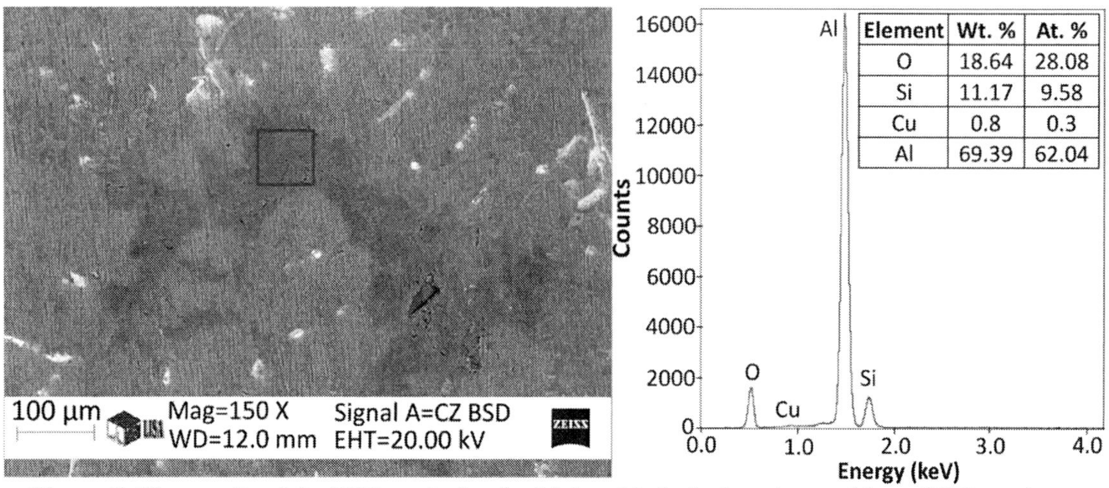

Figure 8. The results of the EDS analysis of a thick oxide inclusion observed in an RPT specimen after fluxing

However, the degree of thickening of the oxide films cannot be explained by only the holding of the melt, as the formation of thick oxide inclusions is only expected after a resting period of several hours [5]. On the other hand, the application of fluxes can remarkably change the rate of oxidation of the alloy. The applied flux contains Na_2SO_4, which is known as an oxidizing compound, which reacts with molten aluminum in a strongly exothermic reaction [31-33]. Based on this, the flux could cause accelerated oxidation of the alloy, which resulted in the formation of thick oxide films whose layers bond together. These oxide films could not open and inflate into pores during the solidification of the alloy, which resulted in lower pore volume fraction and smaller pore size.

4. Conclusion

In this study, two different sampling techniques were compared for the melt quality evaluation of an aluminum casting alloy by the computed tomographic porosity analysis of reduced pressure test pieces. The effect of fluxing treatment on the tendency to pore formation was also investigated. It was found that the type of sampling has a significant effect on the volumetric porosity of the specimens, as by pouring of the liquid metal, a considerable amount of air can be entrapped into the melt, which results in higher pore volume fraction. Fluxing has a notable effect on the structure of oxide films, and in this way on the tendency to pore formation, as the layers of the bifilm defects can bond together due to the application of fluxes.

Acknowledgment

Supported by the ÚNKP-19-3 New National Excellence Program of the Hungarian Ministry for Innovation and Technology.

References

[1] Dispinar D and Campbell J 2011 *Mater. Sci. Eng. A.* **528** 3860–3865 DOI: 10.1016/j.msea.2011.01.084.
[2] El-Sayed M A and Griffiths W D 2014 *Int. J. Cast Met. Res.* **27** 282–287 DOI: 10.1179/1743133614Y.0000000113.
[3] Chiesa F et al 2016 *Int. J. Met.* **10** 216–223 DOI: 10.1007/s40962-016-0029-3
[4] Wang Q G et al 2006 *Metall. Mater. Trans.* B **37** 887–895 DOI: 10.1007/BF02735010.
[5] Campbell J 2015 *Complete Casting Handbook* (Oxford: Elsevier Ltd.) 3-90 341-415 779-783
[6] Campbell J 2006 *Mater. Sci. Technol.* **22** 127–145 DOI: 10.1179/174328406X74248.
[7] Gopalan R and Prabhu N K 2011 *Mater. Sci. Technol.* **27** 1757–1769 DOI: 10.1179/1743284711Y.0000000033.
[8] Campbell J 2012 *Int. J. Met.* **6** 7–18 DOI: 10.1007/BF03355529
[9] Campbell J 2011 *Metall. Mater. Trans. B* **42** 1091–1097 DOI: 10.1007/s11663-011-9575-5.
[10] Dispinar D and Campbell J 2004 *Int. J. Cast Met. Res.* **17** 280–286 DOI: 10.1179/136404604225020704.
[11] Yousefian P and Tiryakioğlu M 2018 *Metall. Mater. Trans. A* **49** 563–575 DOI: 10.1007/s11661-017-4438-6.
[12] Erzi E and Tiryakioğlu M 2019 *Mater. Sci. Technol.* **35** (13) 1656-1659 DOI: 10.1080/02670836.2019.1646010.
[13] Tiryakioğlu M 2018 *Metall. Mater. Trans. A* **49** 5953–5955 DOI: 10.1007/s11661-018-4969-5.
[14] Tiryakioğlu M 2019 *Mater. Sci. Technol.* **35** 738–741 DOI: 10.1080/02670836.2019.1584423.
[15] Tiryakioğlu M 2019 *Shape Casting* (The Minerals, Metals & Materials Society)143–150 DOI: 10.1007/978-3-030-06034-3_14.
[16] Dispinar D and Campbell J 2004 *Int. J. Cast Met. Res.* **17** 287–294 DOI: 10.1179/136404604225020704.
[17] Tiryakioğlu M et al 2018 *Metall. Mater. Trans. A* **49** 5815–5822 DOI: 10.1007/s11661-018-4865-z.
[18] Farhoodi B et al 2014 *J. Mater. Sci. Technol.* **30** 154–162 DOI: 10.1016/j.jmst.2013.09.001.

[19] El-Sayed M A et al 2016 *Int. J. Adv. Manuf. Technol.* **86** 1173–1179 DOI: 10.1007/s00170-015-8240-6.
[20] Shafaei A and Raiszadeh R 2014 *Metall. Mater. Trans. B* **45** 2486–2494 DOI: 10.1007/s11663-014-0135-7.
[21] Samuel A M et al 2017 *Int. J. Met.* **11** 729–742 DOI: 10.1007/s40962-016-0118-3.
[22] Raiszadeh R and Griffiths W D 2006 *Metall. Mater. Trans. B* **37** 865–871 DOI: 10.1007/BF02735007.
[23] Gyarmati G et al 2019 *Mater. Charact.* **157** 109925 DOI: 10.1016/j.matchar.2019.109925.
[24] Fox S and Campbell J 2002 *Int. J. Cast Met. Res.* **14** 335–340 DOI: 10.1080/13640461.2002.11819451.
[25] Fox S and Campbell J 2000 *Scr. Mater.* **43** 881–886 DOI: 10.1016/S1359-6462(00)00506-6.
[26] Lapham D P et al 2002 *Ionics* **8** 391–401 DOI: 10.1039/tf9353101439.
[27] Dispinar D and Campbell J 2006 *Int. J. Cast Met. Res.* **19** 5–17 DOI: 10.1179/136404606225023300.
[28] Uludağ M et al 2018 *Int. J. Met.* **12** 853–860 DOI: 10.1007/s40962-018-0217-4.
[29] Campbell J 2018 *Encyclopedia of Aluminum and Its Alloys* (Taylor & Francis) 275-279 DOI: 10 .1201/9781351045636-140000285
[30] Aryafar M et al 2010 *J. Mater. Sci.* **45** 3041–3051 DOI: 10.1007/s10853-010-4308-8.
[31] Gallo R and Neff D 2008 *ASM Handbook Vol. 15: Casting* (ASM International), 230–239 DOI: 10.1361/asmhba0005300.
[32] Ambrová M et al 2005 *Chem. Papers* **59** 235–239
[33] Strauss K 1970 *Applied Science in the Casting of Metals* (Oxford: Pergamon Press Ltd.) 256-257

IOP Conf. Series: Materials Science and Engineering **903** (2020) 012004 doi:10.1088/1757-899X/903/1/012004

Development of new cladding types for nuclear fuel

Z Hózer[1], T Novotny[1], E Perez-Feró[1], M Horváth[1], A Pintér Csordás[1], P Szabó[1], L Illés[1], M Schyns[2], R Delville[2], D Kim[3], W J Kim[3], M Ševeček [4]

[1]Centre for Energy Research, Budapest
[2]SCK·CEN Belgian Nuclear Research Centre
[3]Korea Atomic Energy Research Institute
[4]Czech Technical University in Prague

E-mail: zoltan.hozer@energia.mta.hu

Abstract. Three different cladding types were tested for nuclear fuel in traditional light water reactors and generation IV gas-cooled fast reactors. Cr coated Zr cladding was tested in steam atmosphere up to 1200 °C to demonstrate moderate oxidation and hydrogen production in accident conditions. 15-15Ti stainless steel alloy and SiC_f/SiC cladding tube samples were treated in helium atmosphere with different impurities for several hours at 1000 °C. Additional mechanical testing and microstructure examinations were carried out with as-received samples and with specimens after high temperature treatments. The experiments results indicated the applicability of the tested materials for reactor conditions in the investigated range of parameters.

1. Introduction

The cladding of the nuclear fuel elements serves as an important barrier against radioactive fission product release from the fuel rod. Furthermore, the cladding tube stabilizes the geometrical arrangement of fissionable materials in the reactor core. Most of current nuclear power plants operates with zirconium cladding tubes. The Zr alloys have low neutron capture cross section, good corrosion and irradiation resistance and high mechanical strength. However, at high temperature the creep of the Zr alloys become significant and in case of incidents or accidents, the steam-zirconium reaction may result in severe degradation of cladding capabilities.

After the Fukushima accident, it was recognized that a hydrogen explosion is one of the major risks of reactor safety during a loss of coolant accident (LOCA) in a light water reactor. The development of accident tolerant fuel (ATF) became a major concern in the nuclear industry. Several ATF cladding concepts are considered [1, 2, 3, 4] and one of the promising technology is to coat zirconium cladding with chromium.

The development of gas-cooled fast reactor technology includes the selection and qualification of fuel components capable to withstand very high temperatures The first core of ALLEGRO will be composed of UOX/MOX pellets in stainless steel 15-15Ti cladding, which was already tested in several sodium-cooled fast reactors for many years. There are very limited data available on the behavior of 15-15Ti cladding in high temperature helium atmosphere [5]. For this reason, a modest experimental program was launched at the Centre for Energy Research (EK).

The second core of ALEGRO could be built using ceramic fuel components. The 850 °C core outlet temperature will need the introduction of new pellets and cladding materials. The candidates are carbide pellets in SiC cladding. The applicability of SiC cladding in gas cooled reactor was addressed in several experimental programmes in the past [6, 7, 8] An important step of this work was the testing of SiC_f/SiC (silicon-carbide fibre reinforced silicon carbide composite) claddings in high temperature helium with the investigation of the effect of different impurities.

2. Coated Zr cladding for light water reactors

The International Atomic Energy Agency (IAEA) launched a Coordinated Research Project (CRP) on the Analysis of Options and Experimental Examination of Fuels with Increased Accident Tolerance (ACTOF) for the period 2015-2019. The CRP objectives included both corrosion and high-temperature oxidation tests. The aim of our studies was to investigate the consequences of high temperature oxidation on the chromium coated E110 type cladding material. (The E110 alloy contains 1% niobium and very low amount of hafnium. This alloy is produced in Russia for structural elements of nuclear fuel assemblies.) The samples were shared between the participating organizations. EK received chromium coated zirconium samples from the Czech Republic for testing at 1100 °C and 1200 °C. The test specimens were Cr-coated samples with about 25 μm coating and reference samples had no coating on the Zr surface (Figure 1.) [4]. The tube specimens were closed by two plugs, so one-side oxidation took place.

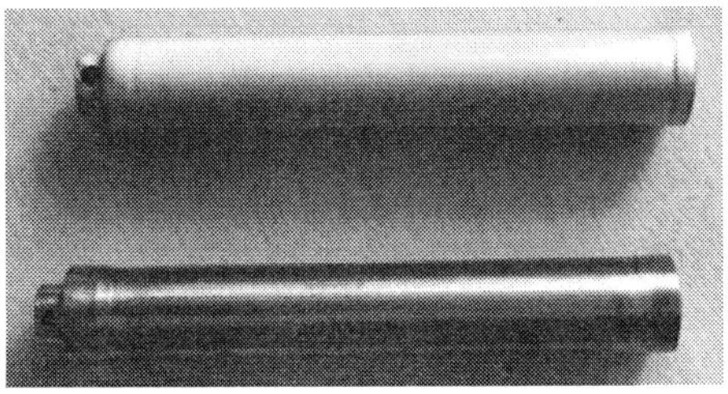

Figure 1. Cr coated (top) and reference Zr (bottom) specimens with 9.1 mm external diameter

The specimens were oxidised in steam mixed with 12 vol.% argon or in pure air atmosphere under isothermal conditions, at 1200 °C and 1100 °C. A high temperature tube furnace with a quartz tube was used for the oxidation of the samples. The experimental setup consisted of a steam generator, a three-zone horizontal furnace with temperature control system and a condensing system. A large vessel (approx. 6 dm^3) filled with cold water was placed under the outlet part of the furnace to quench the oxidised samples. The inlet of the furnace was maintained at 200 °C via a pre-heater to avoid condensation of steam. The outlet gas flow rate was measured by a calibrated bubble gas flow meter. The steam flow rate was about 2 $mg/cm^2/s$. It was determined by measuring the mass of the condensed water collected during the test period, divided by the oxidation time and the cross-sectional area of the quartz tube. After stabilization of the temperature and the steam flow, the quartz boat with the sample was pushed into the heated zone of the furnace. At the end of oxidation, the sample was quenched. The steam oxidation experiments were successfully completed. Table 1 contains the conditions and results of the high temperature oxidation.

Table 1. Results and conditions of the high temperature oxidation

Sample	Type	Atm.	Ox. temp. (°C)	Ox. time (s)	Initial mass (g)	Mass gain (g)	A (cm^2)	Δm/A (g/m^2)
IAEA-02	Zr ref.	steam	1100	3600	11.02454	0.35719	18.64	191.6
IAEA-03	Cr coated	steam	1100	3600	9.96257	0.01582	17.96	8.8
IAEA-04	Cr coated	steam	1100	10800	10.26797	0.02497	18.25	13.7
IAEA-05	Zr ref.	steam	1200	1800	10.98722	0.43748	18.75	233.3
IAEA-06	Cr coated	steam	1200	1800	10.34523	0.02312	18.29	12.6
IAEA-07	Cr coated	steam	1200	2700	9.83425	0.02837	17.88	15.9
IAEA-08	Cr coated	steam	1200	3600	10.37083	0.03443	18.33	18.8

The mass gain of Cr coated samples was very low compared to the reference Zr samples oxidised under similar conditions. The mass gain for Zr samples was 0.3-0.4 g, while the Cr coated samples showed only 0.01-0.04 g increase in steam atmosphere.

After the oxidation tests the tube samples were cut into five pieces. Three 8 mm long rings and two 14 mm long plugs were produced for further testing. From each sample two rings were used for ring compression tests.

Radial ring compression tests were carried out with a standard INSTRON 1195 type tensile test machine at 135 °C. The tensile test machine was equipped with a special furnace in order to keep the high temperature during the test. Each measurement started with a heat-up period (10 min) to establish uniform temperature distribution in the test section. The applied crosshead velocity was 2 mm/min. During the tests, the load-displacement curves were recorded. The results of the ring compression tests showed significant differences between the reference Zr cladding and the Cr coated tubes.

- The Cr coated IAEA-03 sample oxidised at 1100 °C for 3600 s indicated ductile behaviour. The ductile plateau continued after the plastic deformation for a long displacement. The Zr reference sample IAEA-02 reached 450 N maximum load. After that less and less load was enough to cause further deformation, but no clear breakdown was observed for this sample (Figure 2., left)).
- The oxidation at 1200 °C for 1800 s resulted in brittle Zr IAEA-05 samples with maximum force of ≈300 N. The large breakdown at 1.5 mm displacement indicated brittle fracture of the specimen. The Cr coated IAEA-06 rings could withstand loads above 500 N and remained ductile (Figure 2, right).

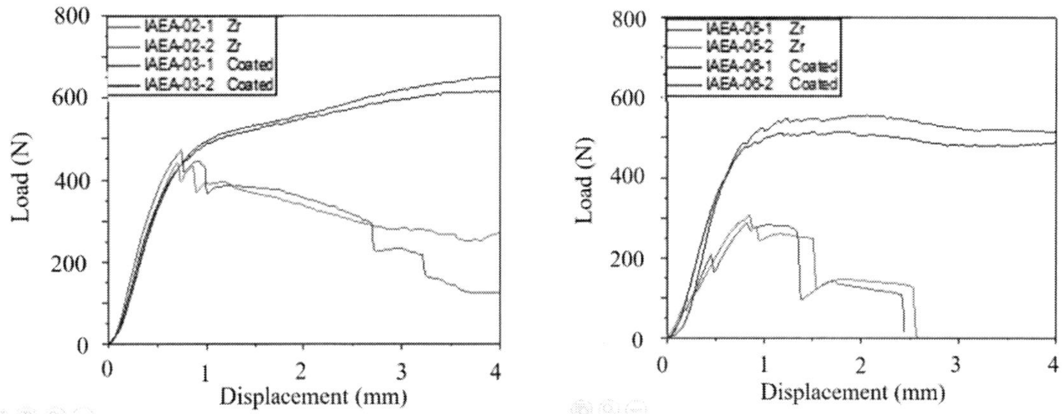

Figure 2. Load-displacement curves of ring compression tests with Zr and coated samples after oxidation at 1100 °C for 3600 s (left) and after oxidation at 1200 °C for 1800 s (right)

SEM examinations were performed to analyse the microstructure of a selected oxidised Cr coated sample. The studied sample was labelled as IAEA-08, oxidised at 1200 °C in steam atmosphere. The oxidation period was 3600 s. It was investigated both in as-received state and after grinding and polishing its cross sectional surface.

A scanning electron microscope (ThermoScientific Scios 2 type) was used for morphological studies. The elemental composition and the distribution of the most interesting elements were investigated with an Oxford energy dispersive X-ray microanalyser (EDX) with a silicon drift detector.

Figure 3 shows a typical Secondary Electrons (SE) image from the outer oxide layer and the neighbouring metal of the as received sample. Also, Backscattered Electrons (BSE) image of the ground and polished sample is represented.

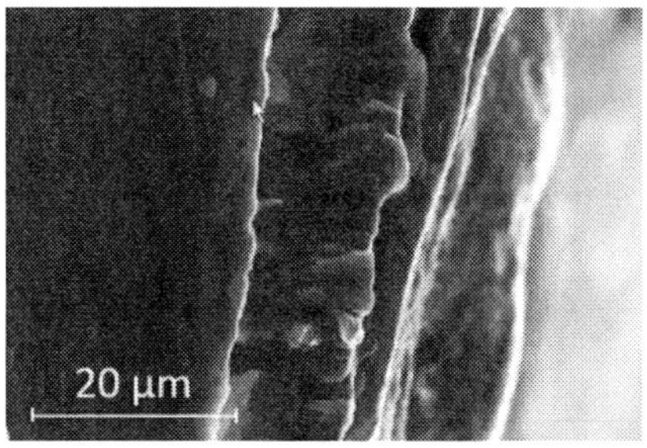

Figure 3. SE and BSE images of IAEA-08 cladding sample taken at 2000 (left) and 3500 (right) times of magnification

On the left hand side part of the SE image the surface of the E110G ring sample can be seen, which has remnants of the cutting process during sample preparation. At the edge of the sample two layers can be distinguished:

- a chromium layer, consisting of columnar crystallites, with thickness between 12 and 16 µm and
- a rather smooth layer of chromium oxide with thickness varying between 2.1 and about 5 µm, which is in good agreement with the data estimated on the basis of mass gain measurements.

On the backscattered electron images (BSE) image shown at right hand side of Figure 3., besides the above mentioned two layers, a third layer between the zirconium and the chromium layer can be recognized. It is brittle intermetallic Cr-Zr layer caused by interdiffusion of Cr and Zr composed of Cr_2Zr phase or $Zr(Fe, Cr)_2$ intermetallic Laves phases with the thickness of about 1 µm.

3. 15-15Ti alloy for the first core of ALLEGRO gas cooled fast reactor

The main objective of the planned experiments was the investigation of the behaviour of titanium stabilized DIN 1.4970 stainless steel claddings under conditions, which are similar to those expected in the ALLEGRO reactor. Furthermore, the high temperature testing in helium was focused on the effect of impurities, which may be present in the atmosphere of the reactor.

The cladding material DIN 1.4970 (15-15Ti) austenitic stainless steel was provided by Studiecentrum voor Kernenergie / Centre d'Étude de l'énergie Nucléaire (SCK·CEN), as support by the MYRRHA project for ALLEGRO. This cladding material was developed as part of the MYRRHA-project. The Ti-stabilized DIN 1.4970 (15-15Ti) austenitic stainless steel cladding tubes were manufactured in 2013 by Sandvik according to specifications from SCK•CEN [5].

The experiments were carried out in a laboratory at EK, in which several high temperature oxidation tests were performed earlier with traditional zirconium claddings [9, 10, 11, 12]. The test conditions were selected taking into account the parameters of GFR reactors and considering the available data on DIN 1.4970 alloy oxidation from other laboratories.

Table 2. Results and conditions of high temperature treatment of 15-15Ti samples

Sample	Atmosphere	Temp. (°C)	Time (s)	Initial mass (g)	Final mass (g)	Mass gain (g)	Mass gain (%)
Allegro–1	100% Helium 6.0	1000	25200	0.56777	0.56873	0.00096	0.3
Allegro–2	90% Helium 6.0 + 10% Hydrogen 6.0	1000	25200	0.56765	0.56768	0.00003	0.0
Allegro–3	90% Helium 6.0 + 10% Methane 4.5	1000	25200	0.56629	0.59128	0.02499	7.7
Allegro–4	90% Helium 6.0 + 10% Nitrogen 6.0	1000	25200	0.56586	0.56696	0.00110	0.3
Allegro-5	as received						

The review of potential impurities for the helium cooled HTR-10 reactor indicated that during normal operation H_2O, CO, N_2, O_2, H_2 and CH_4 components may be present in the atmosphere of the reactor [13]. It was decided to focus on the separate effect of different impurities. Hydrogen, nitrogen and methane were selected for further studies. High content (10%) of impurities were chosen in order to enhance the potential chemical reactions. Testing in pure helium was also a main item in the test programme. The selected test atmospheres are summarised in Table 2.

Treatment at 1000 °C temperature was selected in order to have higher than typical for ALLEGRO reactor temperature in the experimental facility. The treatment time was 7 hours (25200 s) for each test to have comparable conditions in different atmospheres. Furthermore, this period seemed long enough to indicate the potential effects of gases on DIN 1.4970 alloy. The high temperature treatment of DIN 1.4970 steel samples was carried out in a horizontal resistance furnace. The gas mixtures were injected at the inlet of the furnace. The gas flowrate was measured in the outlet section. The cladding samples were placed into a quartz sample holder, which could be moved in and out of the furnace inside a hermetically closed tube.

At the beginning of the tests the furnace was heated up to 1000 °C. When the stabilized temperature profile was reached the samples were moved into the furnace and the treatment was initiated. The gas flow rate was 5.9 $cm^3/(cm^2\ s)$ in the gas flow meter after the quartz tube. At the end of the treatment time (25200 s) the samples were removed from the tube and were cooled down in the external tube. Finally, the samples were taken from the sample holder and the mass changes were determined by weighing.

The change of the masses showed the followings results:

- The pure helium atmosphere resulted in small (0.3%) mass gain.
- The hydrogen content in the atmosphere seemed to have no effect on the mass change of samples compared to pure helium.
- The presence of nitrogen in helium showed 0.3% mass gain for the sample.
- The highest mass gain was measured in the case of methane impurity: the mass of the sample increased by 7.7%. The carbon deposition observed on the investigated samples caused the mass gain. The mass gain in methane atmosphere could be associated to the decomposition of methane at high temperature, which can produce carbon deposit: $CH_4 \rightarrow C + 2\ H_2$.

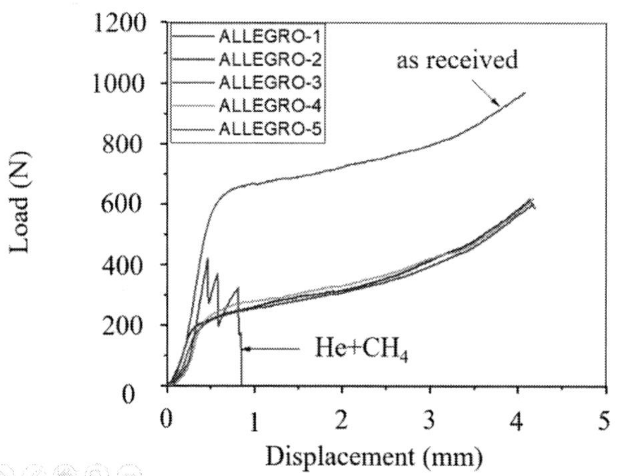

Figure 4. Load-displacement curves of ring compression tests with 15-15Ti samples after thermal treatments in different atmospheres

Ring compression tests were applied to characterize the compression resistance of the samples The as-received and pre-treated 15-15Ti ring specimens were examined in radial compression tests using an Instron 1195 universal test machine. The velocity of the crosshead movement was fixed at 2 mm/min, and the tests were performed at room temperature. In the mechanical tests the samples were loaded until failure or until maximum possible displacement. The high temperature treatment obviously softened the cladding material.

The ring compression tests showed ductile behavior for most of the samples. On the basis of load-displacement curves (Fig. 4.) the following conclusions could be drawn:

- Brittle failure was observed only for the Allegro-3 ring, which was treated in helium+methane atmosphere. It cannot be explained by the simple carbon deposition on the surface of the samples. However, due to the long (7 h) oxidation time and the high temperature (1000 °C) the carbon could interact with the steel and the microstructural changes could cause the brittle behavior.
- The elastic behavior ended at about 600 N force for the as-received sample (Allegro-5), the sample showed ductile behavior up to 4 mm displacement.
- The samples after treatment in pure helium (Allegro-1), in helium+hydrogen (Allegro-2) and in helium+nitrogen mixture (Allegro-4) ended the elastic deformation at about 200 N force. There were practically no differences between these samples in the mechanical behavior.

It was an important observation that the pure helium content did not result in brittle failure. So the suspected helium uptake and its embrittlement effect did not take place.

The failure of 15-15Ti cladding tubes in accident conditions was investigated in burst tests. An electrically heated, three-zone steel tube furnace was constructed and built to keep the steel samples at a constant high temperature. The high pressure argon gas was fed into the cladding samples by a needle valve, in order to maintain a certain rate of internal pressure increment. In the case of very small pressure increments, a 1 dm^3 buffer was put between the needle valve and the sample via a 0.7 mm inner diameter capillary to better regulate the argon flow.

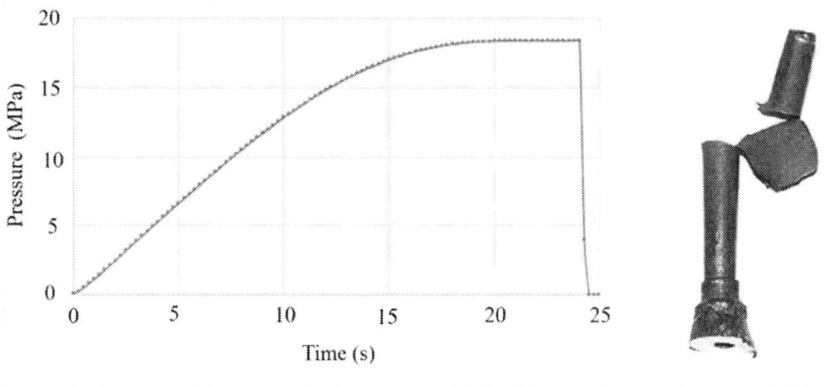

Figure 5. Pressure history and view of the 15-15Ti sample after burst at 1000 °C

The pressure was measured by two Suco 0720 type, linear mechanical– inductive transducers and two MAI-250 type Bourdon-tube manometers to read the pressure of the buffer and the sample during operation. The measurements were performed on 80 mm long stainless steel cladding samples. The pressure of the argon gas in the cylinder used as argon source was around 19 MPa at the beginning of the experiment and therefore this was the highest achievable pressure in the samples. This value has continuously decreased during the experiment as the argon gas has been used up. Burst of the samples with this pressure condition could be reached at 1000 °C temperature. Figure 5. shows the pressure history and the view of the sample after burst.

4. SiC$_f$/SiC cladding for the second core of ALLEGRO gas cooled fast reactor

SiCf/SiC cladding tubes were produced at Korea Atomic Energy Research Institute (KAERI) and their behavior was tested in the framework of extensive experimental series [6, 7, 8]. The work performed at KAERI with SiC$_f$/SiC composites for nuclear applications includes the development of light water reactor (LWR) fuel cladding and in-core components for very high temperature reactors (VHTR). One series of KAERI tests focused on the investigation of behaviour CVD (Chemical Vapour Deposition) SiC and SiC_f/SiC composite in the oxygen containing He and air. In air atmosphere positive mass gains were observed above 1100 °C [6].

Important step of qualification for gas-cooled fast reactor conditions was the testing of SiC_f/SiC claddings in high temperature helium with different impurities. Four Duplex (two-layer composite, cross section shown in the left hand side picture of Figure 6.) and four Triplex (three layer composite with polished external surface, cross section shown in the right hand side picture of Figure 6.) type specimens were treated in following atmospheres: 10% H_2 + 90% He, 10% N_2 + 90% He, 10% CH_4 + 90% He and 100% He. 1000 °C oxidation temperature was selected and the oxidation time was 7 hours. The change of masses showed the followings results:

- The pure helium atmosphere resulted insmall (0.02-0.04%) mass reduction for both Duplex and Triplex samples.
- The hydrogen content in the atmosphere seemed to have no effect on the mass change of samples compared to pure helium. Small mass reduction (0.02%) was detected in both sample types.
- In case of methane impurities the mass gain was measured. The mass of Duplex sample increased with 1.7%, while the Triplex with 0.3%.
- The presence of nitrogen in helium showed 0.03% mass reduction for the Triplex sample and no mass change for the Duplex one.

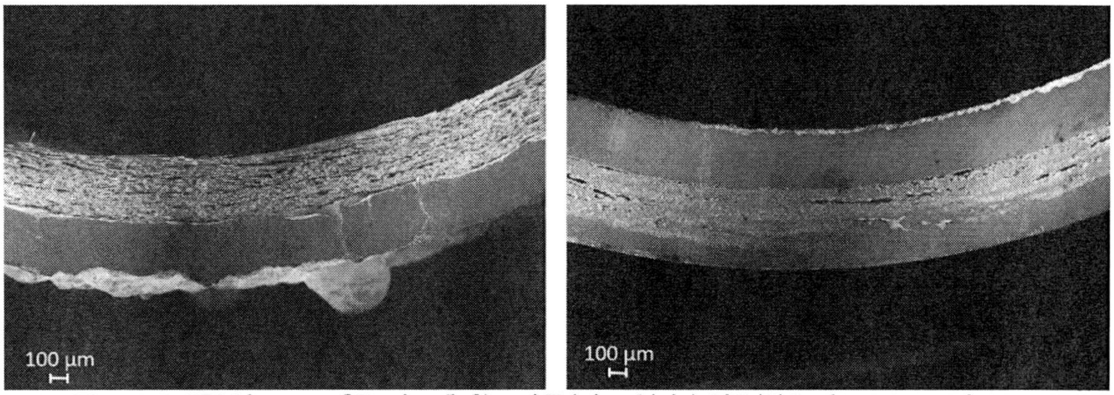

Figure 6. SEM images of Duplex (left) and Triplex (right) SiC_f/SiC tube cross sections

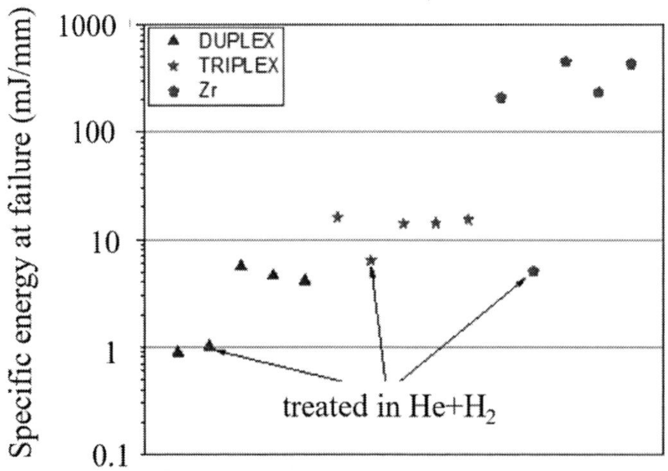

Figure 7. Specific energy at failure for Duplex, Triplex and Zr samples (the order of samples for all three series corresponds to treatments in He, He+H_2, He+CH_4, He+N_2 and as-received condition)

Ring compression tests were applied to characterize the compression resistance of the sample. The as-received and pre-treated Duplex and Triplex ring specimens were examined in radial compression tests using an Instron 1195 universal testing machine. The velocity of the crosshead movement was fixed at 2 mm/min, and the tests were performed at room temperature. The rings were compressed until fracture of the samples. The comparison of Duplex and Triplex samples shows that the Triplex tubes have much higher compression resistance capability.

On the basis of load-displacement curves the specific energy to failure (related to unit length of the ring) was calculated. There were parallel zirconium (E110 type) cladding specimens tested in the high temperature helium environment under the same conditions as the SiC samples. The comparison of calculated specific energies at failure showed that for the SiC_f/SiC samples much less energy was needed to cause failure compared to Zr samples. Figure 7. shows these energy values for Duplex, Triplex and Zr samples in logarithmic scale. The last point of each series is the as-received sample. Figure 7. indicates that the treatment in hydrogen rich helium caused dramatic changes for all three types of samples, while the Duplex sample after treatment in pure He suffered the most impact.

5. Summary and conclusions

Three different new cladding types were tested in high temperature facilities. The investigations included mechanical testing and microstructure examinations, too. As general conclusion, the performed testing confirmed the applicability of all three cladding for reactor conditions in the investigated range of parameters:

- The comparison of Cr coated Zr tubes with the non-coated Zr specimens after oxidation in steam at 1100-1200 °C proved that the coated samples could withstand the high temperature conditions without severe degradation. Only modest oxide scale was formed on the Cr coating and the tube samples remained ductile. The internal metallic phase of coated tube did not suffer such changes, which are typical for uncoated Zr cladding.
- The thermal treatment at 1000 °C softened the 15-15Ti samples. The detrimental effect of methane resulted in brittle failure of the sample at low load. The burst tests proved that the DIN 1.4970 (15-15Ti) cladding tubes can keep their integrity at high temperature even with very high internal pressure. The failure pressure of samples tested at 960-1000 °C was above 18 MPa.
- The ring compression tests in SiC_f/SiC indicated that the compression resistance capability of Triplex type cladding is much better compared to Duplex ones. The comparison with traditionally used Zr claddings showed much less energy to failure for SiC_f/SiC samples.

References

[1] Pint B A *et al* 2013 *J. Nucl. Mat.* **440** 420
[2] Brachet J C *et al* 2019 *J. Nucl. Mat.* **517** 268
[3] Karoutas Z *et al* 2018 *Prog. Nucl. Energy* **102** 68
[4] Ševeček M *et al* 2018 *Nucl. Eng. Technol.* **50** 229
[5] Cautaerts N *et al* 2017 *J. Nucl. Mat.* **493** 154
[6] Daejong Kim *et al* 2013 *Oxid. Met.* **80** 389
[7] Daejong Kim *et al* 2015 *J. Nucl. Mat.* **458** 29
[8] Daejong Kim *et al* 2017 *J. Nucl. Mat.* **492** 6
[9] Király M *et al* 2019 *Nucl. Eng. Technol.* **51** 518
[10] Perez-Feró E *et al* 2016 *Prog. Nucl. Energy* **93** 89
[11] Hózer Z *et al* 2008 *J. Nucl. Mat.* **373** 415
[12] Kozsda-Barsy E *et al* 2018 *J. Nucl. Mat.* **508** 423
[13] Yao R P *et al* 2002 *Nucl. Eng. Des.* **218** 163

IOP Conf. Series: Materials Science and Engineering **903** (2020) 012005 doi:10.1088/1757-899X/903/1/012005

THE SIMULATION OF THE STEADY-STATE CONDITIONS OF ALUMINIUM DIRECT-CHILL CASTING

Daniel Molnar, Viktor Karpati, Imre Budavari

University of Miskolc, Faculty of Materials Engineering, Miskolc-Egyetemvaros, Hungary

E-mail: daniel.molnar@uni-miskolc.hu

Abstract. In this research, the simulation of aluminium direct-chill casting has been achieved focused on the fluid flow and thermal conditions. Control Volume Method is used applying the NovaFlow&Solid simulation software and the JMatPro software was used for the calculation of the material properties. 6 ingot geometries were examined using 4 commercial alloys. A hybrid method is developed for the calculation of the material properties and as a result, the temperature profiles and the liquid phase is analysed.

1. Introduction

The second most economically important metal is aluminium, around 80 million tonnes of aluminium cast every year and more than half is direct-chill cast [1].

Direct-chill castings can be cast from primary metal or secondary scrap materials to create intermediate products for remelting or for forming, such as rolling. The quality of direct-chill castings has an effect on the yield and the properties of the final products used. Typical products are used in transportation, automotive sector, building and electronic industry. The direct-chill casting companies, therefore seek to control the quality of the castings, while minimising costs, maximising capacity, at the same time producing with minimal environmental impact.

2. Modelling and simulation of the direct-chill casting process

Direct-chill casting appears to be a simple process involving the cooling and solidification of liquid metal into an ingot; however, the physical phenomena taking place are complex, with many interactions between heat transfer, fluid flow, and stresses. The development of mathematical and physical models has paralleled the development of technology and continually expanded our understanding of these complex phenomena [2]. From the very beginning of direct-chill casting development, it became clear that the proper choice of process parameters makes all the difference in successful casting. See the examined device in Figure 1. [3].

Figure 1. Direct-chill casting device

Many modelling and simulation methods can be employed for direct-chill casting, such as physical modelling, empirical relationships, statistical methods, analytical solution methods, inverse methods and numerical methods. The most common numerical methods implemented in technological simulation software are Finite Difference Method (FDM), Finite Volume Method (FVM) and Finite Element Method (FEM) [4].

Finite Difference Method is a numerical method where a complex problem is solved by discretizing the complex region of problem (domain) into a finite number of small portions (control volumes). Material properties are assumed to be constant throughout the volume. Therefore, for high accuracy of the results, the domain should be divided into a maximum number of control volumes possible taking into account the computational time. FDM is a differential scheme which is the approximation of Taylor series expansion. Calculations are iterative and done at a predetermined time-step. The results can be stored at the end of each time-step or after a pre-determined number of steps [5].

Unlike FDM, Finite Volume Method is an integral scheme. Although the idea of discretizing the domain into small volumes remains the same, the use of integral formulations is advantageous in treating the Neumann boundary continuous as well as that of discontinuous source terms due to their reduced requirements on the regularity or smoothness of the solution. It is possible to solve to be a control on mesh elements at the borders of the geometry. With cubic elements in the volume and border cells at the boundary, the simulations are much faster and accurate. Also, the method provides filling of a necessary fraction of a cell instead of filling cell by cell [6].

The Finite Element Method discretizes the complete domain of the problem into small pieces (a.k.a. elements). Each element is made up of nodes (corner points) and edges, which store material properties to be used in the computation. The solution is done by using these values to determine a quantity for these specific points (a.k.a. Gauss points) within the elements. The position of these points in elements is a function of the integration applied, initial coordinates of the nodes, and the element shape. Values of variables, which are considered to be constant in FDM/FVM across the elements, are calculated using some interpolation function. The treatment of time in an iterative and step-wise manner is similar to FDM/FVM [7].

2.1 The simulation model

At this point, it should be emphasized that a mathematical model of a technological problem always is an approximation of the original problem no matter the solution method. Thus, an analytical method, although giving the exact solution to the mathematical problem, results in an approximate solution to the problem. It therefore only makes sense to take all the approximations into account if the value or applicability of a solution to a technological problem should be evaluated. Exact consistency between the mathematical solution and the mathematical model is thus not enough to ensure a good solution to the original problem, even though a certain degree of accuracy in the mathematical solution is always desirable [8].

The physical phenomena behind a technological problem should be identified and a mathematical model must be written. This mathematical model must be solved using an analytical or a numerical solution and the physical interpretation of this mathematical solution should be done for the technological solution. Especially in manufacturing processes such as casting, the misinterpretation of otherwise correct mathematical results could lead to wrong conclusions, and hence to no solution of the originating problem [9].

In this paper, the Finite Volume Method is used to solve the material- and heat transport processes using the commercial software NovaFlow&Solid. The concept of simulation experiments can be seen in Figure 2.

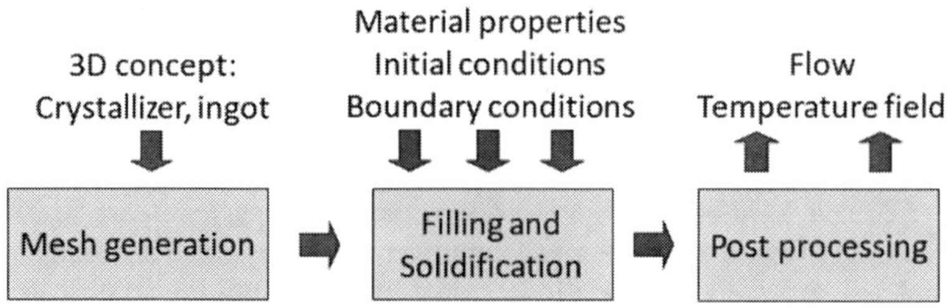

Figure 2. Concept of the simulation experiments

Pre-processing, id est definition of the problem. The first step is to define the geometry of the casting system into a discrete number of segmented volume elements for the subsequent calculations. Before the equations that govern the filling and solidification processes can be solved, the necessary thermophysical material data must be available. Apart from the material data themselves, other relevant process parameters have to be defined. Initial conditions for the unknown quantities and boundary conditions for the unknowns must be defined. Other relevant information also needs to be input, so that all the factors that affect the filling and solidification of the casting can be accounted for.

Main-processing, id est the calculations. After the geometry has been defined and the mesh generation has been performed, the most demanding part of the numerical simulation follows in respect of both the algorithmic development and the requirements for computer capacity, solution of the governing equations. The most usual approach here is to solve all the basic equations, this being a prerequisite for simulating all relevant casting problems of a technical nature. It is clear that these calculations, in which primitive fields such as temperatures, displacements, stresses, velocities, pressure, etc. are determined, require the solution of the governing differential equations.

Post-processing, id est presenting the results. After the computations, the resulting basic fields (temperatures, velocities, pressure, displacements, stresses, etc.) should be presented appropriately. The results are often made more instructive by colour displays and curves.

In the given experiments the material properties, the 3D geometry and the initial conditions were modified for the proper calculation results.

2.2 Material properties

Before the equations that govern the filling and solidification processes can be solved, the necessary thermophysical material data must be available. For the calculation of the temperature field in the casting system, information is required about the densities, specific heat capacities, and thermal conductivities for all of the materials in the casting system (cast metal, mould, cores). Besides, the latent heat of fusion of the cast alloy is required. To calculate velocities and pressures during mould filling, the viscosity of the liquid metal alloy is also needed. Since all of these quantities can vary significantly with temperature, the property variations with temperature must be considered as part of the simulation.

The source of the thermophysical material data can be calculation, literature, data bank and measurement. The examined alloy compositions (wt%) can be seen in Table 1.

Table 1. Chemical composition of the aluminium alloys (wt%)

Alloy	Si (wt%)	Fe (wt%)	Cu (wt%)	Mn (wt%)	Mg (wt%)	Cr (wt%)	Zn (wt%)	Ti (wt%)	Al (wt%)
3117	0.3	0.2	0.8	1.2	<0.1	<0.1	<0.1	0.1	rest
4147	11.8	0.2	<0.1	<0.1	0.2	<0.1	<0.1	<0.1	rest
5083	0.1	0.2	<0.1	0.5	4.4	<0.1	<0.1	<0.1	rest
8006	0.1	1.3	<0.1	0.4	<0.1	<0.1	<0.1	<0.1	rest

The chemical compositions are measured average values, calculated from industrial batches. Based on the chemical composition and the cooling velocity of the ingot, which is calculated by a preliminary simulation (0,0544°C/s), the thermophysical material data are calculated with the JMatPro software. The JMatPro software calculates a wide range of material properties for multi-component alloys. By several iterations, the following material data are calculated as a function of temperature in every 2°C cooling step: heat conduction, specific heat, density, viscosity, heat transfer coefficient and latent heat. This means approx. 1700 material data per alloy. These material properties must be installed to the NovaFlow&Solid simulation software for further calculations.

In the NovaFlow&Solid simulation software, a specific databank is used for the material and heat transfer calculations. A standard aluminium foundry alloy composition (EN AC-46000) is used as a so-called donor alloy, and the values of this, e.g.phase diagram and material properties, are used for the creation of the new alloy properties by the following routine:

1. Study temperature-thermophysical data values of the donor alloy.
2. Determination of the thermophysical data on the NovaFlow&Solid curves.
3. Find these values on the JMatPro curve.
4. If the examined value can be found on the JMatPro curve and the temperature value of it fits the NovaFlow&Solid values, use the data of JMatPro. (See Figure 3.)
5. If the examined value cannot be found on the JMatPro curve or the temperature value of it doesn't fit the NovaFlow&Solid values, use the data of Novaflow&Solid databank. (See Figure 3.)
6. Create the new alloy in the NovaFlow&Solid databank.

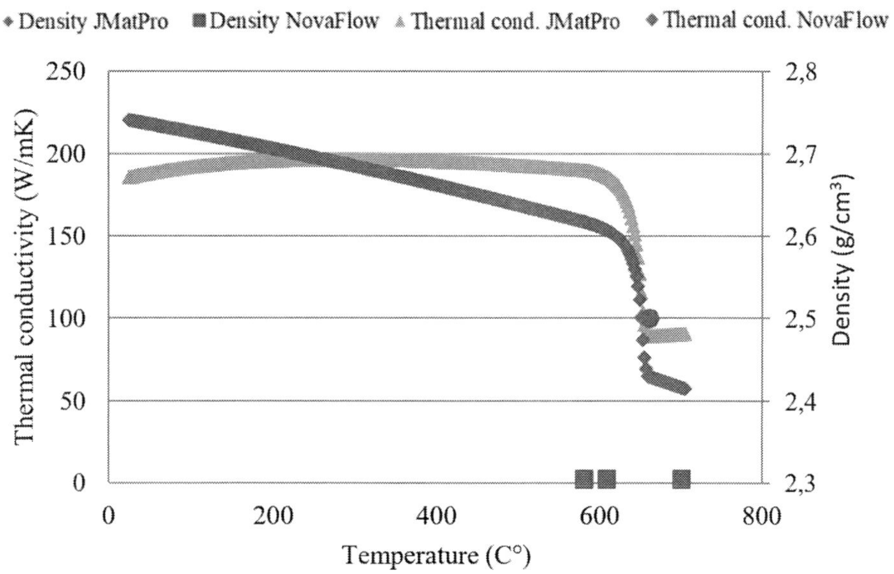

Figure 3. Corresponding values of thermal conductivity and non-corresponding values of density

3. Design of Experiments

First, the filling of the crystallizer is analysed, where the geometry consists of the die, the primary cooling and the starting block in the initial position. Later on, the solidification of the ingot is examined based on several process parameters. In this case, the geometry is completed with the secondary cooling. The mid-line 3D section of the examined CAD models can be seen in Figure. 4.

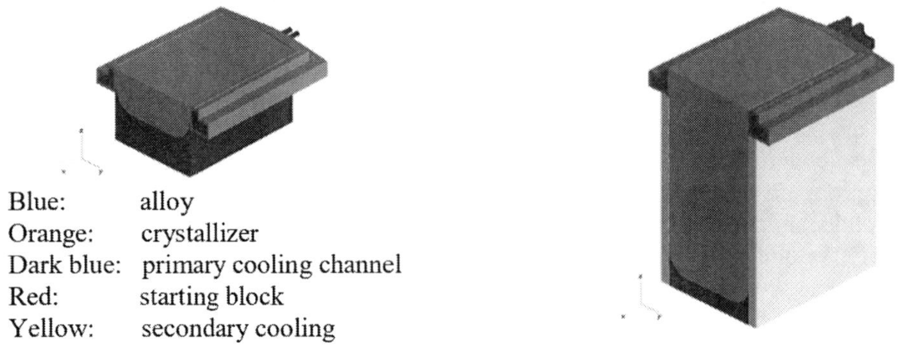

Figure 4. Left side: crystallizer geometry; Right side: ingot geometry

The design of experiments, the main process parameters and the number of iterations (NoI) can be seen in Table 2.

Table 2. Design Of Experiments

Nr.	Action	NoI	Alloy	Ingot mm	Temperature (°C)	Pouring (kg/s)
1	Filling and solidification of the crystallizer	4	4147	520x1590	705	2-27
Nr.	**Action**	**NoI**	**Alloy**	**Ingot** mm	**Alloy temp.** (°C)	**2nd cooling** (°C)
2	Definition of the shrinkage cavity	2	4147	520x1590	705	65
3	Solidification of the divided ingot	1	4147	520x1590	705	65
4	Solidification of the 4147 alloy	3	4147	400x1460	680	65
5	Solidification of the 5083 alloy	3	5083	465x1120 465x1620	675	65
6	Solidification of the 8006 alloy	3	8006	520x1340	705	65
7	Solidification of the 3117 alloy	3	3117	330x1310 400x1310	715	65

4. Simulation results

4.1 Flow analysis

In the filling analysis of the crystallizer, the 4147 alloy is poured to the cavity with the initial temperature of 705 (°C). The alloy is poured to the cavity by a round shape stream, where the radius is 30.0-80.0 (mm), using a 91.5 (mm) defined pressure height and a defined flow. Three cases were examined: slow pouring, fast pouring, and ideal pouring.

In the case of slow pouring, the flow was 2.0 (kg/s), the heat loss of the liquid alloy is too fast, and the metal solidifies to the surface of the crystallizer layer by layer, which causes overlapping and cold flow defects. See Figure 5. left side, liquid phase scale: 5-95(%).

In the case of fast pouring, the flow was 27 (kg/s), the metal flows with a lot of turbulences and splashes in the crystallizer. Turbulence causes air entrapments and surface oxide bifilms which is one of the main sources of shrinkages. See Figure 5. right side, velocity scale: </>0.5 (m/s) [10].

The so-called ideal pouring flow was calculated in the following way. For gravity castings the pouring time (s) can be calculated by the following empirical relationship: pouring time is equal to the square root of the double amount of metal (kg). The relationship is not scientific but works well in case of heavy castings. In the next step, the amount of the poured metal is divided by the pouring time, which gives the pouring flow: 13.7 (kg/s). In case of ideal flow the velocity of the metal is lower than

0.5 (m/s), which means that the movement of the metal is turbulence-free and undisturbed, no pouring defects developed.

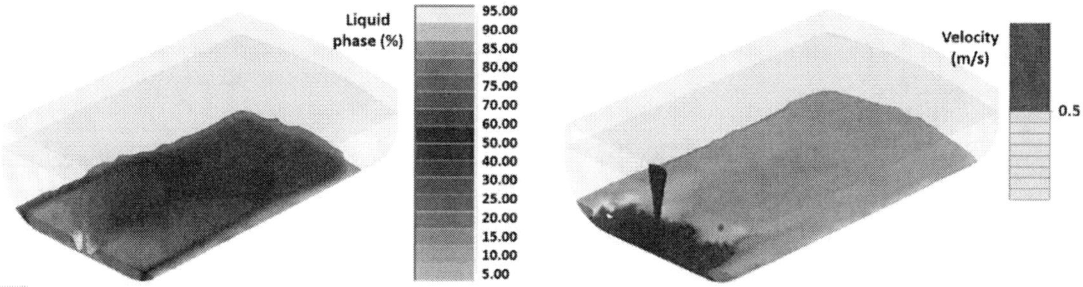

Figure 5. Left side: Cold flow defects (grey) during casting. Filled volume: 10.38%
Right side: Turbulence defects (red) during casting. Filled volume: 13.5%

4.2 Shrinkage analysis
In the solidification calculations, the steady-state conditions were examined, because the moving of the ingot was not possible to observe. In real life, the length of the ingots is 5400 (mm). To determine that which section of the ingots should be examined a so-called shrinkage analysis is implemented where no feeding is applied in normal gravity conditions. The mid-line 3D section of the examined CAD model can be seen on the right side of Figure 4. The initial parameters can be seen in Table 2. Further initial parameters are material of the die and starting block: EN AW-3003. The temperature of the die and starting block: 20 (°C). The flow of primary cooling: 27,7 (l/s). The results of the shrinkage analysis in the mid-line 3D section can be seen in Figure 6.

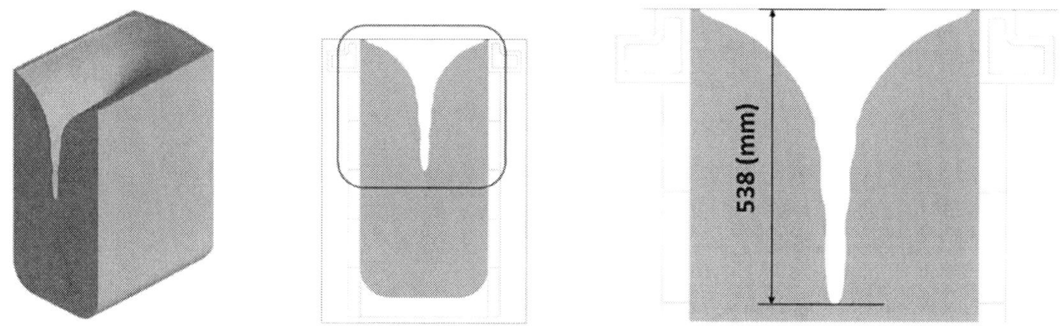

Figure 6. Determination of the shrinkage depth

A calculated depth of the shrinkage hole is 538 (mm). The weight of the poured metal is 1985.3 (kg) and the volume of the shrinkage is 6.3% of the initial geometry.

4.3 Solidification analysis
At the starting point of the shrinkage analysis, the whole ingot geometry had a uniform temperature. To approach the real-life conditions, the partitioning of the ingot is needed. The sections of the ingot are calculated by the parameters of the CAD model. The mid-line 2D section of the divided ingot with the initial temperature of the metal, and the thermal conditions, can be seen in Figure 7. The colours correspond to the earlier models. The ingot geometry is divided as follows: ingot height in the crystallizer is 140 (mm), ingot height in the starting block is (120) mm, the middle part of the ingot is divided into 4 times 200 (mm).

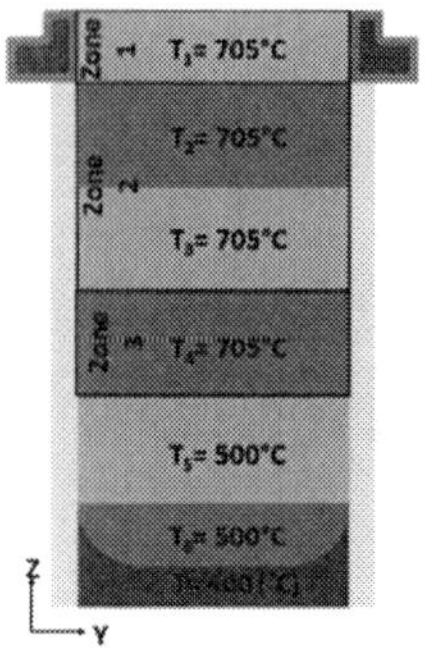

Thermal conditions	
Zone 1	Sideward the heat is added to the crystallizer. In +Z direction the heat is added to the environment. In –Z direction the heat is added to the melt.
Zone 2	Sideward the heat is added to the secondary cooling. In Z direction the heat is added to the melt.
Zone 3	Sideward the heat is added to the secondary cooling. In +Z direction the heat is added to the melt. In –Z direction the heat is added to solidified parts.

Figure 7. Divided ingot geometry with initial temperatures, and the thermal conditions

The solidification of 4 alloys and 6 ingots are examined, here the results of the 520x1340 ingot are presented (alloy 8006). The initial parameters of the calculation can be seen in Table 2. and Figure 7. The solidification of the ingot in the mid-line 3D section can be seen in Figure 8. Liquid phase scale: 5-95%.

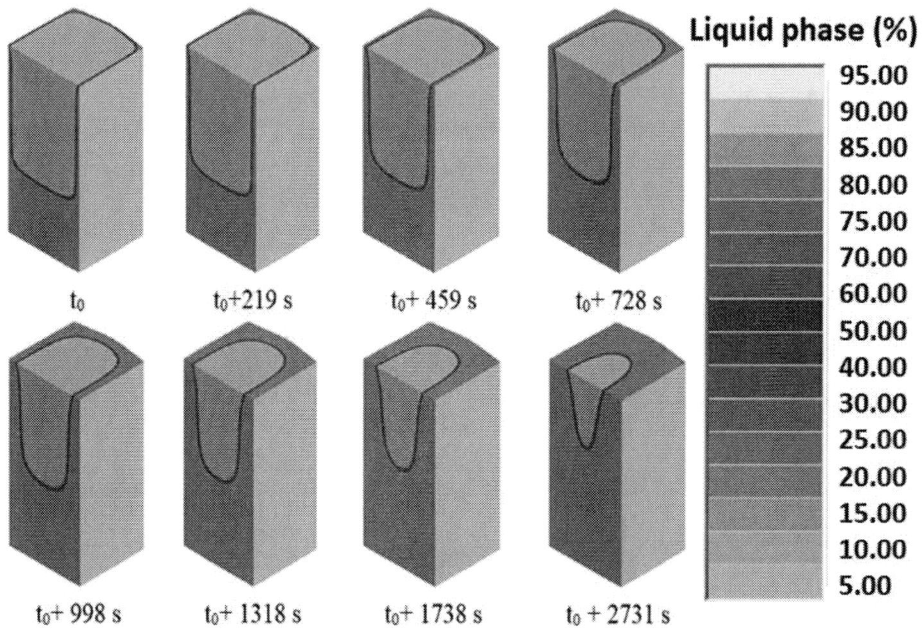

Figure 8. Solidification of the 520x1340 ingot geometry

For further analysis of the solidification process, thermocouples are defined in the mid-line section of the crystallizer with 23 (mm) distances. The cooling curves of the thermocouples can be seen on the left side of Figure 9. The results of the analysis of the phase fraction of the liquid phase can be seen on the right side of Figure 9.

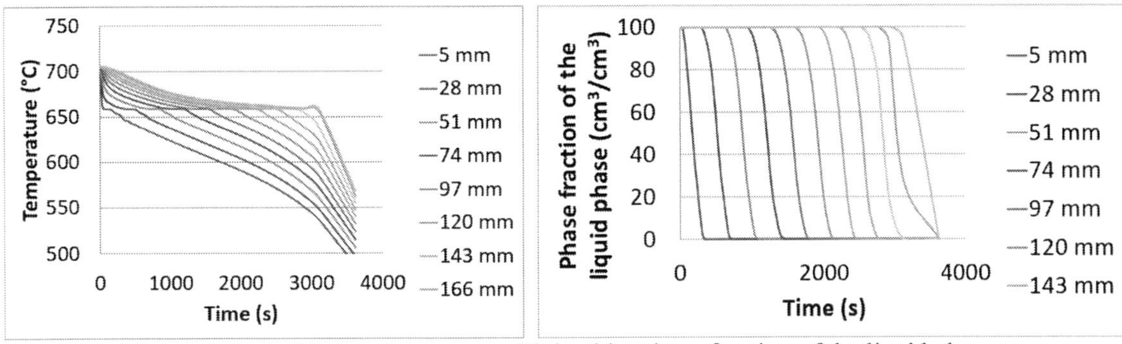

Figure 9. Left side: cooling curves; Right side: phase fraction of the liquid phase

5. Summary

In this research, the simulation of aluminium direct-chill casting has been achieved focused on the fluid flow and thermal conditions. The background of the direct-chill casting process is presented based on the literature and the simulation possibilities are considered. The material properties of the examined alloys are calculated based on industrial data using JMatPro software and the databank of the NovaFlow&Solid simulation software. All the simulation experiments are accomplished with the NovaFlow&Solid platform using Control Volume Method. By the help of fluid flow analysis, optimal casting parameters were developed and the solidification process in steady-state conditions are examined. All examined variables -such as ingot geometry, alloy and cooling conditions- were examined and the specified simulation process is suitable for the analysis of the filling and solidification processes.

Acknowledgements

The described article was carried out as part of the EFOP-3.6.1-16-2016-00011 "Younger and Renewing University – Innovative Knowledge City – institutional development of the University of Miskolc aiming at intelligent specialisation" project implemented in the framework of the Szechenyi 2020 program. The realization of this project is supported by the European Union, co-financed by the European Social Fund.

References

[1] Grandfield J F et al 2013 *Direct-Chill Casting of Light Alloys* (TMS USA) 1-5
[2] Qian M et al 2001 *Journal of Light Metals* **1** 157-165
[3] www.precimeter.com/content/uploads/Precimeter_Slab_casting.jpg (2019 December)
[4] Hakonsen A et al 1995 *Modelling of Casting, Welding and Advanced Solidification Processes VII* (TMS USA) 743-748
[5] Rappaz M et al 1996 *Metall. Trans. A* **27A** 3214-3225
[6] Drezet J M et al 2000 *Metall. Trans.* **31A** 1627-1634
[7] Khan M et al 2018 *International Journal of Simulation Modelling* **17** 197-209
[8] Hattel J 2005 *Fundamentals of Numerical modelling of Casting Processes* (Polyteknisk Denmark) pp 23-28
[9] Novacast Systems AB 2019 *NovaFlow&Solid User Guide*
[10] Campbell J 2011*Complete Casting Handbook* (Butterworth-Heinemann UK) pp 24-28

IOP Conf. Series: Materials Science and Engineering **903** (2020) 012006 doi:10.1088/1757-899X/903/1/012006

The Effect of Primary Phase Grain Refinement of Al-Si alloys on heat cracking

M. Bubenko[1], D. Molnar[1], T. Monika[1], Gy. Fegyverneki[1]

[1]University of Miskolc, Foundry Institute, Miskolc-Egyetemvaros 3515, Hungary

E-mail: daniel.molnar@uni-miskolc.hu

Abstract. In this research thermal analysis was carried out for different alloys which use for gravity die cast cylinder heads. The experiments were implemented under industrial conditions. The aim was to examine the effect of the small amount of supplementary grain refining pre-alloy on primary crystallization for hot-crackings. Different evaluating methods of thermal analysis were tested and compared. Results proved that the dosing small amount of grain refining alloy is favorable in the foundry production practice.

1. Introduction

In the foundry practice, the amount of titanium (Ti) grain refinement is prescribed which is checked by spectrometric analysis. The necessary amount of titanium is set by the supplier of the alloy-block and there is no additional dosing. The aim of the performed tests is to investigate the achievable changes by adding a small amount of grain refining pre-alloy. The starting material for the tests was AlSi7MgCu0,5 (356Cu), AlSi9Cu1 (226L) alloy melt. The thermal analysis of the melt was examined from various foundry casting alloys by gravity mold casting. The effect of the auxiliary small amount of added AlTi5B1, 750 (g/ton) to 800 (kg) of liquid metal, grain refining pre-alloy on primary solidification was investigated. Moreover, various evaluating methods of thermal analysis have applied and compared.

The changing of the charge material, the high ratio alloy/recycled waste, or the charging of scrap castings will result in the change in the state of the grain at the primary solidification of the melt on account of the coarsening of titanium-containing grain forming particles. These effects on primary crystallization can be only detected by thermal analysis. The evaluation and comparison of the characteristics have done, based on the literature survey. The performed tests justify that the qualification of the grain refinement with several parameters together gives a reliable result.

2. The application of thermal analysis to qualify grain refining

In thermal analysis (TA) the temperature of the cooling alloy is detected and analyzed vs. the time. In practice, the simplest method is when we take the cooling curve of a melt which is cooling down in a mold or in a test cup. The evaluation process is based on the fact that all processes occurring during crystallization are influences the shape of the cooling curve. In the case of cast aluminium alloys, the efficiency of grain refinement and the modification of the eutectic can be examined. The shape of the cooling curve is determined by the amount of the latent heat released during solidification and the heat which is given to the environment. Also, the pouring temperature and the amount of the cast melt are determinative [1, 2].

Thermal analysis is the only well-known melt rating method that can effectively characterize the potential of the nucleation of a given alloy, so the efficiency of grain refinement. It is important to emphasize that the parameters determined by thermal analysis are strictly valid only at the cooling rate characteristic of the test. For a given casting, the cooling conditions may be significantly different in a casting part, so the size of the created particles/grains may be different locally than we would be expected from the results of the thermal analysis. However, it can be said that the greater potential of the nucleation in thermal analysis results, the more likely it is that the grain refinement is successful in all parts of the casting [3].

The efficiency of chemical grain refinement can be investigated near the range of liquidus temperature on the cooling curve. Based on the cooling curve and its first-time derivative, the characteristic temperature and the connected time values of the primary α-Al crystallization can be determined (Fig. 1.). $T_{NUC}^{\alpha-Al}$ is the nucleation temperature at which the nucleation of primary α-Al particles begins. At $T_{MIN}^{\alpha-Al}$ temperature, the latent heat released during primary crystallization, and the heat which is released from the mold are come to balance so the measured value of the cooling rate in this point (so the first derivative) is 0 (°C/s). $T_{G}^{\alpha-Al}$ is the temperature which connects to the maximum of warming back which is caused by the released latent heat during crystallization (in literature is named as growth temperature), after which the value of the first derivative is going to be negative again [4].

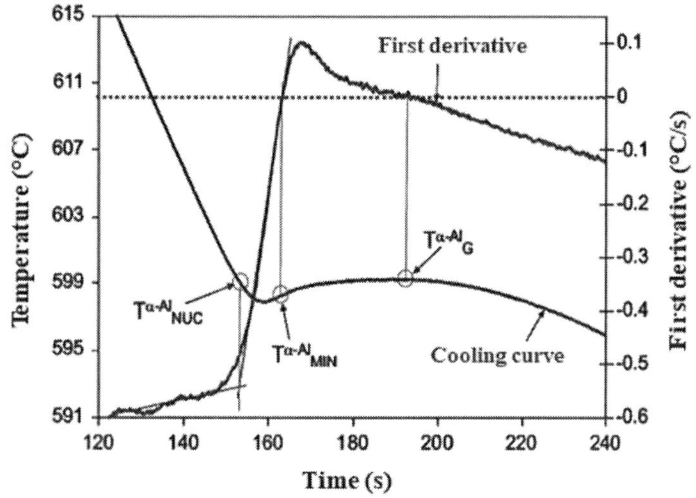

Figure 1. Determining the characteristic temperature of the primary α-Al crystallization

The value of $T_{NUC}^{\alpha-Al}$ the nucleation temperature significantly depends on the efficiency of grain refinement. The more potential heterogeneous nucleating phases are introduced into the melt, the process of nucleation is easier. Thus, the reason for effective chemical grain refinement the nucleation begins at a higher temperature, so the temperature of nucleation is also grown [5]. This is illustrated in Fig. 2., which shows the cooling curves of a grain refined and an unrefined alloy.

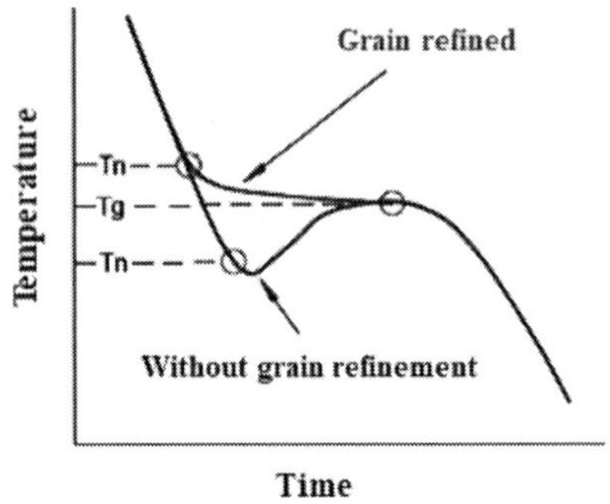

Figure 2. The effect of the grain refinement on nucleation temperature [5]

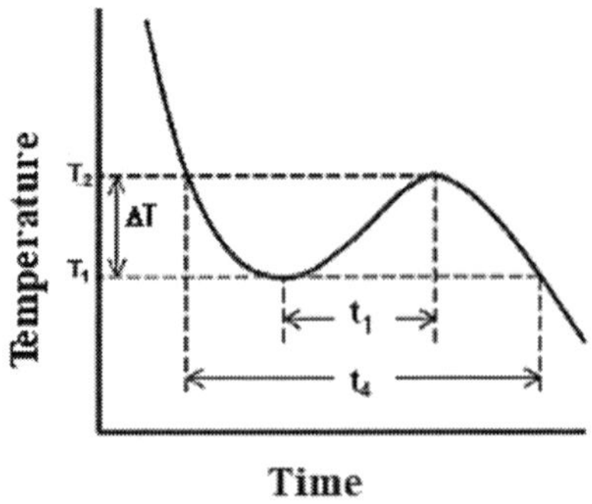

Figure 3. Parameters used to qualify the grain refinement

In Fig. 2. T_N is the (nucleation) temperature of the creation of dendritic nucleation; T_G is the beginner temperature (connecting to the maximum after overcooling) of dendritic crystallization

It is important to note that the small overcooling is necessary, but not a satisfactory criterion for the creation of the properly grain refined structure. Namely, the grain size significantly depends on the particle growth process and on the available time during the growth can happen [6].

Currently, there is no generally accepted evaluating method of the qualification of the grain refinement efficiency by thermal analysis. There are many different opinions in the literature that which parameter describes the best efficiency of the chemical grain refinement. According to Charbonnier [7], the knowledge of two important parameters is necessary for the qualification (Fig. 3.). One is the apparent overcooling which is the difference between the minimum and the maximum temperatures (ΔT) of primary crystallization. The other parameter is the passed time (t_1) between the overcooling and the warming back which is defined by the time between the minimum and maximum temperatures. According to Bekaert and Wettinck [8], the most important qualifying characteristic is the value of t_4 is shown in Figure 3., which with increasing the average grain size is also increasing. The efficiency of grain refinement can be qualified at AlSi7Mg alloys as follows:

- fine grain structure: if $t_4 < 12{,}5$ (s),
- medium-sized grains: if 12,5 (s) $< t_4 < 18{,}8$ (s),
- rough grains: if $t_4 > 18{,}8$ (s).

The qualitative parameters for grain refining, which can be determined by thermal analysis are summarized in Fig. 4. Fig. 4./a) shows only the temperature and in Fig. 4./b) time parameters are displayed. Fig. 4./c) shows qualifying characteristics that are widely used in industrial-scale as part of an automated quick-rating process. KF16 is measured at 2 (°C/s) cooling rate and connecting to the 16. seconds temperature values difference which calculated from the minimum temperature of overcooling. $t_{f,Th-Al}$ is the time from the minimum overcooling temperature to the moment when the temperature reaches again the temperature value connecting to the overcooling minimum point after heating back [9].

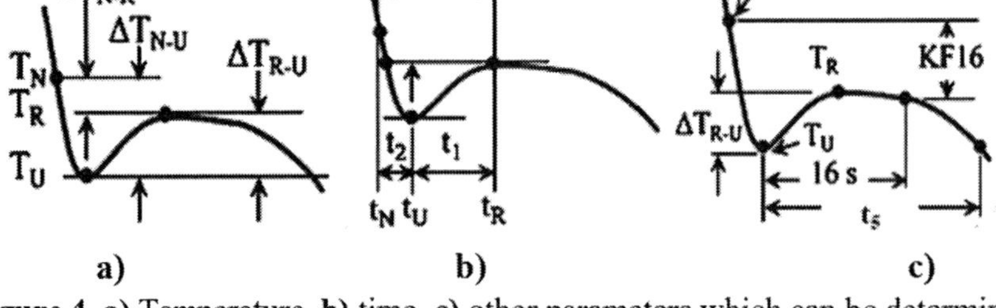

Figure 4. a) Temperature, **b)** time, **c)** other parameters which can be determined by thermal analysis of grain refining [9]

With the improving of grain refinement efficiency ΔT_{N-R} and KF16 are increased, while ΔT_{R-U}, t_1, t_3, $t_{f,Th-Al}$ are reduced. According to recent researches, the time parameters and the connecting KF16 values characterize better the efficiency of grain refinement than the temperature-related characteristics, because of their values dependent more from the energy of primary crystallization [9].

According to R. Döpp et al. [10], the value of KF16 is the temperature in the prevent cooling section of liquidus-transition at the point of 2 (°C/s) cooling rate and this is given by the difference of the temperatures connecting to the followers 16 seconds. An effective qualification can also be provided by the method presented by Bäckerud [11]. It is important to note that similarly to the KF16 described above this method is also based on temperature and time parameters. So it determines the efficiency of the grain refinement based on several measurement results. In the case of time parameters classification, the identity of the tested samples' weight is important. Therefore, the method described by Bäckerud and the results obtained by KF16 can be considered more accurate than the rating methods based on only time or temperature parameters described above [11].

3. The relationship between grain refinement and strength properties

Grain refinement significantly improves the strength properties of castings based on that first the finer microstructure causes a consistent dispersion of secondary phases (intermetallic compounds and porosity) in the material structure [5]. On the other hand, with the increasing grain number, the specific quantity of grain boundaries is also increasing in the metallic matrix which moves dislocations more difficult [12]. The latter effect can be expressed by Hall-Petch correlation [13]. During and after crystallization the occurring mechanical strains are distributed on the grain boundaries in the castings. The smaller average grain size, specifically the greater surface area of the grain boundaries on which strains are dispersed. So grain refinement reduces the stress concentration values within the casting. As a result, with finer grain size the casting is more resistant to mechanical strains, so the hot crack sensitivity will be smaller and the tensile strength will be higher [14].

The finer grain particles do not only have higher tensile strength but result in a higher elongation at rupture, too.

With the help of tensile strength and elongation at rupture, values can be determined by the so-called Quality Index which makes the quality of castings quantitative and comparable based on their strength properties. Calculation of Quality Index [15]:

$$Q = R_m + 150 * \lg A_5 \quad \text{(MPa)} \qquad (1)$$

where Q is the quality index (Mpa), R_m is the tensile strength (Mpa) and A_5 is the elongation (%).

4. Experimental conditions

The basic material of the investigations was AlSi7MgCu0,5 (356Cu), AlSi9Cu1 (226L) alloys, where L means low iron content. The chemical composition is given in Table 1. After melting in the gas-fired Striko furnace during tapping the prescribed amount of AlSr10 auxiliary alloy added to the metal bath in the forwarder ladle to provide the strontium concentration about 200 (ppm) (alloy 1. in Table 1.). In the next step, the melt was poured into an 800 (kg) capacity heat furnace where 750 (g/t) experimental amount of AlTi5B1 pre-alloy is added to liquid metal.

Table 1. Chemical composition of the investigated alloys (weight%)

Element (%)	356Cu	356Cu+Ti	226L	226L+Ti
Si	6,93	7,10	9,80	9,74
Fe	0,091	0,097	0,110	0,110
Cu	0,468	0,491	0,967	0,958
Mg	0,388	0,393	0,424	0,418
Ti	**0,1194**	**0,1229**	**0,1109**	**0,1131**
Sr	0,0235	0,0251	0,0239	0,0238

Before casting the melt was held for 1 hour at 730 ± 5 (°C). The hydrogen content of the alloy was checked by the density-index method and thermal analysis was used/done for the melt before and after adding the pre-alloy. The investigations were performed on three doses per alloy.

5. Experimental results

The cooling curves of investigated alloys show differences in the chemical composition. In the primary crystallization phase of all two alloys, overcooling and heat back are occurred. Fig. 5-6. show the cooling curves of the effect of a small amount of grain refining pre-alloy addition.

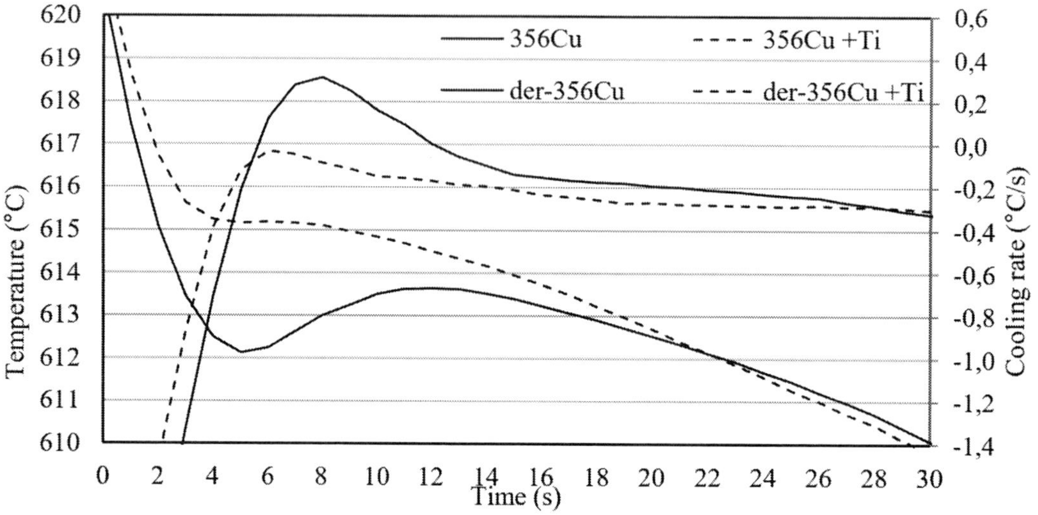

Figure 5. Effect of the grain refining pre-alloy added to melt of AlSi7MgCu (356Cu) alloy on the primary crystallization of the cooling curve

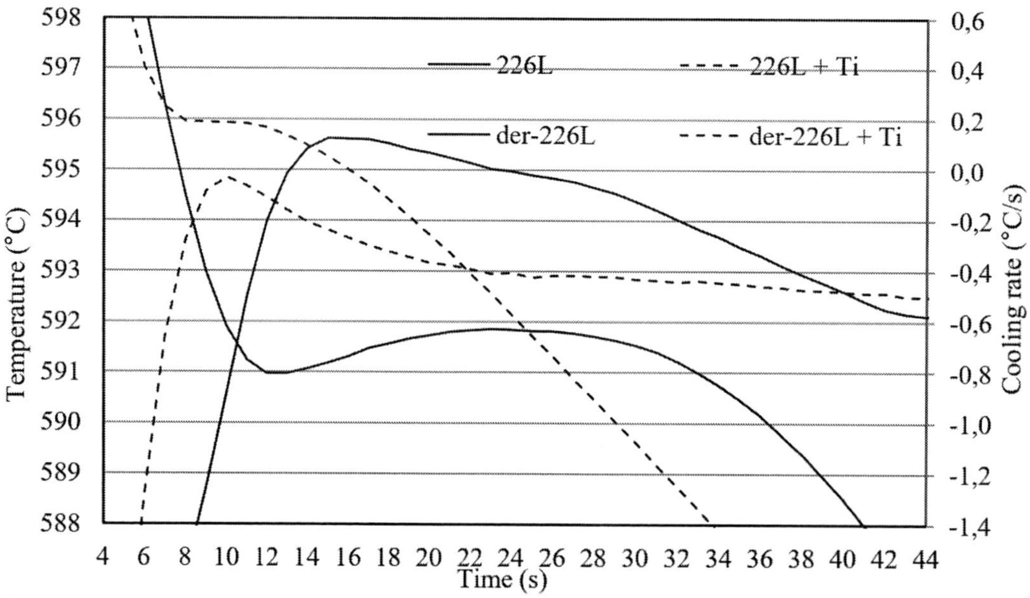

Figure 6. Effect of the grain refining pre-alloy added to melt of AlSi9Cu1 (226L) alloy on the primary crystallization of the cooling curve

We examined the microstructure of the casting samples after Barker etching to count the grain numbers. The specimens were examined by optical light microscopy at 25X amplification. The results of before and after titanium addition were averaged and shown in Fig. 7. The results of the diagram show that the microstructure was refined after the titanium addition.

We have also selected those samples in which mechanical properties were the best. Results of samples without titanium addition and with titanium addition were shown in Table 2. The effect on the mechanical properties was not detectable due to other factors influencing the strength properties (melt inclusions, directional freezing, etc.) The standard deviation is significant.

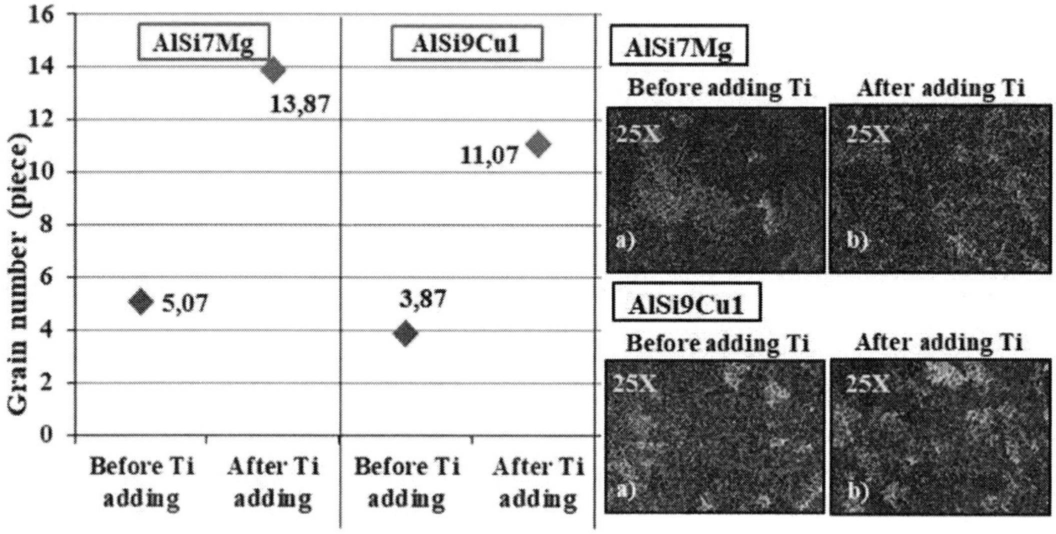

Figure 7. The effect of the 750 (g/t) grain refining pre-alloy on the primary crystallization stage of the cooling curve which was added to the alloy melt, a) before pre-alloy addition, b) after pre-alloy addition

Table 2. Average mechanical results of tensile rods machined out from experimental castings

Samples	Experimental castings			
Marking	AlSi7Mg - 356		AlSi9Cu1 – 226L	
	Before Ti adding	After Ti adding	Before Ti adding	After Ti adding
R_m (MPa)	237,2	236,1	321,6	314,8
$R_{p0,2}$ (MPa)	163,1	168,6	260,1	256,1
A_5 (%)	10,5	8,9	3,8	3,9
Q (MPa)	390,4	378,5	408,6	403,5

We investigated with stereo microscopy the test bars and found different sized aluminium-oxide films and porosities on the fracture surfaces at both alloys which generate the decreasing values also before and after titanium additions (Fig. 8.).

AlSi7Mg - 356 AlSi9Cu1 – 226L

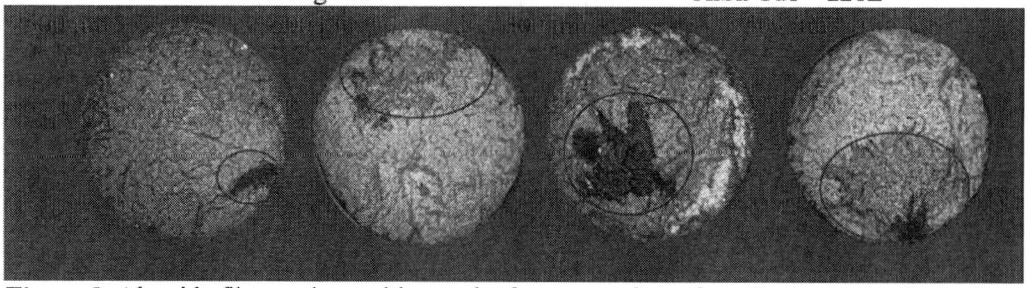

Figure 8. Al-oxide films and porosities on the fracture surface of tensile test bars at both alloys

Significant differences are shown between the cooling curves and their first derivative curves (cooling rate) at all three alloys by the effect of a small amount AlTi5B1 grain refinement pre-alloy are added. Typical differences at the auxiliary titanium addition:

- primary crystallization initial temperatures, the liquidus maximum (T_{LR}) and liquidus minimum (T_{LU}) temperature values are higher,
- the temperature difference is lower ($\Delta T = T_{LR} - T_{LU}$) at heat back in the phase of primary crystallization,
- the grain refining rating number (KF16) is higher,
- the duration of the primary crystallization stay in the initial phase (t_4 and t_5) is shorter,
- in the initial phase of primary crystallization, the area of positive (above 0) values of the first derivative curve is smaller.

The rating indicators of primary crystallization at the tested alloys are determined based on the evaluation of cooling curves (Fig. 4., [9]). The rating data of prepared experimental doses are summarized in Table 3. with the addition of a small amount of AlTi5B1grain refinement pre-alloy at the foundry and degassing treatment.

Table 3. Measured and calculated data for the thermal analysis of investigated alloys

Rating	AlSi7Mg	AlSi7Mg+Ti	AlSi9Cu1	AlSi9Cu1+Ti
T_{LU} (°C)	**612,3**	**615,6**	**590,1**	**595,9**
ΔT_{LR}-T_{LU} (°C)	1,31	0,00	0,88	0,00
KF16 [9] (°C)	**3,50**	**9,47**	**4,06**	**6,87**
KF16 [10] (°C)	**2,74**	**7,58**	**3,94**	**5,23**
t_1 [3] (s)	7,0	0,0	12,0	0,0
t_4 [9] (s)	18,5	0,0	24,0	0,0
t_5 [9] (s)	16,0	0,0	21,5	0,0

The determination of characteristic temperature and time data with the grain refining of the foundry aluminium alloys allows the formation of multi-parameters rating methods which help reliably estimate the modification and the primary crystallization processing.

6. Conclusion

Thermal analysis of various foundry alloys is performed and the effect of the AlTi5B1 grain refining pre-alloy auxiliary addition is investigated.

In the foundry practice, thermal analysis is performed with a target device developed for this purpose. The rating of primary crystallization of the melt based on the defined parameters by the associated program connecting to the device.

The evaluation and comparison of characteristics in literature have done connecting to the thermal analysis of primary crystallization. The examinations prove that the rating of grain refinement gives a reliable result with the consideration of several parameters.

In the foundry practice, the change of the used deposit material, the high recyclable waste ratio, or the melting of scrap, the long term heat keeping of melting at primary crystallization induces the change of nucleation phase because of the roughness of titanium-containing nucleation forming compound particles. Their effect on primary crystallization can only be detected by thermal analysis.

The test results confirm the beneficial effect of using 0,75 kg/t AlTi5B1 grain refining pre-alloy for melt which meets the foundry requirements for titanium-content according to the regulations.

Acknowledgement
The described article was carried out as part of the EFOP-3.6.1-16-2016-00011 "Younger and Renewing University – Innovative Knowledge City – institutional development of the University of Miskolc aiming at intelligent specialisation" project implemented in the framework of the Szechenyi 2020 program. The realization of this project is supported by the European Union, co-financed by the European Social Fund.

References

[1] Stefanescu D M 2015 *International Journal of Metalcasting* **9(1)** 7–22
[2] Djurdjevic M et al 2014 *Revista de Metalurgia* **50(1)** 1–12
[3] Spittle J A 2006 *International Journal of Cast Metals Research* **19(4)** 210-222
[4] Robles-Hernandez F C et al 2017 *Al-Si Alloys - Automotive, Aeronautical, and Aerospace Applications* (Springer USA) pp 17-33, 209-213
[5] Sigworth G K et al 2007 *International Journal of Metalcasting* **(1)** 31–40
[6] Murty B S et al 2002 *Intern. Mat. Reviews* **47(1)** 3–29
[7] Charbonnier J 1984 *AFS Transactions* **92** 907
[8] Bekaert F et al 1996 *Aluminium* **72(6)** 442-447
[9] Ratvik A P 2017 *Light Metals 2017* (TMS Springer) 1027-1034
[10] Menk W et al 1992 *Giesserei* **79(4)** 125–134
[11] Bäckerud L et al 1990 *American Foundrymen's Society* (AFS USA) 256
[12] Guan R et al 2017 *Acta Metallurgica Sinica* **(30)** 409-432
[13] Ghassemali E et al 2017 *Procedia Engineering* **207** 19–24
[14] Majo J F 2008 *ASM Handbook Vol. 15* (ASM International, USA) 1059-1084
[15] Sigworth G K 2011 *AFS Transactions* **119** 211-227

Investigation of the formability of polymeric elements in different software environments for the automotive industry

Péter Zoltán Kovács[1] and Bence Tóka[2]

[1]Associate professor, University of Miskolc, Institute of Material Science and Technology, 3515 Miskolc, Miskolc-Egyetemváros, Hungary
[2]BSc student, University of Miskolc, 3515 Miskolc, Miskolc-Egyetemváros, Hungary

E-mail: metkpz@uni-miskolc.hu

Abstract. Market-leading injection molding simulation software enables product designers, designers, injection molding tool manufacturers to predict the outcome of their work, making injection molding technology and end product quality much more reliable. This article describes how Moldex3D, with its advanced networking and computational mechanism helps simulate injection molding processes across multiple application areas. In addition, it optimizes product design and manufacturability, thus shortening the time to reach the market and maximizing the return on investment (ROI) of the product.

1. Introduction

In the 21st century, thanks to the exponential increase in computing power, many design and simulation programs help engineers avoid global design losses due to poor design / manufacturing.

The most significant simulation program (CAE application) for the plastics industry is Moldex3D. Within the program can be performed simulations on a given workpiece according to various plastic machining technologies and evaluate the result. Technologies include: injection molding (conventional, reactive, combination (bi - 2 channel, 2 component or co - 1 channel, 2 component), gas injection, water injection), foaming, injection molding, precision casting, PIM, MIM.

The Moldex3D program can be divided into four large sections during the study: product geometry, product design, mold design, material selection and injection machine injection molding technology.

With Moldex3D, due to advanced computational methods, can be determined the temperature and angle, and the tear stress, of the collision lines in order to avoid strength problems. Helps to optimally position the sprues, and how the polymer will behave in future applications. Inspection of air bubbles and welding / collision lines may help with aesthetic changes. In addition, the proper adjustment of the tool temperature for warp testing greatly contributes to optimum warping, and examining percentage shrinkage results will assist in the subsequent optimization process.

2. Optimization studies in Moldex3D to avoid / reduce errors due to different material selection, machine setup and structural design.

The properties of the ABS (Terluran GP-22) chosen as the base material of the spacer (Figure 1.) made by us as an automotive supplier are represented in Figure 2.

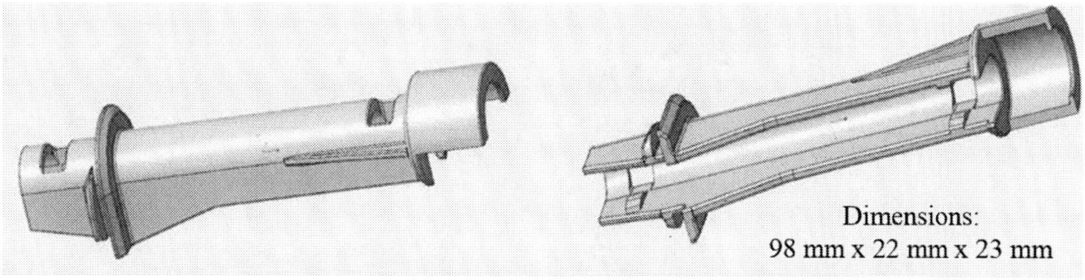

Figure 1. Modell of spacer

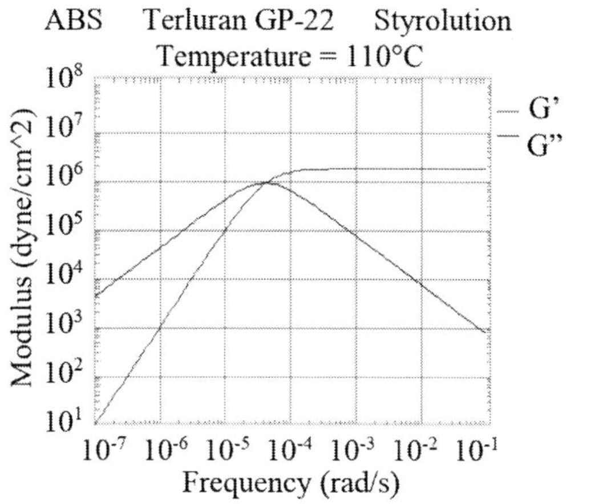

Polymer	ABS
Grade Name	Terluran GP-22
MFI (220,10)	19.6 g/10min
MVR (220,10)	19 cm³/10min
Mechanical properties	Isotropic properties
Elastic modulus	2.3e+010 (dyne/cm^2)
Poisson's Ratio	0.38 (-)
CLTE	9.5e-005 (1/K)
Process condition	
Melt temperature (min.)	220 °C
Melt temperature (norm.)	250 °C
Melt temperature (max.)	280 °C
Mold temperature (min.)	30 °C
Mold temperature (norm.)	50 °C
Mold temperature (max.)	60 °C
Ejection temperature	110 °C
Freeze temperature	130 °C

Figure 2. Properties of Terluran GP-22 (ABS) manufactured by Styrolution in Moldex3D

Various comparative tests were performed on the component (selecting ABS + PC material instead of ABS, gate and other channel locations), focusing on air bubbles and weld lines. Because the component is relatively small at a maximum distance of 49 mm from the barrier, it is difficult to show a significant difference, but small changes tend to show some tendency.

2.1. Etalon

Initially, the simulation ran based on the configuration of a particular machine (Table 1.) (optimal sprue and channel placement, and ABS (Terluran GP-22) material that was manufactured in-house) at a constant pressure of 600 bar. Keep this setting for now called etalon.

Table 1. Standard machine settings

Injection time (sec)	5.5
Melt Temperature (°C)	250
Mold Temperature (°C)	50
Maximum injection pressure (MPa)	240
Injection volume (cm^3)	47.9704
Maximum packing pressure (MPa)	240
Cooling time (sec)	13
Mold-Open Time (sec)	5
Eject Temperature (°C)	110
Air Temperature (°C)	25
Cycle time (sec)	23.5
Mesh file	Tavtarto+0.6% zsugor-ABS
Material file	ABS_TerluranGP-22_1.mtr

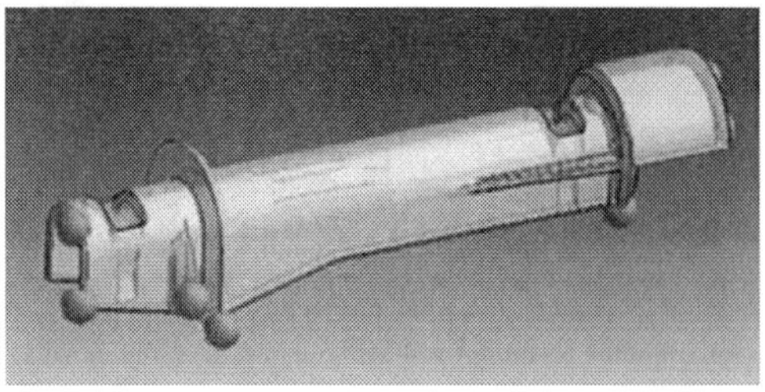

Figure 3. Air bubble number and position next to standard settings

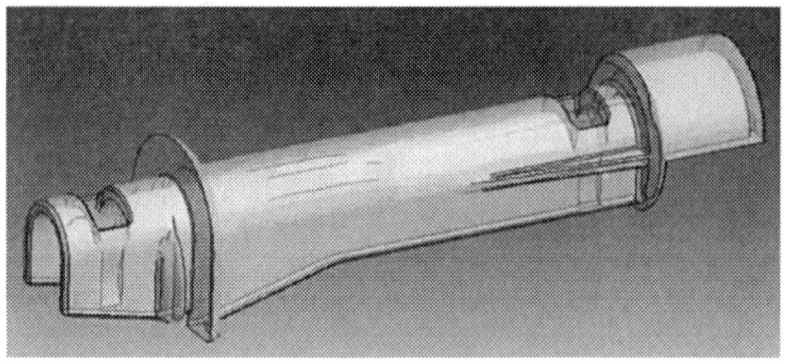

Figure 4. Conflict line location at standard settings

Can be seen during filling the standard setting with 12 possible air bubble locations (Figures 3. and 5.), which are all on the surface (possible ventilation points for the tool), this is a better case because if there is a greater chance leads to burn-in and cannot be solved with ventilation, other construction (eg.: geometry conversion, sprue, channel placement and size) is required. In addition, 10 clash lines are formed, each of which, globally, is concentrated in 2 locations (Figures 4. and 5.). Finally, the program indicates that 2% of the piece is subjected to high tear stress (Figure 5.).

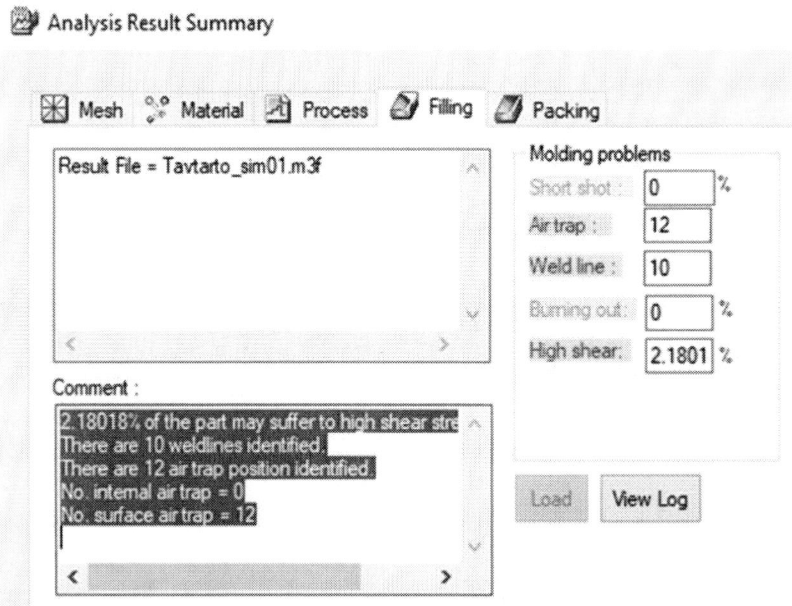

Figure 5. Percentage distribution of potential problems during filling with benchmark settings

2.2. Different sprue and channel arrangement

On the next run, the sprue and channel will be bypassed from the program-recommended location at one end of the component, but the machine settings will be the same as those mentioned in the benchmark. It can be observed with the other sprue placement that although the number of air bubbles decreases (Figures 6. and 8.), the number of collision lines increases (Figures 7. and 8.). It is primarily due to the suboptimal location of the barrier, of course, due to the small volume of the cavity, growth is not significant but is present. So proper placement of the barrier in the mold cavity is very important to avoid future problems.

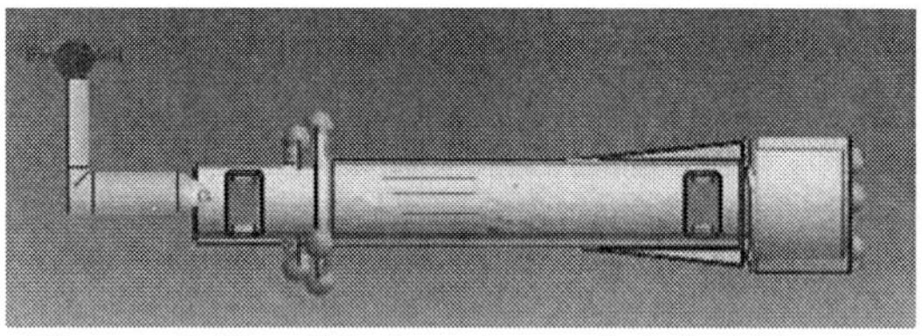

Figure 6. Possible locations of air bubbles for other barrier constructions

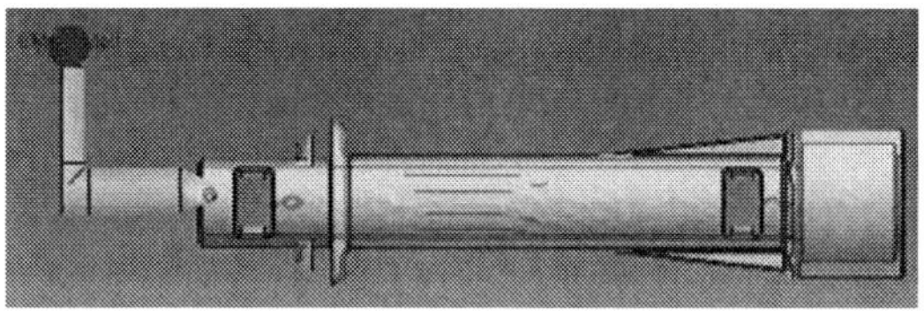

Figure 7. For other sprue constructions, possible locations of collision lines

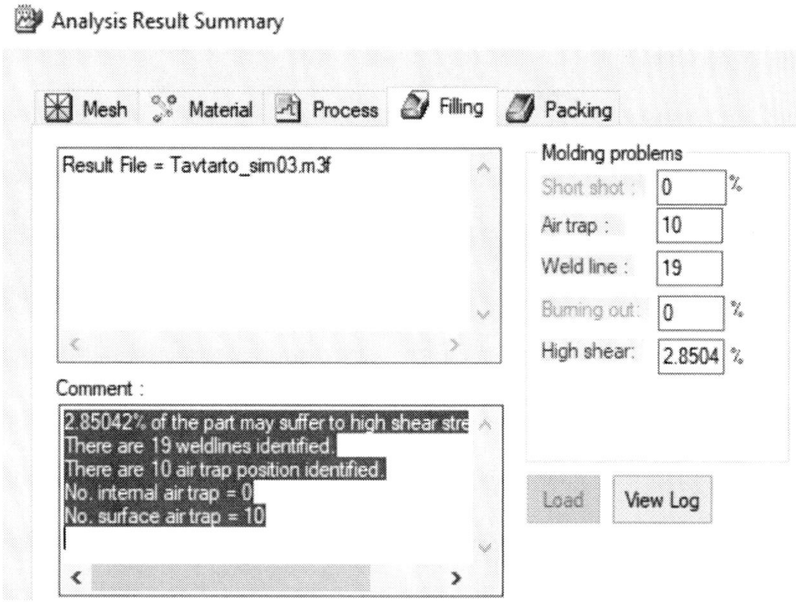

Figure 8. For other sprue constructions, the percentage distribution of errors is possible

2.3. Short shot

If I change the injection molding time in the setup (Table 2.), leaving the barrier in the standard and ignoring the Moldex3D recommendations can easily discern any problems. While using the same material, changing the injection time (lowering below the lower limit) results in a short shot (The melt drips into the mold cavity before filling it completely).

Table 2. Bad machine settings

Filling time (sec)	0.3
Melt Temperature (°C)	250
Mold Temperature (°C)	50
Maximum injection pressure (MPa)	60
Injection volume (cm^3)	35.5575
Packing time (sec)	4
Maximum packing pressure (MPa)	60
Cooling time (sec)	13
Mold-Open Time (sec)	5
Eject Temperature (°C)	110
Air Temperature (°C)	25
Cycle time (sec)	22.3
Mesh file	Tavtarto+0.6% zsugor-ABS
Material file	ABS_TerluranGP-22_1.mtr

Looking at the spread of the melt front over time (Figure 9.), the program will produce a very representative figure to illustrate the fact of the short shot, and it is clear that the inlet pressure reached the set maximum very quickly (Figure 10.) time was too short, resulting in a short shot. There are several ways to eliminate this problem (leaving the injection time unchanged): changing the barrier diameter, increasing the injection pressure and choosing another material.

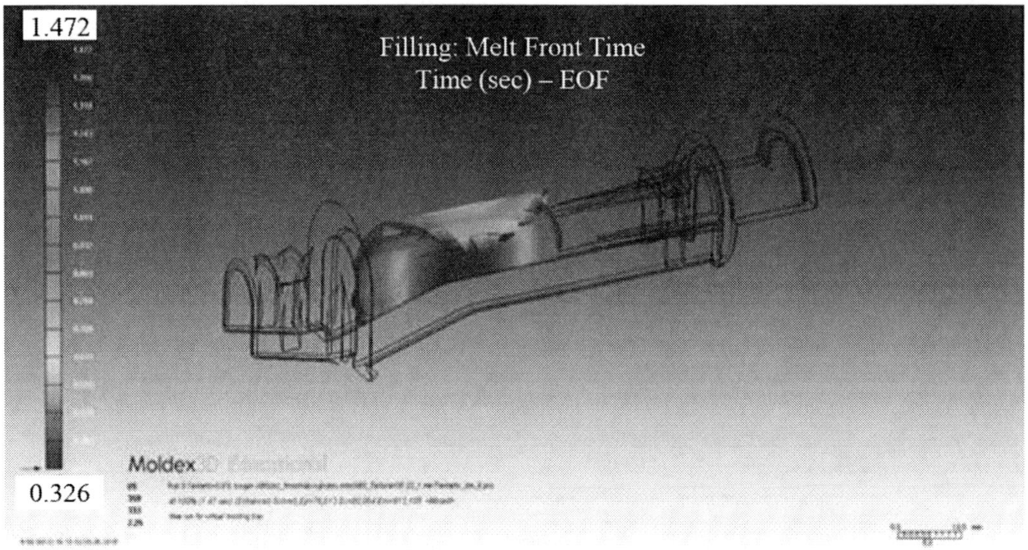

Figure 9. Flame frontal propagation in the case of short-shot

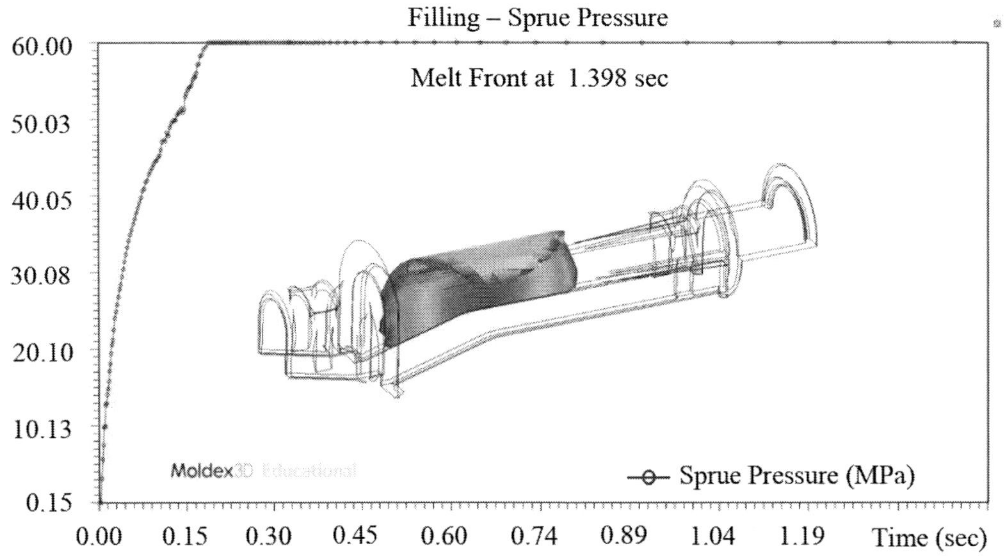

Figure 10. Pressure versus time inlet

2.4. Choice of other material

In the third case, I changed the ABS material to ABS + PC from an arbitrary supplier (Table 3.). If you use the standard machine settings for this material, you will find problems with the material's temperature conditions and warping.

Table 3. Material properties of selected ABS + PC

Item name	Item data
Material type	Thermoplastic
Generic name	PC
Supplier	SABIC (LNP)
Trade name	Thermocomp DX06313l
MFI	MFI(300.5)=30 g/10min
Fiber percent	30.00 (%)
Melt temperature range	315 – 340 (°C)
Mold temperature range	80 – 115 (°C)
Ejection temperature	123 (°C)
Freeze temperature	150 (°C)

The recommended temperature for melting the raw material is between 315 and 340 °C. At these machine settings, however, at elevated temperatures, air bubbles are more likely to burn (Figure 11.).

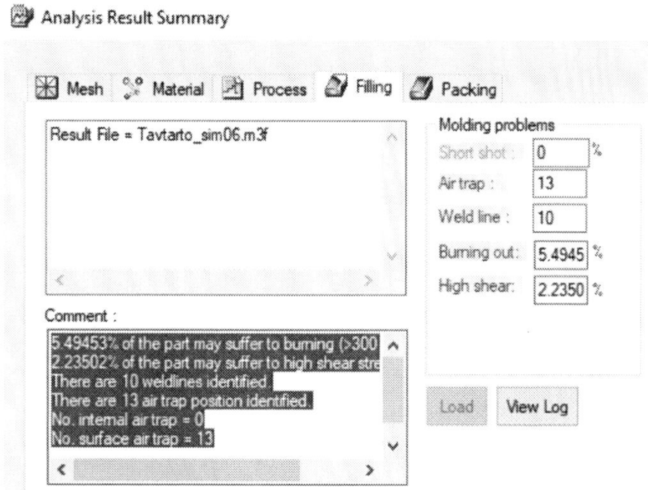

Figure 11. Possible Problems with ABS + PC Material at Machine Settings

The elevated temperature results in a much greater volume shrinkage (Figure 13.) during filling than in the case of plain ABS material (Figure 12.), thereby increasing the formation of suction and cavities. This is why it is recommended for this material, if you want to keep the preset geometric parameters unchanged, to modify the cooling system (to reduce later warping) and to adjust the postpressure accordingly.

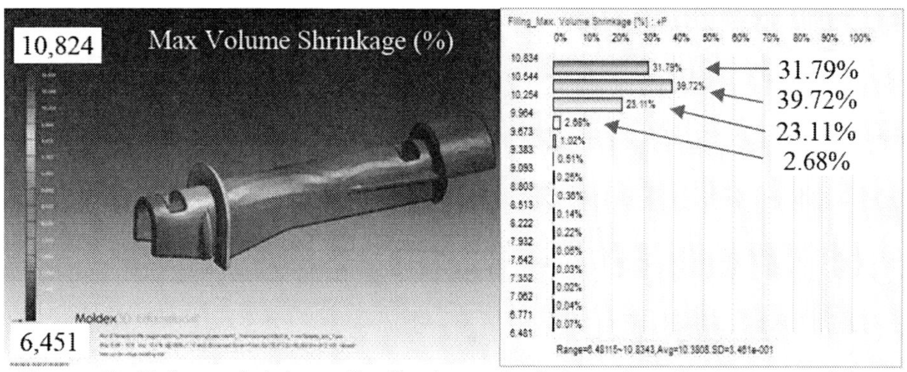

Figure 12. Volume shrinkage distribution and tabular values for ABS material

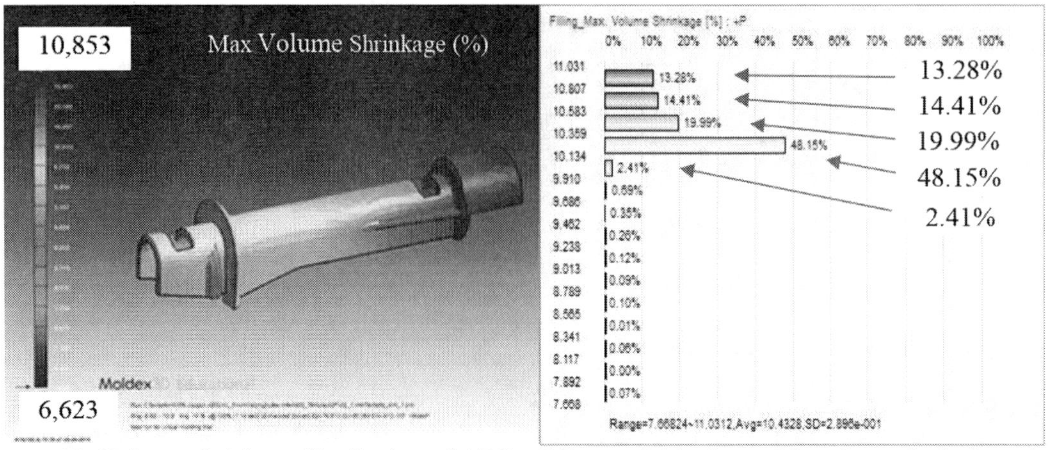

Figure 13. Volume shrinkage distribution of ABS + PC material in the mold cavity and tabular values

3. Summary of received results

It can be seen from the obtained simulation results that the standard barrier setting, barrier arrangement, machine setup has actually been optimized and, with different settings, the possible errors and their probability. Personally, I have experienced how much time it takes for a machine to set up a machine during my internship, and in light of this, it can be said and proven that Moldex3D can do a great deal to achieve optimal settings in terms of cost and time.

Acknowledgements

This research was supported by the European Union and the Hungarian State, co-financed by the European Regional Development Fund in the framework of the GINOP-2.3.4-15-2016-00004 project, aimed to promote the cooperation between the higher education and the industry. The authors would also like to thank PSM Hungary KFT. for their assistance in providing Moldex3D with the help of Bay Zoltán Nonprofit Ltd. for Applied Research.

References

[1] József Péterfalvi, Péter Primusz, Péter Szabó, Computer-based modelling systems (in Hungarian), online document, 2011, https://docplayer.hu/8006522-Szamitogepes-modellezo-rendszerek.html

[2] Tibor Czvikovszky, Péter Nagy, János Gaál, The Basics of Polymertechnology (in Hungarian), online document, 2007, https://regi.tankonyvtar.hu/hu/tartalom/tkt/polimertechnika-alapjai/index.html

[3] József Gábor Kovács, Design and Simulation of Injection-molded products (in Hungarian), Budapest University of Technology and Economics, Faculty of Mechanical Engineering, PhD dissertation, online document, 2007, https://repozitorium.omikk.bme.hu/bitstream/handle/10890/556/ertekezes.pdf?sequence=1&isAllowed=y

[4] Dr. Balázs Mikó, Design and Production of Injection-molded products (in Hungarian), Óbuda University, Donát Bánki Faculty of Mechanical and Safety Engineering, online document, 2006, http://old.bgk.uni-obuda.hu/ggyt/targyak/seged/bagms15nnk/sztervgyart.pdf

[5] László Molnár, Basics of CAD (in Hungarian), online document, 2012, https://regi.tankonyvtar.hu/hu/tartalom/tamop412A/2010-0017_43_cad_alapok/adatok.html

[6] Moldex 3D application: https://www.moldex3d.com/en/products/software/moldex3d/

Influence of 3D printing parameters

P Ficzere[1] and N L Lukács[2]

[1]Senior lecturer, Budapest University of Technology and Economics, Faculty of Transportation Engineering and Vehicle Engineering, Department of Vehicle Elements and Vehicle-Structure Analysis, Budapest, Hungary
[2]BSc Student, Budapest University of Technology and Economics, Faculty of Transportation Engineering and Vehicle Engineering, Department of Vehicle Elements and Vehicle-Structure Analysis, Budapest, Hungary

Email: ficzere@kge.bme.hu

Abstract. In this study, the influence of printing parameters have been investigated. The inner structure of 3D printed specimens were captured by a scanning electron microscope and these records were investigated by a CAD software. The mechanical properties of 3D printed parts can be determined by the size changing of failures between layers. Additive manufacturing technologies have become widely used, for this reason the influence of printing parameters must be well known. Optimisation of production time and mechanical strength is possible with this method.

1. Introduction

Additive technologies like Fused Deposition Modelling (FDM) became popular in the last decade. Because of the spreading of the technology the low-cost machines are affordable for everyone and by the help of communities the technology has been developing fast [1]. It can be used as a rapid prototyping technology and might can be used directly in custom prostheses production [2], [3] or in non-series manufacturing [4], [5]. The influence of printing parameters like temperature, speed, or layer height to the structure of the printed parts has been tested in many ways [6]. The optimal range of printing temperature is determined by the manufacturers, usually it means the best temperature range for the reliable production process. The time impact of the layer height also well known (in this case, the impact of layer height for the surface quality is not important) but the mechanical influence of these properties have not been exactly determined [7], [8]. For this applications the exact influence of printing parameters must be well known.

2. Method

In this study the layer height was investigated. AM technologies like FDM use thin layers of different molten polymers (FDM), photopolymers (SLA/DLP) and powder (SLS) to produce three dimensional bodies and parts. Connection between layers - intralayer bonding - cannot be perfect because of the imperfection of the technology.
Amabel et al. looked for an ideal test method for 3D printed specimens and observed significant failures between layers [9].

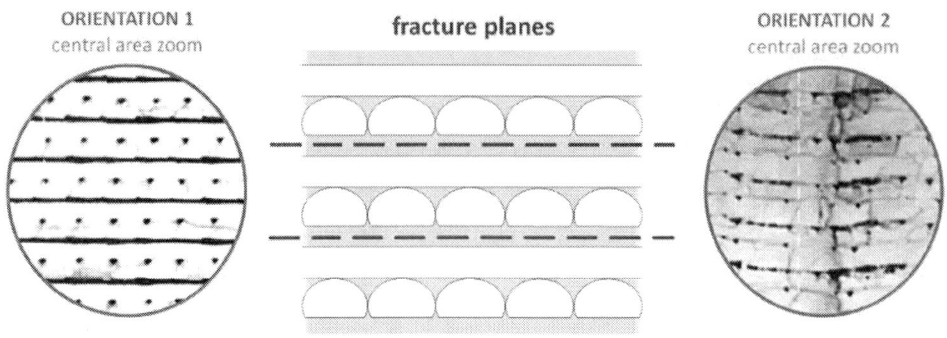

Figure 1. Air gaps between layers [9]

Air gaps and failures are observable between the rasters. Mouhamadou et al. studied the extruded material diameter and proved a high dependency on velocity and temperature.
The higher the layer, the faster the molten polymer is [10] and it can affect to the diameter of the lines also. This phenomenon can cause to shrink the lines and produce bigger air gaps. Tested material was a biopolymer named PLA. This is a biodegradable semi-crystalline polymer, which can be also produced from renewable resources like corn or sugarcane. It is the most popular 3D printer filament type because it is easy to print with it. It does not warp and it is relatively hard (its tensile strength is about 50 MPa). Its printing temperature is around 190-230 °C and heated bed is not required. This material is not an engineering polymer but the results can be implemented and used to other materials. The 3D printer was an affordable commercial FDM 3D printer, the Hephestos Prusa I3 [11]. It has many advantages and disadvantages against a professional one, for example the printing parameters are changeable separately. However, this machine is less precise (±0,2 mm in x and y directions). Layer height was changed during the manufacturing process and the other parameters were always the same.

3D printing parameters			
Number	Temperature (°C)	Layer height (mm)	Speed (mm/s)
1.	220	0,05	40
2.	220	0,1	40
3.	220	0,2	40
4.	220	0,3	40
5.	220	0,4	40

Table 1. Printing parameters

For this investigation a simple test specimen was designed by us. It had to be easy to reproduce and brake. The normal test-bars would have been needlessly oversized and they would have produced unnecessary spoilage. After the manufacturing and fracture making process, pictures were taken by SEM (Scanning Electron Microscope) and CT. Figure 2. shows a CT record from a specimen and it shows failures inside of the printed parts. These failures can reduce the mechanical strength of 3D printed parts. CT records were also useful to check if the size of gaps were influenced by the fracture making method.

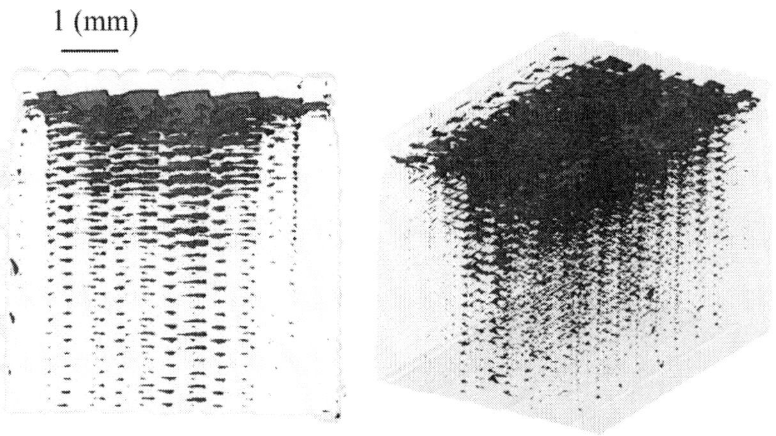

Figure 2. The inside of the test specimen

Fractures were measured and exactly determined by a CAD software named Solid Edge. Modern CAD software use Non Uniform Rational Bezier Splines (NURBS) and the failures between the layers could be approximated by them. We had the possibility to set the SEM pictures in a draft document and then measure them by Solid Edge. In the Figure 3. a failure and its size can be seen.

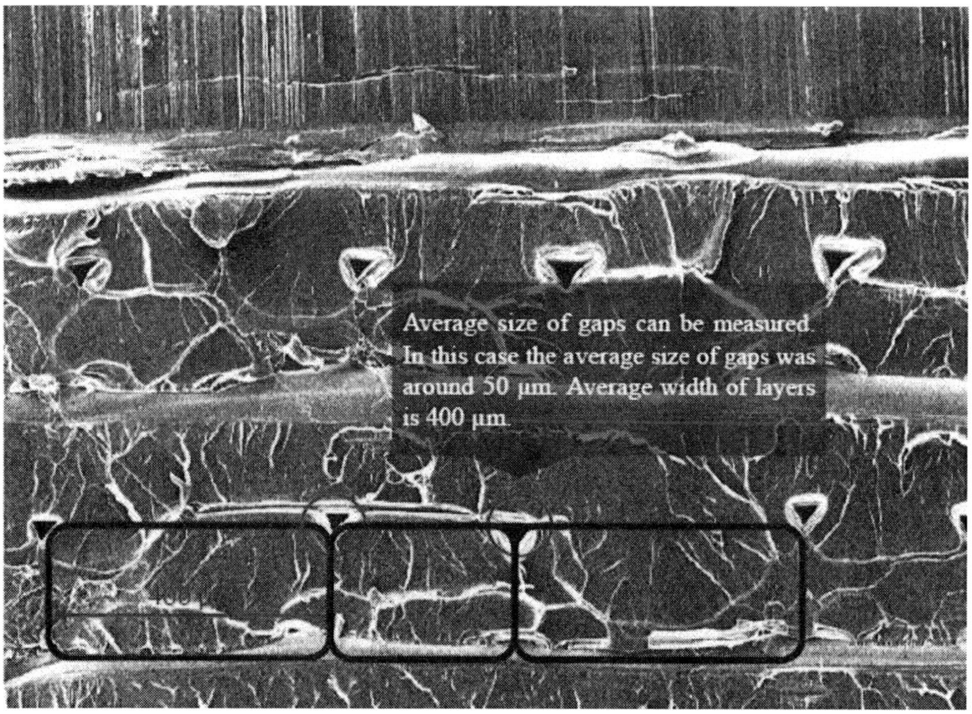

Figure 3. Measuring the size of gaps by Solid Edge (layer height: 0,2 mm)

As can be observed SEM pictures show similar results as it was expected based on Figure 1. The failures are relatively big compared to the layer height. The size of the failures has an effect on the printed parts and maybe they can influence the anisotropic properties.

Figure 4. SEM picture shows gaps between the layers [12]

Small gaps were compared to the nozzle diameter. By this method the relative failure size can be determined. This relation is more clear and understandable. Figure 5. shows the connection between layers.

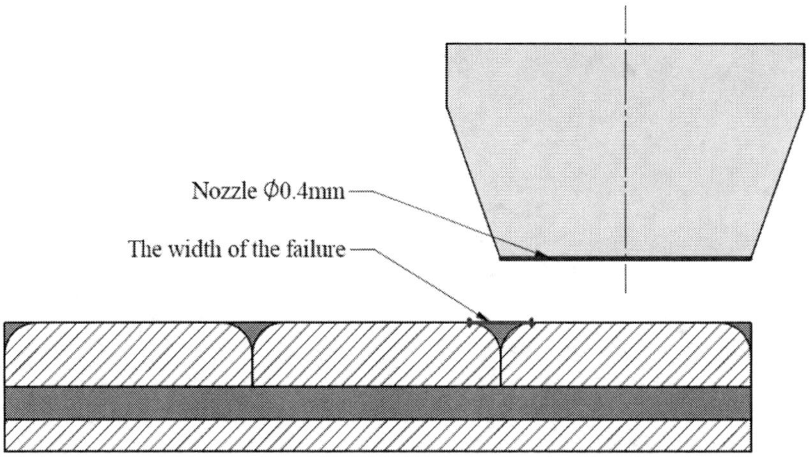

Figure 5. Failures sizes were compared to the nozzle.

3. Results

Our presumption was that the size of gaps will change if the printing parameters are changing. Every specimen have been analysed and more than 100 gaps have been measured by Solid Edge. The results are represented on this diagram. The statistical dispersion of the results is significant because the failures have been directly measured. Statistical dispersion of conventional measurements like tensile test is less significant because in this case just the average failures can be measured.

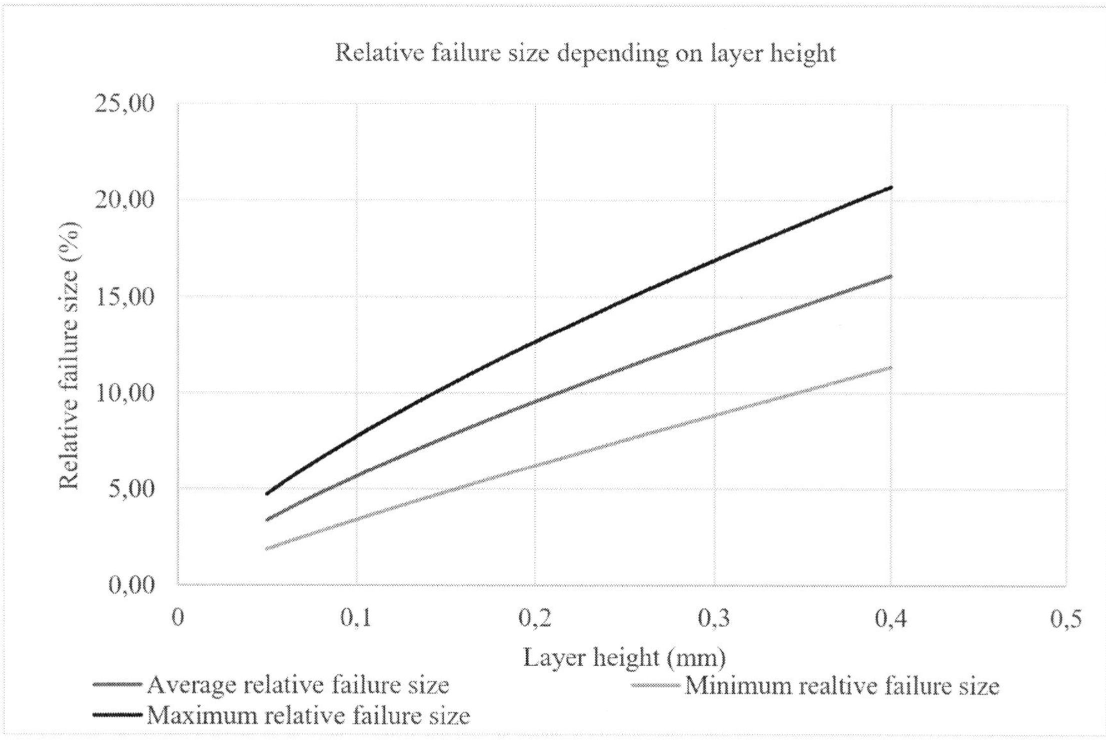

Figure 6. Relative failure size according to the layer height

This diagram shows the changing of gap sizes according to the layer height. Layer height and failure size are in a relation of proportionality. Reducing layer height always increases the printing time and it can negatively affect costs [13]. Increasing layer height can lead to production of weaker parts, furthermore the gaps can raise the risk of infection in case of medical devices like orthoses and prostheses [2]. The highest recommended layer height is around 80% of the nozzle diameter [14]. In this case it is 0.32mm.

4. Discussion

Influence of printing speed and temperature also can be significant. Printing speed is usually a physical constrain, most of commercial printers can print around 80mm/s or less. Theoretically the smaller printing speed is better. The temperature is depending on the used material. Melting point of PLA is around 150-160°C [15], but the printing temperature is much higher, it is around 190-230°C. Different applications can require different printing temperatures at the same material. Influence of printing parameters can be determined by this method, but for the supervision a conventional tensile test needed.

Our study was about the influence of layer height. Spreading of additive technologies can help us to produce cheaper and better products. These technologies often used in production of custom orthoses and prostheses. In this case the price, strength and weight are really important [16]. The number of people who use orthoses or prosthetics expected to increase from 5.6 million in 1995 to 7.3 million by 2020 [1]. Nowadays the conventional prostheses are not perfect: "The prostheses are rigid and heavy. Installation is needlessly hard, this technology has been used for 20 years. The usage of bandage is not modern, hard and break down fast. The hand itself works well. Consequently, the hand is great but the case is a nightmare." – said Szabolcs Kollinger [17], who uses prostheses. For this reasons to investigate new materials and methods are important and necessary.

References

[1] Ficzere P. 2020 Experimental Dynamical Analysis and Numerical Simulation of the Material Properties of Parts Made by Fused Deposition Modelling Technologies Periodica Polytechnica Transportation Engineering.
[2] Jin Y et al 2015. *CIRP 25th Design Conference Innovative Product Creation*
[3] Ficzere P, 2018. *Proceedings of the 4th International Interdisciplinary 3D Conference* pp. 57-67.
[4] https://www.bugatti.com/media/news/2018/world-premiere-brake-caliper-from-3-d-printer/ (2020.04.26.)
[5] https://www.aerospacemanufacturinganddesign.com/article/stratasys-to-3d-print-airbus-a350-xwb-parts/ (2020.04.26.)
[6] Ficzere P et al, 2018. *Materials Today: Proceedings* **5** 13 pp. 26489-26494.
[7] Ficzere P, Borbás L, 2018 *Transactions of Famena* **42** : 1 pp. 27-36.
[8] Ficzere P, Borbás L, 2010 *Youth Symposium On Experimental Solid Mechanics* pp. 120-122.
[9] Amabel García-D. et al 2019 Considerations on the Applicability of Test Methods for Mechanical Characterization of Materials Manufactured by FDM
[10] Mouhamadou et al, 2018 Extruded diameter dependence on temperature and velocity in the fused deposition modeling process
[11] https://reprap.org/wiki/Calibration#Layer_height (2020.04.26.)
[12] Sophia Z., Maryvivian O., Constance W. Z., 2018. Tensile and fatigue behavior of layered acrylonitrile butadiene styrene
[13] Ficzere P et al, 2013 *International Journal for Traffic and Transport Engineering* **3** : 3 pp. 344-350.
[14] https://reprap.org/wiki/Calibration#Layer_height (2020.04.26.)
[15] Hajba S. 2018. Politejsav és társított politejsav kristályszerkezetének műszaki tulajdonságokra gyakorolt hatásának elemzése
[16] Ficzere P et al, 2018 *Design of Machines And Structures* **8** : 2 pp. 5-12.
[17] https://eletemszavai.blog.hu/ (2020.04.26.)

IOP Conf. Series: Materials Science and Engineering **903** (2020) 012009 doi:10.1088/1757-899X/903/1/012009

The effect of titanium dioxide on the moisture absorption of polyamide 6 prepared by T-RTM

Orsolya Viktória Semperger[1,2], András Suplicz[2]

[1] HD Composite Zrt., Újszász utca 45, 1165 Budapest, Hungary; orsolya.semperger@hd-composite.com

[2] Department of Polymer Engineering, Faculty of Mechanical Engineering, Budapest University of Technology and Economics, H- 1111 Budapest, Műegyetem rkp. 3, Hungary; suplicz@pt.bme.hu

Email: orsolya.semperger@hd-composite.com

Abstract. In recent years, the replacement of metal components with composites is becoming more and more common in the automotive industry. Thermoplastic composites are recyclable, which meets requirements. Impregnating the reinforcing material is difficult but caprolactam is a good choice because in melt state, it has similar viscosity to that of water. Polyamide 6 can be produced from caprolactam by ring opening in-situ polymerisation. We produced PA 6 samples from caprolactam using T-RTM. We examined the effects of the manufacturing parameters on morphology on residual monomer content and crystalline proportion. Then we filled the samples with TiO_2 and examined the same parameters. Finally, we tested the effects of crystallinity, residual monomer content and the filler on moisture absorption.

1. Introduction

Nowadays, composites have a significant role in the manufacturing of technical components, due to their low weight (because of their low density) and their excellent mechanical properties. Being recyclable, thermoplastic matrix composites are increasingly used because of their excellent heat resistance, stiffness and high strength. Thermoplastic materials have high viscosity (10-100 Pas) in the melted state, therefore they cannot properly impregnate a continuous reinforcing material. This problem can be solved by caprolactam, which, in the melted state, has a viscosity (3-5 mPas) similar to that of water. Caprolactam can be used to produce polyamide 6 if combined with the appropriate initiator and activator, in a closed mould by ring opening in-situ polymerisation. These polyamides belong to the group of technically significant polymers. They have excellent mechanical properties, but their hydrophilic attribute limits their application. T-RTM (Thermoplastic Resin Transfer Moulding) is a highly efficient automated processing technology for in-situ polymerisation, whereby ring opening polymerization is carried out at low pressure with a sufficiently short cycle time in the shaping mould [1-4].

We produced polyamide 6 samples from caprolactam by T-RTM using TiO_2 as filler. The aim of our study is to analyse the effect of technological parameters, the filler and the environmental conditions on the moisture absorption of the samples. Furthermore, we studied the effect of the morphological structure of the specimens on water uptake.

2. Materials and Methods

2.1. Materials

The polyamide 6 specimens were prepared by anionic ring-opening polymerization from a system of ε-caprolactam. The system contained ε-caprolactam (AP-Nylon), hexamethylene-1,6-dicarbamoyl caprolactam (Brüggolen C20P) as activator and sodium caprolactam as iniciator (Brüggolen C10) which was purchased from L. Brüggemann GmbH and Co. KG, Germany. The filler used was titanium dioxide (Titan Dioxide KTR 600, KOLTEX COLOR s.r.o., Czech Republic) which size is 45 μm.

2.2. Sample preparation

The 290 mm × 510 mm × 5.5 mm specimens were produced below the melting point of polyamide 6 (130–170 °C), by anionic ring opening in-situ polymerisation with the use of the T-RTM (Thermoplastic Resin Transfer Moulding) technology of KraussMaffei Technologies GmbH. The equipment consisted of a DU (Dosing Unit) and a hydraulic press, which is the shaping mould carrier (Figure 1). The DU, whose temperature is 110 °C, is suitable for melting the raw materials in an inert nitrogen atmosphere. One DU has two tanks for melting the CL + C10 compound and the CL + C20P + filler separately. In our case, the filler was TiO_2.

The DU is connected to the forming mould installed on the press with the use of heated pipes, one for CL + C10 and one for the CL + C20P + TiO_2 blend. The pipes are connected to a mixing head, whose temperature is also 110 °C. The mould was filled with the melt through them. The components come into contact with each other directly after the mixing head in the cavity. Proper mixing of the components is ensured by the design of the mixing head and the flow rate of the components. Furthermore, there is vacuum in the mould for the best possible filling and the proper temperature ensures polymerisation. After polymerisation, the sample was removed.

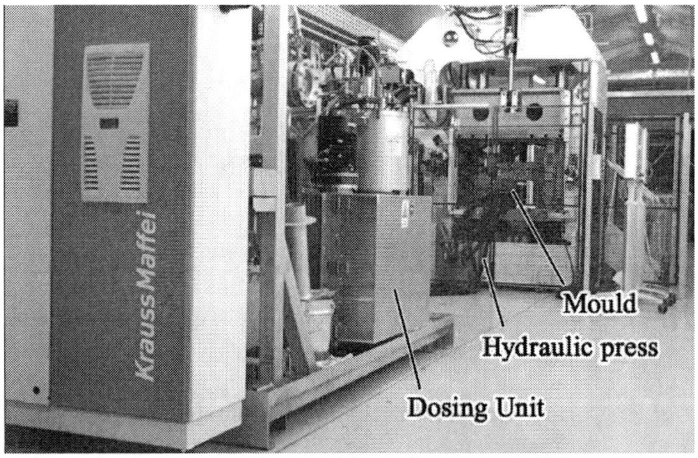

Figure 1. T-RTM machine of KraussMaffei

2.3. Design of experiments

Based on our literature research the most suitable polymerisation time is 3–5 min and the most suitable mould temperature range is 140–190 °C. The proposed mixing ratio is 94% from caprolactam, 3% from initiator, and 3% from activator [7-11]. We examined the crystalline ratio and residual monomer content with two different mould temperatures (150 °C and 175 °C), and three different residence times (120 s, 180 s and 240 s). Then we prepared TiO_2 filled specimens with the same production parameters and investigated their effect on the crystalline ratio and residual monomer content. The amount of filler added to the system was 0 m/m%, 1 m/m% and 3 m/m%. With more filler, significant sedimentation was observed. We also studied the moisture uptake of samples produced with different mould temperatures and with TiO_2. Five samples were produced, then crystallinity, residual monomer content and their effects on moisture absorption were examined. Table 1. and Table 2. show the manufacturing and test parameters of the specimens.

Residence time means the time from the injection of the material into the cavity until take out the finished sample. Mould temperature is the temperature of the tool during production. Filler content is TiO_2 concentration. During production, these parameters determine the morphological structure of the specimens (molecular weight, crystalline structure and proportion, residual monomer content, etc.).

Table 1. Manufacturing parameters

Mould temperature	Residence time		
	120 (s)	180 (s)	240 (s)
150 (°C)	x	x	x
175 (°C)	x	x	x

Table 2. Test parameters

Filler content	Relative humidity	
	50(%)	under water
0 (m/m%)	x	x
1 (m/m%)	x	x
3 (m/m%)	x	x

2.4. Characterization of the samples

The samples were produced with a mould temperature of 150 °C and 175 °C and a residence time of 120 s, 180 s and 240 s and they contain different amounts of TiO_2. Residual monomer content was determined with TGA tests on a TGA Q500 instrument (TA Instruments, New Castle, USA) at temperatures from 50 °C to 350 °C with a heating rate of 10 °C/min. The crystalline proportion of the samples was defined with a TA DSC Q2000 device (TA Instruments, New Castle, USA). We taked the measurement in a heat-cool-heat cycle from 25 °C to 250 °C with a heating and cooling rate of 10 °C/min. The first heating gave the information about the production parameters, and the second heating described the properties of the material. The crystalline proportion of the specimens was determined as follows:

$$X = \frac{\Delta H_m - \Delta H_{cc}}{\Delta H_f \cdot (1-\varphi)}, \quad (1)$$

where ΔH_{cc} is the enthalpy of cold crystallization, ΔH_m is the enthalpy of melting, and ΔH_f is the melting enthalpy of a theoretically fully crystalline polymer (ΔH_f = 188 J/g [12] in the case of PA 6) and φ is the mass fraction of the filler.

The water uptake was tested on 40x15x2 mm sized, rectangular samples. They were placed in a desiccator with 50% relative humidity and under water for analysis. Before the tests, we dried the samples in a dryer at 80 °C for 24 hours to remove moisture. During the tests, the weights were measured at predetermined intervals. In the first step, we removed the samples from the desiccator and wiped off the water and vapour on them. Then we measured the mass of the specimens with an Ohaus balance. The mass-specific moisture uptake was determined with Vas-Nagy equation as follows [13]:

$$m(t) = m_\infty \cdot \sqrt[3]{1 - e^{-\sqrt{\left(2t\frac{c}{m_\infty}\right)^3}}}, \quad (2)$$

where m_t (t) is the mass increase of the sample at the time of water uptake relative to the sample mass before water uptake, m_∞ is the saturation moisture content of the sample, t is the time of water uptake and c is a constant. The moisture content of the samples was determined as follows [13]:

$$N_{calculate,up} = \frac{m_v(t) - m_{ve}}{m_v(t)} \cdot 100 = \frac{m(t)}{m_v(t)} \cdot 100, \quad (3)$$

where $N_{calculate,up}$ is the measured moisture content of the sample, m_{ve} is the mass of the sample before water absorption, $m_v(t)$ is the mass of the sample measured at the time the water was taken.

3. Results

3.1. Monomer concentration

Residual monomer is the monomer that remains unreacted during polymerisation. It reduces the final conversion and it acts as a plasticizer inside the material, over 4%. It can condense onto the surface of the product, which prevents coating it. In the case of a composite material, it can reduce the interfacial bonding of the matrix and reinforcing material, thereby impairing mechanical properties.

The residual monomer fraction was determined by TGA [14]. Our tests showed that the decomposition temperature of caprolactam is around 200 °C, which is below the melting temperature of PA 6. Figure 2. shows a typical TGA curve of PA 6, where three different temperature ranges can be separated: the first is when the water evaporates from the sample at a temperature of about 80-100 °C, and the second is between 100-250 °C, where the solvents and small molecules (monomers, oligomers) are removed. Finally, the molecular chains begin to break up, and the polymer degrades when heated above the melting temperature.

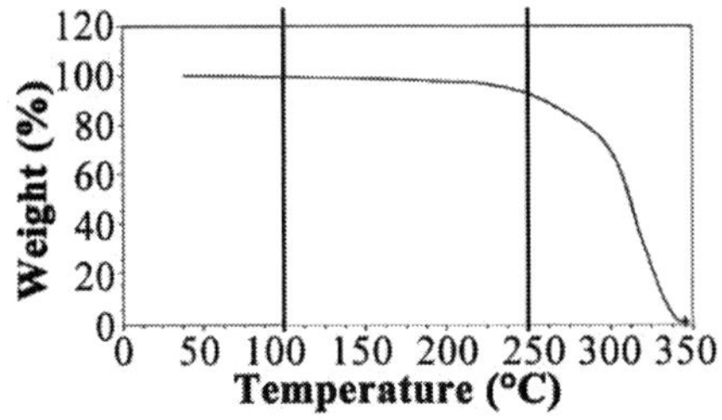

Figure 2. TGA curve of polyamide 6 sample

Conversion is 95-97% at a mould temperature of 150 °C (Figure 3. a)). Monomer ratio decreases when mould temperature increases. This may be the reason why the thermodynamic equilibrium of the monomer and polymer shifts toward the polymer at lower temperatures. A further reason can be that the crystalline phase means a full conversion, hence when crystalline fraction increases, conversion increases also. Residence time does not significantly influence conversion.

We also measured the residual monomer content of the TiO_2 filled samples. The results show that an increase in the amount of TiO_2 (Figure 3. b)) causes the residual monomer content of PA 6 to decrease. Perhaps TiO_2 can stabilize or induce the polymerization process.

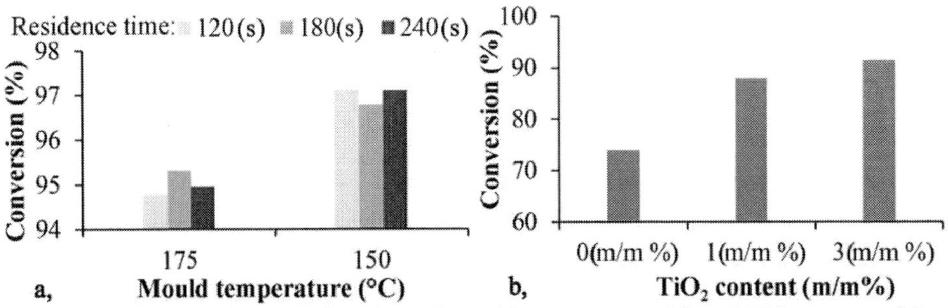

Figure 3. Conversion as a function of mould temperature (a) and TiO_2 content (b)

3.2. Crystalline proportion

We performed DSC tests on the polyamide-6 based specimens to determine the crystalline proportion. Figure 4. shows the crystalline proportion as a function of mould temperature. The specimens prepared with a mould temperature of 150 °C have higher crystallinity than the specimens prepared with a mould temperature of 175 °C. The crystalline proportion is approx. 40–43% in samples produced at the lower mould temperature, and it decreases linearly with increasing mould temperature (at 175 °C it is 20%). Residence time does not significantly influence crystalline proportion.

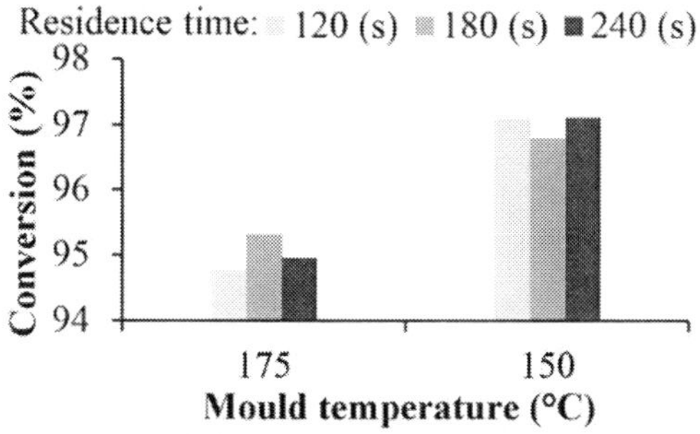

Figure 4. Crystallinity as a function of mould temperature

Figure 5. shows the results of the DSC test. The tests were performed on PA 6 specimens filled with 0 m/m%, 1 m/m% and 3 m/m% TiO_2. Samples of around 5 mg were used for the tests.

The curves show the effect of TiO_2 as a nucleating agent in PA 6 specimens produced by in-situ polymerisation and PA 6 manufactured by a melt process (like injection moulding, extrusion or compression moulding). In in-situ polymerisation, the TiO_2 slightly reduces crystallinity, while it increases the crystalline proportion in the specimen produced from melt. In the first heating stage, increasing TiO_2 content widens the crystallization range and the peaks shift to lower temperatures. The first cooling section shows the nucleation effect of TiO_2 during production from melt. In this case, increasing amounts of TiO_2 cause narrower crystallization ranges and the peaks shift to higher temperatures. Consequently, TiO_2 is an effective nucleating agent for polyamide 6.

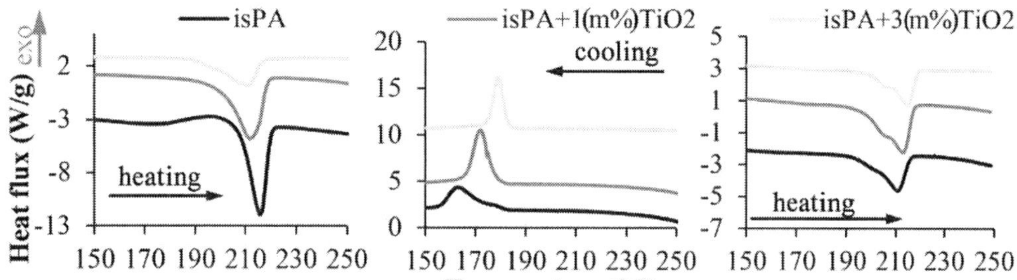

Figure 5. Heating curves for the DSC test: First heating (a), First cooling (b), Second heating (c)

The results of evaluation are shown in Figure 6. which clearly shows that increasing the amount of TiO_2 in the specimens produced by in-situ polymerisation reduced the crystalline proportion. In contrast, the crystalline proportion is significantly increased when the amount of TiO_2 is increased in specimens produced from melt. This may be important in the recycling of in-situ polymerised products, because the fillers in the surface coating can increase the crystalline content of the recycled material, thereby reducing moisture absorption.

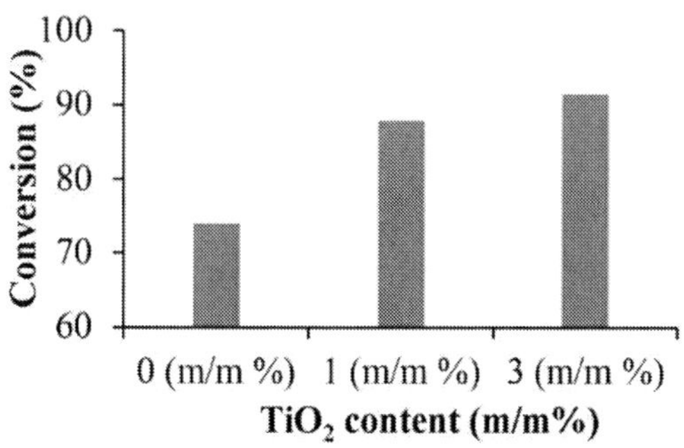

Figure 6. Crystallinity as a function of TiO_2 content

3.3. Moisture absorption

Due to hydrophilic properties of PA 6, it can absorb water, which also has a major influence on its mechanical properties. Moisture has a softening effect, thereby reducing elastic modulus, strength and hardness. It causes a significant increase in the size of specimens due to the swelling of the polymer. This must be taken into account with strict tolerance for the reproducibility of the product [15].

Figure 7. a) shows as a function of time the measured and calculated moisture content of specimens produced with a mould temperature of 150 °C, and Figure 7. b) shows the same for 175 °C. The DSC tests show that the specimens produced with the higher mould temperature have a lower crystalline ratio; at 175 °C crystalline ratio is 21.2% and at 150 °C it is 41.7%. Samples produced with a mould temperature of 175 °C absorbed more water in all test environments than samples produced with 150 °C. This is due to their lower crystalline proportion, because only the amorphous phase is able to absorb water in semicrystalline polymers [15]. We proved that the moisture uptake formula can describe the whole process well; the measured and calculated results correlate well.

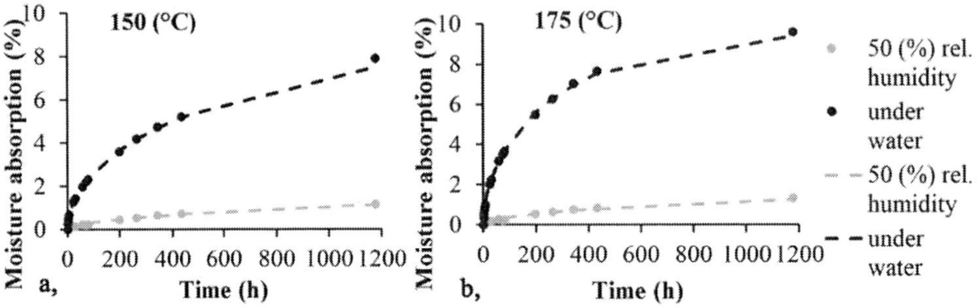

Figure 7. Moisture content as a function of residence time on mould temperature of 150 °C (a) and 175 °C (b)

The specimens filled with different amounts of TiO_2 were produced with a mould temperature of 150 °C and a residence time of 240 s. At 50% humidity, the measured and calculated moisture content of the specimens as a function of time is shown in Figure 8. a). The DSC tests show that increasing the amount of TiO_2 from 1 m/m% to 3 m/m% reduces the crystalline proportion from 40% to 37.1% in in-situ polymerisation specimens. Increasing TiO_2 content decreases water uptake because only the PA can absorb moisture; the TiO_2 does not. In the case of 3 m/m% TiO_2, the decrease is 2.4%, while it is around 2.9% in unfilled specimens. Figure 8. b) shows the measured and calculated moisture content of the specimens under water, as a function of time. During the construction of the model, the values marked with the red ellipse were ignored. This can be explained by the less than 100% conversion rate. The specimens have a residual monomer content of 4-5%, which is dissolved by the water, therefore the weight of the specimen is reduced. Contrary to expectations, the 0 m/m% TiO_2 specimens have a lower water content, due to its higher monomer content and consequently the monomer is dissoluted by the water. The measured and calculated results are in good agreement, so the proposed formula can also be used to evaluate TiO_2 filled specimens.

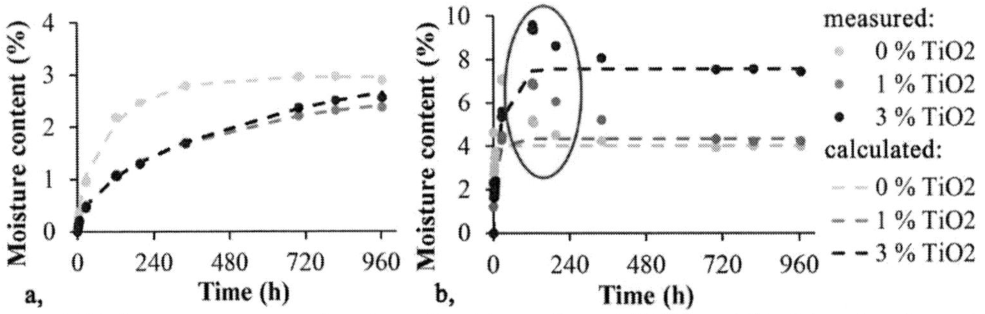

Figure 8. Moisture content as a function of residence time at 50% humidity (a) and under water (b)

Table 3. shows the result of the moisture absorption.

Table 3. Result of the moisture absorption

		Crystallinity	**Moisture absorption**	
			50 % rel. hum.	under water
Mould temperature	175 °C	21.2 %	1.3 %	9.6 %
	150 °C	41.7 %	1.15 %	7.9 %
TiO_2 content	0 m/m %	40 %	2.9 %	4.01%
	1 m/m %	37.3 %	2.37 %	4.24 %
	3 m/m %	37.1 %	2.56 %	7.45 %

4. Conclusions

We produced polyamide 6 specimens from caprolactam by in-situ polymerisation, using T-RTM. Mould temperature and residence time was varied and TiO_2 was used as filler. We studied the effects of mould temperature, residence time and TiO_2 concentration on crystalline ratio, residual monomer content and moisture absorption, and also measured the effect of the crystallinity and residual monomer content of the specimen on moisture absorption.

We found that crystalline proportion decreased (by about 3%) with increasing amount of TiO_2 in the specimens. TiO_2 inhibits the formation of crystals. In addition, PA 6 produced by anionic ring-opening polymerisation has a very high (37-40%) crystalline content. Crystalline proportion significantly increased when the amount of TiO_2 was increased in the samples produced from melt. In this case, the TiO_2 behaves as a good nucleating agent. This may be important in the recycling of in-situ polymerised samples, because the fillers in surface coating can increase the crystalline proportion of the recycled material, thereby improving its mechanical properties and reducing its water uptake. We examined the residual monomer content of the samples. The result was that the monomer content of polyamide 6 decreased with increasing TiO_2 content, and the curves shifted upwards.

Finally, we tested the water absorption of the samples produced with mould temperatures of 150 °C and 175 °C. These samples had different crystalline proportions; the samples produced at 150 °C have 41.7% crystallinity, while the samples produced at 175 °C have 21.2% crystallinity. Samples with a higher crystalline proportion absorbed less moisture. Furthermore, we proved that the TiO_2 reduces the water uptake of PA 6 at 50% relative humidity, but we found the opposite tendency under water.

Acknowledgements: The project is funded by the (NKFIH) National Research, Development and Innovation Office (NVKP_16-1-2016-0046, 2017-2.3.7-TÉT-IN-2017-00049, 2018-1.3.1-VKE-2018-00001). This paper was supported by the János Bolyai Research Scholarship of the Hungarian Academy of Sciences. Supported by the ÚNKP-19-4 New National Excellence Program of the Ministry for Innovation and Technology and by the National Research, Development and Innovation Fund (TUDFO/51757/2019-ITM, Thematic Excellence Program).

References

[1] Karger-Kocsis J et al 2018 *Polymers.* **10** 357
[2] Czigány T et al 2018 *J. Mater. Sci.* **48** 2530-2535
[3] Mochane M J et al 2019 *Express Polym. Lett.* **13** 159-198
[4] Semperger O V and Suplicz A 2020 Materials. **13** 4-15
[5] Librelon W et al 2018 *Express Polym. Lett.* **12** 1106-1116
[6] Kéri O et al 2019 *Period. Polytech. Chem.* **63** 378-387
[7] Ben G et al 2011 ICCM 201: *In Proceedings of the 18th International Conference of Composite Materials*, p 6
[8] Van Rijswijk K et al 2006 *Compos. Part A Appl. Sci. Manuf.* **37** 949-956
[9] Yan C et al 2013 *Materials and Design* **46** 688-695
[10] Tunc D 2014. *Ph.D. Thesis*
[11] Barhoumi N et al 2013 *Express Polym. Lett.* **7** 76-87

[12] Mark J E 1999 *Polymer Data Handbook.* 181
[13] Tábi T and Kovács J G 2009 Műanyag és Gumi **46** 189-192
[14] Cai-Liang Zhang et al 2006 *J. Appl. Polym. Sci.* **101** 1972-1981
[15] Monson L et al 2008 *J. Appl. Polym. Sci.* **107** 355-363

IOP Conf. Series: Materials Science and Engineering **903** (2020) 012010 doi:10.1088/1757-899X/903/1/012010

Mechanical property evaluation of closed and open-cell foam structures with finite element method

K Kulcsár[1], J Kónya[2] and I Zsoldos[3]

[1] Development engineer, Dent-Art-Technik Ltd., Győr, Hungary
[2] Head development engineer, Dent-Art-Technik Ltd., Győr, Hungary
[3] Professor, Széchenyi István Egyetem, Győr, Hungary

E-mail: kulcsar.klaudia@dentarttechnik.hu

Abstract. In this study, different lattice structures were examined with finite element method. Mechanical stress evaluations were performed on porous structured lattices that consisted simple cubic, face-centered cubic, body-centered cubic, and diamond unit cells. Finite element method simulations showed that structured foams with open and closed cells produced significantly different stress levels depending on their packing fraction. The aim of this study is to provide a basis for the design of an implant system that would promote the process of osseointegration.

1. Introduction

Estimates show that 70 to 80 percent of medical implants are made from metallic materials [1]. Medical implants are utilized to replace hard tissues and are mostly made from pure titanium or titanium alloys. Their most important characteristics are outstanding mechanical, physical and biological properties [2]. As a result of these, titanium-based alloys soon became widely-used for medical applications. Porous titanium alloys possess excellent mechanical properties and have low modulus of elasticity. Moreover, porous titanium structure and rough surface provides improved biological stability and biocompatibility compared to other porous materials [3,4]. In recent past, different technologies were studied to manufacture porous titanium alloys including precision casting [5], titanium powder sintering [6], rapid prototyping [7], and mixed titanium powder sintering [8]. In recent times additive technology has been chosen to manufacture porous structures. Exceptionally precise and computable structures can be created that have numerous advantages over conventional technologies [9-13]. Chen *et al.* manufactured open-cell porous structures with selective laser melting from Ti-6Al-4V titanium alloy. They studied lattices with 40 to 80 percent porosity and pore sizes between 600-1000 μm. Their aim was to design open-cell porous structures that best resembled human bone tissue. Specimens were designed with CAD software and were printed with SLM (Selective Laser Melting) technology. Lattices in CAD environment were compared to the printed specimens, and the largest difference was found at 80% porosity. The rest of the specimens showed minimal difference. Elastic moduli of printed lattices were studied, and they concluded that specimens with 67% porosity were the best compatible with the elastic modulus of human bone [14]. Ahmadi *et al.* studied six different lattice structures manufactured with additive technology. Material testing was conducted on printed specimens to determine maximum stress, elastic modulus, and energy absorption. Type of studied lattice structures were simple cubic, diamond, truncated cube, truncated cuboctahedron, rhombic dodecahedron, and rombicuboctahedron [15]. Finite element method (FEM) is a calculation technique used in engineering practice to approximate solutions in boundary value problems. A boundary value problem is a mathematical problem in which one or more variable must fulfil differential equations within the known boundaries of independent variables and must fulfil special conditions on the boundaries themselves [16]. Steps of finite element method-based calculation process are the following: modelling, pre-processing, calculation, post-processing, and evaluation [17].

2. Presentation of evaluated structures

The purpose of this study is to create an implant system that makes osseointegration faster. Osseointegration is a structural and functional connection between bone tissue and implants. Studies previously mentioned in the introduction part were investigating different foam-like structures, but porous structures resembling crystal structures have not been found in literature. That is why we started to investigate these configuration types. The starting body was a *20x20x20* mm cube for each simulation.

2.1. Simple cubic structure

In simple cubic structure, atoms sit in the corner points of the single cell. In this work, porous structures were designed as negatives of crystal structures. It means we have cut the spheres from the corner points of the solid cell. Figure 1. represents phases of volume reduction for simple cubic structure. Volume was first reduced by increasing the diameter of these spheres. Volume was further reduced by rounding sharp edges. The lattice with the simple cubic alignment remained a closed-cell structure after these reduction steps. As spherical diameter increased, strut thicknesses decreased.

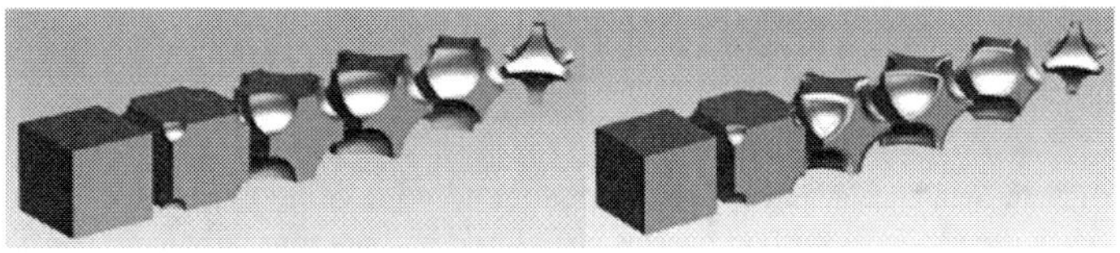

Figure 1. Phases of volume reduction for simple cubic structure

2.2. Face-centered cubic structure

In face-centered cubic structure, atoms sit in the corner points and, also in the centre of each side of the cube. Porous structures were designed as negatives of the crystal structure. It means we have cut the spheres from the lattice points of the solid cell. Figure 2. represents phases of volume reduction for face-centered cubic structure, which was achieved by increasing spherical diameter and rounding edges. Due to growing cut-out sphere size, the structure changed from closed-cell to open-cell lattice. It is notable that face-centered cubic alignment allowed extensive volume reduction.

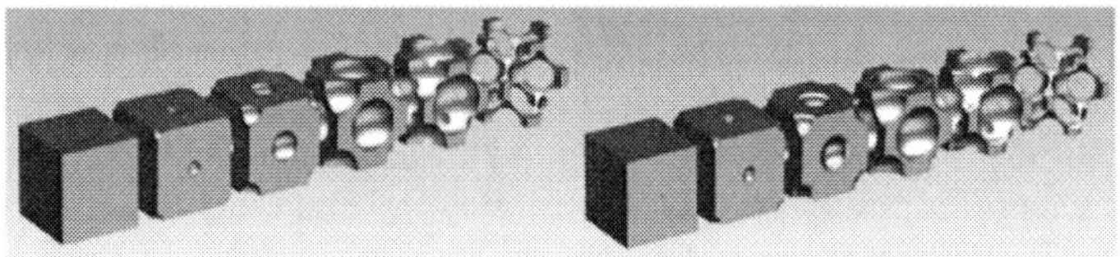

Figure 2. Phases of volume reduction for face-centered cubic structure

2.3. Body-centered cubic structure

In body-centered cubic structure, atoms sit in the corner points and in the middle of the body diagonal. Porous structures were designed as negatives of the crystal structure. It means we have cut the spheres from the lattice points of the solid cell. Figure 3. represents phases of volume reduction for body-centered cubic structure, which was achieved by increasing spherical diameter and rounding edges. Due to increasing cut-out sphere diameters, the structure changed from closed-cell to open-cell lattice. Extensive volume reduction can also be observed here.

Figugre 3. Phases of volume reduction for body-centered cubic structure

2.4. Diamond structure

Diamond structure can be derived from face-centered cubic structure. Atoms sit in the corners, face-centres, and mid-octan centres of the cube. Porous structures were designed as negatives of the crystal structure. It means we have cut the spheres from the lattice points of the solid cell. Figure 4. represents phases of volume reduction for diamond structure, which was achieved by increasing spherical diameter and rounding edges. Due to growing cut-out sphere size, the structure changed from closed-cell to open-cell lattice. Three phases in the change of cell openness could be identified. We could also achieve extensive volume reduction with this cell type.

Figure 4. Phases of volume reduction for diamond structure

3. Results of finite element analysis

Finite element simulations were carried out in Ansys software. Starting body of the previously presented structures was always a cube with the size of *20* mm x *20* mm x *20* mm. First, spherical diameter and edge rounds were increased by *0.5* mm. Smallest rounding radius was *0.1* mm for each structure. They were then increased in *0.1* mm increments until it was geometrically allowed for each structure. Due to the transition from closed-cell to open-cell structure, sphere diameters reduced by *0.1* mm were evaluated as well. Constraint and load definitions remained the same for every test. Distributed force acting on the bodies was 500 N, and friction-free fixation was chosen. Maximum stress levels were evaluated during result analysis, irreal results were neglected. Maximum stress level was the highest equivalent stress in the entire body. These stresses arised in different areas of the structures depending on their geometry. In our simulations, a sample data representative of Ti-6Al-4V (Additive Manufacturing Materials) was considered, and a corresponding isotropic material model was used by the software.

3.1. Results for simple cubic structure

1408 calculations were conducted with different spherical diameters and round edge radii for simple cubic structures. The largest cut-out sphere diameter was *19.5* mm. Figure 5. represents maximum stress levels in simple cubic structures. It is notable that up until 80%, result distribution is very dense. Simple cubic structure did not transform into open-cell structure. Due to its closed-cell structure, high volume fraction could not be achieved. In this case, volume reduction did not reach even 50%. Maximum stress was around 17 MPa. The diagram shows correlation between maximum stress levels and the extent of volume reduction. Thus, cut-out sphere diameters were increased in *0.1* mm increments and calculations were performed again without rounding edges. Calculation results can be seen in Figure 6. The tendency is clear, an outlier at *20* mm diameter is observable when comparing it with to the rest of the results. Calculated stress results were low, but they came along with low levels of volume fraction. The curve starts at the volume fraction of 100% with gentle slope. Curve gradient slowly increases after the volume fraction of ca. 60%.

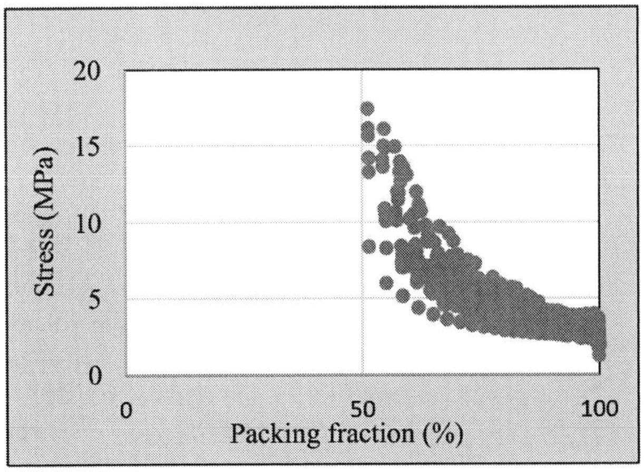

Figure 5. Summary diagram for simple cubic structure

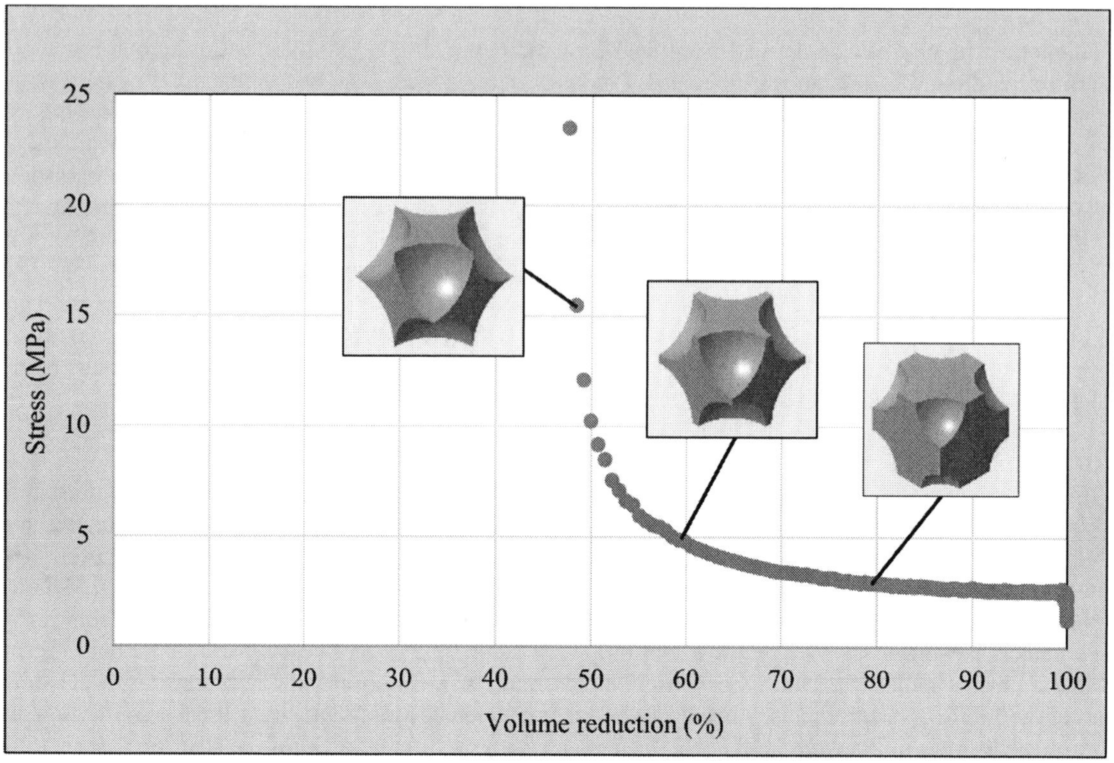

Figure 6. Simplified diagram for simple cubic structure

3.2. Results for face-centered cubic structure

761 calculations were made with increasing spherical diameters and round edge radii for face-centered cubic structure. The largest cut-out sphere diameter was *15.5* mm. Figure 7. represents maximum stress levels in face-centered cubic structures. It is notable that up until 70%, result distribution is very dense. Face-centered cubic structure transformed from closed-cell to open-cell architecture by reducing volume fraction. Compared to simple cubic structure, low volume fraction values were achieved during our simulations. Result points were distributed sparsely below 55% volume fraction. Maximum stress was 382 MPa at the volume fraction of 13%. Cut-out sphere diameters were increased in *0.1* mm increments due to the transition from closed to open-cell structure, which can be seen in Figure 8. Result point distribution was very dense until 60% volume fraction, then result data points became sparser. Until 28% volume fraction, the structure remained closed, and at 25% it was perforated and turned into an open-cell structure. Meanwhile, maximum stress level was raising above 85MPa. From this threshold value, maximum stress gradient was increasing: at 14% volume fraction maximum stress

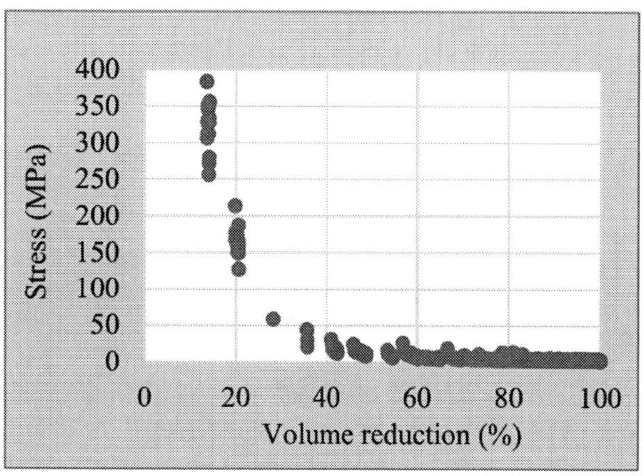

Figure 7. Summary diagram for face-centered cubic structure

was 293 MPa. Results here can be considered reasonably good. High fraction volumes could be reached and even at 30% fraction volume, maximum stress remained at 44 MPa.

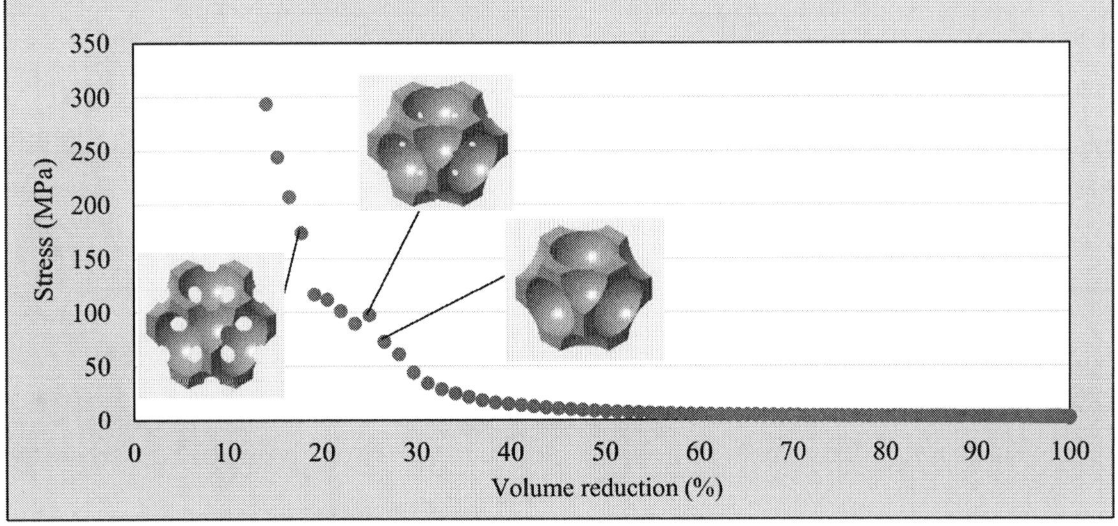

Figure 8. Simplified diagram for face-centered cubic structure

3.3. Body-centered cubic structure

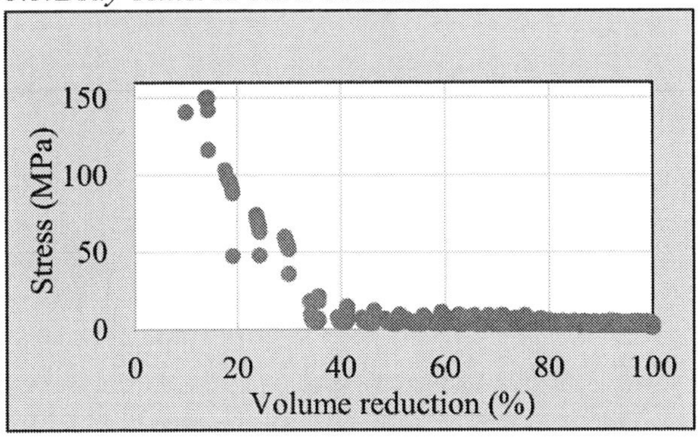

Figure 9. Summary diagram for body-centered cubic structure

1268 calculations were carried out with increasing spherical diameters and round edge radii for body-centered cubic structure. The largest cut-out sphere diameter was *19* mm. Figure 9. represents stress levels in body-centered cubic structures. It is notable that up until 48%, result distribution is very dense. Body-centered cubic structure transformed from closed-cell to open-cell architecture. Result points were distributed densely below 48%, but then distribution got more intermittent. 10% volume fraction could be achieved. Stress intensity was 150 MPa at 14% volume fraction. From 100 to 35% volume fraction, the curve had low gradient. This architecture provided the best results so far. New calculations were carried out for body-centered cubic structure without round edges. They are presented in Figure 10. Cut-out sphere diameters were increased in *0.1* mm increments here as well. It is notable that until 32% volume fraction, the set of result points was dense and evenly distributed. There, the architecture remained closed. At 31% volume fraction, stress intensity increased sharply. It is well-defined in the graph where closed-cell architecture opens up. After this transition, stress intensity decreased then started to rise again as volume was reduced.

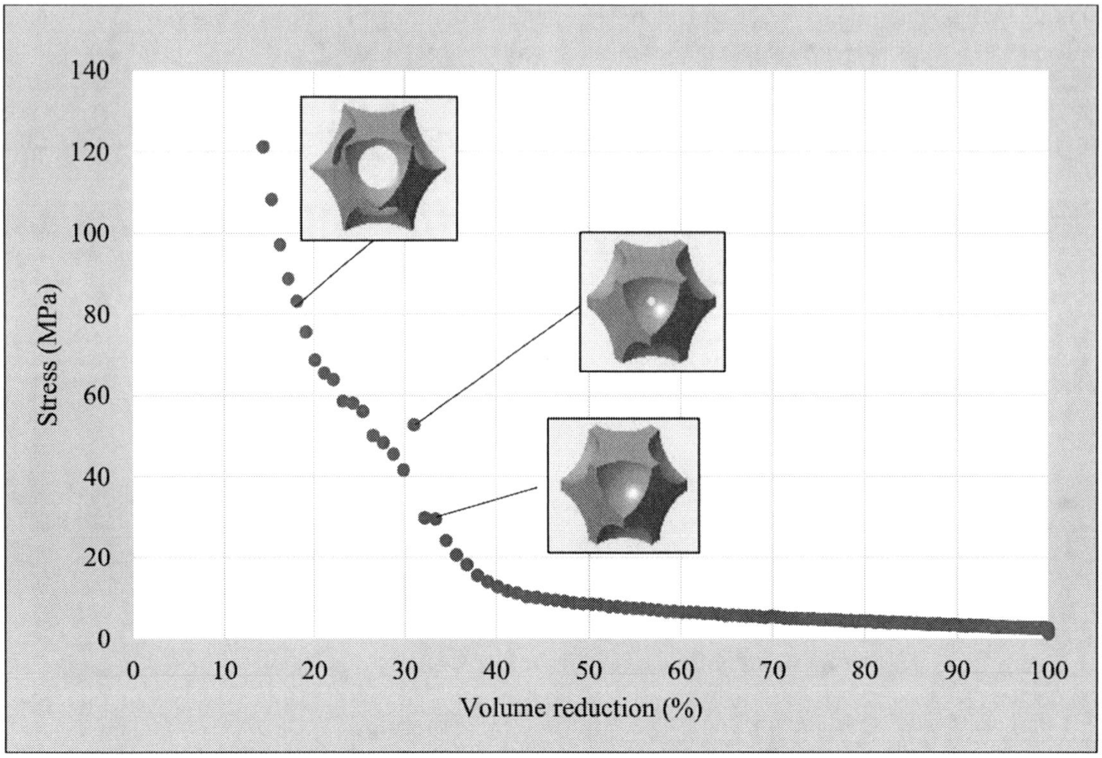

Figure 10. Simplified diagram for body-centered cubic structure

3.4. Diamond structure

589 calculations were carried out with increasing spherical diameters and round edge radii for diamond structure. Diameter of the largest cut-out spheres was *13.5* mm. It was the most complex structure to analyse due to the spheres located in mid-octans of the body. Simulation results can be seen in Figure 11. Result distribution was very dense until the volume fraction of 68%, then there were no result points until 60%. This tendency remained all along. Several quasi-separate set of result points emerged because the open-cell structure became multiphase. To evaluate phases of the open-cell structures, new calculations were carried out where cut-out sphere diameter was increased in *0.1* mm increments with no round edges used. Figure 12. demonstrates results for cut-out sphere diameters increased by *0.1* mm. Calculations were conducted with non-rounded edges. Until volume fraction of 84,9%, the structure stayed closed, the cut-out sphere from the mid-octan appeared at 84.25%. Due to this, stress intensity increased sharply, then plummeted back. Until the volume fraction of 72%, stress level remained quasi-constant, then as cut-out sphere diameter increased at the mid-octans, result values increased again. For the blue-coloured body, which had a volume fraction between 59 and 66%, false results were gained (thousands of MPa-s) due to the unrealistically thin walls formed during the transition to open-cell structure. At the volume fraction of 59%, stress intensity decreased as thin walls were completely wiped out from the structure.

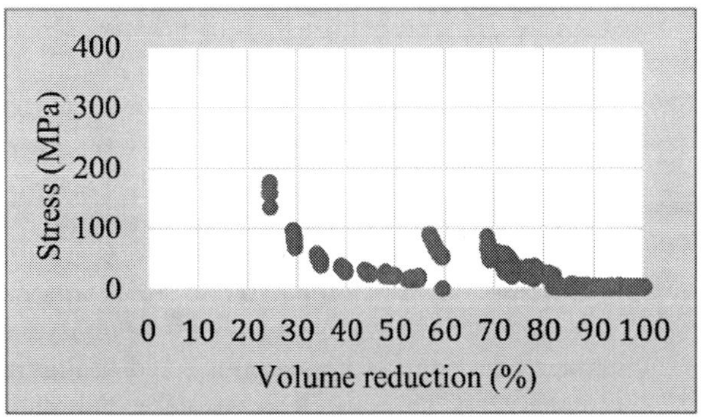

Figure 11. Summary diagram for diamond structure

This process lasted until the volume fraction of 40%, where spheres at the mid-octans met those at the face-centres. Due to this, stress increased sharply, then plummeted and increased again. Six stages can be identified altogether for diamond structure.

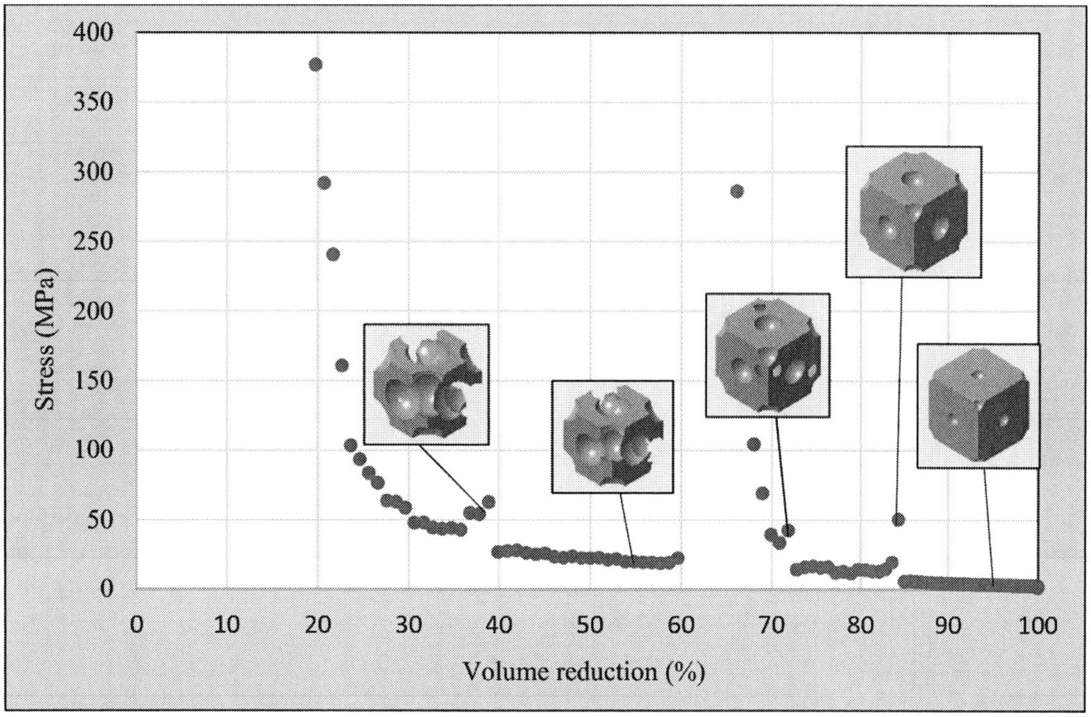

Figure 12. Simplified diagram for diamond structure

4. Summary

Diagrams presented in the previous sections are summarized in Figure 13. This diagram contains the set of results of both simple cubic, face-centered cubic, body-centered cubic, and diamond structures. Set of result points were attained from spherical dimensions decreased by *0.1* mm. In the summarizing diagram, result points for simple cubic structure are covered by the rest of the results. The lowest volume reduction could be achieved with this type of structure in spite of the low stress levels that remained quasi-constant even until high volume fractions. Neither did the structure transformed to an open-cell architecture. Stress intensity in diamond structures were generally higher at lower volume fractions compared to the other two structure types. Face-centered and body-centered cubic structures produced almost identical results until the volume fraction of 27%. Afterwards, stress levels in face-centered cubic structure started to rise and much higher results were attained. In general, the best result was achieved by body-centered cubic structure if we consider stress levels and volume fractions. Curve discontinuities marked the transition phase to open-cell structure. Namely, open-cell structure comes along with reduced wall thicknesses, which produce extremely high stress levels.

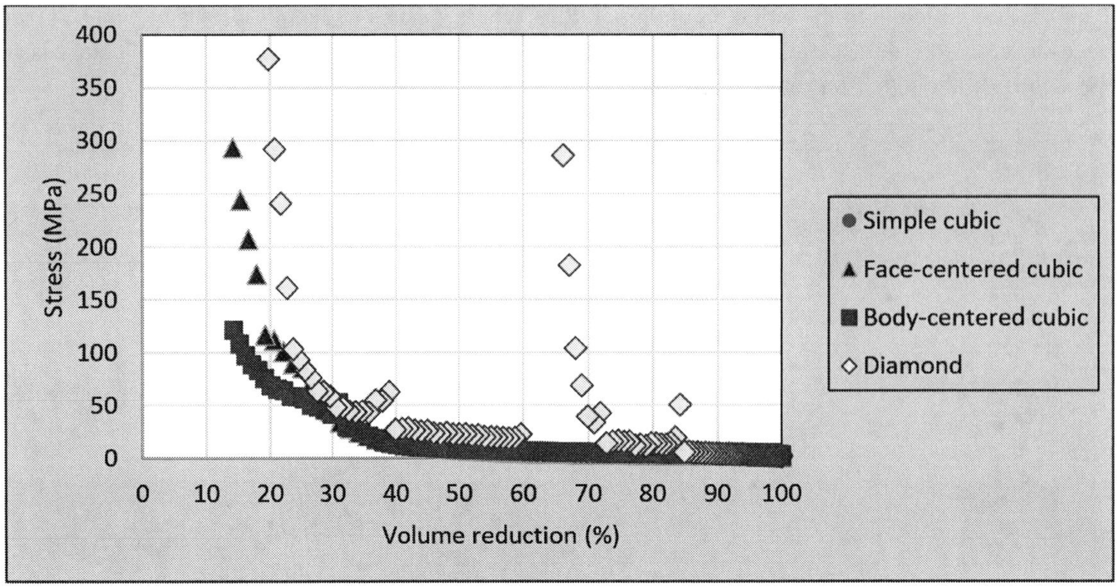

Figure 13. Integrated set of results

Our research is going to be continued. In the following stages, environmental behaviour of different porous structures are going to the evaluated with great emphasis on stress levels and volume fractions. We are going to investigate structures that make up an entire lattice. Calculations are going to be validated empirically using additively manufactured test specimens.

Acknowledgement

"EFOP-3.6.1-16-2016-00017 Internationalization, initiatives to establish a new source of researchers and graduates, and development of knowledge and technological transfer as instruments of intelligent specializations at Szechenyi University"

References

[1] Chao J et al 2014 *Mater.* **7** 1709-1800
[2] Wen C E et al 2002 *J. Mater. Res.* **17(10)** 2633-2639
[3] Lopez-Heredia M A et al 2008 *Biomater.***29** 2608-2615
[4] Yamada Y et al 2000 *Adv. Eng. Mater.* **2** 184-187
[5] Oh I H et al 2003 *Scr. Mater.* **49(12)** 1197-1202
[6] Li J P et al 2006 *Biomater.* **27(8)** 1223-1235
[7] Bartolo P et al 2012 *CIRP Annals* **61(2)** 635-65
[8] Gu D D et al 2012 *Mater. Rev.* **57(3)** 133-164
[9] Mironov V et al 2009 *Biofabr.* **1(2)** 1-16
[10] Murr E et al 2010 *Phil. Trans. R. Soc. A* **368** 1999-2032
[11] Hatos I et al 2018 *Stroj Vestn-J Mech E* **64** 121-129
[12] Chen S Y et al 2017 *J. Alloys Compd.* **713** 248-254
[13] Ahmadi S M et al 2018 *Mater.* **8(4)** 1871-1896
[14] Hutton D V *ISBN: 0-07-239536-2*
[15] Gyimóthy Sz 2003 *Ph.D BME VIK*

IOP Conf. Series: Materials Science and Engineering **903** (2020) 012011 doi:10.1088/1757-899X/903/1/012011

Detection of neutron irradiaton induced degradation of reactor steel by magnetic method

G Vértesy[1], A Gasparics[1], I Uytdenhouwen[2], I Szenthe[1], F Gillemot[1] and R Chaouadi[2]

[1]Hungarian Academy of Sciences, Centre for Energy Research; 1121 Budapest, Hungary
[2]SCK•CEN Belgian Nuclear Research Centre; 2400 Mol, Belgium

E-mail: vertesy.gabor@energia.mta.hu

Abstract. A novel nondestructive method, based on step by step measurement of minor magnetic hysteresis loops was used to determine the neutron irradiation induced degradation of nuclear reactor pressure vessel steel material. This method is called Magnetic Adaptive Testing (MAT), and two types of reactor steel material were investigated: base material type 22NiMoCr37 and weld material type 18MND5. Charpy geometry samples were irradiated by neutrons at the BR2 in SCK•CEN. The neutron irradiation generated structural changes in the material. As a result of our measurement, it was shown that this degradation could be detected by MAT measurements. Charpy impact tests were also done on the investigated samples and the results of these measurement were compared with the nondestructively determined magnetic quantities. It was shown that a closely linear correlation existed between the ductile-to-brittle transition temperature (DBTT) and MAT parameters.

1. Introduction

Pressurized water reactors are applied in majority the nuclear power plants. The reactor pressure vessel (RPV) is the most important part from point of view of safety. The lifetime of the power plant is determined by the lifetime of RPV. It is exposed at high pressure to neutron radiation and is operated at elevated temperature. Reactor pressure vessel material is degraded during service because of these effects. This material is a high quality low alloyed ferromagnetic CrNiMo or CrMoVsteel, and its properties must be monitored during the whole lifetime of the power plant. The forged rings and their welds at the core level are the most critical parts. Regular inspection of reactor pressure vessel material is performed by the so called surveillance programmes. During this programme surveillance samples (Charpy samples) placed inside of the reactor are exposed to the same conditions (neutron radiation and thermal conditions) as the pressure vessel walls. Charpy impact, tensile stress and fracture toughness tests are performed on surveillance samples. All these tests are destructive, and sooner or later there are no more samples (especially if the life time of nuclear power plants is extended) for mechanical tests.

The most frequently applied method is the well-known, conventional, mechanical Charpy impact test, which is performed on 55x10x10 mm³ size samples. Samples are periodically taken out from the reactor pressure vessel and inspected. These tests have been used as a standard testing method of material toughness for a long time [1]. The fixed standard size and shape specimens to be tested are

broken by a swinging hammer, and the absorbed energy by the fracture of the sample is a typical property of the material. This energy depends on the temperature at which the test is being performed. As a definition, ductile-to-brittle transition temperature (DBTT) is the temperature where the absorbed energy is 41 J. One single sample is not enough for the reliable determination of DBBT, usually several (at least a dozen) of surveillance specimens are used for one single DBTT characterization.
The surveillance specimens became highly radioactive during irradiation. It means that they can be measured only in hot cell laboratories. This type of inspection is very costly. Because of this application of nondestructive tests instead, or at the beginning parallel with the present surveillance techniques is very important.

There exist numerous nondestructive methods, and those which are based on electromagnetic measurements are encouraging for potential practical applications. Among them magnetic hysteresis methods are rapidly developing and occupy a special position. A close correlation was found between modification of the microstructure of steels due to different influences (e.g. mechanical load, thermal treatment, fatigue, corrosion, etc.) and between magnetic behaviour, which is reflected the best in the magnetic induction of samples under an applied magnetizing field. Domain wall movements – which influence the shape of hysteresis curve – and dislocations in the ferromagnetic material are both affected by the micro-structural defects. This direct correlation between magnetic and mechanical hardness of the same ferromagnetic material is well-known, and also well understood.

Magnetic methods are not expensive, mobile and technically simple. A review of magnetic methods in nondestructive testing is given in [2]. An advantage of these methods is that they can be applied via electrical cables and coils, such a way it is possible to place the necessary electronic equipment outside the protected space where the irradiated, and due to irradiation radioactive specimens are kept. Some measurements were published, which proved the practical applicability of magnetic methods in the inspection of steel degradation. For example cold rolling, special thermal processing and fast neutron irradiation effect on nuclear pressure vessel steels and welds were tested, see e.g. [3-11].

A novel method, Magnetic adaptive testing (MAT) has been developed recently and this technique seems to be a promising candidate for nondestructive inspection of steel degradation. In MAT method series of minor magnetic hysteresis loops are measured [12,13], permeability and/or hysteresis matrices are calculated and the matrix parameters of degraded material are compared with the same parameters of virgin samples. Such a way so called magnetic descriptors are generated, which reflect any modification of material structure with high sensitivity. This procedure is multi-parametric and robust. Details of technique is described detailed in [13].

In a previous work of us [14], different samples, made of JRQ, 15CH2MFA and 10ChMFT type steels were investigated by MAT. The samples were irradiated by E>1 MeV energy fast neutrons with total neutron fluence of 1.58×10^{19} – 11.9×10^{19} n/cm^2. Regular correlation was found between the optimally chosen MAT degradation functions and the neutron fluence in all three types of the materials. In the present work the applicability of MAT is shown for the detection of the degradation of the reactor steel, caused by neutron irradiation. Our final purpose is to compare the magnetic parameters measured by this non-destructive way with the Charpy impact testing shifts on irradiated samples.

2. Sample preparation

Standard Charpy specimens from 22NiMoCr37 type base and from 18MND5 type weld materials were measured. These steel grades are members of the typical RPV steel groups: Mn-Ni-Mo steels (western RPV design). The geometry of samples is shown in Fig. 1 and the chemical compositions of the samples are given in Table 1.

The samples were prepared, irradiated and measured at the SCK•CEN Belgian Nuclear Research Centre. The samples were irradiated in the BR2 reactor at 260 °C temperature by E>1 MeV energy fast neutrons with total neutron fluence in the range of 4.02×10^{19} – 8.95×10^{19} n/cm^2. Since irradiated samples are radioactive, they can be measured only in hot cell laboratories. Charpy impact testing (ASTM-23-16b [15]) was used to determine the 41 Joule transition temperature (T_{41J}). In this paper the ductile-to-brittle transition temperature shift (ΔDBTT) is the difference between the neutron irradiated T_{41J} and the as-received (base-line) T_{41J} which is a measure for the embrittlement of the material.

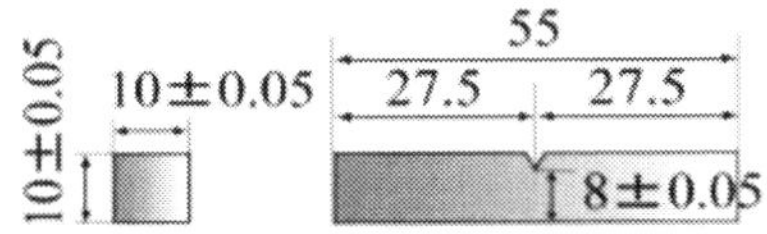

Figure 1. Dimensions (in mm) and shape of the measured Charpy samples. (Each sample had a V-notch at one face.)

Table 1. Chemical composition as measured by Optical Emission Spectroscopy of the measured materials (values are given in wt%)

Material	C	Si	Mn	P	S	Cr	Mo	Ni	Cu
18MND5 weld	0.09	0.23	1.21	0.018	0.009	0.12	0.49	0.96	0.13
22NiMoCr37 base	0.20	0.25	0.87	0.009	0.007	0.39	0.49	0.85	0.06

3. Magnetic adaptive testing

Set of minor hysteresis loops is measured (starting from zero magnetizing field, then magnetizing field is increased step by step up to closely saturated state). This measurement is performed on each sample within the given set of samples. A specially designed equipment is used, which is controlled by a PC. A magnetizing yoke (a C-shaped laminated Fe-Si transformer core) is placed to the sample surface. Size of the yoke fits geometry of the specimen. In our case the cross-section is S=10x5 mm^2, the total outside length is 18 mm, and total outside height of the bow is 22 mm. There are two coils, both of them wound on the yoke. The magnetizing coil has 200 turns and the pick-up coil has 75 turns. The photo of the sample holder, which was built specially for hot cell measurement is shown in Fig. 2. The magnetizing yoke is placed below the sample, the samples can be replaced within the hot cell by manipulator. The whole electronic unit, which controls the measurement and manages the data acquisition is placed outside of the hot cell.

Figure 2. The sample holder with a Charpy geometry sample to be measured.

An evaluation program processes the measured data. Experimental noise is filtered by the program and the experimental data are interpolated to a square grid of elements, $\mu \equiv \mu(h_a,h_b)$, of a μ-matrix (or permeability matrix), where h_a is the actual magnetic field value and h_b is the amplitude of the actual minor loop. These μ-elements are designated as "MAT-descriptor" and they characterize the material structure variation of the investigated specimens.

Another evaluation program processes the matrices. All matrix elements are divided by the corresponding element of the reference matrix (matrix elements of the reference sample). As a result, normalized $\mu(x)$-degradation functions are generated, which reflect the magnetic response of the material to any kind of degradation. x can be any parameter, which characterize material degradation. In our case these are the neutron fluence on one side, and the shift of the transition temperature (ΔDBTT) on the other side, determined independently in the samples. Taking into account the accuracy of the measurement, and analysing the measured MAT parameters, the error of the MAT descriptors is estimated about 5%.

4. Results and discussion

MAT measurements were performed on two series of the above described reactor steel samples before and after neutron irradiation. The μ matrix elements from the reference (not irradiated) samples were used for normalization. In the graphs (Figs. 3 and 4) shown below the modification of the MAT descriptors due to neutron irradiation are presented (normalized MAT descriptors) as functions of two independent parameters.

In Fig. 3 the optimally chosen MAT descriptor can be seen as a function of the neutron fluence for both materials. The optimal μ matrix elements were calculated from the measured permeability loops. Optimal means that these descriptor has the largest sensitivity with respect to the independent parameter (neutron fluence and ΔDBTT), which characterizes the material degradation, and at the same time they are reliable and well reproducible The exact description on how these optimal descriptors are determined is given in Ref. [13]. The optimally chosen MAT descriptor is characterised by μ (h_a=500mA,h_b=1150mA) values. A linear correlation was found between magnetic parameter and neutron fluence which is monotonously increasing. A 30% increase of the MAT descriptor was detected due to 6×10^{19} n/cm^2 neutron fluence in case of the base material, compared with the not irradiated sample; and close to 40% increase due to 8.5×10^{19} n/cm^2 neutron fluence in case of the weld material.

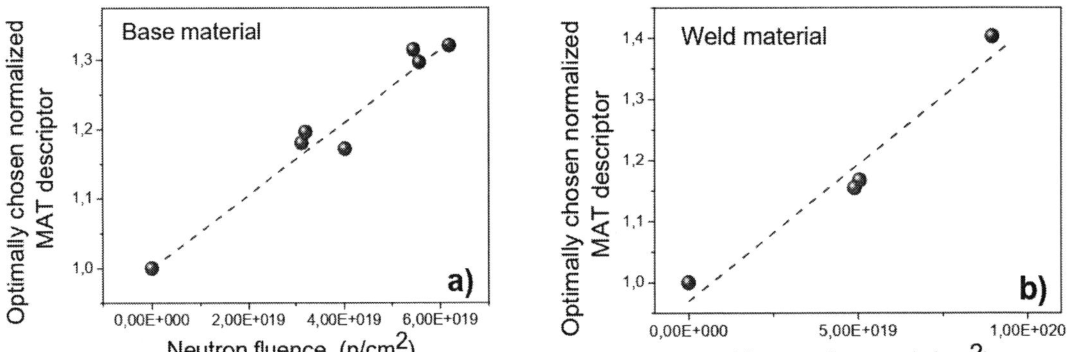

Figure 3. Optimally chosen normalized MAT descriptor (μ degradation functions) as function of neutron fluence for 22NiMoCr37 type base material (a) and for 18MND5 type weld material (b).

As a second step of evaluation, the optimally chosen MAT descriptors were considered as a function of the transition temperature shift ΔDBTT (see Fig. 4) for both materials. The optimally chosen MAT descriptor is characterized by μ(h_a=500mA,h_b=1150mA) values in this case, too.

It can be seen that a similar correlation was found between the MAT descriptors on the one hand and the transition temperature shift and the neutron fluence on the other hand. In this case, an averaged value of the transition temperature shift is used for samples that were irradiated with similar (but not exactly same) neutron fluence.

The correlation between the magnetic parameters and the neutron fluence shows that the material properties were changed due to the neutron irradiation. In other cases, see e.g. [14], also a monotonous, but not linearly increasing correlation was found. In general, the neutron fluence values alone are useless for the characterization of the material degradation, because the radiation damage is measured through the effect of the irradiation on the mechanical properties (amount of hardening and embrittlement). But in this case, for the two investigated series of samples, a similar correlation was found between the optimally chosen MAT descriptors; and the neutron fluence and transition temperature shift. It is not surprising, because if the ΔDBTT is considered as a function of the neutron fluence, the correlation is also linear, as shown for the two materials in Fig. 5. For the correct interpretation of the material embrittlement the modification of the transition temperature is the most important parameter.

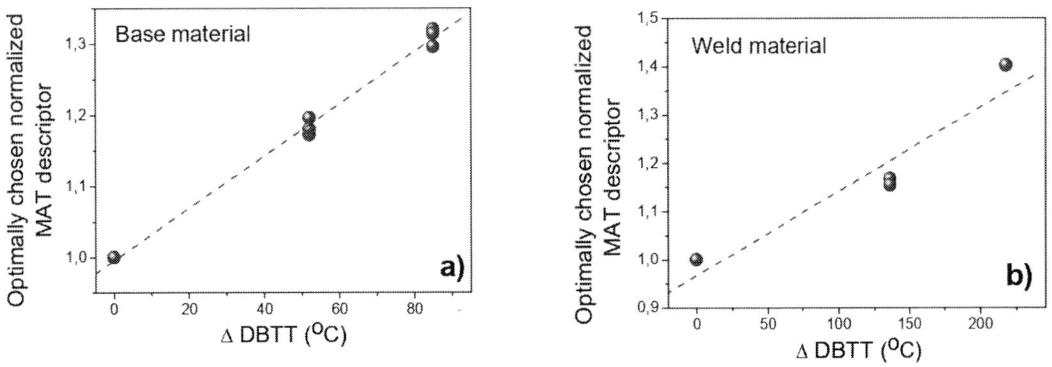

Figure 4. Optimally chosen normalized MAT descriptor (μ degradation functions) as a function of the transition temperature shift for a 22NiMoCr37 type base material (a) and for a 18MND5 type weld material (b).

The linear fit on the measured points is to lead the eye to reflects more visible the character of the correlation. It does not reflect a strict mathematical fitting, because of the limited number of measured points.

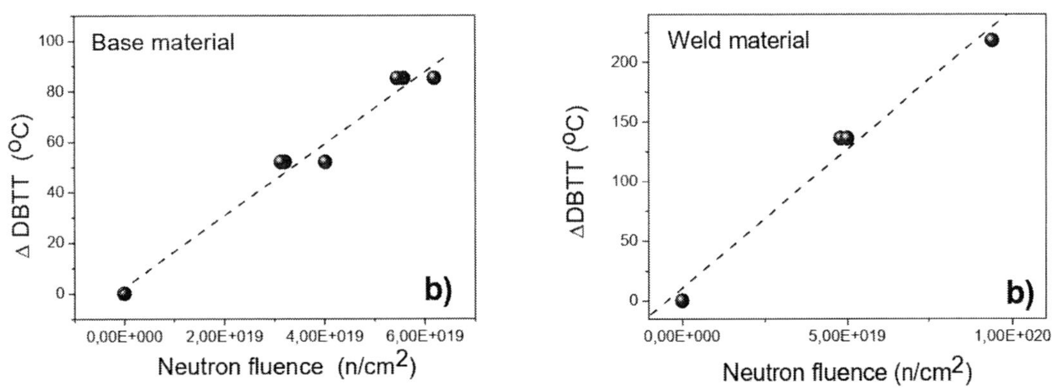

Figure 5. Correlation between the transition temperature shift and the neutron fluence for a 22NiMoCr37 type base material (a) and for a 18MND5 type weld material (b).

5. Conclusions

It was shown that the Magnetic Adaptive Testing method can characterize the structural changes in a reactor pressure vessel material in a sensitive and reliable way. A definite and monotonous correlation was found between the neutron fluence and the MAT descriptors for both types of investigated materials. The modification of the transition temperature due to neutron irradiation can also be followed by this nondestructive method. A linear correlation was found between the MAT descriptors and the Charpy 41 Joule transition temperature.

As a general conclusion, it was shown that the MAT method is suitable for the nondestructive structural inspection of reactor pressure vessel steel: degradation functions characterize structural changes of this material easily and reliably. Our presently described results are in good coincidence with previous results, performed on other types of reactor steel.

MAT is also an advantageous method, from point of view of measurement, because there is not necessary to saturate the sample magnetically, which is almost impossible in case of an open magnetic circuit. As has been proved above, it is also possible the measurement of highly radioactive specimens. The evaluated magnetic descriptors are not absolute parameters, but MAT is powerful tool to perform comparative measurements, and such a way to detect changes in the material behaviour due to any type of degradation. However, calibration of MAT procedure should be done on a reference series of samples (by comparing magnetic descriptors with independently, destructively measured parameters) and then the degradation of any unknown sample can de determined

Acknowledgements

The authors gratefully acknowledge the technical testing team at SCK•CEN for the Charpy impact tests: J. Knaeps, R. Vanuytven, G. Berkmans, J.L. Puzzolante. The hot-cell operators Y. Florenty and G. Cools and the workshop for the manufacturing of the samples under the supervision of J. Verlinden.

The work was supported by "NOMAD" project. This project (Nondestructive Evaluation System for the Inspection of Operation-Induced Material Degradation in Nuclear Power Plants) has received funding from the Euratom research and training programme 2014-2018 under grant agreement No 755330.

References

[1] Ferreño D et al 2011 http://www.intechopen.com/articles/show/title/non-destructive-testing-for-ageing-management-of-nuclear-power-components
[2] Jiles D C 2001 Encyclopedia *of Materials Science and Technology*, Elsevier Press, Oxford, K.H.J. Buschow et al., Ed.,p.6021
[3] Takahashi S et al 2006 *J. Appl. Phys.* **100** 113908
[4] Takahashi S et al 2006 *J. Appl. Phys.* **100** 023902
[5] Dobmann G et al 2008 *Electromagnetic Nondestructive Evaluation (XI),* p.54, IOS Press
[6] Vandenbossche L 2009 "*Magnetic Hysteretic Characterization of Ferromagnetic Materials with Objectives* towards *Non-Destructive Evaluation of Material Degradation"*, Gent University, Gent, Belgium, PhD Thesis
[7] Gillemot F and Pirfo Barroso S 2010 *Proc. 8th International Conference on Barkhausen Noise and Micromagnetic Testing (ICBM8)*
[8] Pirfo Barroso et al 2010 *Nuclear Engineering and Design* **240** 722
[9] Kobayashi et al 2012 *Journal of Nuclear Materials* **421** 112
[10] Minov B 2012 „Investigation *of the hardening in neutron irradiated and thermally aged iron-copper alloys, on the basis of mechanical and magnetic relaxation phenomena*", Gent University, Gent, Belgium, PhD Thesis
[11] Dobmann G 2011 http://www.intechopen.com/articles/show/title/non-destructive-testing-for-ageing-management-of-nuclear-power-components - 11.10.2019
[12] Tomáš I 2004 *J.Mag.Mag.Mat.* **268/1-2** 178
[13] Tomáš I and Vértesy G 2012 http://www.intechopen.com/articles/show/title/magnetic-adaptive-testing - 22.11.2018
[14] Tomáš I et al 2013 *J. Nuclear Materials*, **432** 371
[15] DeVan MJ 1993 *6th Int. Symp. ASTM STP 1175 American Society for Testing and Materials, Philadelphia,* 26

IOP Conf. Series: Materials Science and Engineering **903** (2020) 012012 doi:10.1088/1757-899X/903/1/012012

Semi-continuous casting and microstructure investigation of the AlSi12 alloy

Agota Kazup[1], Viktor Karpati[2], Balazs Hegedus[3], Zoltan Gacsi[4]

[1, 2, 4] Institute of Physical Metallurgy, Metalforming and Nanotechnology; University of Miskolc; Miskolc-Egyetemvaros 3515 HU

[3] Institute of Energy and Quality; University of Miskolc; Miskolc-Egyetemvaros 3515 HU

E-mail: femagota@uni-miskolc.hu

Abstract. Semi-continuously cast AlSi12 alloy was investigated by our research group at the University of Miskolc in the Institute of Physical Metallurgy, Metalforming and Nanotechnology under laboratory condition. Our goal was to define correlation and tendency between casting parameters and solidified microstructures. The examined alloy is used as filler material, thus cover soldering technology and exchangers are made from it, like car coolers, radiators, air conditioners. During the investigation optical microscope was used for sight of ingot microstructure, secondary dendrite arm spacing was measured in this way and even the amount of primary α-phase.

1. Introduction

Aluminium and its alloys are often applied as industrial commodity in many ways. Besides high strength conventionally as cast Al alloys with more alloying elements (for examples AlSi, AlMg, AlSiCu, AlZnMg) and lower strength wrought Al alloys with less alloying elements (for examples AlMgMn, AlMn, AlMgZn) aluminium is used as a filler material because of its low melting point and sufficient strength for consumption with several amount of Si, perhaps with other micro alloying elements too [1, 2]. The technology conditions of casting greatly influence the solidifying microstructure [3]. At the University of Miskolc in the AluForm project our group carried out semi-continuous castings of different alloys by Indutherm CC3000 equipment [4]. AlSi12 was produced by this way in a laboratory at small size, which alloy is applied for making cooler/heater systems by soldering technology in the industry at large size [5]. The aim of the research was in order to observe the casting parameters for the solidifying microstructure. These kind of parameters are the secondary water-cooling and the casting speed, which is combined by the pulling speed and the pause function during the process. Microstructure directly influences attribution of the material [6, 7], that plays significant role in the working up of the raw material and usage of the end product. Primary solid solution is a soft texture element in the alloy, so it can be formed easily. Its shape is a branching dendrite and it has got primary and secondary arms. The secondary dendrite arm spacing (sDAS) defines the refinement of the microstructure. The smaller this distance, the greater the strength of the material. At higher solidification rate smaller dendrite arm spacing comes into being [8]. From the usage perspective the most appropriate microstructure of the AlSi12 alloy is homogeneous, formable with the soldered alloy together and melts easily in the furnace.

2. The AlSi alloy

Aluminium constitutes eutectic system with silicon as shown in Fig. 1. The eutectic temperature is 577°C, and the eutectic point is at 11-13% of silicon content. As the silicon content changes in the alloy several phases form in case of equilibrium solidification process, which are introduced below:

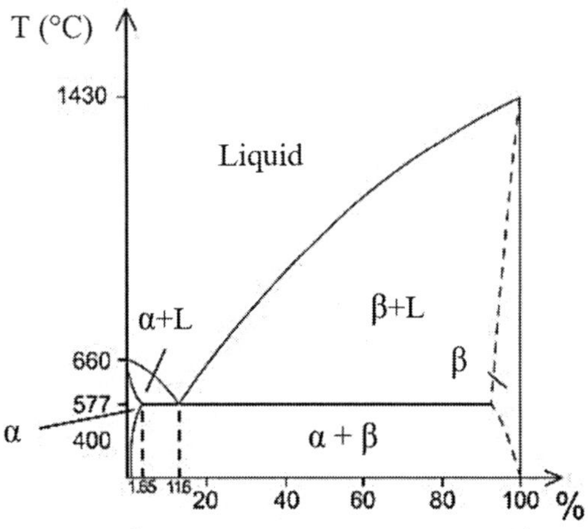

Figure 1. Al-Si phase diagram [9]

(a) From 0% to 1.65% Si content: From the molten metal primary α-solid solution solidifies, then under the solvus temperature β-silicon precipitates.
(b) From 1.65% to 11-13% Si content (hypoeutectic alloy): After the solidification of dendrite shaped primary α-solid solution, eutectic α+β phases solidify from the remainder melt.
(c) 11-13% of Si content (eutectic alloy): In this case eutectic is created if it is an absolutely equilibrium solidification. In practice it contains other texture elements.
(d) More, than 11-13% of Si content (hypereutectic alloy): Firstly primary silicon solidifies, then eutectic comes to be from the remainder melt. The primary silicon occurs with different Si contents in the texture if the solidification is not equilibrium.

This is not advantageous for working the raw material up because of its rigidity. It can be refined by phosphorus in melt condition [10].

2.1. AlSi alloy, as a filler

As a filler AlSi can be used between different type of materials to bind, like Al-Al, TiAl-TiAl, Al-steel and even ceramics to make bond [1, 2, 11, 12]. The most often used alloy is the AlSi12 melting at the lowest temperature and it has enough good viscosity. These criterions are needed so that Al fills strait gaps during soldering, but higher Si content can induce porosity in the bond, which influences strength to negative line [5].

Sheets are got with each other during covering solder technology, so that filler raw material is between the sheets and these are rolled. After this process rolled sheets get heat treatment in a furnace, the filler (solder) melts and fills gaps [5].

3. Experiments

Semi-continuous casting means casting of the molten metal into a water-cooled die solidifying an ingot, which is pulled out from the die by rolls [4]. Limited ingot length is produced by this method, its cross-section size is from few centimeters even to some meters, which geometry is circular or square. The cast ingot is a raw material for rolling, extruding and forging technologies.

Casting equipment is shown in Fig. 2., which was applied for the AlSi12 alloy casting. Its chemical composition is 12.7% Si, 0.2% Fe and 0.1% > Mg, Mn, Cu, which are weight percents. Principle of operation of the equipment had been introduced in another paper [4]. Cross-section was 10 x 100 mm and 3-3 pieces of samples were taken from every section, where centre and edge parts were investigated. These are shown in Fig. 3-4. In all 5 pieces of experiments were done and one piece of casting parameter was changed in each case. We mark these in Table 1. Comparisons were made in pairs illustrated by colours: secondary cooling by blue, casting speed without secondary cooling by green and pause with secondary cooling during casting by orange. Parameters were chosen taking into account of ability the technological solidification of AlSi12 alloy. During the casting process the speed of solidification frontline depends on the casting speed through the mushy zone (contains liquid and solid phases) geometry [3]. As mushy zone is getting deeper, the higher difference of rate of solidification velocity is between edge and centre of the ingot. Depth of the zone depends on the heat energy quantity of the system. Quantity of energy increases when casting speed is lower and secondary water-cooling is not used. In case of higher solidification rate secondary dendrite arm spacing will be smaller, so causes finer microstructure. The amount of α-solid solution depends on local concentrations during solidification also in contact with the rate of solidification.

Figure 2. The Induterm CC3000 equipment

3.1. Measurements

Microstructure and homogeneity of the polished and Barker etched samples were investigated by optical microscope in 50x magnitude without polarized light.

Table 1. Casting parameters

#Experimental matrix	$T_{,melt}$ (°C)	$T_{,die}$ (°C)	$v_{,pulling}$ (mm/s)	Pulling pause (s)	$v_{,casting}$ (mm/s)	Secondary cooling
#1	750	450	0.1	0.1	0.091	Not used
#2	770	450	0.1	0.1	0.091	Used
#3	770	450	0.3	3.0	0.075	Not used
#4	740	450	4.0	0.0	4.000	Used
#5	770	450	4.0	9.9	0.370	Used

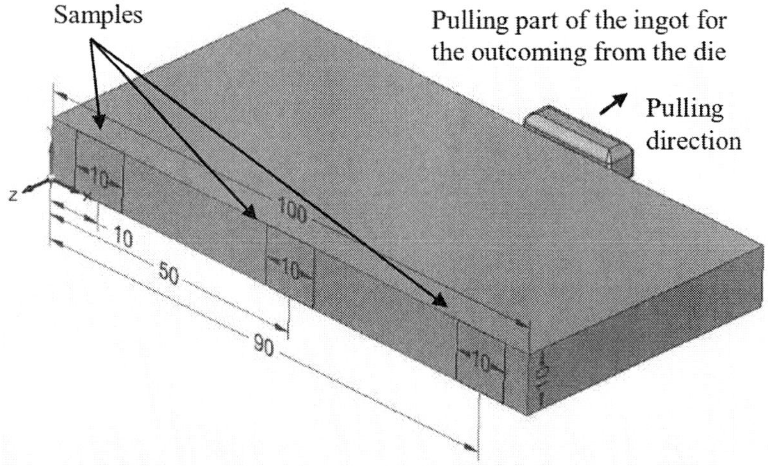

Figure 3. Size of the ingot and pulling direction of the samples

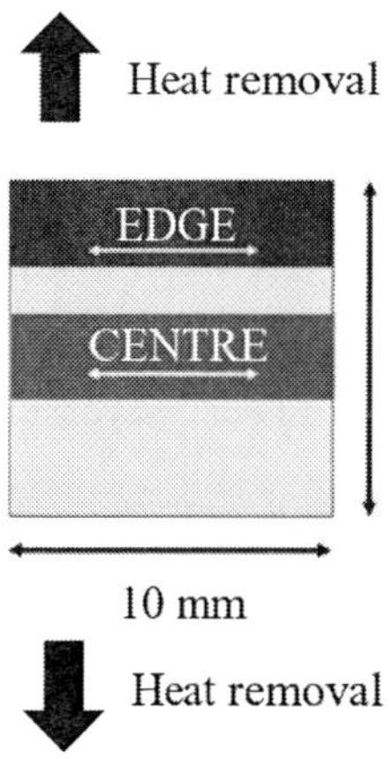

Figure 4. Size of a sample

3.1.1. Secondary dendrite arm spacing. Secondary dendrite arm spacing is measured from centres of two arms, then the encased length was divided with the number of "taken arms-1", shown in Fig. 5. The resulted number is the average sDAS. In case of AlSi alloy primary α-solid solution phase solidifies from the melt getting dendrite structure and gives the refinement of its microstructure (Fig. 6.). sDAS = $\frac{L}{P-1}$ (μm), where L – measured distance between two secondary arms (μm), P – numbers of arms (-).

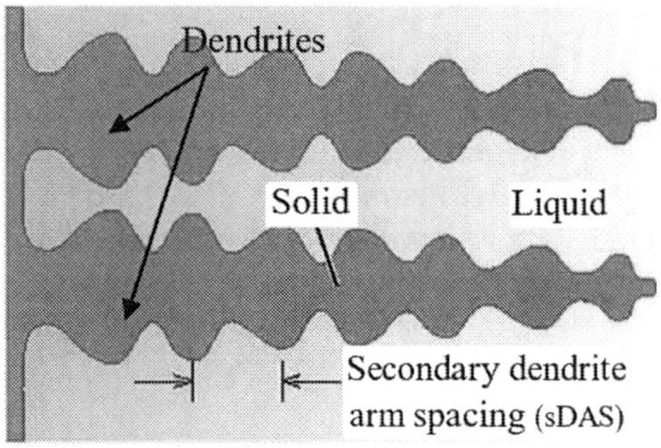

Figure 5. Measurement of the sDAS [13]

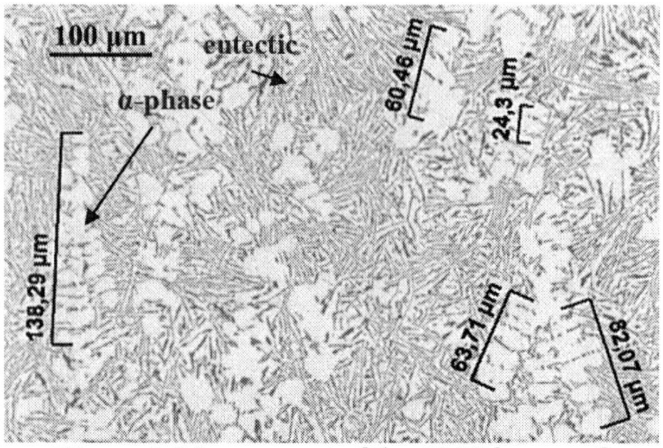

Figure 6. Measurement of the sDAS in case of practice, HF etching, magnification: 200x

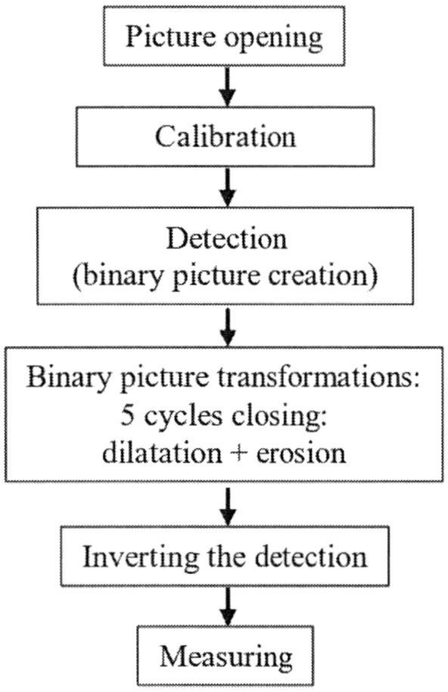

Figure 7. Process of the picture analysis for measuring the amount of primary α – phase area

3.1.2. Picture analysis. Black and white images were taken about microstructure with microscope in the right magnitude. After reading into Leica QWin software calibration was done through scale bar in order to define how many pixels mean a micrometer. In the next steps dark parts of images were detected and the level of the gray-shade were given from 0 to 255 (-), which exactly determines the dark phases (Si) according to the given image, thus in this way new, binary pictures were generated. Morphological binary picture transformations were accomplished on all binary pictures like: by 5 cycles of closing in the eutectic light coloured secondary α-phase was also detected. 5 cycles of closing means consecutive of 5 steps of dilatation and erosion. Dilatation means increasing the edge of the detected object with 1 pixel in every direction, furthermore erosion means decreasing of its edge. Inverting of the detection the area of the primary α-phase (A_α) was immediately able to measure and we calculated the amount of it (A_A). $A_A = \frac{A_\alpha 100}{A_{all}}$ (%), where A_α – the area of primary phase (μm^2) and A_{all} – the all area of the image (μm^2).

4. Results

4.1. Microstructure

Barker electrolytic etched microstructures of samples of experiments (5 pieces) were investigated from different positions through cross-section of the ingots. In this paper just centre structures of all experiments are introduced in Fig. 8. in 50 mm sample position along the ingot width (Fig. 3-4.).

In case of same casting speed without using secondary cooling microstructure is more disperse in the edge part of the ingot. Structure of the centres have not changed (Fig. 8. a and b). With using secondary cooling in case of lower casing speed homogeneous and fine microstructure has been reached in both of edge and centre. Ordered dendrite structure has disappeared. These are shown in Fig. 8. a) and c).During the casting process using of pause function can be more advantageous for the solidifying microstructure. Its result is an unified and ordered microstructure through the cross-section (Fig. 8. d and e).As number #3 experience presented, lower casting rate (0.075 mm/s) with no secondary cooling (assuming it had the lowest solidification rate) has resulted the most homogeneous microstructure. As number #5 experience displayed, casting speed can be increased if secondary cooling is used for a good result. Using this method casting process can be accelerated getting advantage for the industry.

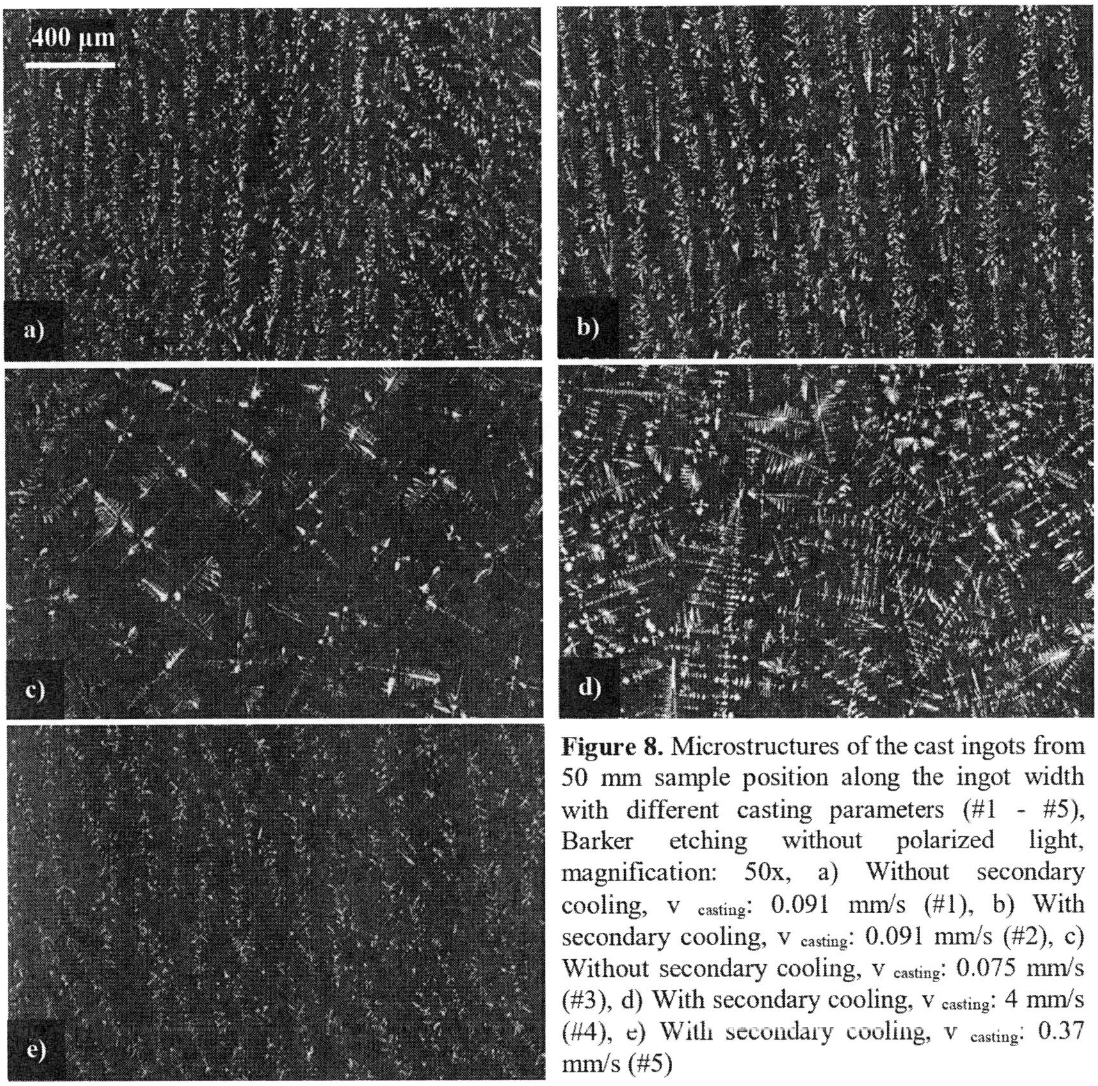

Figure 8. Microstructures of the cast ingots from 50 mm sample position along the ingot width with different casting parameters (#1 - #5), Barker etching without polarized light, magnification: 50x, a) Without secondary cooling, $v_{casting}$: 0.091 mm/s (#1), b) With secondary cooling, $v_{casting}$: 0.091 mm/s (#2), c) Without secondary cooling, $v_{casting}$: 0.075 mm/s (#3), d) With secondary cooling, $v_{casting}$: 4 mm/s (#4), e) With secondary cooling, $v_{casting}$: 0.37 mm/s (#5)

4.2. Results of secondary dendrite arm spacing of cast ingots

Results of sDAS measurement are shown in Fig. 9. Secondary cooling has had positive effect.

In the edge sDAS has decreased uniformly from 15-35 μm to 15-16 μm (Fig. 9. a and b). sDAS of the centre part has not changed using secondary cooling, it is 13-15 μm. This is due to the cooling, because of the depth of mushy zone has decreased in the ingot during casting process. In the edge the rate of solidification velocity has increased. At lower casting speed (0.075 mm/s) sDAS is uniformly 13-15 μm through the cross-section of the ingot as shown in Fig. 9. a) and c). The depth of mushy zone has decreased, and same solidification rate is assumed in both of edge and centre. Using the pause function during casting has not had significant effect to sDAS, it is 10-23 μm (Fig. 9. d and e).

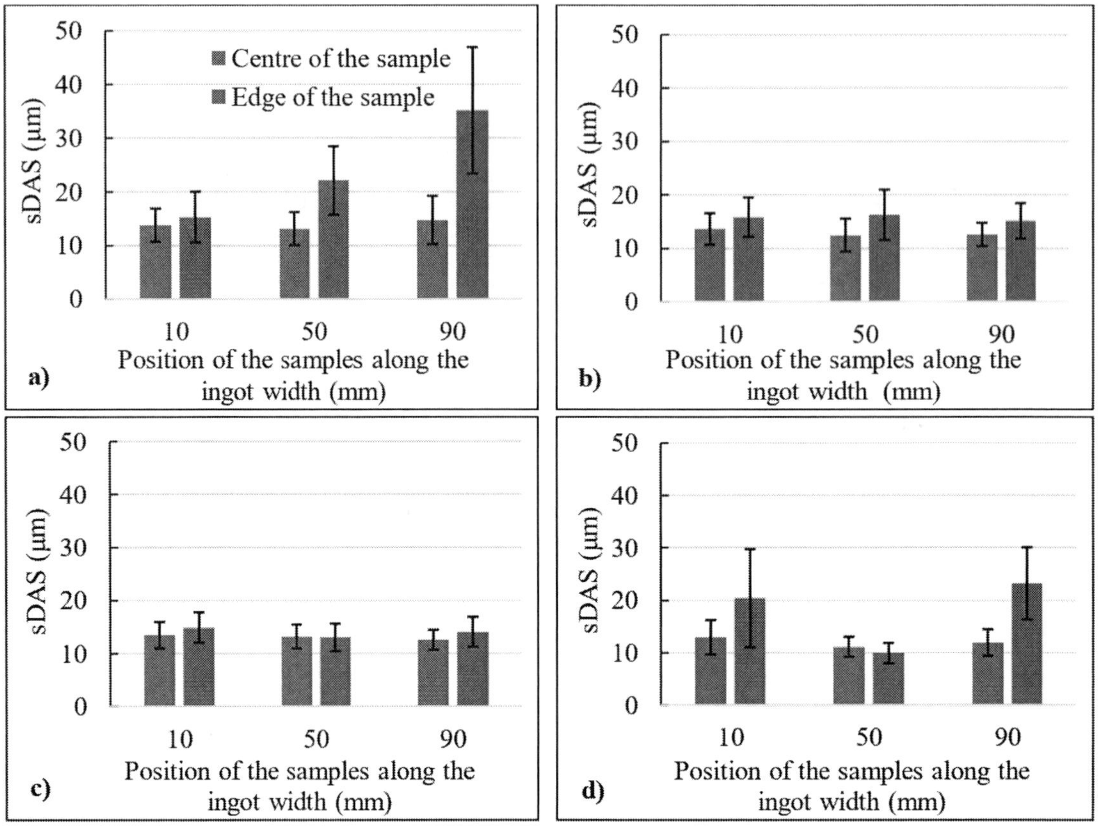

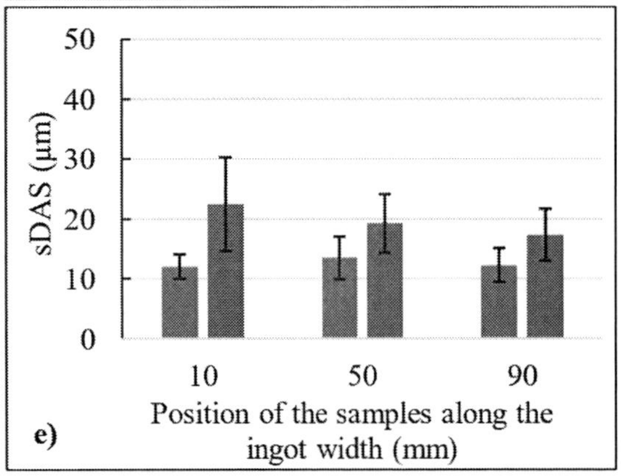

Figure 9. Secondary dendrite arm spacing (sDAS) of the cast ingots with different casting parameters (#1 - #5): a) Without secondary cooling, $v_{casting}$: 0.091 mm/s (#1), b) With secondary cooling, $v_{casting}$: 0.091 mm/s (#2), c) Without secondary cooling, $v_{casting}$: 0.075 mm/s (#3), d) With secondary cooling, $v_{casting}$: 4 mm/s (#4), e) With secondary cooling, $v_{casting}$: 0.37 mm/s (#5)

4.3. Average area amount of the primary α–solid solution

In the solidified microstructure the calculated amount of primary solid solution has shown in Fig. 10.

With secondary cooling the amount of α-phase has decreased from 23-33% to 19-22% in the edge approaching 18% like in the centre, which has not changed. Results are shown in Fig. 10. a) and b). At lower casting speed (0,075 mm/s) the area amount of α-phase has decreased in the edge, so uniformly is has been 20% through cross-section of the ingot (Fig. 10. a and c). Using the pause function during casting (0.37 mm/s) the amount of α-phase has decreased to 17-18% in the centre, which is same value to the edge's (Fig. 10. d and e).

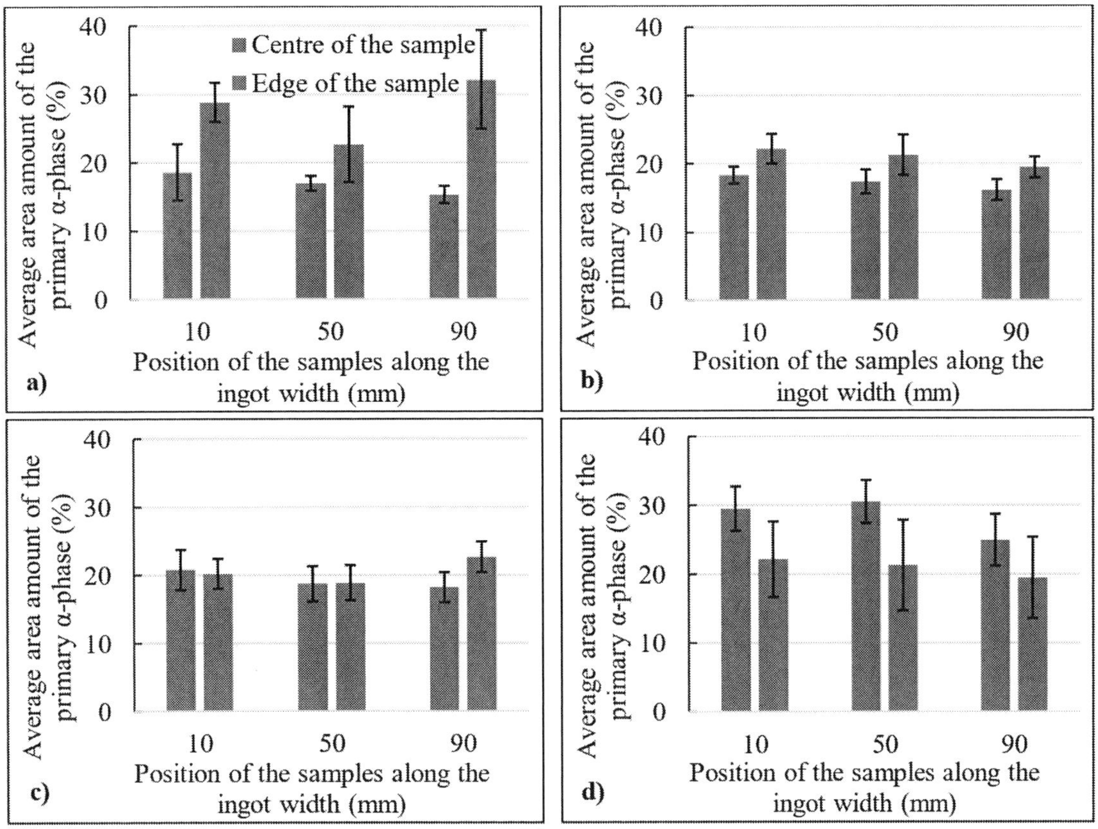

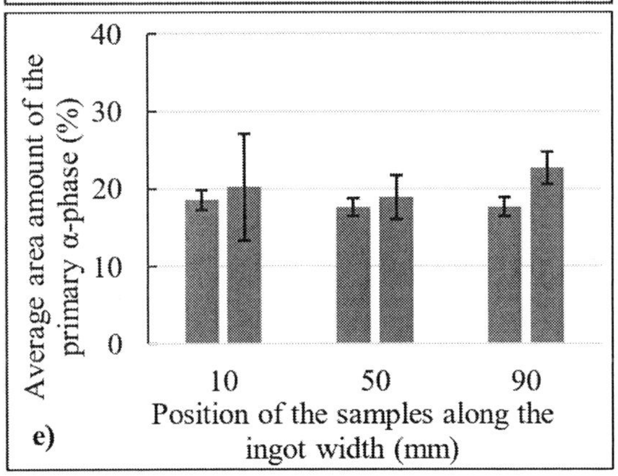

Figure 10. Average area amount of the primary α-phase of the cast ingots with different casting parameters (#1 - #5): a) Without secondary cooling, $v_{casting}$: 0.091 mm/s (#1), b) With secondary cooling, $v_{casting}$: 0.091 mm/s (#2), c) Without secondary cooling, $v_{casting}$: 0.075 mm/s (#3), d) With secondary cooling, $v_{casting}$: 4 mm/s (#4), e) With secondary cooling, $v_{casting}$: 0.37 mm/s (#5)

5. Conclusions

The aim of our research was to investigate microstructure of semi-continuous cast AlSi12 alloy affected by changing casting parameters, like rate of casting speed and using of secondary cooling.

Due to our results we take the following statements:

(a) Secondary water-cooling has significant and advantageous effect to the solidified microstructure, because using it the cooling rate is intensive, thus structure becomes refined and homogeneous. In this case secondary dendrite arm spacing is 15 μm also in the edge part of the ingot, the amount of α-phase has decreased 12%.

(b) Decreasing the casting speed (0.075 mm/s) homogeneous dispersion of sDAS has formed in the microstructure through the cross-section (13-15 μm). The amount of α-phase has decreased from 23-33% to 20%.

(c) During the casting process using the pause function has great effect to the solidification of microstructure. Across the cross-section of AlSi12 ingot solidification rate becomes equal due to the less depth of mushy zone [3], so microstructure solidifies homogenously.
(d) Consequently stated how homogeneous microstructure can be reachable: in case of lower casting speed (0.075 mm/s) without using secondary cooling or in case of greater casting speed (0.37 mm/s) using secondary cooling.

Acknowledgments
In this paper the published research is supported by the ÚNKP-19-2 New National Excellence Program of the Ministry for Innovation and Technology and the EFOP-3.6.1-16-2016-00011 "Rejuvenating and Renewing University – Innovative Knowledge City – part of the University of Miskolc intelligent specialization aided institutional development" project – in the Szechenyi 2020 – by the assistance of European Union, realized by the European Social Fund co-financing.

References
[1] Z Niu et al Effects of germanium additions on microstructures and properties of Al–Si filler metals for brazing aluminum, *Trans. Nonferr. Met. Soc.* China **26** (2016) 775-782
[2] X Chen et al Interfacial structure and formation mechanism of ultrasonic-assisted brazed joint of SiC ceramics with Al-12Si filler metals in air, *Journ. of Mat. Sci. & Techn.* **33** (2017) 492-498
[3] R Nadella et al Macrosegregation in direct-chill casting of aluminium alloys, *Prog. in Mat. Sci.* **53** (2008) 421–480
[4] Á Kazup et al High purity primary aluminium casting by Indutherm CC3000 semi-continuous casting equipment, *Multiscience - XXXIII. Microcad Intern. Multid. Sci. Conf.* (2019) ISBN 978-963-358-177-3
[5] T. Markovits CO_2 laser soldering of aluminium alloys, PhD dissertation, Budapest, 2004 (in Hungarian: Markovits T Alumíniumötvözetek CO_2 lézeres forrasztása, PhD értekezés, Budapest, 2004)
[6] D G Eskin Physical metallurgy of direct chill casing of aluminium alloys, 2008
[7] L Y Zhang et al Effect of cooling rate on solidified microstructure and mechanical properties of aluminium-A356 alloy, *Journ. of Mat. Proc. Techn.* **207** (2008) 107–111
[8] B Benjunior et al Effect of different cooling rates condition on thermal profile and microstructure of aluminium 6061, *Proc. Engin.* **184** (2017) 298 – 305
[9] tudasbazis.sulinet.hu/hu/szakkepzes/mezogazdasag/muszaki-alapismeretek/vas-szen-otvozetek-egyensulyi-diagramjai/ontottvas-szovetelem-diagramjai, 2019.11.17.
[10] P. Jónás Knowledges of light castable metals, National Textbook Publisher, 2009 (in Hungarian: Jónás P Könnyűfém öntészeti ismeretek, Nemzeti Tankönyvkiadó, 2009)
[11] R K Shiue et al Infrared brazing of TiAl using Al-based braze alloys, *Intermetallics* **11** (2003) 661–671
[12] D Narsimhachary et al Mechanical and microstructural characterization of laser weld-brazed AA6082-galvanized steel joint, *Journ. of Mat. Proc. Tech.* **263** (2019) 21–32
[13] V. Mertinger Presentation of structure of metal materials, University of Miskolc, 2019.05.05. (in Hungarian: Mertinger V Fémes szerkezeti anyagok előadásanyag, Miskolci Egyetem, 2019.05.05.)

IOP Conf. Series: Materials Science and Engineering **903** (2020) 012013 doi:10.1088/1757-899X/903/1/012013

Fracture mechanical analyses of high strength steels applying experiments and simulation

Zs Koncsik[1] and Zs Lukács[1]

[1]associate professor, Institute of Materials Science and Technology, University of Miskolc, Hungary

E-mail: zsuzsanna.koncsik@uni-miskolc.hu

Abstract. The paper contains information about the experimental and simulation test results of thermomechanically rolled high strength steel. The S960TM material grade was produced by a modern technological process, which provide longer estimated life time, better performance, and due to higher strength enable smaller applied wall thickness, which require less welding activity. The only disadvantage of this material grade is that they can contain material discontinuities in their microstructure. The characterisation of the response of these materials against load can be analysed by fracture mechanical investigation. Fracture mechanical experiments were carried out with three-point bending method, calculating fracture toughness value of crack-tip opening displacement (CTOD). To reduce the complicated real experiment in fracture mechanics, the simulation possibility of fracture mechanical behaviour of the investigated material was applied, using DEFORM software, which gave an acceptable narrowing to the measured phenomenon.

1. Introduction

Innovative steel production processes have resulted in novel material grades, with improved quality and enhanced performance. One of the significant driving forces behind the development is the requirement for lighter-weight structures. Light-weight structures provide several advantages, like smaller applied wall thickness, easier carriage, savings the required amount of welding, etc. The appearance and wider application of the weldable high strength steels, produced by novel technologies is the outcome of their characteristic features, i.e. higher strength and toughness [1], [2].

Based on the novelty, there is incomplete information about the detailed properties of these material grades in different application areas, which hinder its applicabilities in the engineering structures. However, there is relatively less data available for structure design engineer to accomplish the mechanical controls based on the fatigue or fracture mechanical approaches for design and operate structures built from high strength steels [3]. The Eurocode 3 [4], chapter 12. gives information about standard investigation results for steel grades up to a guaranteed yield strength of 700 MPa. The measures for fracture mechanical control are only available for grades with lower strength [5].

There exist several methods to increase the strength properties of steel materials [6]. One of them is thermomechanical rolling process, which is carried out in two main steps [7]. Firstly, the base material is heated to 1100°C, following that the rolling of the slab is carried out in austenitic state. In the next step the plate cools down on calm air to "as rolled" condition. Thermomechanical rolling is following, which is individually designed (rolling temperature and the number of rolling steps) by the producer.

The result of this manufacturing process is an extremely fine grain size with reduced carbon and alloying element content which further improved weldability due to cleaner steel composition.

Beside the several advantageous properties, the microstructure, produced by this process can contain material discontinuities, that influences the behaviour of the engineering structure under loading condition. Aiming to get information about the response of a thermomechanically rolled high strength steel containing material discontinuities (supposed crack = artificial failure), fracture mechanical investigations were carried out [8]. Due to tough behavior of the investigated advanced high strength steel and the complicacy of the experimental work [9] the possibility of application of simulation was also analysed.

2. Experimental work

2.1. Investigated materials

The experimental work was carried out on thermomechanically rolled ALFORM 960M grade steel produced by Voestalpine Anarbeitung GmbH. The chemical composition, according to the producer's certificate, with the alloying elements of the investigated samples are given in Table 1.

Table 1. Chemical composition of the investigated material, (wt%)

C	Si	Mn	P	S	Al	Cr	Ni	Mo	Cu	V	Ti	Nb	N	B
0.084	0.329	1.65	0.011	0.0005	0.038	0.61	0.026	0.29	0.016	0.078	0.014	0.035	0.006	0.0015

The yield strength according to the producer's material certificate reached 1051 MPa, while the measured value showed only 958 MPa.

For fracture mechanical tests samples with 15 mm wall thickness were applied. The average surface roughness changed between 0.8…1.6 µm. The lower value belongs to the surface containing the notch.

The microstructure of the base material was analyzed in longitudinal section. This section contained elongated grains due to forming during rolling and as a result of fast cooling the microstructure contained predominantly martensite and bainite beside ferrites [11].

2.2. Experimental circumstances

Advanced high strength steels have an excellent combination of high yield strength and high toughness even at lower temperatures. In consequence of this the determination of general fracture toughness value is not unambiguous. Therefore, crack tip opening displacement (CTOD or COD) measurements were carried out on three-point bend specimens, worked out in T-L and L-T orientation from the sheet base material. For the reliable investigation result 5-5 specimen were cut out in each orientation. The geometry of the samples was 13 mm thickness, 26 mm width, and the machined notch depth varied between 11.46…11.52 mm. The specimens were prepared according to the ISO 12135 standard [10].

For fatigue precracking a maximum bending load of 8560 N was applied, which was reduced by degrees, in 7 steps, to the lowest load of 3230 N, as the crack at the notch tip arose and started to propagate. The fatigue precracking was carried out on MTS 810.23 type electro-hydraulic testing machine, with R=0.1 asymmetry factor. The planned crack length after precracking was selected as *"longer than 1.3 mm"*. Completing the fatigue precracking, the length of initial cracks on specimen surfaces (length of machined notch + crack from fatigue precracking) changed between a= 12.73…12.92 mm. The real crack length could be measured after opening broken the specimen at the final stage of the experimental procedure.

For fracture mechanical tests the above mentioned three-point bending test apparatus was selected, and the experiments were carried out at room temperature with the support distance of 104 mm. The lower support and the upper bending radius were 5 mm, the applied velocity during the test was 0.05 mm/s, and the applied load was static increasing. The crack tip opening displacement was measured by a strain gauge type MTS 632.02C attached to a clip placed between the two accurately positioned knife edges at the mouth of the machined notch. The controlled variable was the piston displacement. During investigation the piston displacement, the crack tip opening data and the applied load were registered

IOP Conf. Series: Materials Science and Engineering **903** (2020) 012013 doi:10.1088/1757-899X/903/1/012013

and stored. After the three-point bending test the samples were marked by heat tinting in a chamber at 250°C for an hour, then they were broken open and the crack propagation was observed on the fracture surfaces.

2.3. Experimental results

Examining the recorded load-notch opening displacement diagrams three different types could be observed, related to the standard [10]. One type is a so-called maximal diagram, containing elongated plateau, as maximum value for load, shown in Figure 1., the other type of the recorded diagrams contains sudden springing, the so-called pop-in as shown in Figure 2. One type of a diagrams contained only one pop-in, while the other type of the diagrams contained two pop-ins. From these diagrams two basically different types are presented on Figure 1. and Figure 2., one with maximum value, and the other one with one significant pop-in.

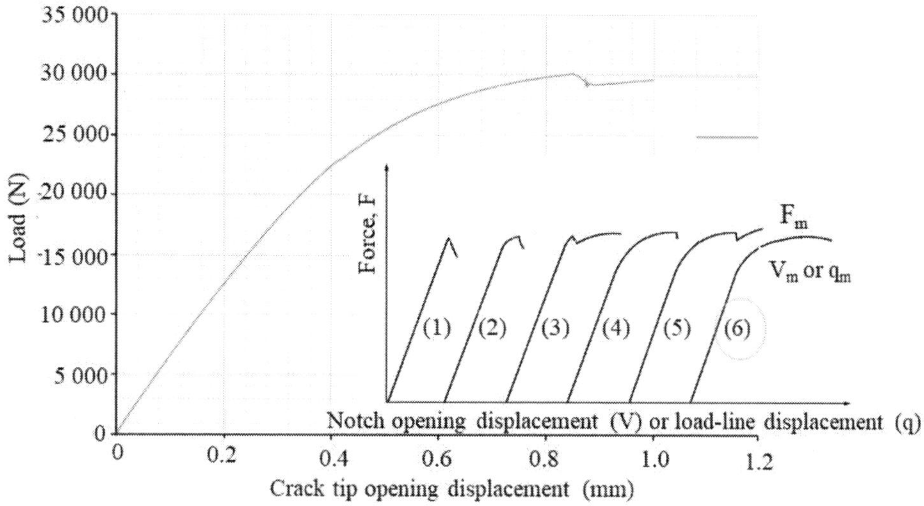

Figure 1. Recorded Load – Crack tip opening displacement diagram, type 1.

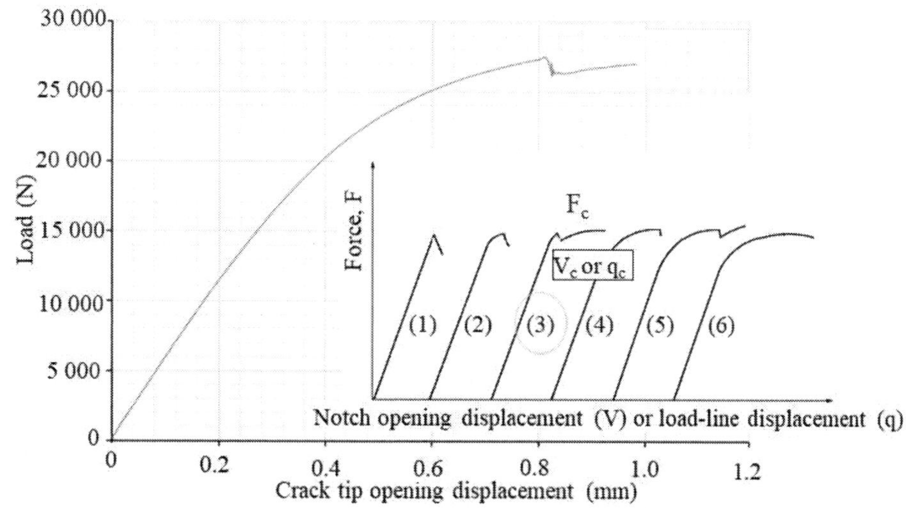

Figure 2. Recorded Load – Crack tip opening displacement diagram, type 2.

In some cases, there are more pop-ins to observe on the recorded diagram. According to the test evaluation procedure [10], the first significant pop-in have to be considered for the calculation of CTOD value. In the case illustrated in Figure 1. F and V values were taken on that place, where the diagram reached a maximum force plateau prior to fracture, without a significant pop-in. In these cases, the indices of F and V values are *'m'*. In the case, illustrated by Figure 2., a significant pop-in could be identified. It has no importance, which number of pop-ins was the first significant, for these cases the F and V values get an index *'c'*.

The CTOD values (δ_0) are calculated according to the standard [10]. The calculation details and the detailed results and conclusions for all measurement are already published in [11]. For the calculation, the real accurate crack length is also necessary, which was measured after heat tinting, on nine locations on the broken open surfaces as shown in Figure 3. The calculated CTOD values with additional pop-in data and the maximum forces are given in Table 2.

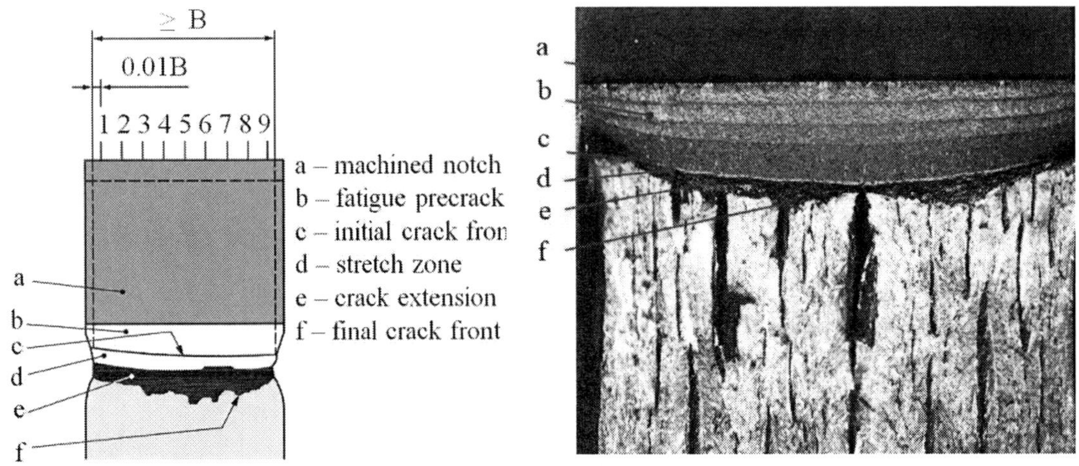

a) b)

Figure 3. a) Crack length measurement on the broken open surfaces [10]; **b)** Fracture surface of one investigated sample, TM1_1

Table 2. CTOD results

Measured value / Sample number	Significant pop-in	Orientation	F	δ_0	Type of F, V, and δ
	(number)	(mm)	(N)	(mm)	-
TM1_1	no pop-in	L-T	29402	0.200	m
TM1_2	no pop-in	L-T	30102	0.161	m
TM1_3	2. pop-in	L-T	29908	0.113	c
TM2_1	1. pop-in	T-L	27400	0.140	c
TM2_3	no pop-in	T-L	27805	0.228	m
TM2_4	no pop-in	T-L	29103	0.288	m
TM2_5	2. pop-in	T-L	29101	0.155	c

From the planned measurements three successful results for L-T orientation and four results for T-L orientation are given in Table 2. In four cases no pop-in phenomena could be observed on the diagram, in one case the first pop-in was the significant, while in two cases the second pop-ins were significant.

In L-T direction one δ_c and two δ_m values, while in T-L direction two-two δ_c and δ_m values could be considered, at the investigated thickness of 13 mm.

As conclusion to the test results given in Table 2. it can be stated, that lower δ_0 value, which correspond to δ_c, belongs to brittle behaviour, while higher δ_0 value, which becomes δ_m belongs to more plastic behaviour, especially in the vicinity of crack tip.

The several requirements regarding the initial crack length, crack and notch location, surface roughness, the measuring equipment, precracking procedure, were observed and controlled during the complete investigation. It can be stated that they correspond to the requirements of the standard [10], except one minor difference:

- according to the requirement, given in the standard: no part of the fatigue precrack front shall be closer to the starter notch than 1.3 mm or 2.5%W, whichever is larger. This requirement is only valid for the seven interior cracks, the a_1 and a_9 crack lengths are usually closer to the crack starter notch than 1.3 mm, see also Figure 3.

3. Simulation with DEFORM

DEFORM is a Finite Element Method based process simulation system, designed to analyze various forming and heat treatment processes used by metal forming and related industries [12]. By simulating deformation processes on computer, it allows the designers and operators to reduce a real physical experimental need, and to get faster information about the material or structure behaviour possibly also as prediction. Nowadays it is increasingly applied for the deformation analyzes of advanced high strength steels in various conditions, thought there is no data available for the material in the software [13].

As it was mentioned before the fracture mechanical physical test is a complex investigation with several influencing parameters and several requirements. The simulation of this process needs partially different input information as the real physical investigation. The main benefits of using the simulation of this damage process could be time, material and cost saving. While the real experimental tests last approximately for weeks for one batch, the simulation can run in some hours with the appropriate input data and adjustment.

3.1. Input parameters

During the simulation the following input parameters were set:

- applied model: 2D model – supposed plain strain;
- applied material law: elastic-plastic;
- material characteristics:
  - for elastic deformation part: measured elastic modulus, E=195000 MPa;
  - for plastic deformation part: flow-curve, measured and converted data for true stress – true strain curve for true strain 0 - 0.2 value, and approximation applying Ludwik's equation for the strain from 0.2 – 1., see Figure 4.:

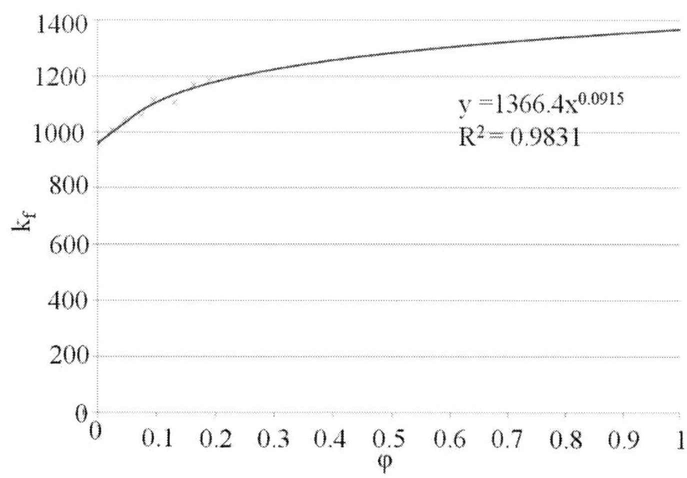

Figure 4. Measured flow-curve

- applied mesh: rectangular with remeshing;

- applied damage model: Cockroft & Latham model [14], the critical fracture criteria: 0.5:
  - $\int_0^{\varepsilon_f} \bar{\sigma}\left(\frac{\sigma^*}{\bar{\sigma}}\right) d\bar{\varepsilon}$
  - where
    - ε_f: fracture strain;
    - $\bar{\sigma}$: equivalent stress;
    - $\bar{\varepsilon}$: equivalent strain;
    - $\frac{\sigma^*}{\bar{\sigma}}$: non-dimensional stress concentrating factor, representing the effect of the highest tensile stress, σ^*.
- for the simulation the accurate sample geometry data was selected of TM2_5 sample;
- on Figure 5. the meshed sample is illustrated, with locally enriched meshing parts on places, where deformation and damage data are more interesting.

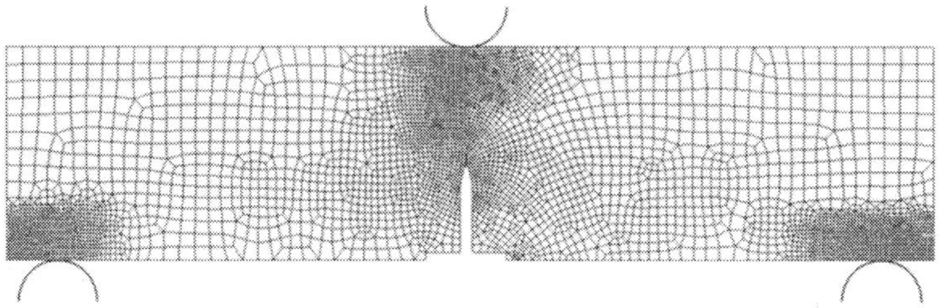

Figure 5. The meshed sample

3.2. Simulation results

With the application of finite element simulation several information can be extracted after the simulation is completed. For the investigated case the most important data were the load-deflection curve and the crack tip opening information.

The simulated load-deflection curve is illustrated in Figure 6. with red line. For comparison between simulated and real test diagram, the real testing diagram is also presented for the single investigated specimen, marked with blue line.

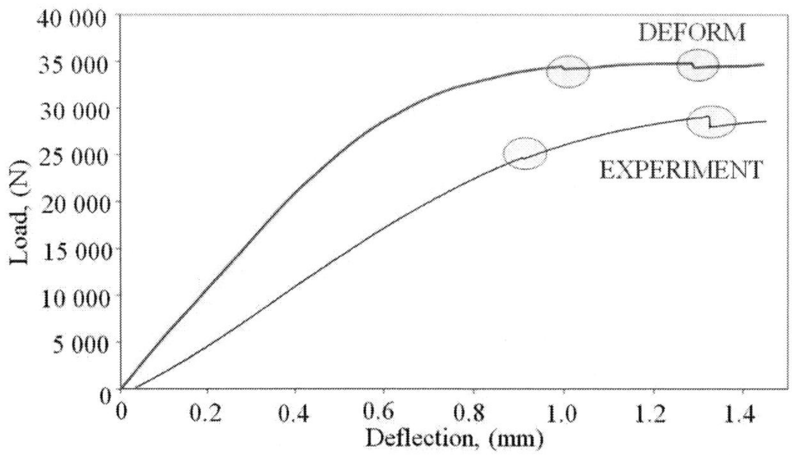

Figure 6. Comparison of load-deflection curve derived from experiment and simulation

In real physical experiment, the load-deflection curve contained two pop-ins, at load = 24700 N, deflection = 0.9 mm and at load = 29101 N, deflection = 1.33 mm. The running down of the curve of simulation follows the real experimental results, but the slope of the curve is higher, therefore the load values at the pop-ins are higher. The location of the pop-in in the x axes narrows the real results.

The crack tip opening was measured by the displacement changes during the investigation of two points at the top of the notch (P1 and P2 in Figure 7.), as it is measured also in the physical test. It is only possible to record a displacement-time diagram in the software. At the point of the first pop-in the process time is 19 s, while at the second pop-in the process time is 26 s. The crack tip opening value was measured at that times, as it is illustrated in Figure 7. At the location of the first pop-in the crack tip opening value is 0.64 mm, in the physical test it was 0.48 mm. At the location of the second pop-in the crack tip opening value is 0.82 mm, which was the same value also for the physical test.

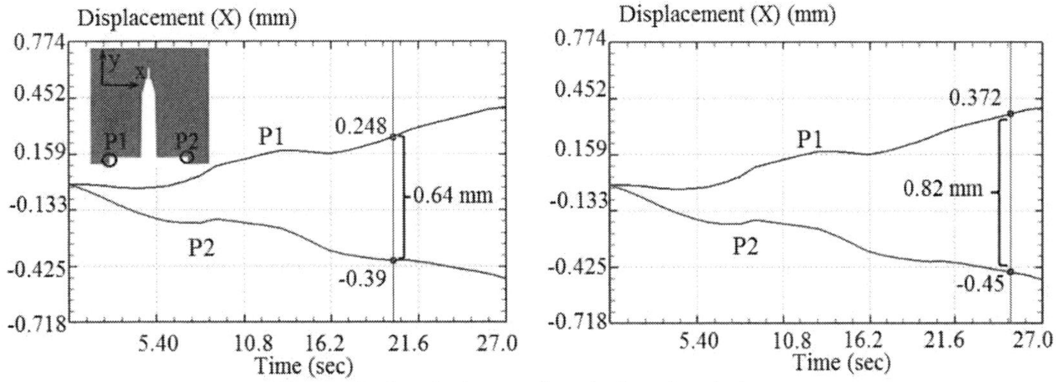

Figure 7. Crack tip opening during simulation

Analyzing the damage rate, Figure 8., as the test/simulation is being proceed, two changes in the gradient of damage diagram can be observed in Figure 8.a. The damage slope changed first at 19 s of the test, close to the first pop-in observed in physical test. Here supposedly the crack started to propagate, so the damage got higher. Furthermore, the slope of the damage curve changed at 26 s, close to the location of the second pop-in in physical experiment. Both locations of slope changes in the simulated damage curve narrow the locations of pop-ins experienced during the physical investigation (dotted lines in Figure 8.a). Analyzing the damaged picture of the sample, Figure 8.b), the highest damage can be observed in the vicinity of the crack tip.

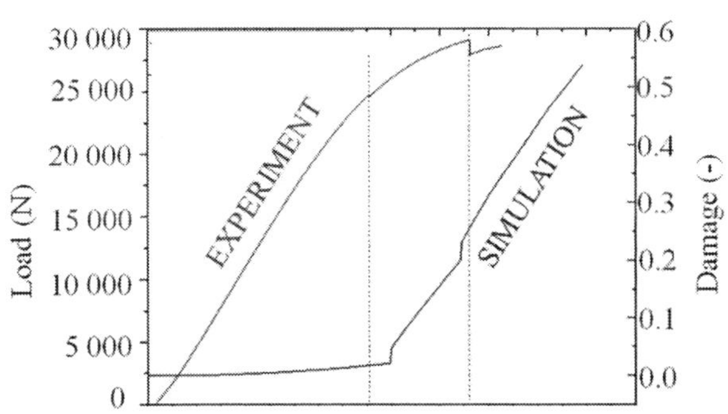

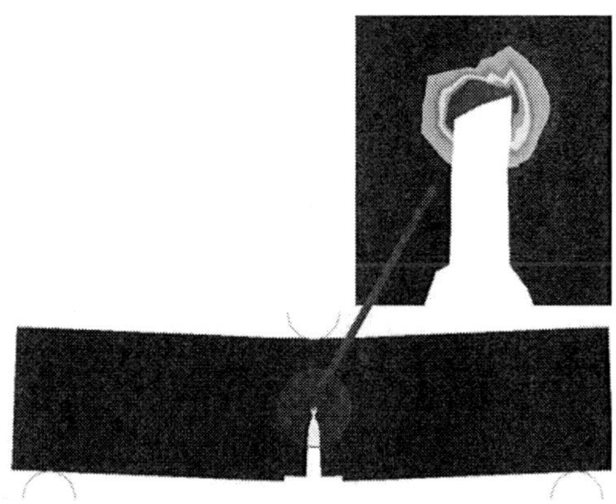

a) b)

Figure 8. a) Comparison of experimental result (load-deflection curve) and simulated damage curve b) Damaged sample after completing the simulation

Concluding the simulation results it can be stated that DEFORM software simulation for fracture mechanical test, applying three-point bending test method gave a comparable approximation. Further refinement of the simulation can be tested, by for example choosing a different damage model, or considering a different fracture criterion.

4. Conclusion

In this paper the behaviour of thermomechanically rolled, high strength steel grade, ALFORM 960M was investigated applying real physical test and finite element modelling. During the experiment an artificial material discontinuity (notch and fatigue precrack) was machined and initiated in the samples. This discontinuity was also geometrically given to the FEM model.

In the real physical test crack tip opening displacement method was applied for the evaluation of fracture mechanical behaviour. The average value of the calculated δ_0 measure is 0.183, at the investigated thickness of 13 mm, which value is lower for *'c'* type measures, possessing more brittle character, and higher for *'m'* type measures, possessing more ductile behaviour.

DEFORM simulation was carried out to simulate the response, of the investigated high strength steel containing material discontinuity, to load. The simulation result showed, that the extracted load-deflection diagram had an acceptable correspondence to the measured one, and even the values for the crack tip opening were close, or same as it was measured in the reality. For a reliable result in the simulation further refinements should be tested concerning for example damage model, damage criterion. The advantage of this simulation is, that the unknown fracture mechanical behaviour of the investigated sample can be predicted, even if the advanced high strength steel properties are not involved in the software database.

Acknowledgement

The described article was carried out as part of the EFOP-3.6.1-16-2016-00011 *"Younger and Renewing University – Innovative Knowledge City – institutional development of the University of Miskolc aiming at intelligent specialisation"* project implemented in the framework of the Széchenyi 2020 program. The realization of this project is supported by the European Union, co-financed by the European Social Fund.

References

[1] Lukács J, Dobosy Á and Gáspár M 2018 *Adv. Mat. Res.* **1146** pp 44-56, doi: 10.4028/www.scientific.net/AMR.1146.44
[2] Dobosy Á, Gáspár M and Lukács J 2018 *Adv. Mat. Res.* **1146** pp 73-83, doi: 10.4028/www.scientific.net/AMR.1146.73
[3] Glodez S, Knez M, Jazaernik N and Kramberger J 2009 *Eng. Fail. Anal.* **16** pp.2348-2356, doi: 10.1016/j.engfailanal.2009.03.023
[4] European Standard EN 1993-1-1 2009 *Eurocode 3: Design of steel structures*
[5] Hagedorn K E and Eckel M 1992 *Nucl. Eng. & Design* **137** pp.343-353
[6] Willms R 2009 *Proc. Nordic Steel Construction Conference* pp 597-604, (Malmö, Sweden) ISBN: 9171270582
[7] Bandyopadhyay P S, Ghosh S K, Kundu S and Chatterjee S 2011 *Metall. and Mat. Transact. A*, **42A**, pp 2742-2752, doi: 10.1007/s11661-011-0711-2
[8] Barsom J M and Rolfe S T 1999 *Fracture and Fatigue control is Structures: Application of Fracture Mechanics* Third Edition, (American Society for Testing and Materials, Philadelphia) ISBN 0-8031-2082-6
[9] Zhu X-K and Joyce J A 2012 *Eng. Fract. Mech.* **85** pp 1-46 doi: 10.1016/j.engfracmech.2012.02.001
[10] International Standard ISO 12135 2016 *Metallic materials – Unified method of test for the determination of quasistatic fracture toughness*
[11] Koncsik Zs 2020 *Proc. of the 1st Int. Conf. on Eng. Sol. for Sust. Dev.; Solutions for Sustainable Development* ed. Tóthné K Sz, Jármai K and Voith K (CRC Press, London) ISBN 978-0-367-42425-1, pp. 316-324. doi: https://doi.org/10.1200/9780367824037
[12] http://home.zcu.cz/~sbenesov/Deform2Dlabs.pdf, *DEFORM User Manual,* available 10th of January 2020
[13] Tisza M and Lukacs Zs 2015 *Acta Metall. Sinica Engl. Lett.* **28** pp. 1471-81.
[14] Cockroft M G and Latham D J 1968 *J. of the Inst. of Metals* **96**, pp. 33-39.

IOP Conf. Series: Materials Science and Engineering **903** (2020) 012014 doi:10.1088/1757-899X/903/1/012014

Measurement possibilities of water content in polyamide

D V Terjék[1] and E Kókai[2]
[1] MSc Student, John von Neumann University, GAMF Faculty of Engineering and Computer Science Department of Materials Technology, 6000 Kecskemét Izsáki Str. 10, Hungary
[2] Research Fellow, John von Neumann University, GAMF Faculty of Engineering and Computer Science Department of Materials Technology, 6000 Kecskemét Izsáki Str. 10, Hungary

E-mail: kokai.eszter@gamf.uni-neumann.hu

Abstract. Water content of polyamide is an important parameter, because it has great effect on the elongation, toughness and impact strength. The aim of our research was to determine moisture in polyamide-6 and polyamide-6.6 granules with different pretreatment of the base material. The specimens were prepared and conditioned in distilled water or non-conditioned. Thermogravimetric analysis, moisture analysis and coulometric Karl-Fisher indirect titration were used for the determination of humidity content. We used polyamide granules and flakes, and polyamide nanocomposites filled with montmorillonite for the water content measurements. First, we compared the technics and chose the most applicable to measure the water content in the samples, and made further measurement of polyamide nanocomposites.

1. Introduction

Automotive industry widely used polyamide (PA) and their composites in automotive structure applications [1–3]. Angioplasty ballon also made from polyamide, used as medical application [4]. In polymer processing moisture content is a distraction, so polymer granules need to be dried before the process. If the basic material is polyamide, it is necessary to make a conditioning procedure after polymer processing, like injection moulding or extrusion. The water absorption of polyamide strongly influences on its mechanical properties [3, 5, 6]. The amide bond can bind the water by H-bond. Our previous research investigated polyamide nanocomposites, reinforced with montmorillonite (MMT) nanoclay [7]. MMT can reduce the water absorption in polyamide [8], and in the same time it can increase mechanical properties, like tensile and impact strength [9, 10]. In our experience, MMT do not always improve the tensile and impact strength. We used pure polyamide-6 and polyamide-6.6 reinforced with 30% glass fiber (GF) materials, and made 1 wt%, 2 wt%, and 5 wt% of MMT PA-nanocomposites. It showed that the MMT could not cause more reinforcing properties besides glass fiber reinforcement.

IOP Conf. Series: Materials Science and Engineering **903** (2020) 012014 doi:10.1088/1757-899X/903/1/012014

2. Experimental

2.1. Materials
The examined raw materials were PA-6 (Grilon BS produced by EMS Chemie), and PA-6.6 with 30% GF reinforced (Zytel 70G30HSL produced by DuPont). The MMT was Nanomer I.30P (produced by Nanocor USA) a surface modified nanoclay. The surface modifier was octadecyl ammonium.

2.2. Measurement methods and technics of moisture absorption in polyamide

2.2.1. Method 1: PA-6 was used in all measurements to investigate the measurement technics to determine the moisture content. One of the PA-6 base material was flattened (it means the normal granules was flattened with a mill) and the other was normal granules. The two different shape granules were conditioned by distilled water for 2 days or non-conditioned. The non-conditioned granules were stored in air for 2 days.

2.2.2. Method 2: Six new nanocomposites were prepared with surface modified MMT. The PA-6 and PA-6.6 with 30% GF were mixed with 1 wt%, 2 wt% and 5 wt% MMT content [7]. A small amount (ca. 10 g) of each composite granulate were conditioned in Weiss Umwelttechnik GmbH WK11-340/40 climate chamber for seven days on a temperature of 70±1 °C with relative humidity of 62±1%.

2.2.3. Measurement technics 1: TGA Q50 thermogravimetric analyser by TA Instruments was used first to determine the moisture content in flakes and normal granules. 5−50 mg of granulates were tested on 110 °C for 5h.

2.2.4. Measurement technics 2: Radwag MAC 50/1/NH type gravimetric moisture analyser was also used to determine the moisture content in the two different shaped granules. Ca. 10 g of granulates were tested on 150 °C.

2.2.5. Measurement technics 3: The amount of absorbed water in the granules and the six new nanocomposites was measured with Karl-Fisher (KF) titration. The testing was carried out with a Metrohm 899 coulometer, and a Metrohm 885 oven. Ca. 50 mg of granulates were tested on 170 °C.

3. Results and discussion

3.1. Comparison of the three measurement technics

3.1.1. Results of moisture content with TGA
TGA Q50 thermogravimetric analyser was used to determine moisture content in PA-6 granules. The results were shown on the Figure 1.

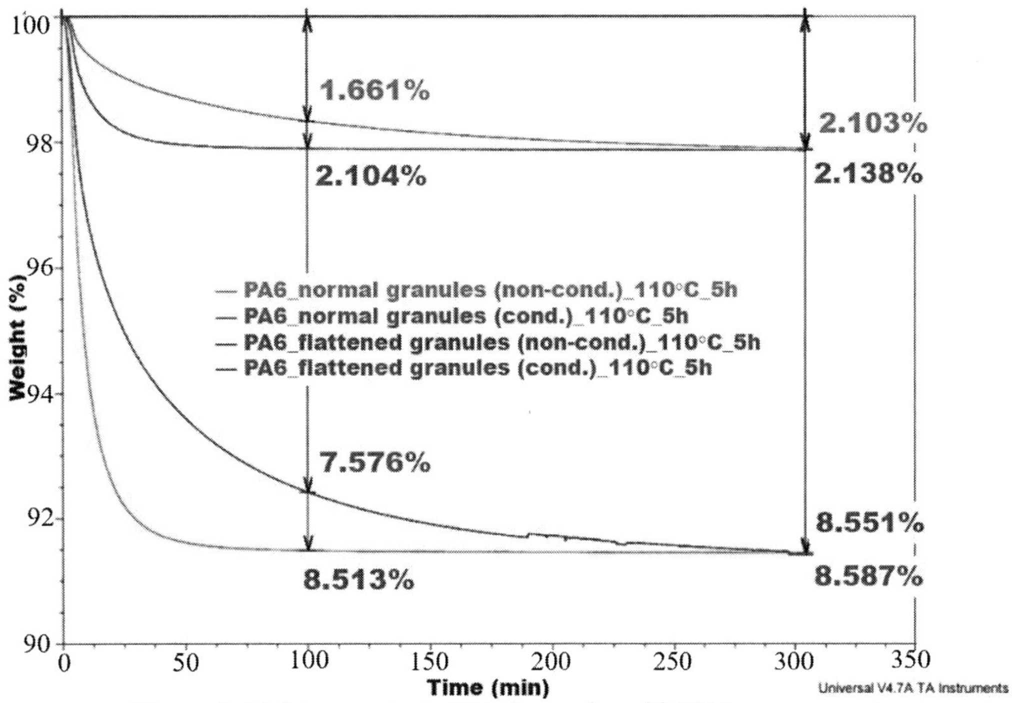

Figure 1. Moisture content of PA-6 granules with TGA measurement

It is showed that the flattened granules could absorb more moisture content and the granules could absorb much more water in distilled water than in the air during 2 days. The flattened granules has bigger surface, and on the surface the PA can easier absorb water then the sample inside.

3.1.2. Results with gravimetric moisture analyser

The moisture content of the samples measured by gravimetric moisture analyser. The results was showed in Table 1.

Table 1. Results of moisture content measured by gravimetric analyser

Material: PA-6	Measured moisture content (%)
Non-cond. granules	1.929
Non-cond. flattened granules	2.067
Cond. granules	7.780
Cond. flattened granules	8.184

The moisture content results with gravimetric analyser were similar to TGA results.

3.1.3. Results of the water absorption measurements by Karl-Fisher titration

The amount of absorbed water in the granules were measured by Karl-Fisher coulometric indirect titration on the conditioned and non-conditioned flattened and normal granules. The results were showed in Table 2.

Table 2. Results of water content measured by coulometric KF titration

Material: PA-6	Measured water content (%)
Non-cond. granules	1.75
Non-cond. flattened granules	2.33
Cond. granules	7.19
Cond. flattened granules	9.26

Compared the three measurement technics KF indirect titration was chosen to determine the water content in the six nanocomposites. The most important parameter was water content and only this measurement technics can determine the water content excluding other solvents.

3.2. Water measurement results of PA nanocomposites

The prepared PA based nanocomposites were measured by coulometric KF titration and the results were showed in Table 3.

Table 3. Results of Karl-Fisher titration in the six nanocomposites and the base materials

Material	Absorbed water (%)	Water absorption of commercial material (%)*
PA-6	1.06	9
PA-6 + 1% MMT	1.59	9
PA-6 + 2% MMT	1.35	9
PA-6 + 5% MMT	1.38	9
PA-6.6 + 30% GF	0.95	6
PA-6.6 + 30% GF + 1% MMT	0.72	6
PA-6.6 + 30% GF + 2% MMT	0.76	6
PA-6.6 + 30% GF + 5% MMT	0.81	6

*According to the data sheet of the materials by the producers

The climate chamber was used for 7 days with 70±1 °C and with 62±1% relative humidity. It is showed that PA-6-MMT nanocomposites absorbed more water than the base material, but in the other case, PA-6.6, which has 30% GF reinforcement, absorbed less water content than PA-6.6 base material.

4. Conclusion

Three measurement technics investigated to determine moisture or water content in polyamide-6 flattened and normal granules, which were conditioned in distilled water or air. In all measurement technics the granules conditioned in water, absorbed more water. Compared the three technics we chose the Karl-Fisher indirect titration which was the most accurate technics to measure the water content in polyamide samples. We investigated the water content in six polyamide nanocomposites reinforced with montmorillonite, and the water content results showed that the nanocomposites absorbed less water next to glass fiber reinforcement than unfilled polyamide material.

Acknowledgement

The project has been supported by the European Union, co-financed by the European Social Fund. EFOP-3.6.1-16-2016-00014.

References

[1] Abbasi H, Antunes M and Velasco J I *Progress in Materials Science*Volume 103, 2019, Pages 319-373.
[2] Ho Q B, Osazuwa O, Modler R, Daymond M, Gallerneault M T and Kontopoulou M *Composites Science and Technology*2019, https://doi.org/10.1016/j.compscitech.2019.03.024
[3] Ferreno D, Carrascal I, Ruiz E and Casado J A *Polym. Test.* 30 (2011) 420–428.
[4] Halim K A A, Farrell J B andKennedy J E *Materials Chemistry and Physics* 143 (2013) 336–348.
[5] Chaichanawong J, Thongchuea C and Areerat S *Advanced Powder Technology* 27 (2016) 898–902.
[6] Chaichanawong J, Thongchuea C andAreerat S *Advanced Powder Technology* 27 (2016) 898–902
[7] Terjék D V, Manová A and Kókai E *Perner's Contacts Special Issue 2* Vol. XIX, May 2019, p. 291–297.
[8] Faghihi K, Taher M and Hajibeygi M *Arabian Journal of Chemistry* 2012, http://dx.doi.org/10.1016/j.arabjc.2012.03.010.
[9] Fereydoon m. et al *Polymer Engineering and Science* 2013 Vol. 54. 2617-2631.
[10] Fereydoon m. et al. *Journal of Plastic Film and Sheeting* 2014 Vol. 31. 1-33.

IOP Conf. Series: Materials Science and Engineering **903** (2020) 012015 doi:10.1088/1757-899X/903/1/012015

Investigation of the tribological properties of nano-scaled ZrO_2 and CuO additive in automotive lubricants

Á. D. Tóth[1], J. Knaup[1]

[1] Széchenyi István University, Department of Internal Combustion Engines and Propulsion Technology, Győr, Hungary

E-mail: toth.almos@sze.hu

Abstract. To improve the fuel efficiency and the lifetime of the internal combustion engines, the lubricants and their additives have to be developed further. One of the possible future engines lubricants can be the nano-sized ceramic particles, which can provide positive tribological properties also in the presence of non-metallic surface materials. This paper presents the results of investigations with the help of ZrO_2 and CuO nano-sized ceramic particles. To define the tribological properties of these additives, lubricant samples with different additive-concentrations were prepared and tribologically analysed. The frictional losses of these lubricant samples were analysed by a ball-on-disk sliding friction machine. The worn surface on the test specimens was analysed by different high-resolution microscopes. To define the functional mechanisms of the nano-additives, the worn surfaces were investigated by high resolution scanning electron microscopes. The ZrO_2 additive has experimentally shown an excellent wear reduction property (over 40% wear reduction compared with the neat Group 3 base oil) at the optimum mixing concentration of 0.4wt%. Both frictional and wear reduction properties could be determined at the application of CuO additive (15-15% friction coefficient and wear scar diameter reduction) at its optimum concentration (0.5wt%). A copper-yellow layer can be seen on the worn surface of the disc specimens with CuO, which indicates the mechanism of chemical transformation to elementary copper from the cupric-oxide nanoparticle and this elementary copper can be melted on the surface, because of the applied high temperature and high loads during the experiments.

1. Introduction

The emission regulations in Europe are becoming continuously strict which requires the development of the different components of the vehicles on the road. One of the most important development topics is the reduction of the different losses in the internal combustion engines, e.g. thermal and mechanical losses including frictional losses. During the last years, the engineers and researchers have elaborated several solutions to increase the efficiency of the engines, e.g. low-viscosity lubricants, low-friction coatings, etc. To take advantage of the maximum potential of these solutions, the different solutions have to be suited together.

The lubricants are one of the most important components of the piston engines. Their roles are to separate the rigid surfaces, prevent them from wear, corrosion, and decrease the frictional losses. To fulfill these challenges, the modern engine lubricants have to be formulated with the use of different additives. The lubricant additives of today are usually function due to polarity difference and they can form a protective nano-scale layer on the rubbing surfaces. In the last years, the compatibility of the lubricant additives and the exhaust gas after-treatment systems became vital to the harmful emission products. Some additives can form ash during their burning process and this ash can seal the channels in the particle filters, or they can cause damages in the catalyst systems of the vehicles.

These tribological losses will lead to different energy losses and failures in the mechanical systems due to friction and wear, and these losses can strongly influence the fuel consumption and component lifetimes of the engines and vehicles. Different statistics are existing, which show a third of the world's primary energy consumption due to friction, and the wear is responsible for 60% of the damage of machine parts. Besides the lubrication failure can cause more than 50% of accidents involving machinery equipment. [1]

One of the possible solutions for the future lubricant additive can be the nano-particles. These particles can be made of a huge variety of materials. According to their material, they can offer quite different properties. However, this huge amount of variety requires research activities to map their properties and their optimal mixing concentration in different types of lubricants. These types of particles can have lots of influence factors regarding tribology, e.g. particle size, material, the property of building agglomerations, concentration, or attendance of surface-active solving agents.

2. Nano-sized ceramic particles as lubricant additives

The existence of powder form nano-materials can be very useful for different applications: they can be the starting material for different coatings (e.g. atmospheric plasma spray) or they can be sintered into new and complex material and geometry. The research activities to use this kind of additives in lubricants of machines have started in the recent few years. During this research time plenty of nano-particle material, size, and form were investigated by many institutes which shows until now encouraging.

The most widely investigated size of these particles is the nano-scale with spherical form, which means the average particle size is between 1 and 100nm [2]. These size of additives results in the best homogenization properties: between the particles, there are always secondary Van der Waals forces which can be separated via e.g. ultrasonic homogenizer resulting in a homogeneous lubricant-additive mixture. Besides, these secondary forces can also increase the adhesion property of the nano-particle additives to the rubbing surfaces [3].

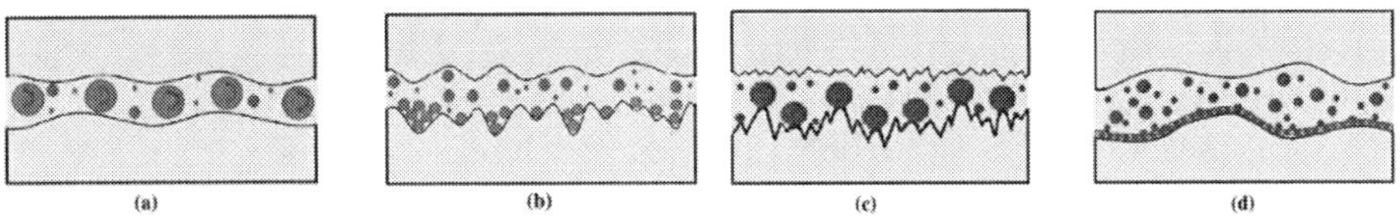

Figure 1. Different working mechanisms using nanoparticles, a) rolling mechanism, b) mending mechanism, c) polishing mechanism, d) protective film [4] [5]

Different working mechanisms were reported in the literature to describe the investigated results. Zhang et al. [4] summarized the different mechanisms for nanoparticles (Fig. 1):

- Rolling (or ball bearing) mechanism: the nanoparticles will work as nano-scale balls to roll between the rubbing surfaces and change the sliding friction to sliding-rolling one.
- Mending mechanism: the nanoparticles can be collected in the grooves of the rubbing surfaces resulting in a smoother contact surface.
- Polishing mechanism: the nanoparticles will polish the roughness peaks of the surface leading to a smoother surface, which reduce the running-in phase of the rubbing surfaces
- Protective film mechanism: the nanoparticles can attach to the rubbing surfaces (e.g. with secondary forces or via polarity difference) and they can form a protective tribofilm between the sliding surfaces.

3. Investigation method

The aim is this paper is to analyse the friction and wear influencing properties of the selected ZrO_2 and CuO nano-particle. The average size of the particles is between 15-25 and 30-50nm respectively. To

IOP Conf. Series: Materials Science and Engineering **903** (2020) 012015 doi:10.1088/1757-899X/903/1/012015

characterise the tribological behaviour of these nano-ceramic particles, a homogeneous particle-lubricant sample has to be prepared. To the characterisation, a base oil (classification Group 3, kinematic viscosity 4cSt) without any additives was chosen. The chosen lubricant is worldwide used to produce engine lubricant with relatively low viscosity class (SAE 0W-20). The nanoparticles with the proper concentration have to be mixed and homogenised and for this task, a two-phase method was used, which contains a 30-minute homogenisation phase (ultrasonic homogeniser) and a 30-minute magnetic stirring.

The friction and wear influencing behaviour of the nanoparticles was investigated with an Optimol SRV5 friction machine. This tribometer enables to produce the necessary oscillation movement pattern between the used testing specimens and measures the most important tribological values (e.g. friction coefficient of the system). As testing specimens, ISO-standardised 10mm balls and 24mm discs were used (material: 100Cr6, Ra 0.020 and 0.047μm respectively) [6]. For the tribological test, a self-developed testing method and setup parameters were used [7], which can be seen in Tab.1.

Table 1. Investigation parameters used for the tribological tests

Parameter	Stroke	Frequency	Specimen T	Oil T	Oil circulation speed	Load	Time
Step 1	1mm	50Hz	100°C	100°C	225ml/h	50N	30s
Step 2	1mm	50Hz	100°C	100°C	225ml/h	100N	2h

After the tribological experiments, the worn surface was analysed with both digital- and scanning electron microscopes to define the wear behaviour of the investigated nanoparticles. The recordings made with these microscopes can carry important information to understand the working mechanisms.

4. Investigation results

During the investigation, the following three main measured values are considered as a comparison base:

- Coefficient of Friction (COF): The maximum value of the friction coefficient in one sliding stroke, measured by the Optimol SRV5 tribometer
- Friction Absolute Integral (FAI): the absolute integral value of the measured friction coefficient in the function of the stroke, measured by the Optimol SRV5 tribometer
- Wear scar diameter (WSD) from the ball specimen, measured by a digital microscope

4.1. Results with the reference lubricant sample

For the evaluation of the measurements, a neat Group 3 base oil was chosen as the reference sample. This base oil is widely used in the lubrication industry to produce low-viscosity engine lubricants for modern passenger cars. This base oil enables to produces lubricants according to the SAE 0W-20 viscosity classification and the engine manufacturers (OEMs) are also specify this kind of lubricants as official engine oil in their products. The reference oil was delivered by the MOL-Lub Kft. from Hungary.

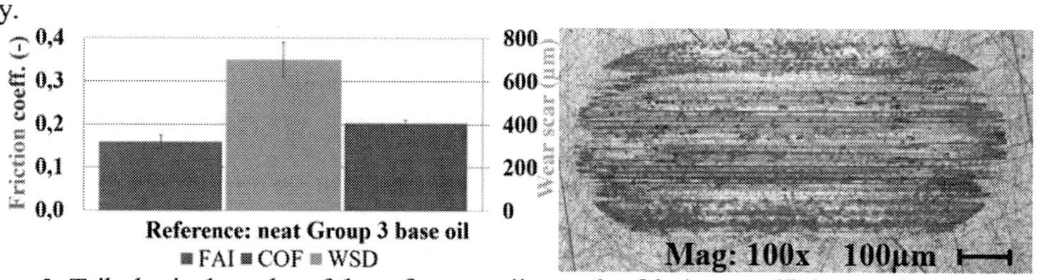

Figure 2. Tribological results of the reference oil sample: friction coefficient and wear results (left) and the wear scar on the disc specimen (right)

Fig. 2 demonstrates the experimented and evaluated results of the used reference oil sample. The evaluation of the reference sample was carried out with 8 independent experiments and their average value and their deviation were stastistically calculated. The measured frictional coefficient values have shown a normal operation of the system. The wear scar image on the disc specimen presents high wear with heavy abrasion and adhesion. The signs of burned lubricant cannot be considered on this worn surface, which can be explained with this high wear velocity: the lubricant may have burned on the surface but the high abrasive and adhesive wear have removed these oil molecules.

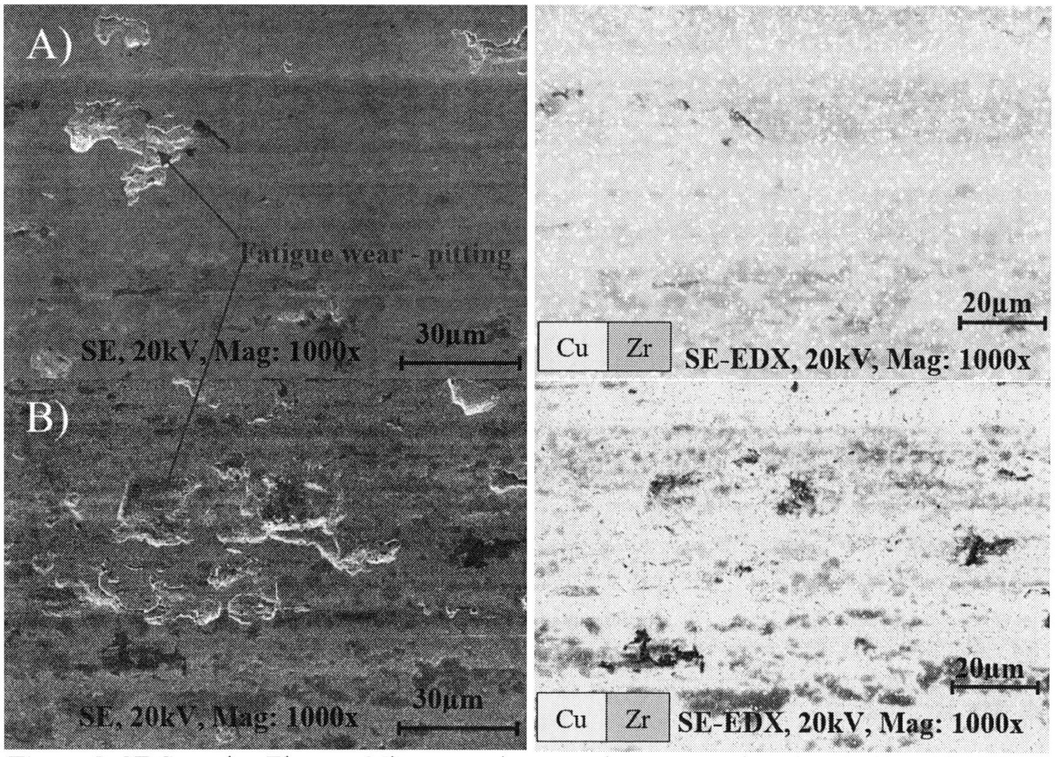

Figure 3. SE-Scanning Electron Microscope image and EDX Mapping picture about the surface of disc specimen with neat Group 3 reference oil, A) middle-stroke area, B) dead-centre area

The SEM-images of the disc specimen with reference oil sample (neat Group 3 base oil) can be seen on Fig. 3. The SEM-images can show a high amount of fatigue wear, which is called also pitting. This demonstrates the lack of the tribological additives of the used oil sample: there are no additives which can protect the rubbing surfaces against overload and high wear. These pittings can be established by the following process: the high load causes micro-cracks in the surface-near material of the specimens, resulting in a grid of cracks in the material. Further load and sliding of these surfaces produce wider and deeper cracks, which can lead to a solid material particle departure from the surfaces resulting in these craters on the surface [8]. The amount of the pitting is higher at the dead centre areas, because of the direction change in these areas. The EDX-analyse of the surfaces shows a small amount of Cu and Zr on the surface, which can be considered as the natural background spectrum of the measurement method. According to the quantitative analysis of the electron microscope, this amount can be considered as zero.

4.2. Results with ZrO_2 nano-particle

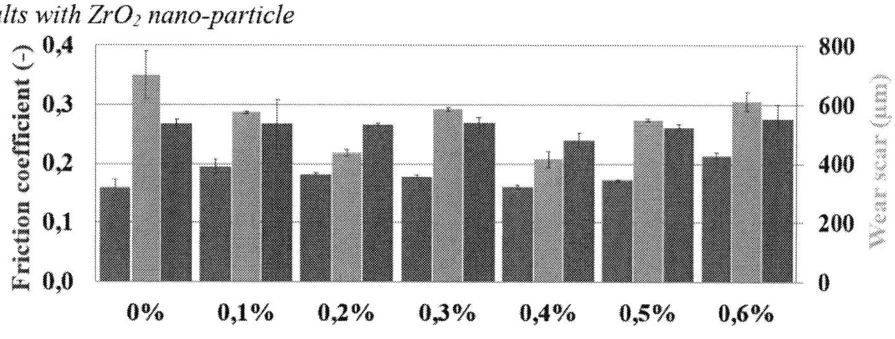

Figure 4. Tribological results of nano-sized ZrO_2 additive

IOP Conf. Series: Materials Science and Engineering **903** (2020) 012015 doi:10.1088/1757-899X/903/1/012015

Fig. 4 illustrates the results of the investigated ZrO_2 additive. This additive could decrease the WSD, up to 40% at the concentration of 0.4wt%. However, the friction coefficient values have shown a slightly increasing tendency. Only the optimum 0.4wt% concentration was able to reach the coefficient of friction value of the reference sample without any additive.

The wear scars on the disc and ball specimen (Fig. 5.) show a relatively smooth surface with a low level of surface damages. Even some roughness valleys can be seen on the worn surface. It can also be considered that on the edges the colour of the wear is different compared with the middle-areas. A very low amount of burned lubricant can also be seen on the surface of the disc specimen, with this blue-colour area in the middle of the width.

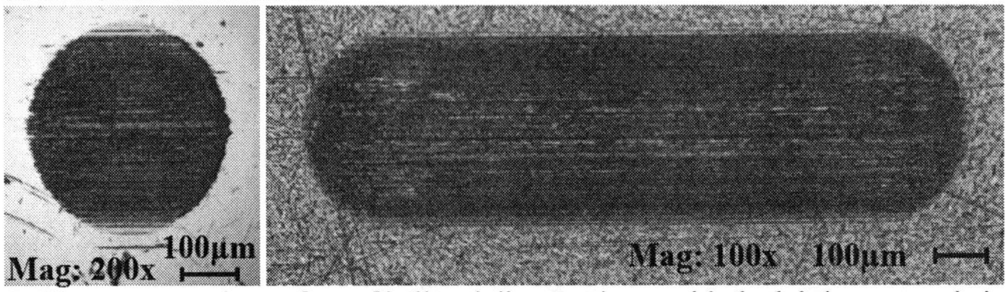

Figure 5. Wear scars on the surface of ball and disc specimen with the lubricant sample including 0.4wt% ZrO_2

The worn surfaces were also analyzed by a scanning electron microscope located in the laboratory of the department in Győr. Two different areas were selected for the analysis: one middle-stroke and one dead-centre area. The images can be seen in Fig. 6. On the worn surface, slight abrasive wear scars and tiny fatigue craters (pitting) can be found. The relatively small amount of heavy abrasion wear can show the existence of a protective tribological layer between the rubbing surfaces and the pits are the signs for the reciprocating movement pattern and the fatigue of the surface. On the EDX images, the amount of the founded Zr and O material can be seen with the colour of red and green. On the images from the dead-centre areas can be measured less particle which means that the dead-centre areas were loaded more heavily because of the slowing-stopping-restarting movement pattern of the oscillation.

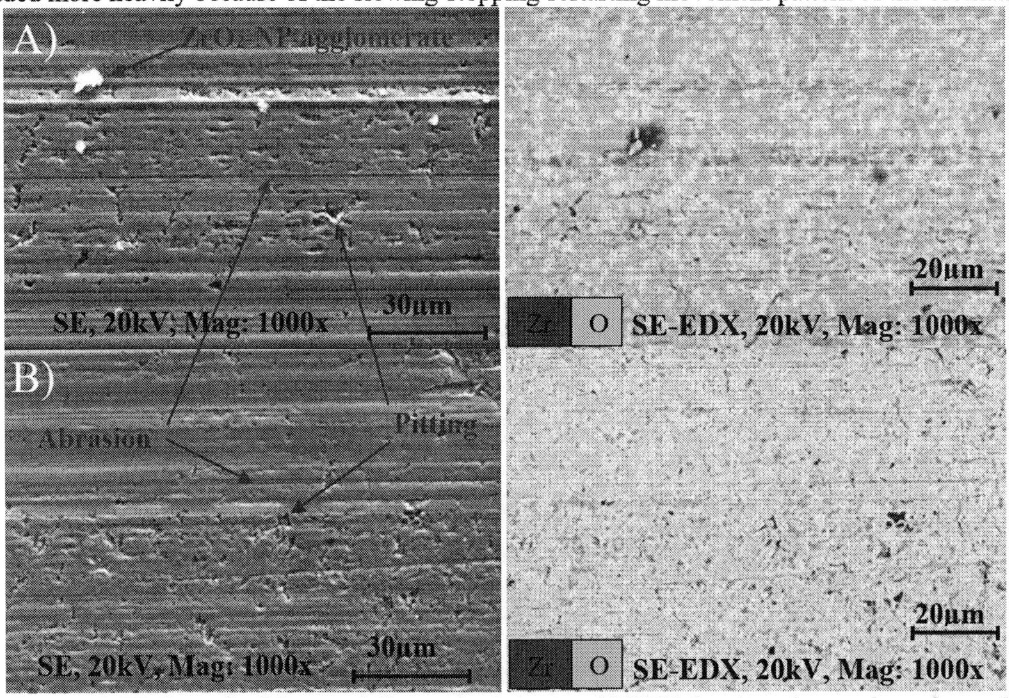

Figure 6. SE-Scanning Electron Microscope image and EDX Mapping picture about the surface of disc specimen with ZrO_2 nano-additive, A) middle-stroke area, B) dead-centre area

The ZrO_2 nano-particle has the hardness value of 1200 HV, that's why it is very difficult to decrease the average size of the particles during the loading and rubbing phase of the tribosystem. These particles are also not able to be solved into the used mineral base oil (G3, 4cSt). The EDX-SEM images show the existence of the Zr and O material on the worn surface even after a thorough ultrasonic cleaning in brake cleaner medium. This shows the sign of the existence of secondary bonding force (van der Waals) between the nano-particles and the rubbing surface. The ZrO_2 nanoparticles can be collected in the roughness valleys, fill them up resulting a smoother rubbing surface with the lower surface distribution of the load. Besides they can be pressed into the metal-matrix via deformation of the metal. During the loaded and rubbing phase on the surfaces of the two testing specimens, this protective layer can be formulated and that's why the specimens are only in contact with each other via this layer. Of course, as from the SEM-images of the middle-stroke and dead-centre areas can be seen, the higher tribological load results in less particle on the surface at the end of the test.

4.3. Results with CuO nano-particle

The tribological properties of CuO nano-particle were also investigated according to the same process. Fig. 7 describes the results of this type of additive. This chart shows a slight reduction in friction coefficients and WSD as well. Both values have been decreased by 15%, compared with the reference oil. The optimum concentration can be established at 0.5wt%. After 0.5wt%, a slight increase in friction and wear could be measured.

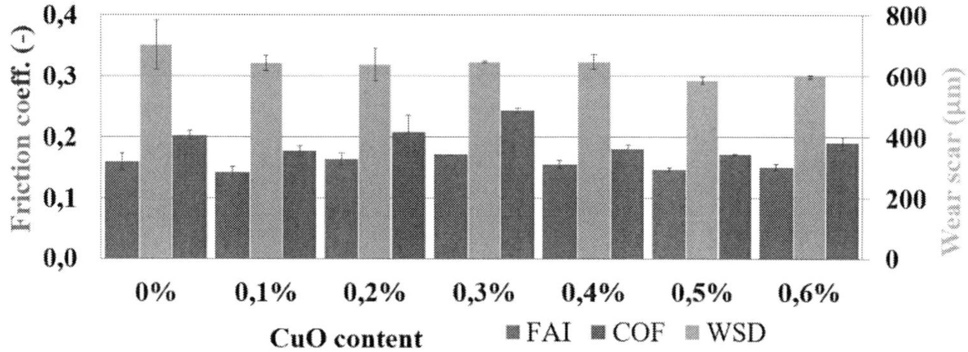

Figure 7. Tribological results of nano-sized CuO additive

The wear scars on the disc and ball specimen (Fig. 8.) illustrate the signs of a middle-loaded surface. There are some valleys with relative higher depth, but the whole worn surface can be described as normal wear of these systems. The speciality of this wear scar is the cooper-yellow colour. This colour may be formated from the CuO additive. Burned lubricant cannot be seen on the surfaces.

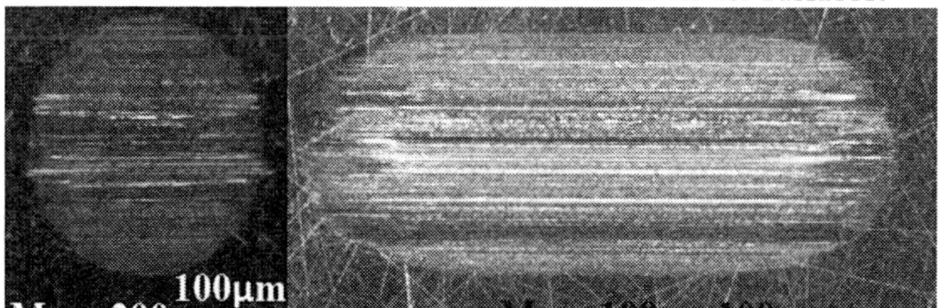

Figure 8. Wear scars on the surface of specimens with the lubricant sample including 0.5wt% CuO

The worn surface of the disc specimen was also analysed via a scanning electron microscope (Fig. 9.), both at middle-stroke and dead-centre area. The surfaces show slightly worn areas by abrasion wear and polished areas as well. The material content of the surfaces was also analysed via EDX SEM and the results can show that there are 2 different areas: one with higher abrasion wear and higher Cu content and one with smoother, polished surface with lower copper on the surface. It can be defined that in the dead-centre areas less amount of Cu and O material was found which can be explained with the higher tribological loads and with the worse lubrication properties because of the oscillation motion.

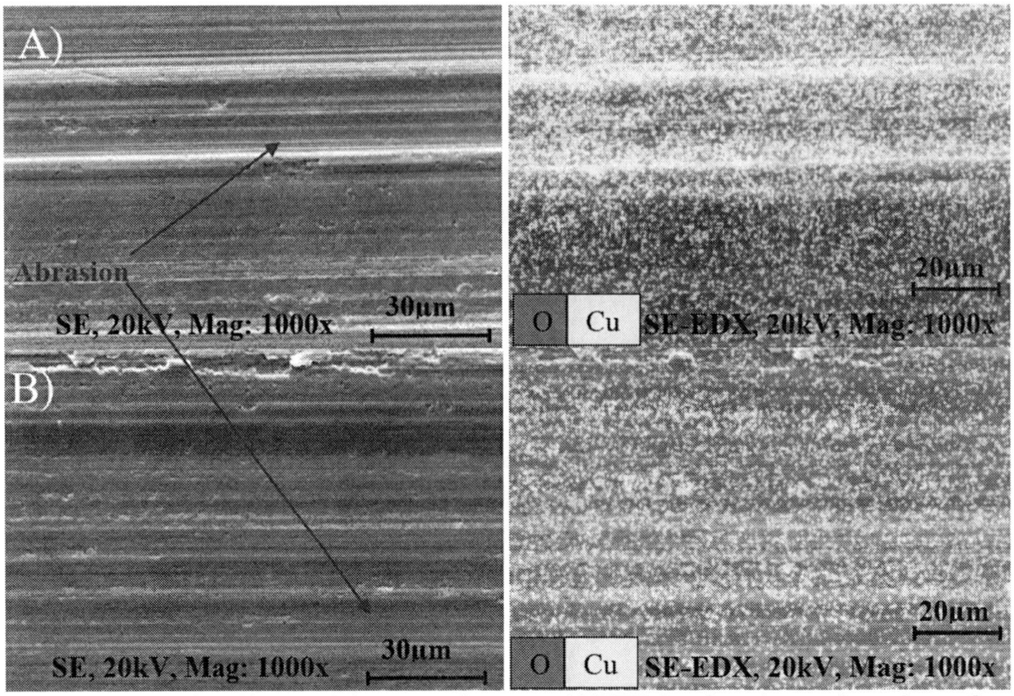

Figure 9. SE-Scanning Electron Microscope image and EDX Mapping picture about the surface of disc specimen with CuO nano-additive, A) middle-stroke area, B) dead-centre area

The CuO additive has significant lower hardness value compared with the ZrO_2, which enables the formation of the spherical additives. After the homogenization of the CuO and the base oil, the colour of the sample was black, but the colour of the additive on the worn surface is significantly different, copper-yellow. To formulate this elementary copper, the following hypothesis was founded: the cupric-oxide can be reducted with the carbon-hydrogen content of the base oil (the Group 3 base oil is formulated from different carbon-hydrogens between C20 and C50. With these long-chain molecules, the cupric-oxide may react, resulting in CO, CO_2 or H_2O. One possible reaction equation can be presented with the decane ($C_{10}H_{22}$), which is widely used in fuels and low-viscosity lubricants [9][10]. During the reaction, the decane can be reduced to nonane (C_9H_{20}):

$$CuO + C_{10}H_{22} \xrightarrow{reduction} Cu + CO + H_2 + C_9H_{20}$$

Further analysis is required to verify this chemical reduction equation which can be proved with the calculation of Gibbs free enthalpy values.

In case the copper has already been formulated near the rubbing surfaces, this material is enough soft to be mended on the surface, forming a tribological protective sliding film between the two rubbing specimens. This copper-layer can fill up the roughness valleys resulting smoother contacting surfaces (see on Fig. 1, d) and this layer provides an increase of the contacting area which decreases the load pressure of the ball and disc specimen. To verify the hypothesis, further investigation has to be carried out in the future via FIB-SEM microscope or XPS method. Besides the valid transformation equation and the starting temperature have to be clarified.

5. Summary

This paper presents the investigation results with the ZrO_2 and CuO nanoparticles as lubricant additives. The tribological properties were analysed via Optimol SRV5 tribometer and the worn surfaces via digital and scanning electron microscopes.

The following results can be summarised:

- Both the ZrO_2 and CuO nanoparticles have shown positive tribological properties, compared with the neat Group 3 base oil: friction decrease by 3 and 15%, wear reduction by 40 and 15%, respectively. These values belong to the optimum concentration values, 0.4 and 0.5wt% respectively.
- The worn surface after the investigation with ZrO_2 0.4wt% shows a significant smooth character. With SEM analysis a relatively huge amount of Zr was found on the worn surface. The ZrO_2 NP could adhere to the rubbing surfaces via mending mechanism, which resulted in a protective tribological layer between the specimens.
- On the surface after the CuO 0.5wt% experiment Cu was also found, however, in a different form: it was melted in the valleys of the worn surface. This melted layer could be formulated by the reaction of CuO and the carbon-hydrogen content of the Group 3 base oil.

To understand the exact working mechanisms of these additives, further investigation is necessary, with both tribometers and microscopes. Further experiments with these additives and lubricants with higher additive content (both active and inactive additives) is also obligatory.

Acknowledgement
This research was supported by the EFOP-3.6.1-16-2016-00017 Internationalization, initiatives to establish a new source of researchers and graduates, and development of knowledge and technological transfer as instruments of intelligent specializations at Széchenyi University.

The authors would like to thank Dr József Perger, Mr Ádám István Szabó, Ms Dóra Olimpia Csepreghy and Ms Boglárka Nagy for their general support.

References

[1] Wang W, Xie G and Luo J 2018 *Friction* **6** pp 116-142
[2] Mohan N, Sharma M, Singh R and Kumar N 2014 *SAE Technical Paper* **2014-01-2781** p 9
[3] Hernández Battez A, González R, Viesca J L, Fernández J E, Díaz Fernández J M, Machado A, Chou R and Riba J 2008 *Wear* **265** pp 422-428
[4] Zhang Z, Simionesie D and Schaschke C 2014 *Lubricants* **2014** pp 44-65
[5] ShafiW K, Raina A and Ul Haq M I 2018 *Tribology – Materials, Surfaces and Interfaces* **12** pp 27-43
[6] ISO 19291:2016(E), Lubricants – Determination of tribological quantities for oils and greases – Tribological test in the translatory oscillation apparatus, Switzerland, 2016
[7] Tóth Á D and Paulovics L, et al 2017 *58. Tribologie-Fachtagung 2017 Reibung, Schmierung und Verschleiß Forschung und praktische Anwendungen* Gesellschaft für Tribologie e.V., *58.* Aachen vol 2, pp P8/1-P8/11
[8] H. Czichos and K.-H. Habig 2010 *Tribologie-Handbuch, Tribometrie, Tribomaterialien, Tribotechnik* Viewer+Teubner Verlag Wiesbaden vol 3 p 757
[9] Jones H A and Taylor H S 1923 *J. Phys. Chem.* **27** pp 623-651
[10] Kim J Y, Rodriguez J A, Hanson J C, Frenkel A I and Lee P L 2003 *J. Am. Chem. Soc.* **125** pp 10684-10692

Examination of welded joint of titanium alloy used in oral surgery

J Kónya[1], K Kulcsár[1], H Hargitai[2]

[1] Dent-Art-Technik Kft., Győr, H
[2] Széchenyi István University, Győr, H

E-mail: hargitai@sze.hu

Abstract. The additive manufacturing of customized Ti-6Al-4V (Grade 23) implants, which in the present research concerns the dental and maxillofacial surgical field, may require complex manufacturing technology due to their complexity and assembly. In practice, additively manufactured titanium implant elements are supplemented by threaded sleeves that allow disassembly. These turned elements made from Grade 5 material quality rolled preform are joined by laser micro-welding with Grade 1 material. This special process for implant production is still not widespread in current manufacturing practice, and there is no technical recommendation for the manufacturing parameters of these implants. The aim of our research is to explore the possibilities and limitations of additive process in the manufacture of custom-made implants and to provide guidelines for optimal manufacturing and welding parameters. In the initial phase of the research, preliminary experiments were conducted with the conventional and additive manufacturing of cylindrical test specimens and with the production of combined products by micro-welding. The specimens were subjected to tensile testing, the weld was examined by CT, and the fracture surfaces were studied by optical microscope and scanning electron microscope.

1. Introduction

Various materials are used in dental applications, such as metals, ceramics and polymers. Among metallic materials, commercially pure titanium and titanium alloys are becoming more and more widespread due to their relatively simple handling and high strength, their excellent corrosion resistance and their biocompatibility [1]. The most widely used titanium alloy is Ti-6Al-4V, which is generally chosen for medical devices, prostheses and various implants, as well as the material for bone replacement. Until the 1980s, manufacturing technology was divided into two classes: metal forming through plastic deformation, without adding or removing material, and removal processes, and those technological solutions, in which the final product is obtained by removing excess metal from the stock, thus results in waste.

By the end of the 20th century, a new manufacturing process became available due to advances in digital technology. In this technology, the desired geometry is obtained from digital data and then builds the object layer by layer. These processes are referred to as "Additive Manufacturing (AM)" [2,3]. Due to its many advantageous features, not only products with special geometry (mould inserts with conformal cooling) can be manufactured, but also components with special mechanical properties can be produced by using materials with unique composition by mixing powders [4,5].

Computer-aided design and manufacturing (CAD-CAM) became widespread in dental applications in the 1980s [6]. Powder bed fusion technology allows the creation of porous-structured products based on digital models. Due to several advantages of this technology, not only fields, such as tooling and of These can be used as dental implants that fulfil both mechanical and physical requirements [7, 8]. Porous structures can improve osseointegration between the bone and implant. There is an increasing number of studies in this field [9-11].

Titanium, the main base material of the implant, requires great attention in all areas of processing, especially during welding [12]. Welding technology is complex for titanium as the material is highly reactive with atmospheric gases, such as oxygen and nitrogen in liquid phase and above 550 °C. Contamination can be caused by impurities in the shielding gas, poor protection of the welding zone, and improper preparation or cleaning of welding and filler materials before or during welding [13].

There are numerous welding processes available for pure titanium and titanium alloys, such as laser welding, gas tungsten arc welding, and brazing [14]. For metallic materials, there are currently three different techniques investigated based on welding materials: laser welding [15-17], shaped metal deposition [18-20], electron beam deposition [21]. Numerous studies have investigated welded zone microstructure in commercially pure titanium and titanium alloys and have especially focused on analysing the effect of contamination on mechanical properties [19]. Two different welding processes – solid state laser welding and gas tungsten arc welding – have been compared for additively manufactured Ti-6Al-4V alloys [20]. Welding tests were conducted on (Grade 2) titanium alloys and it was proved that discoloration at the weld zones was correlated to the oxygen and nitrogen content of the weld zone, the absence of protective atmosphere, and the mechanical properties of the weld joint [19].

In our present research we used Ti-6Al-4V alloy and produced cylindrical specimens by conventional way with machining and by powder bed fusion. We made pieces by joining the differently made parts with micro-welding. The specimens were subjected to rupture, the weld was examined by CT, and the fractograph were analysed by optical microscope and scanning electron microscope.

2. Experimental

2.1. Materials

For the experiments different grades of titanium and alloys were used. Ti-6Al-4V titanium alloy, Grade 5 (Bibus Metals AG) was used to produce tensile test specimens by machining rods. Ti-6AL-4V ELI powder, or Grade 23 (LPW Technology Ltd), which is the higher purity version of Ti-6Al-4V (medical titanium grade) was used for powder bed fusion. We used wire (Ø 0.2 mm) made of commercially pure titanium Grade I for welding experiments. Table 1. shows the nominal chemical composition and density of applied materials.

Table 1. Chemical composition and density of applied titanium and titanium alloys

Titanium grade	Chemical composition (wt%)								Density
	Al	V	Fe	O	C	N	H	Ti	(g/cm^3)
Grade I	-	-	0.20	0.18	0.08	0.03	0.015	res	4.51
Grade 5 ELI	5.5-6.5	3.5-4.5	0.25	0.13	0.08	0.05	0.012	res	4.47
Grade 23	5.5-6.5	3.5-4.5	0.25	0.13	0.08	0.03	0.012	res	2.49

2.2. Manufacturing of specimens

Cylindrical Ø6 mm tensile test specimens were produced by turning from rod and by laser metal fusion (LMF) using a SISMA mysint100. Specimens were produced by welding in the following steps:

1. Fixing of the conventionally produced and additively manufactured parts by spot welding in argon atmosphere.
2. Filling the seam with filler material based on the set parameters of the laser welding machine (Dentaurum Basel Laser Desktop).
3. Stress relief heat treatment process in argon (600°C)

The technical characteristics of welding machine are summarized in Table 2. Two different configurations (W1 and W2) were used when setting the parameters of the welding machine:

- W1: 295 V; 9.5 m/s; 2 focuses
- W2: 280 V; 7.5m/s; 0 focus

Table 2. Main technical characteristics of laser welding equipment

Electrical connection	200 – 240 V, 50-60 Hz, 10 A
Max. power	2.2 kW
Average power	50 W
Wavelength	1064 nm
Pulse shape	4 pre-formed pulse shapes
Pulse energy	50 J
Pulse peak power	5 kW
Pulse duration	0.5-20 ms
Pulse frequency single pulse	25 Hz

2.3. *Test methods*

For characterization of the parts tensile tests (3-3 pieces of homogenous specimens and 2-2 pieces of welded parts) according to MSZ EN ISO 6892-1:2016 method B were carried out at room temperature with Instron 5582 equipment (test speed 10 mm/min). Welded specimens were examined by computed tomography before and after the tensile test by YXILON Modular CT equipment. Fractographs of the specimens were studied by Zeiss Stereo Discovery V20 stereo-microscope and by HITACHI 3400 scanning electron microscope (SEM).

3. Results

The machined parts (M), additively manufactured specimens (AM) and their combinations joined by laser welding (W1, W2) were tested under tensile loading and the average values of tensile strengths can be seen in Figure 1.

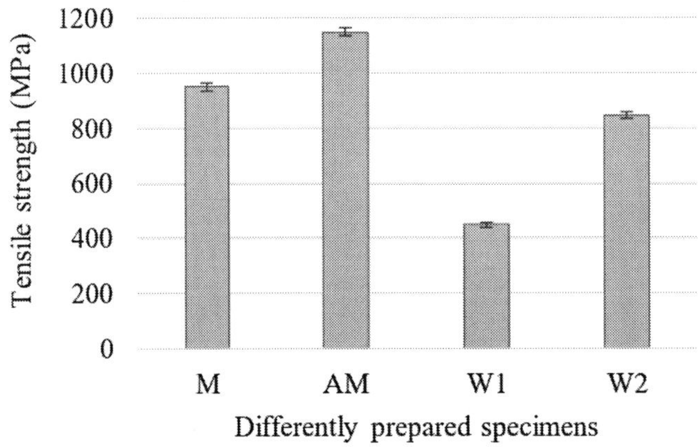

Figure 1. Tensile strength of Ti-6Al-4V parts made by machining (M), additively manufactured (AM) and their combination joined by laser welding (W1, W2)

In the case of tensile strength values, much lower values were obtained for the welded pieces than for the homogenous specimens. Machined and AM parts showed different behaviour under tensile loading, which is indicated also by the percentage elongation values after fracture (Figure 2).

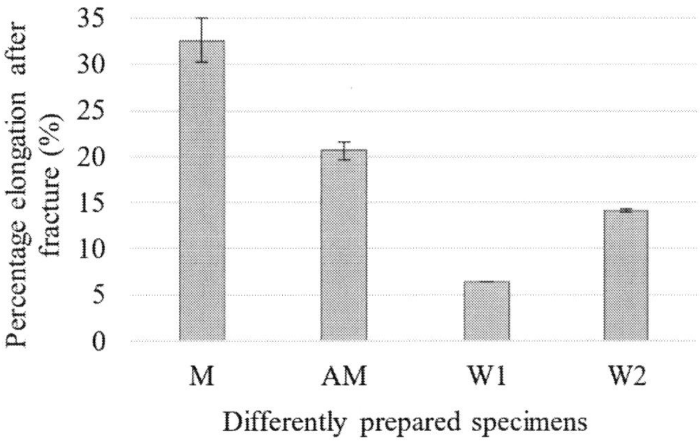

Figure 2. Percentage elongation values after fracture of Ti-6Al-4V parts made by machining (M), additively manufactured (AM) and their combination joined by laser welding (W1, W2)

Additively manufactured parts showed almost 40% less elongation after fracture under tensile loading than machined parts, where the neck formation can be clearly seen. Images made by stereo-microscope show the fractured surface of the tested parts (Figure 3).

a) b) c) d)

Figure 3. Images (stereo-microscope) of typical fractured surface of tensile test specimens (a) machined, b) additively manufactured, c) W1 welded, d) W2 welded

As it can be seen on Figure 3.c the W1 test specimen fractured in the AM part, while W2 welded parts broke in the seam. The results show that the welding technology was not carefully carried out. The high porosity of the weld and the lack of fusion resulted in a weak connection.

Examination with high magnification of a fracture surface is critical of a failure investigation, because the topography and fracture features can help to determine the fracture mode.

The scanning electron microscope is very important in the proper evaluation and classification of a fracture surface. The greater depth of field provided by the SEM allows to make high resolution images of surfaces after fracture at higher magnifications. Fractured surface of broken parts in case of machined, additively manufactured and welded (W2) showed quite different structure (Figure 4.).

IOP Conf. Series: Materials Science and Engineering **903** (2020) 012016 doi:10.1088/1757-899X/903/1/012016

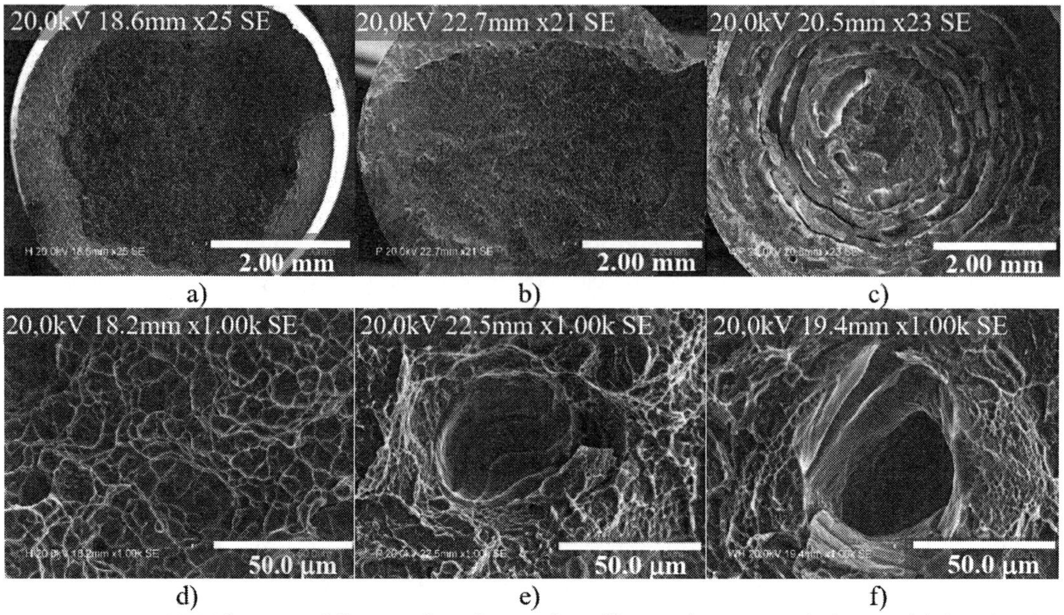

Figure 4. SEM images of fractured surfaces of tensile specimens, made by machining (a, c), additively manufactured (b, e) and W2 welded parts (c, f) (scale bar: 2.00 mm (top) and 50.0 µm (bottom))

Ductile fractures are characterized by extensive plastic deformation just prior to fracture compared to brittle fractures. Ductile fractures often initiate at voids, or inclusions, forming distinct elongated dimples such as shown both in case of homogeneous parts, shown in Figure 4. In case of welded parts, there are several areas when the lack of fusion (red arrows) can be easily explored (Figure 5). Compared the SEM images with the CT tomographs (Figure 5) the same patterns can be observed which confirms the lack of fusion.

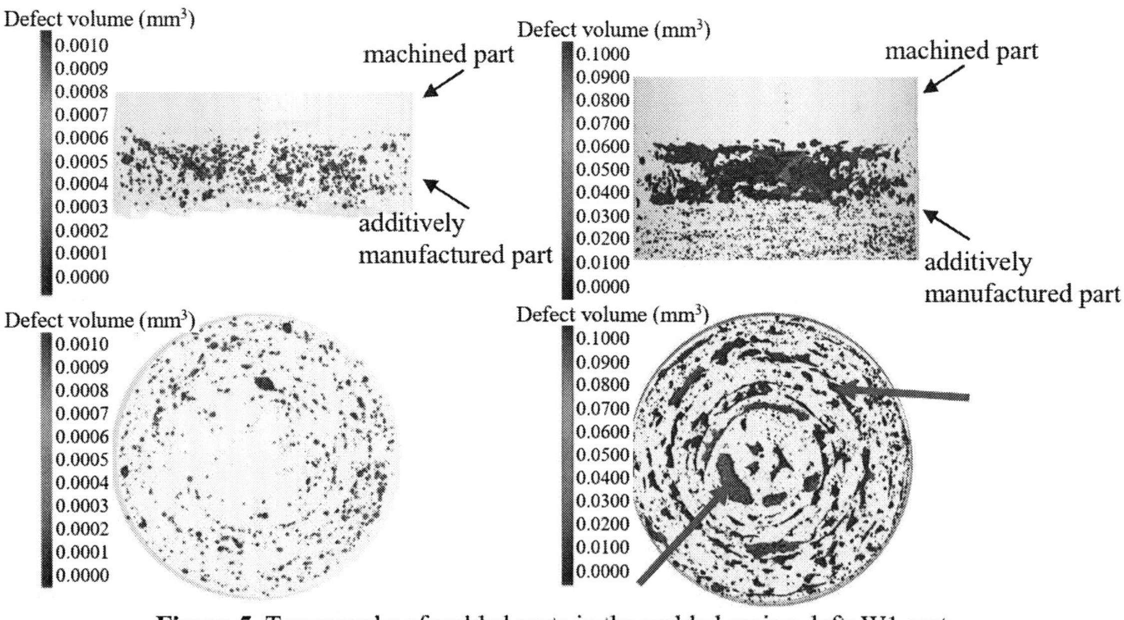

Figure 5. Tomographs of welded parts in the welded region, left: W1 part, right: W2 part (scale: defect volume (mm^3))

4. Conclusions

In our present experiments cylindrical tensile test specimens were prepared from Ti6Al4V alloy by conventional technology (cutting) and by powder bed fusion, and the differently made parts were joint by micro-welding with two parameter settings. The welding quality was evaluated by tensile test, CT examination, and the fractographs were studied by stereo-microscope and scanning electron microscope.

Based on the results, the following conclusions were made:

- The conventionally manufactured specimens have higher deformation capacity (based on the percentage elongation after fracture) and lower tensile strength than the additively manufactured parts.
- The mechanical properties of the welded specimens could not reach those of base materials, so welding parameters should be chosen more carefully with further experiments.
- Both specimens produced by different technologies and welded parts exhibit tough behaviour based on SEM images.

A further aim of the research is to map the effect of laser micro-welding parameters on the properties of combined specimens from conventionally made and additively manufactured pieces and to provide recommendations for the application of optimal technological parameters in the implant production.

Acknowledgement

The research presented in this paper was supported by EFOP 3.6.1-16-2016-00017 'Internationalization, initiatives to establish a new source of researchers and graduates, and development of knowledge and technological transfer as instruments of intelligent specializations at Széchenyi István University'. We would like to thank Imre Fekete, Dóra Harangozó and Tamás Ibriksz for their help in the investigations.

References

[1] Brunette DM et al (Eds.) 2001 *Titanium in medicine, Material Science, Surface Science, Engineering, Biological Responses and Medical Applications*, (Springer, Berlin, Heidelberg) pp. 25-51
[2] Hatos I and Zsoldos I 2012 *Proceedings of Factory Automation* (University of Pannonia (Eds.)) 42-45
[3] Gibson I et al 2009 *Additive Manufacturing Technologies* (Springer-Verlag New York) pp. 19-42
[4] Hatos I et al 2018 *Stroj Vestn-J Mech E* **64(2)** 121-129
[5] Kocsis B et al 2020 *J Magn Magn Mater* **501** 166425
[6] Al-Mesmar HS et al 1999 *J Prosthet Dent* **82 (1)** 15-21
[7] Arvidsson A et al 2015 *J Biomed Mater Res B Appl Biomater* **103** 12-20
[8] Yang F et al 2017 *Sci Rep* **7** 45360
[9] Maniatopoulos C et al 1986 *J Biomed Mater Res* **20** 1309-1333
[10] Unger AS et al 2005 *J Arthroplasty* **20 (8)** 1002-1009
[11] Wu S et al 2013 *Artif Organs* **37 (12)** 191-201
[12] Saresh N 2007 *J Mater Process Technol* **192-193** 83-88
[13] Klas W 2003 *Welding Processes Handbook* (Woodhead Publishing Ltd) pp. 148-170
[14] Xie J and Safarevich S 2003 *Proceedings of the Materials and Processes for Medical Devices Conference* (Anaheim, CA, ASM International) 25-30
[15] Kelly SM and Kampe SL 2004 *Metall Mater Trans A* **35(6)** 1869-1879
[16] Qian Let al 2005 *Mater Sci Tech* **21(5)** 597-605
[17] Dinda GP et al 2008 *Metall Mater Trans A* **39(12)** 2914-2922
[18] Balasubramanian M et al 2008 *Mater Design* **29(1)** 92-97
[19] Li X et al 2005 *J Mater Sci* **40(13)** 3437-3443
[20] Brandl E et al 2010 *Phys Procedia* **5 Part B** 595-606
[21] Nowotny S et al 2007 *J Therm Spray Tech* **16(3)** 344-348

IOP Conf. Series: Materials Science and Engineering **903** (2020) 012017 doi:10.1088/1757-899X/903/1/012017

Semi-continuous casting of aluminium alloys

Viktor Karpati[1], Balazs Hegedus[2], Agota Kazup[3], Tibor Ferenczi[4], Daniel Molnar[5], Zoltan Gacsi[6]

[1, 3, 6] Institute of Physical Metallurgy, Metalforming and Nanotechnology
[2] Institute of Energy and Quality; University of Miskolc
[4] Institute of Metallurgy
[5] Institute of Foundry
[1-6]University of Miskolc; Miskolc-Egyetemváros 3515 HU

E-mail: femkv@uni-miskolc.hu

Abstract. The target of our experiments is to simulate a complex production process under laboratory conditions. An industrial aluminium production line is examined using a semi-continuous casting device, running preliminary and trial experiments to find the optimal technological parameters for the physical simulation. Be the determination of the correct casting parameters the technological industrial process can be optimized.

1. Introduction

The semi-continuous casting technology has been known and used since the 19th century mainly for copper alloy production. In the vertical semi-continuous casting process the molten metal is fed from a crucible to the mould where it starts to solidify. The outer layer of the metal is solidified, while the inner part of still molten. The partly solidified geometry, the so-called strand can be drawn from the crystallizer, using additional cooling [1, 2]. During the solidification process, there are three phases in the system: liquid, semi-solid and solid [3]. The mould, where the metal shell solidifies is water-cooled, which is called primary cooling, while for strand is cooling using secondary cooling, which is usually a water spray [4]. Through the solidification process, numerous parameters are of key importance such as drawing speed and the intensity of the secondary cooling. The casting process can be only influenced indirectly, and the various parameters are all affecting solidification and must be aligned for the desired results. The melting- and solidification temperatures of the examined alloys, were determined using Thermo-Calc software, while the solidification characteristics were determined with NovaFlow&Solid software [5, 6, 7].

2. The semi-continuous casting device

As a part of a research, an Indutherm CC3000 semi-continuous casting device is installed in the laboratory of the University of Miskolc, which is mainly used for non-ferrous and precious metal casting. The target of the experiment is to create an experimental method to simulate the industrial process in a laboratory environment. By the help of the Induterm CC3000 casting device (Fig. 1) slabs can be cast with varying sizes [8]. In the device, the metal is melted by an induction heating unit in a graphite crucible.

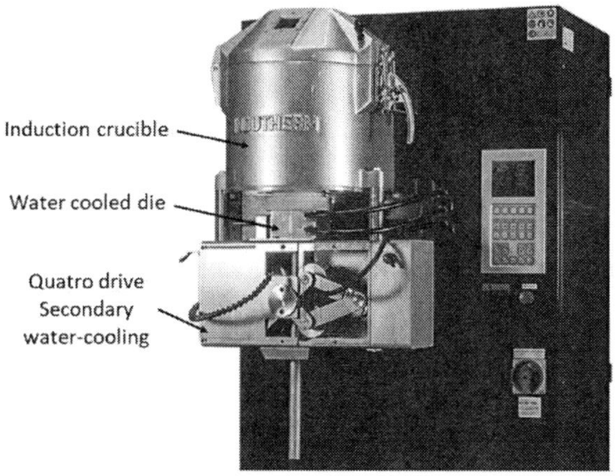

Figure 1. The Indutherm CC3000 semi-continuous casting device [9]

Above the molten metal, the atmosphere can be vacuumed or a protective atmosphere can be applied. Further parts of the melting zone are the inductor coil, the ceramic refractory, the thermocouples in the graphite crucible and the plug. The graphite plug is responsible for closing the crucible, having a 10 mm wide hole on its bottom, through of it the molten metal flows to a graphite mould surrounded by the water-cooled brass die. The temperature of the graphite mould cannot be controlled directly, but with changing the temperature of the molten metal, the drawing speed and the intensity of the secondary cooling.
The temperature of the water-cooled die can be regulated from 400 to 800 (°C) with the water flow rate of the primary cooling. The solidified metal can be drawn from the die with a starter strand.

Under the die, there are four pneumatic ribbed rolls for drawing the starter and the casted strands. The drawing speed is composed of two parameters: a drawing length, which can be 0.1-4.0 (mm), and an ensuing pause, which can be 0.1-9.9 (s). During the casting process, the strand can be stopped and even pushed back in the opposite direction with the reversing function. The secondary cooling is attached to the pneumatic roll unit.

To begin the casting process, the starter strand must be inserted into the graphite mould then the pneumatic roll is closed to hold it. For the experiments, a 10x100 (mm) graphite mould is used, and the strand has got the same size [9].

3. Experiments

The experimental semi-continuous casting machine has not been used for aluminium casting so far and the main goal was to simulate the casting process of different aluminium alloys (Al99.5, AlSi, AlFe) which are used by the industrial process.

As the first step in the research plan, the exact compositions of the alloys were determined using inductive coupled plasma mass spectrometry (ICP). Based on the exact compositions, using the Thermo-Calc software, the temperature-dependent liquid phase ratio was calculated. The calculated curves are shown in Fig. 2. The Thermo-Calc software can be used to solve thermodynamic and kinetic calculations even for multi-phase systems. The base concept of thermodynamic equilibrium is, while the pressure, temperature, and composition are constant, the number, the ratio and the quality of the phases can be calculated minimizing the Gibbs free energy:

$$G_m = \sum_{\Phi}^{F} y_\Phi \cdot \sum_{i}^{K} x_{i(\Phi)} \cdot G_{m,i(\Phi)} \rightarrow min \qquad (1)$$

where G_m is the molar Gibbs energy (J/mol), F is the number of present phases, K the number of components, y_Φ is the phase ratio of the Φ phase, $x_{i(\Phi)}$ is the mole fraction of the component i in Φ phase, and $G_{m,i(\Phi)}$ is the partial Gibbs energy of i component in Φ phase [5, 9-10].

The casting experiments were carried out by pouring three different alloys: Al99.5, AlSi12.8%, and AlFe. Using the liquid phase ratio curves the melting and solidifying parameters of the alloys can be determined.

The second step of the research was the simulation experiments. The NovaFlow&Solid simulation software was used, running Control Volume method. By the help of the simulation mass and heat transfer in three dimensions can be calculated [6, 7]. The geometric model of the casting device was created using Solid Edge V20 CAD program. The created models were installed to the NovaFlow&Solid program where initial and boundary parameters were defined and the solidification was calculated as can be seen in Fig.3.

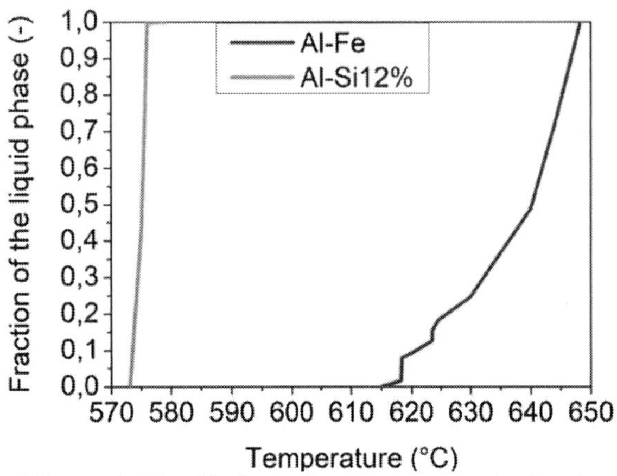

Figure 2. Liquid phase fraction curve calculated with the Thermo-Calc software [10]

Four solidification cases were examined defining the melt temperature between 720-850 (°C). Parallel with the physical experiments 40 simulations were calculated to find the optimal casting parameters. With the help of computer modeling, the solidification time of alloys in the critical cross-section of the strand was determined, without drawing. In the case of 720 (°C) initial temperature the solidification time was 24 (s), while in the case of 850 (°C) it was 36 (s).
By the help of the physical and simulation experiments, the optimal casting parameters that can be used to reproduce the casting of the Al99.5 alloy were determined. The initial temperature of the last experiment was 850 (°C) which value was used later as the ideal graphite mould temperature.

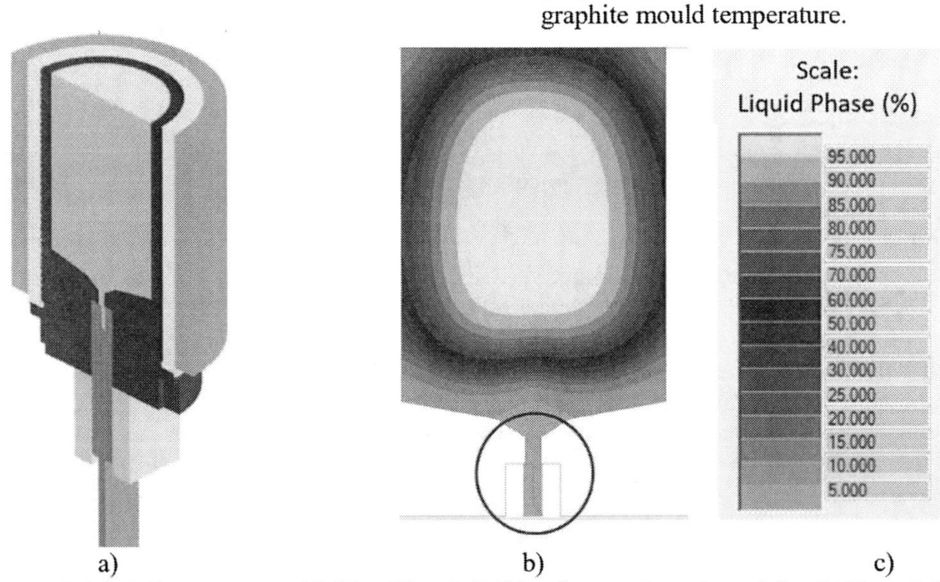

Figure 3. Modeling process with NovaFlow&Solid software a) section of the CAD model, b) the liquid phase ration in the examined cross-section using 850 (°C) initial temperature, the critical cross-section is signed with red c) liquid phase ratio scale for *b)*

If the temperature of the graphite mould is higher than 680 (°C) secondary cooling is necessary. To prevent overheating of the graphite mould the temperature of the water-cooled die was modified from 600 to 450 (°C). By the end of the experiment, the drawing speed and the secondary cooling intensity were set to the maximum value, which resulted in the best surface quality of the aluminium slabs [11].

IOP Conf. Series: Materials Science and Engineering **903** (2020) 012017 doi:10.1088/1757-899X/903/1/012017

4. Results

The Design of Experiments, which summarizes the technological parameters can be seen in Table 1.

Table 1. Design of experiments and technological parameters

Parameters	**Experiments**		
Amount of poured alloy	3 (kg)		
Protective atmosphere	argon		
Pressure of the protective atmosphere	1 (bar)		
Primary cooling	distilled water		
Primary cooling temperature	25-50 (°C)		
Primary cooling flow	0,07 (l/min)		
Secondary cooling	water		
Primary cooling temperature	20 (°C)		
Primary cooling flow	0,2 (l/min)		
Alloy	Al99.5	AlSi	AlFe
Calculated liquidus temperature	-	579.2 (°C)	648.3 (°C)
Calculated solidus temperature	-	573.1 (°C)	614.9 (°C)
Solidification range	-	6 (°C)	33.4 (°C)
Pouring temperature	850 (°C)	770 (°C)	840 (°C)
Crystallizer temperature	450 (°C)	450 (°C)	450 (°C)
Initial drawing speed	0.3 (mm)	0.3 (mm)	0.3 (mm)
Initial pause	3 (s)	3 (s)	1 (s)
Drawing length	4 (mm)	4 (mm)	4 (mm)
Technological pause	9.9 (s)	9.9 (s)	7 (s)

4.1. Solidification of Al99.5 alloy

The casting process started with the optimized parameters. The measured temperature of the graphite mould was not higher than 680 (°C), therefore the secondary cooling was not utilized. Based on the preliminary experiments, the initial drawing lengths was 0.3 (mm) with 3 (s) pause, which greatly reduced the risk that the metal solidifying prematurely and the strand tearing.

Figure 4. The Barker etched microstructure of Al99.9 aluminium was investigated with optical microscope [9, 11]

During the experiment, the drawing length was increased to 4 (mm) with a 9.9 (s) pause, which resulted in adequate surface quality. The optical microscope image of the Barker etched Al99.9 aluminium sample from the casted strand is shown in Fig.4. It represents the grain structure of the shell (finer) and the central (larger) strand parts. The experiment was repeated multiple times with the inspected technological parameters. The surface quality could be reproduced, which confirmed so that these settings are fixed to use for the Al99.5 alloy [9].

4.2. Solidification of the AlSi alloy

Using the Thermo-Calc software the melting point, 580 (°C), and the melting range, 8 (°C) of the AlSi alloy was determined. The melting range, like in case of Al99.5 alloy, is negligible, so the experiments started with the following initial values: liquid metal initial temperature is 770 (°C) and initial die temperature is 450 (°C). The initial drawing parameters were the same as in the case of Al99.5 alloy. During the casting process, secondary cooling was used. In this case, as well it proved to be true that for the successful casting the temperature of the graphite mould must be above the melting point of the alloy. The result was a product with exceptional surface quality. Fig.5. a) shows the casted AlSi strand and b) shows its Barker etched cross-section microstructure. The more and finer dendrites are in the shell part of the strand [9].

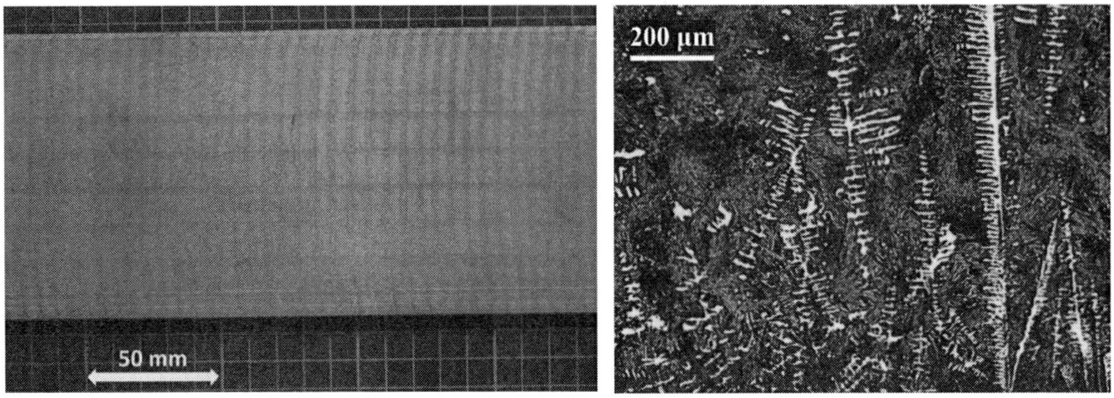

a) b)

Figure 5. a) Casted AlSi strand b) Barker etched microstructure investigated with optical microscope [9, 10]

4.3. Solidification of the AlFe alloy

In the case of the AlFe alloy, the initial liquid metal temperature was 840 (°C), while the initial die temperature was 450 (°C). Based on the first results the casting parameters were optimized and for finally the drawing speed was set to 4 (mm) with a 7 (s) pause. With these settings reproducible and good surface quality was achieved.

It was proved to be true as well that the temperature of the graphite mould has key importance and has to be higher than the melting point of the alloy. Fig.6. a) shows the casted AlFe strand and b) shows its Barker etched cross-section microstructure. Its texture mainly contains dendrites, which is a primary solid solution.

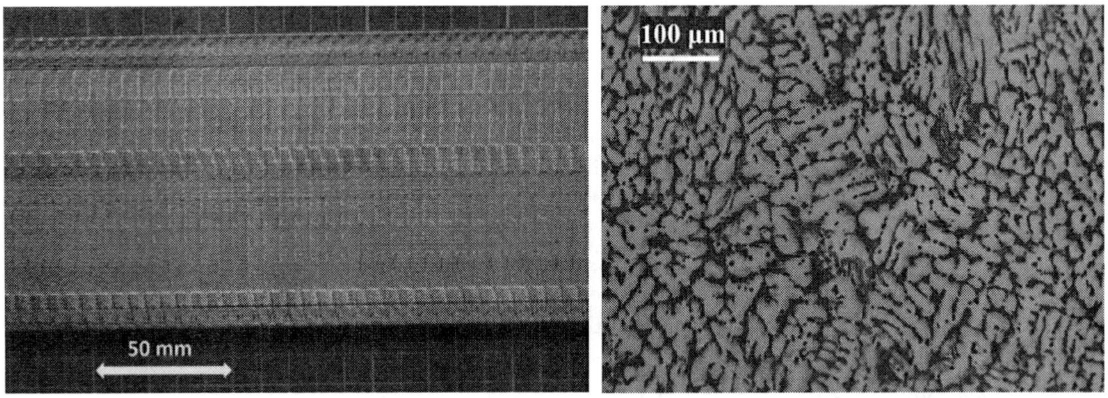

a) b)

Figure 6. a) The casted AlFe strand b) Barker etched microstructure investigated with an optical microscope

IOP Conf. Series: Materials Science and Engineering **903** (2020) 012017 doi:10.1088/1757-899X/903/1/012017

5. Conclusions

The target of this research was to simulate an industrial process under laboratory conditions. Based on the results of the experiments the optimal casting parameters can be determined for the Al99.5, AlSi and AlFe alloys. One of the importance of the results is, that this is the first time when the Indutherm CC3000 semi-continuous casting device is used for aluminium alloy casting. Additionally from the experiments, we could deduce the following results:

1. The Thermo-Calc calculations correspond to the values experienced during casting.
2. The NovaFlow&Solid software is capable to determine the time needs for the solidification of the metal in the mould which can be used for drawing time calculation.
3. In case of the examined alloys, the liquid metal must be overheated with ~190 (°C), while the temperature of the die, which responsible for the primary cooling, must be set to 450 (°C).
4. To prevent surface flaws, additional to the optimal liquid metal and die temperature, the drawing speed has key importance. In case of the AlSi alloy, the ideal value is 4 (mm) drawing with 9.9 (s) pause.
5. In case of higher drawing length, the leakage of the liquid metal can be avoided using longer pause and secondary cooling.

Acknowledgements

Thank you of the Aluform GINOP 2.2.1-15-2016-00018 project for the support of the research.

In this paper the published research is supported by the ÚNKP-19-2 New National Excellence Program of the Ministry for Innovation and Technology and the EFOP-3.6.1-16-2016-00011 „Rejuvenating and Renewing University – Innovative Knowledge City – part of the University of Miskolc intelligent specialization aided institutional development" project – in the Szechenyi 2020 – by the assistance of European Union, realized by the European Social Fund co-financing.

Thank you for the help of Mr Adam Vegh for the Thermo-Calc calculations and Dr Oliver Banhidi for the ICP examinations.

Thanks to the NovaCast Systems AB company for providing the NovaFlow&Solid software.

References

[1] Müller H R et al 2015 *Stranggießen von NE-Metallen Syposium* pp 1-14
[2] Roller E et al 1996 *International Wrought Copper Council, Technical Seminar* 3/95
[3] Eskin D G 2008 *Physical Metallurgy of Direct Chill Casing of Aluminium Alloys* (Taylor and Francis) pp 79-103
[4] Sommerhofer H et al 2010 BHM Berg- und Hüttenmännische Monatshefte **155** 12-16
[5] Kaptay Gy 2011 *Equilibrium of Materials* in Hungarian (Univ. Press, Miskolc) pp 63-68
[6] Hattel J 2005 *Fundamentals of Numerical Modelling of Casting Processes* (Polyteknisk, Denmark) pp 23-28
[7] Bagyinszki Gy et al 2007 *Introduction to the Informatics of Material Technologies* in Hungarian (EME, Cluj Napoca) 84-102
[8] www.bluepower-casting.com, (2020 April)
[9] Kárpáti V et al 2019 *BKL Kohaszat Journals* **152** 5-6
[10] Kárpáti V et al 2019 *Student Research Societies Miskolc* pp 33-35
[11] Kazup A et al 2018 *Student Research Societies Miskolc* pp 16-31

IOP Conf. Series: Materials Science and Engineering **903** (2020) 012018 doi:10.1088/1757-899X/903/1/012018

Characterization of chemically foamed poly(lactic acid)

K Litauszki[1] and Á Kmetty[1,2]

[1] *Budapest University of Technology and Economics, Faculty of Mechanical Engineering, Department of Polymer Engineering, 1111 Budapest, Műegyetem rkp. 3.*
[2] *MTA–BME Research Group for Composite Science and Technology, 1111 Budapest, Műegyetem rkp. 3.*

E-mail: litauszkik@pt.bme.hu

Abstract. This paper presents the characterization of poly(lactic acid)-based chemically foamed foam structures. We used extrusion grade, high molecular weight poly(lactic acid) as matrix material and azodicarbonamide as foaming agent. The resulting foams were characterized by foam density, cell population density, scanning electron microscopy, and based on the fracture surface of the foam cross-section, we calculated cell size distribution. The obtained distributions were fitted with the use of log-normal distribution in each case. We presented the effect of chemical blowing agent content (0.5 wt%, 1 wt%, 2 wt%, 4 wt% and 8 wt%) on the PLA-based foam structure and the effect of processing temperature (190 °C, 210 °C and 230 °C) on cell size distribution.

1. Introduction

The characterization of foam structures at a microstructural level can be approached from four aspects. The first approach is based on the amount of material in the foam structure (density, relative density and void fraction). The foam structure can be characterized (2) by the ratio of open and closed cells, (3) cell size distribution and (4) cell structure anisotropy. Nowadays, cell size distribution is typically investigated along a surface, such as a cross-section (2D). The surface can be examined with a digital optical microscope or a scanning electron microscope (SEM). The cross-sectional surfaces used for observation are typically produced with a microtome or are cryogenic fracture surfaces. The number of cells, their size, their distribution and the degree of anisotropy can be determined from these images [1]. Based on the literature, cell size distribution is typically described with a log-normal distribution [2-5].

The applicability of chemical blowing agents is very different from the usage of thermally expandable microspheres (EMSs). Depending on the chemical blowing agent (CBA) content, different phenomena can be observed in the foamed polymer structures. Effective gases evolved during the decomposition of CBA affect the number of cells formed and their size [6]. The number of cells formed is affected by the amount and type of dissolved effective gas [7]. We determined how the resulting cell size distribution changes when foaming agent content and processing temperature are increased, using PLA, a promising biopolymer, as matrix [8, 9].

2. Experimental

2.1. *Materials*

The polymer matrix material was selected from the NatureWorks LLC (Minnetonka, MN, USA) Ingeo PLA product line. We selected extrusion type Ingeo 2003D. The D-lactide content of this PLA is 4.3 mol% [8], its melting temperature is 150.9 °C (determined by DSC), its Melt Flow Index is 2.0 g/10 min (CEAST 7027.000, 2.16 kg, 190 °C). We determined its number average molecular weight (100,422 g/mol), its weight average molecular weight (180,477 g/mol), and its polydispersity index (1.79) [10].

We used a chemical blowing agent (CBA) for the extrusion foaming of PLA. The CBA was Tracel IM 3170 MS, provided by Tramaco GmbH (Tornesch, Germany). Tracel IM 3170 is an azodicarbonamide-based, exothermic foaming agent. The effective gases are mainly carbon dioxide, but some carbon monoxide is also generated, cyanic acid is formed as well, but it will further decompose to nitrogen and ammonia. The maximum intensity of gas generation measured by thermogravimetric analysis is at 183 °C, and the decomposition range is 147-212 °C, according to the dTG curve (measured in air) [10].

2.2. *Foaming process*

Foaming was carried out in a Teach-line ZK25T (Collin GmbH, Ebersberg, Germany) twin-screw extruder, which has an L/D ratio of 24 and a screw diameter of 25 mm. The processing parameters were 155 °C at the feed section (Z1), 160 °C (Z2), 175 °C (Z3), 190 °C (Z4) and 190 °C at the rod die (rod diameter: 3 mm). The rotation speed of the screw was 10 rpm. Before dosing, the PLA and CBA were dry mixed, the amounts of CBA were 0.5 wt%, 1 wt%, 2 wt%, 4 wt% and 8 wt%. Prior to extrusion, the PLA was dried in a drying oven for 6 hours at 80 °C.

We also investigated the effects of higher processing temperature. The foaming process was also carried out at 175/180/195/210/210 °C, and 195/200/205/230/230 °C. The rotation speed of the screw was also 10 rpm, and all the other parameters were the same. The amounts of CBA that we used were also 0.5 wt%, 1 wt%, 2 wt%, 4 wt% and 8 wt%.

3. Methods and analysis

3.1. *Density measurement*

Density was calculated according to Equation 1. The weight of the specimens was measured with an Ohaus Explorer (Nänikon, Switzerland) balance (accuracy 0.0001 g). The medium was distilled water.

$$\rho = \frac{m_{sa}}{m_{sa} - m_{sl}} \cdot \rho_{dw} \tag{1}$$

where ρ (g/cm^3) is the density of the sample, m_{sa} (g) is the mass of the specimen measured in air, m_{sl} (g) is the mass the sample measured in distilled water and ρ_{dw} (g/cm^3) is the density of distilled water at room temperature.

3.2. *Scanning electron microscopy*

The cell structures were investigated by scanning electron microscopy (SEM). The microscope was a JEOL (Japan) JSM 6380LA. The samples were coated with a gold-palladium alloy with a JEOL 1200 device. The accelerating voltage was 10 kV. We determined cell sizes using the ImageJ software.

3.3. *Foam characterization*

Cell size distribution can be described as the density function of the log-normal distribution (2.) [11]:

$$f(x) = \frac{1}{x\sigma\sqrt{2\pi}} \cdot e^{-\frac{(lnx-\mu)^2}{2\sigma^2}} \tag{2}$$

IOP Conf. Series: Materials Science and Engineering **903** (2020) 012018 doi:10.1088/1757-899X/903/1/012018

where x is the random variable, μ is the mean value of the normally distributed probability variable, σ is the standard deviation of the normally distributed variable.

3.4. *Calculation of cell population density*

Cell population density was calculated according to Equation 3, where n is the number of cells, A [cm^2] is the area of the sample, M [-] is the magnification factor, and V_f [-] is the void fraction [12].

$$N_c = \left(\frac{n*M^2}{A}\right)^{\frac{3}{2}} \cdot \frac{1}{1-V_f}, \quad (3)$$

4. Results and discussion

4.1. *The effect of the chemical blowing agent content on the PLA-based foam structure*

Scanning electron microscope images of foam structures produced with different amounts of the chemical blowing agent are shown in Figure 1. A small number of cells nucleated at small content (0.5 wt% and 1 wt%) and the size of the cells was in the order of 300 µm. At medium dosing, the number of nucleated cells increased dramatically and the cells typically had a smaller size. That results in a more favourable, homogeneous, lower density PLA foam, thus providing greater porosity than lower content. At a high content of exothermic chemical blowing agent (4 wt% and 8 wt%), the number of nucleated cells is still high, but the phenomenon of cell collapse appears (Figure 1. e). This is unfavourable—although large cells reduce density, they also decrease mechanical and insulating properties [13, 14].

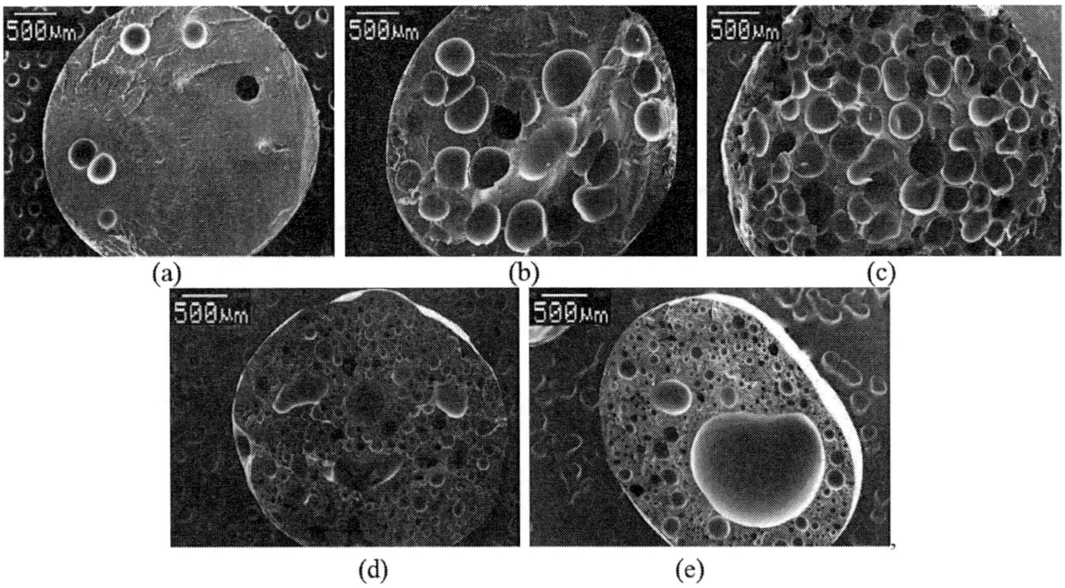

(a) (b) (c)
(d) (e)

Figure 1. SEM images of PLA foams produced with a) 0.5 wt%, b) 1 wt%, c) 2 wt%, d) 4 wt%, e) 8 wt% exothermic chemical blowing agent at the same manufacturing temperature (190 °C); magnification is 30x

The phenomenon that we described based on the SEM images earlier were characterized by measuring cell sizes and cell size distribution was calculated according to Chapter 3.3. Figure 2. a) shows the resulting density function as a function of cell size in the case of 8 wt% CBA. Cell size distributions were plotted against foaming agent content for Ingeo 2003D polylactic acid (Figure 2. b). In the case of chemical foaming, cell size distribution is shifted towards smaller cells as the content of the chemical blowing agent was increased. Larger cells formed by the collapse of the cell walls appear in the case of higher (4 wt% and 8 wt%) CBA content. By performing the χ^2 test, we found that the observed cell size differences are significant (Table 1.).

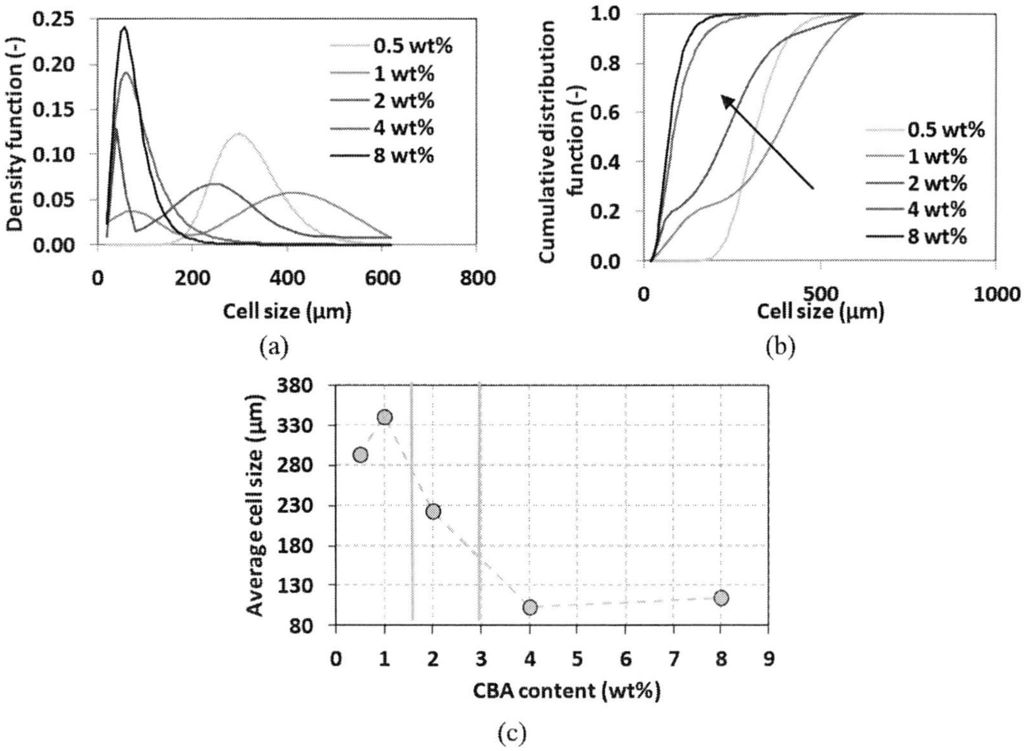

Figure 2. Calculated values of foam structures containing exothermic CBA a) density function as a function of cell size (8 wt% CBA), b) Cumulative distribution function and c) average cell size as a function of CBA content

Table 1. P-values of the χ^2 test for the exothermic chemical blowing agent; the different concentrations compared to each other

CBA content	PLA_CBA$_{exo}$ p-value
0.5 (wt%)-1 (wt%)	0.00024
1 (wt%)-2 (wt%)	0.00000
2 (wt%)-4 (wt%)	0.00000
4 (wt%)-8 (wt%)	0.00001
8 (wt%)-0.5 (wt%)	0.00000

The results indicate that the number of cells formed in the case of small amounts (0.5 wt% and 1 wt%) of chemical blowing agent is low, typically less than 10 cells. In this case, the gases evolved in the polymer matrix are not used for cell nucleation, but typically the growth of the nucleated cells. At 2 wt% CBA, cell population density increased to 1.32 x 10^9 cells/cm^3 and the distribution is bimodal. The reasons of the bimodal cell size distribution is the phenomenon of secondary nucleation [15]. At higher contents of CBA (4 wt% and 8 wt%), the phenomenon of cell collapse occurs, due to cell roughening [13, 16], therefore the large number of nucleated cells cannot grow further.

4.2. *The effect of processing temperature on the foam structures produced*

Not only the CBA content, but production temperature also has paramount importance in the case of polymer foam processing [2, 16]. Both the amount of effective gases produced during the decomposition of CBA and the solubility of the gases depend on the temperature and pressure of the polymer melt. Therefore, processing temperature parameters are important [13]. In the case of chemical foaming, low viscosity of the polymer melt typically adversely affects the resulting foam structure. We described the cell size distribution of the cells at increasing production temperatures (die temperatures were 190 °C, 210 °C and 230 °C). SEM images of the produced PLA-based foams are shown in Figure 3. The foam structures show that fewer cells were nucleated at 190 °C at the same CBA content (2 wt%) than at 210 °C. At a higher processing temperature the number of nucleated cells is higher (at 210 °C 3.96 x 10^{10} cells/cm^3), but the phenomenon of cell collapse appears (Figure 3. a) and b). This is also unfavourable in accordance with the decreasing mechanical and insulating properties [13, 14].

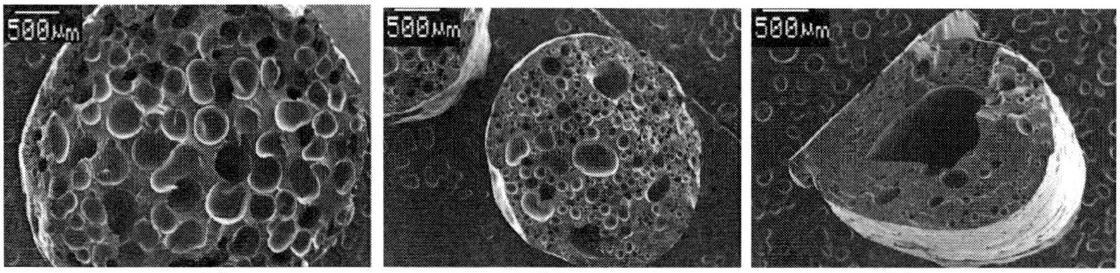

(a) (b) (c)

Figure3. SEM images of PLA foams with 2 wt% exothermic CBA at different processing temperatures a) 190°C, b) 210 °C, c) 230 °C; magnification 30x

Based on the images in Figure 3., we also evaluated cell size distribution according to Chapter 3.3. Cell size distribution (Figure 4.) is shifted towards smaller cell sizes as processing temperature is increased. As chemical blowing agent content is increased, larger cells are produced from the collapse of cell walls.

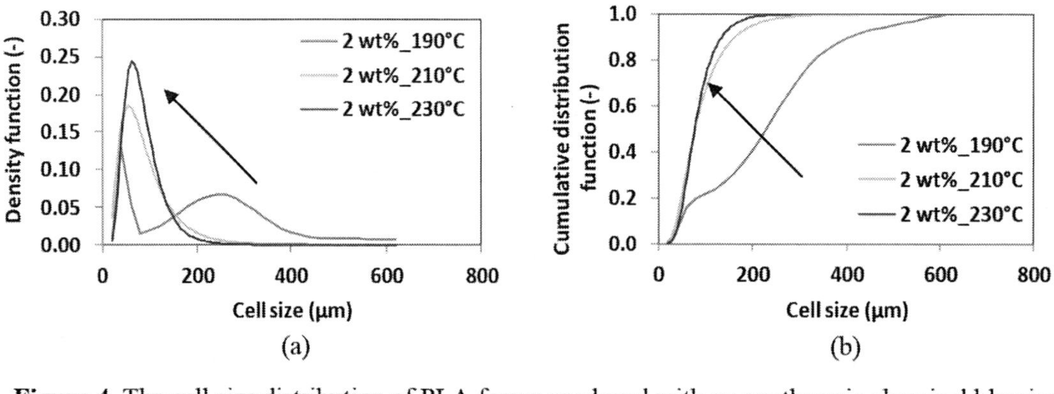

Figure 4. The cell size distribution of PLA foams produced with an exothermic chemical blowing agent at different processing temperatures (190 °C, 210 °C and 230 °C)

There were significant differences in cell size distribution, which was confirmed by the χ^2 test (Table 2.). The increasing addition of chemical blowing agent reduced the gas retention capacity of the polymer. Therefore, more than 2 wt% CBA addition will not decrease further the density (Figure 5.). The foam production limit at 190 °C it is 8 wt%, means that the pressure inside the extruder die decreased to a level, where continuous product can not be extruded, With the increasing temperature, the foam production limit is reduced from 8 wt% to 4 wt% at 210 °C, and 1 wt% at 230 °C. Expansion and density reduction are more favourable for samples made at a lower (190 °C) temperature.

Table 2. P-values of the χ^2 test for the exothermic chemical blowing agent at 2 wt% CBA for samples produced at different processing temperatures (190 °C, 210 °C, 230 °C)

Processing temp.	**PLA_2wt%_CBA$_{exo}$**
	p-value
190 (°C)–210 (°C)	0.00000
210 (°C)–230 (°C)	0.00000
230 (°C)–190 (°C)	0.00000

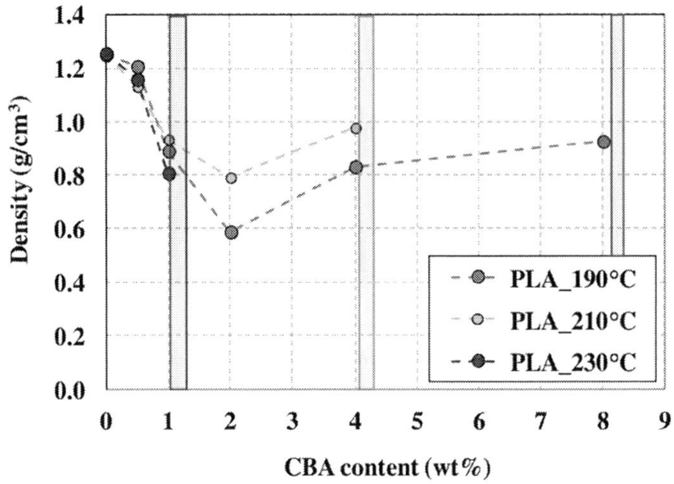

Figure 5. The density of foams as a function of CBA content at different production temperatures (190 °C, 210 °C and 230 °C); blue line: production limit at 190 °C; orange line: production limit at 210 °C; red line: production limit at 230 °C

5. Conclusions

In this study, we characterized chemically foamed poly(lactic acid)-based foam structures. We measured the density of foam samples calculated their cell population density, examined them by SEM and calculated the log-normal cell size distribution. First, we investigated the effect of chemical blowing agent content (0.5 wt%, 1 wt%, 2 wt%, 4 wt% and 8 wt%). The SEM images show that a small amount of cells nucleated at low CBA content (0.5 wt% and 1 wt%), and cell size is around 300 μm. At medium dosing, the number of nucleated cells increased considerably, and the cells are typically smaller. At high contents (4 wt% and 8 wt%) of CBA, the number of nucleated cells is still high, but cell collapse sets in. Cell size distributions were plotted against foaming agent content. Cell size distribution shifted towards smaller cells as CBA content was increased. We found that the observed cell size differences are significant. Second, presented the effect of processing temperature (190 °C, 210 °C and 230 °C). The cell size distribution of PLA foams shifted towards smaller cells when processing temperature was increased. With an increase in chemical blowing agent content, fewer larger cells were produced due to the collapse of cell walls—we found that there were significant differences in cell size distributions. With the increasing temperature, the foam production limit is reduced from 8 wt% to 4 wt% at 210 °C, and 1 wt% at 230 °C. To achieve the largest density loss in the case of examined processing temperatures and CBA content, we recommend the use of 190 °C and 2 wt% CBA.

Acknowledgements
This paper was supported by the National Research, Development and Innovation Office, Hungary (K 132462). This project was supported by The National Research, Development and Innovation Office (NVKP_16-1-2016-0012). This research was also supported by the ÚNKP-19-4 New National Excellence Program of the Ministry for Innovation and Technology; and the János Bolyai Research Scholarship of the Hungarian Academy of Sciences.

References

[1] Gendron R 2004 *Thermoplastics foam processing, Principles and Development* (Boca Raton: CRC Press)
[2] Julien J, Bénézet J, Lafranche E, Quantin J, Bergeret A and Lacrampe M 2012 *Polymer* **53** 5885–5895
[3] Chen Y, Das R and Battley M 2015 *Int. J. Solids. Struct.* **52** 150–164
[4] de Icaza M, Ceniceros J A and Castaño V M 1995 *J. Appl. Phys.* **76** 7317
[5] Du L, Prokop A and Tanner R D 2003 *J. Colloid. Interface. Sci.* **259** 180–185
[6] Shaft M A, Joshi K and Flumerfel R W 1996 *Chem. Eng. Sci.* **52** 635-644
[7] Matuana L M, Faruk O and Diaz C A 2009 *Bioresour. Technol.* **100** 5947–5954
[8] Standau T, Zhao C, Murillo C S, Bonten C and Altstädt V 2019 *Polymers* **11** 306
[9] Hajba S and Tábi T 2019 *Period. Polytech. Mech. Eng.* **63** 270–277
[10] Kmetty Á, Litauszki K and Réti D 2018 *Appl. Sci.* **8** 1960
[11] Prékopa A 1974 *Valószínűségelmélet műszaki alkalmazásokkal* (Budapest: Műszaki könyvkiadó)
[12] Xu X, Park C B, Xu D and Pop-Iliev R 2003 *Polym. Eng. Sci.* **43** 1378-90.
[13] Lee S T and Ramesh N S 2004 *Polymeric foams - Mechanisms and Materials* (Boca Raton: CRC Press)
[14] Szabó V A and Dogossy G 2020 *Period. Polytech. Mech. Eng.* **64** 81-87
[15] Lee S T, Park C B and Ramesh N S 2007 *Polymeric Foams- Science and Technology* Polymeric Foams Series (Boca Raton: CRC Press)
[16] Wu H, Zhao G, Wang J, Wang G and Zhang W 2019 *Express. Polym. Lett.* **13** 390-405

IOP Conf. Series: Materials Science and Engineering **903** (2020) 012019 doi:10.1088/1757-899X/903/1/012019

Deformability Tests of Pure Niobium

V Mertinger[1], I Aviles Santillana[2], M Benke[1], A Gallifa Terricabras[2], A Hlavács[1], D Janovszky, V Kárpáti, Gy Krállics[1], T Mikó[1], A B Palotás[1], M Sepsi[1], G Szabó[1], P Szobota[1], M Szűcs[1]

[1]University of Miskolc, Faculty of Materials Science and Engineering, Miskolc, Hungary
[2]CERN, Geneva, Switzerland

E-mail: femvali@uni-miskolc.hu

Abstract. A research team at the University of Miskolc's Faculty of Materials Science and Technology has signed a cooperation agreement with the Geneva-based European Organization for Nuclear Research (CERN) for testing of the materials employed in the Crab Cavities will be installed in the next generation of the LHC (the so-called High Luminosity Large Hadron Collider – HL-LHC). At the University of Miskolc, high purity niobium rolling experiments were carried out in conventional (unidirectional) and cross-rolled manners in order to increase the deep drawability of the final sheet. The deformability of niobium was measured by Watts-Ford and compression tests. The microstructure and anisotropy (texture) results of the initial material and the straight-rolled products are reported.

1. Introduction

Crab Cavities, SRF cavities of non-axisymmetric shape, made of ultra- high purity Nb are essential parts of the HL-LHC project. These components are submitted to complex metal sheet forming processes, especially deep drawing [1]. The initial material of deep drawing is always a cold rolled product whose technical characteristics and crystalline texture determine its deep drawability. Ideally, the properties of the plate during deep drawing are direction-independent and isotropic.

Unidirectional rolling is one of the bulk metal forming processes, where metal is plastically deformed by allowing it to flow between two rollers which rotate in opposite senses. The rolling process not only reduces the thickness associated with elongation in length, but also enhances material properties like strength, toughness and texture. The purpose of several researchers in this field is to improve the quality of the product through optimizing the process parameters [2-6]. For unidirectional or straight rolling, where it is considered as a plane strain deformation, the sample has an orthorhombic symmetry. Symmetry is something that will bring a sample into coincidence with itself when certain operations are performed it. Orthorhombic symmetry means all the three (rolling, transverse, and normal) directions of the sample are axes of twofold symmetry. For cross rolling, where the sample is rotated by 90° after each pass, the sample symmetry becomes tetragonal, having one fourfold axis which is parallel to the normal direction [7]. That is why cross rolling is a way of reducing the directional dependency of the different properties of the rolled plate. This method can be advantageous in the case of avoiding the earing during deep drawing. [8-10].

During plastic deformation process of polycrystalline metals, the crystallographic orientation of each grain changes. This is known as texture evolution. After cold working the texture remains and it has a significant effect on the anisotropy of the properties, so the texture evolution needs to be controlled [11].

In case of unidirectional rolling, strong textures, which may lead to significantly different properties in the rolling and transverse directions, are formed in the direction of the rolling. In cross rolling, due to continuous change in rolling direction, the microstructure, the texture and the plastic anisotropy will be different than that obtained in unidirectional rolling.

Texture, generally characterized by a crystallographic plane and a crystallographic direction, can be represented graphically in pole figures and orientation distribution functions (ODF). Basically, this representation indicates the intensity of an orientation or texture component, which depends on the number of grains having same orientation [12].

The aim of the present research is to determine the deformability, reveal the developed microstructure and texture and recrystallization temperature of a niobium sheet during unidirectional rolling. The critical levels of deformation steps had been calculated by finite elements analysis as a representative of the effective plastic strain seen by the niobium sheets during deep drawing operation. The obtained results will serve as fundamental data for the upcoming cross rolling process of the examined niobium sheets.

2. Experimental

3 pieces of high purity niobium sheets of 4 mm nominal thickness, 50 mm width and 100 mm length provided by CERN (and supplied by Ningxia Orient Tantalum Industry Co. Ltd, (OTIC)) were employed in the rolling experiments. Their composition of impurities, given by certificate is presented in Table 1. In the as received condition, the sheets have moderate grain size and distribution (Figure 1.). The sampling for different tests and the sample identification are shown in Figure 2. Samples with identification 1W2 and 1W2 were subjected to a Watts-Ford test and subsequently to microstructure investigation. The cylindrical samples with diameter of 4 mm, represented by yellow circles, were subjected to multi-step compression tests. The remaining part (20mm x 50mm) of sample ID 1, went true in unidirectional rolling as an initial test sample.

Table 1. Chemical concentration of the impurities by certificate (ppm) [1]

Ta	Zr	Fe	Si	W	Ni	Mo	Hf	Ti	Cr+Co	H	N	O	C
300	5	5	10	<5	<5	<5	10	<5	<10	1	8	7	10

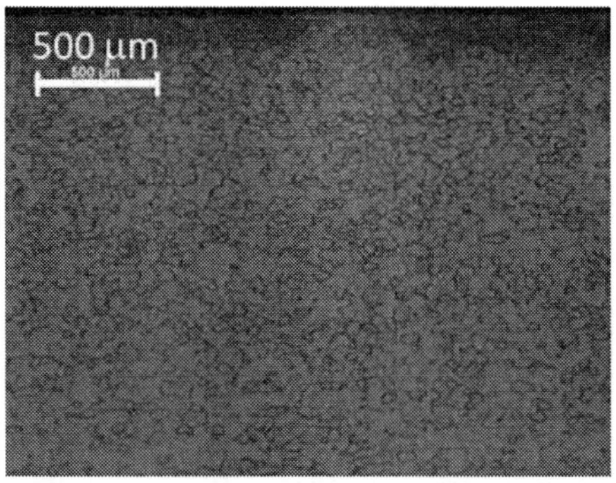

Figure 1. Microstructure of the initial niobium sheet in the as received condition

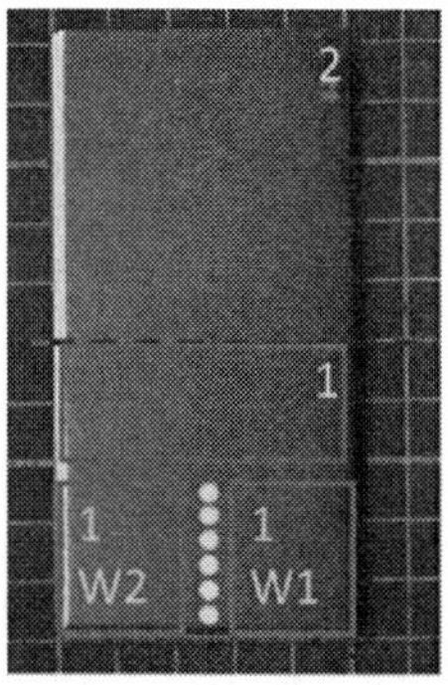

Figure 2. Sample identifications

The rolling process was performed with a Von Roll rolling mill employing two high rolling modes at room temperature. The roll diameter and its length were 220 mm. Prior to each pass, the rolls were lubricated with a mineral oil-based lubricant (PetroPal) through nozzles. The rolling speed was kept constant at 30 m/min throughout the entire rolling process and the force and the torque were recorded for each pass. To reach the final thickness the rolling process was interrupted by intermediate samplings.

A Watts-Ford test (plain strain compression test) was performed to achieve the flow stress curve of niobium under compression. The geometric parameters and the applied tool can be seen in Figure 3. During the test the following conditions were fulfilled:

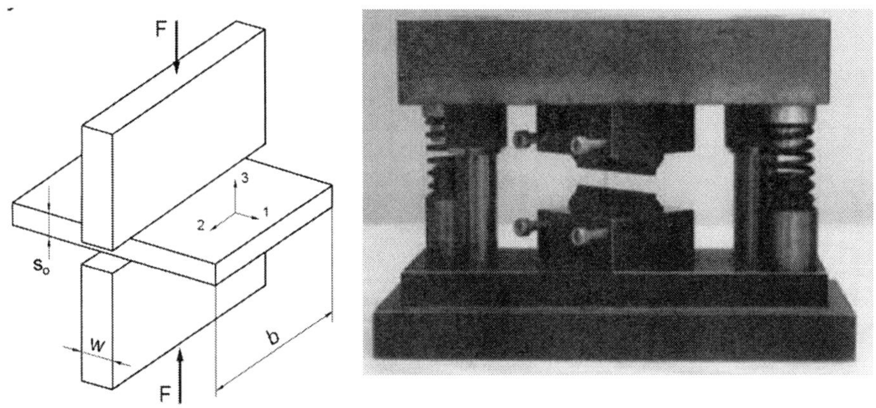

$$\frac{b}{w} \geq 5 \qquad (1)$$

w= 6.0; 3.6; 2.2 mm,
b=min 31 mm

$$2 < \frac{w}{s} < 4 \qquad (2)$$

Figure 3. Geometry for Watts-Ford test (s_0 – initial thickness of specimen, b – initial length of specimen, w – width of dies, F- force) and the inserted die in holder

Based on the results of the Watts-Ford test, we used finite element simulation (MSC.Marc software) to calculate the equivalent Von Mises Stress during rolling. Multi-step compression test on three cylindrical specimens by 0.5 mm reduction steps were performed by Instron 5982 type universal mechanical tester. The maximal (D1) and the minimal (D2) diameter of the samples were measured by an optical system. A ratio of D1/D2 was calculated and plotted as a function of reduction.

According to the supplier, the initial sheets have seen a light levelling operation after the last annealing treatment the texture investigation was also done not only on the rolled but also on the initial sheets. A Bruker D8 Advance diffractometer, equipped with a Eulerian cradle with CoKα radiation and line focus of 20 mm, was used for texture measurements. The {200}, {310} and {110} pole figures of the initial and rolled sheets were measured. Ferrite powder was used for defocusing correction. The pole figure recalculation was done using triclinic deformation while ODF calculation using the software, TEXEVAl, of the equipment.

Thermal analysis, in a Netzsch STA 449 F1 Jupiter Simultaneous TG-DTA/DSC (Netzsch Ltd.) apparatus at a heating rate of 0.16 K/s, was performed to determine the recrystallization temperature. The samples for these measurements were kept in closed Al_2O_3 crucibles and were under a continuous 99.9999% Ar flow. Optical microscopy was used to characterize the rolled and the recrystallized (after the DSC cycle) microstructure.

3. Results and discussion

The measured and recalculated pole figures with texture numbers and calculated α fibers for both sides of the 1W1 initial samples (as received) can be seen in Figure 4. Due to the shape and the asymmetry of the curves it is obvious that a strong shear component is present in the texture. The pole figures were determined for both sides of the sheet (sides A and B) and the comparison of the pole figures shows that the initial sheet was probably cross rolled, but rolling in one direction caused more deformation than in the other direction (Figure 4). This is apparent on the α fibers as well (Figure 5). The asymmetric lubricant condition during unidirectional rolling can also be a reason for this strong shear component.

Figure 6 shows the optical view of one sample after multi step compression tests and the evaluation of D1 and D2 diameter of three parallel samples. The D1/D2 value of two samples shows a maximum in two cases. Maximum means that in the initial phase of the reduction the deformation is strongly anisotropic (ellipsoidal shape of the disk), and after the maximum the ovality decreases. This kind of behavior refers to the state of the inhomogeneous texture of the initial plate confirming the results obtained from the pole figures. This type of behavior should be kept in mind in any future interpretation of the rolling texture.

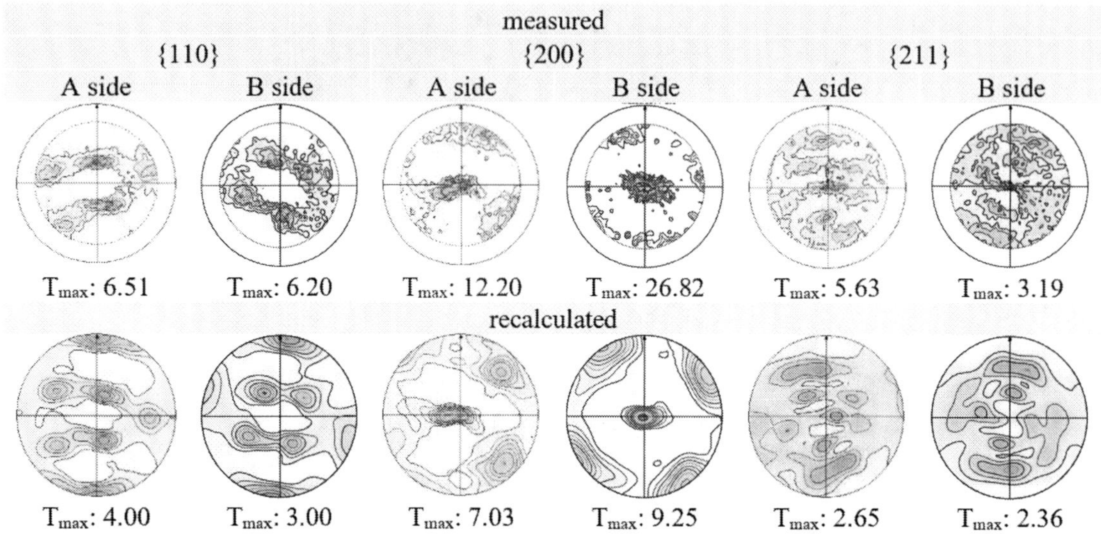

Figure 4. Measured and recalculated pole figures of the initial sample ID 1W1, sides A and B

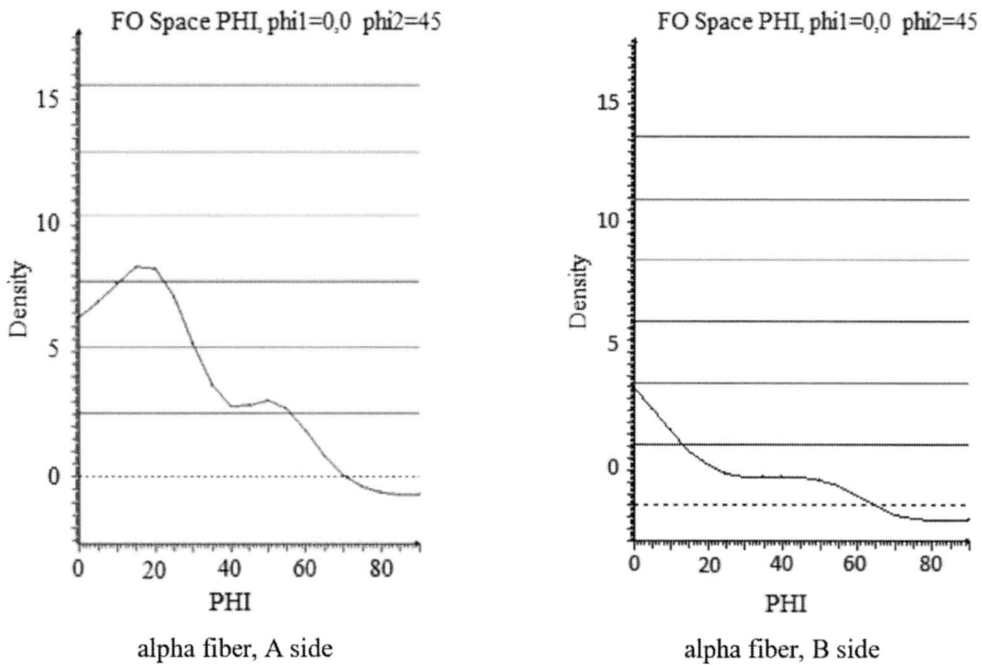

Figure 5. Alpha fiber of the initial sample ID 1W1, sides A and B

The fitted flow curve of niobium, determined by the Watts-Ford test and Lademo fitting, can be seen in Figure 7. Based on the measured equivalent plastic strain, a finite element simulation of stress distribution in the longitudinal section of niobium during one rolling pass was performed. The two-dimensional static model includes a rigid roll and the sheet sample defined as a linear elastic and nonlinear plastic material. The body of niobium sheet meshed with four noded, quad type element and the total number of finite elements was 900. The flow stress data provided by plain strain compression test. The Kudo's frictional model was applied to simulate the friction between roll and sheet material. The difference between the values of measured and calculated force was minimized by modifying the friction factor. The average values of measured and calculated force in the first pass are respectively Fm=61.021 kN and Fc=64.989 kN. The result of the simulation for the upper half part of the niobium section is showed in Figure.8.

To test the boundary states of rolling and the friction problems, a unidirectional rolling experiment was performed on a small sample (ID1). After 10 steps of rolling with a deformation of 2.667 φ., the final thickness that was obtained was 0.28 mm. The resulting ribbon was wavy, and this is due to the small sample size (Figure 9.).

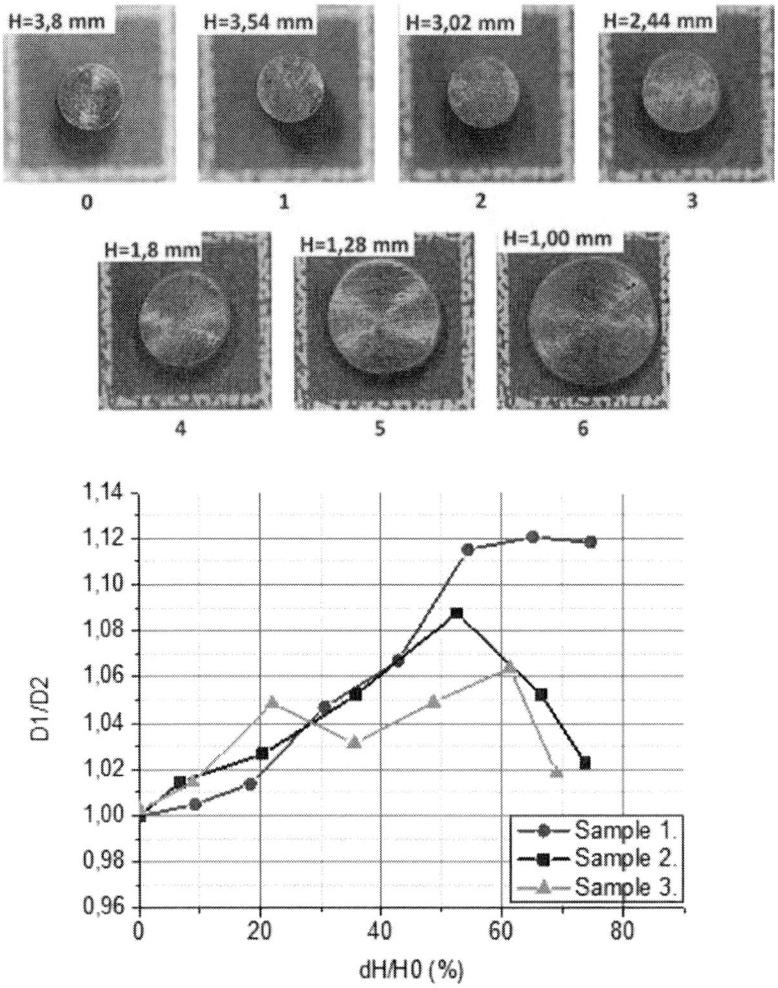

shape of the sample after different reduction, H- the actual high

H0 - the initial high, D1- the maximum, D2-the minimum diameter at each step

Figure 6. Results of multi-step compression test. Samples 1, 2 and 3 are parallel tests.

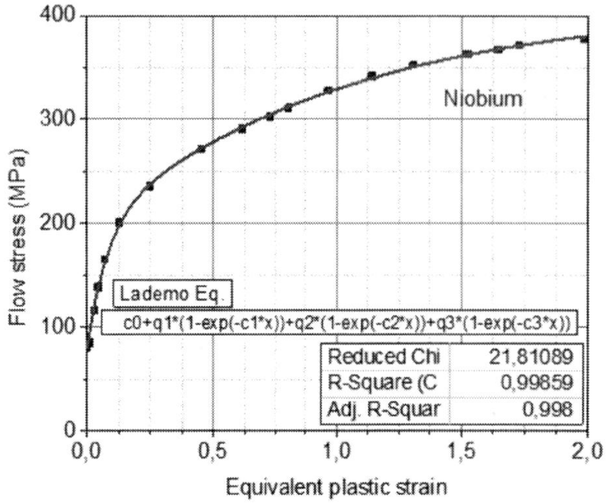

Figure 7. Flow curve of niobium based on Watts-Ford test.

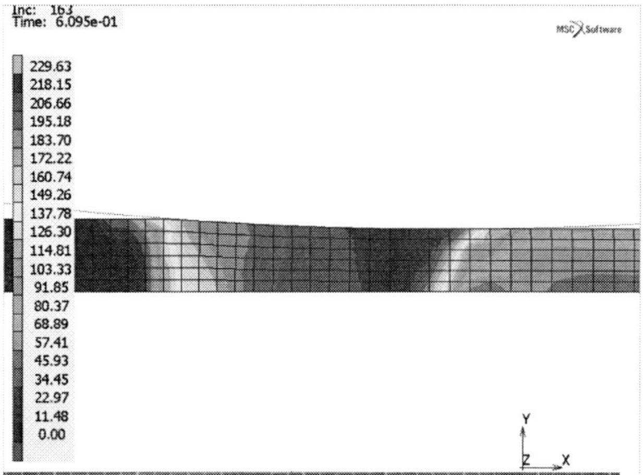

Figure 8. Finite element simulation of the equivalent Von Mises Stress during rolling in the half longitudinal section of niobium strip.

Figure 9. Rolled niobium strip.

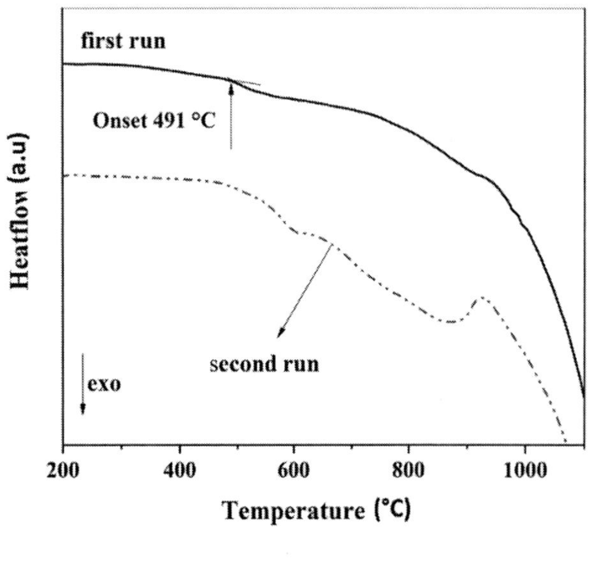

Figure 10. DSC curve of sample 1.6 after 6 steps rolling, φ=1.435.

DSC analysis showed that recovery, stress relaxation and recrystallization started at 491° C (Figure 10). An effect of the oxidization process is also visible even though shielding gas was used. The microstructures formed after the rolling and DSC are shown in Figure 11. The microstructure after the DSC cycle shows not only recrystallization but also grain coarsening.

In Figure 12, the rolling texture evaluation can be clearly seen on the pole figures and α fibers (Figure 13) of the samples subjected to 6 (φ=1.435), and 10 passes (φ=2.667). It is also apparent that the texture is sharper after 10 passes compared to 6 passes.

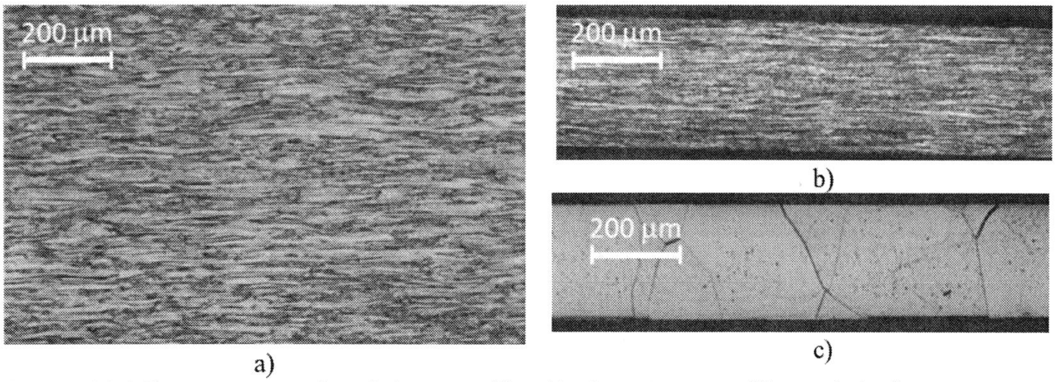

Figure.11. Microstructures after a) 6 steps rolling b) after 10 steps rolling and c) after DSC test.

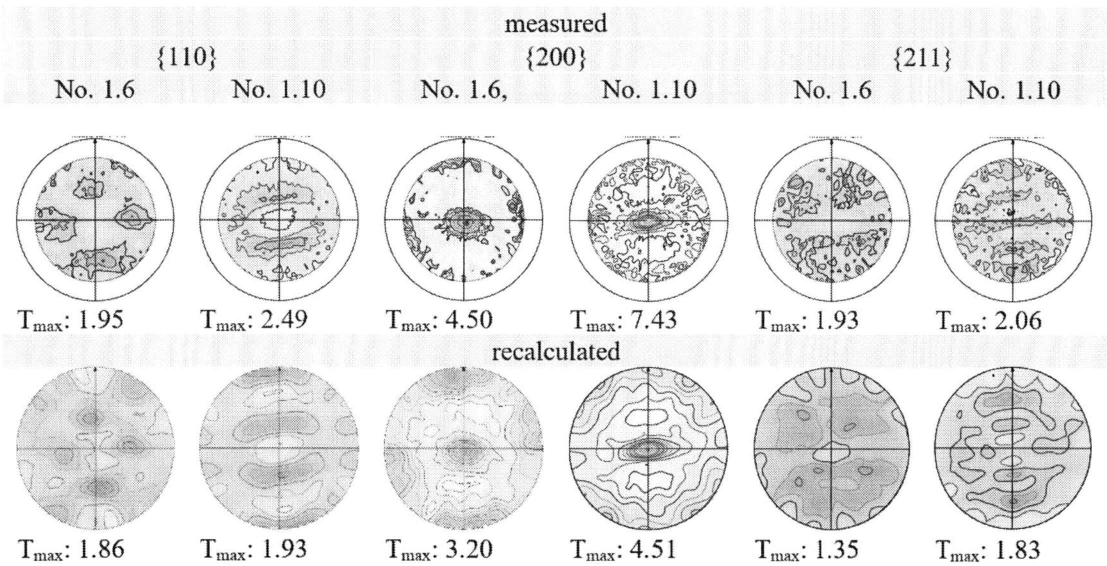

Figure.12. Measured and recalculated pole figures of the rolled No. 1.6 (φ=1.435) and No. 1.10 (φ=2.667) samples.

4. Conclusion

Based on the mechanical experiments, the rolling plan of the examined niobium sheet was established. The rolling experiment was performed in 10 steps up to φ=2.667 deformation at room temperature. Although the resulting ribbon was wavy, due to the small sample size, it was revealed that at room temperature niobium can be rolled very well. It was also found that the recrystallization of the rolled sheet started at 491° C. Grain coarsening also occurred during the DSC examination.

Acknowledgement

The research was carried out at the University of Miskolc, within the framework of the Thematic Excellence Program funded by the Ministry of Innovation and Technology of Hungary. The present paper was supported through the National Research, Development and Innovation Office – NKFIH K119566 project and the new national excellence program of the ministry of human capacities. Author say thanks to Department of Materials Science and Engineering (BME) for Watts-Ford test.

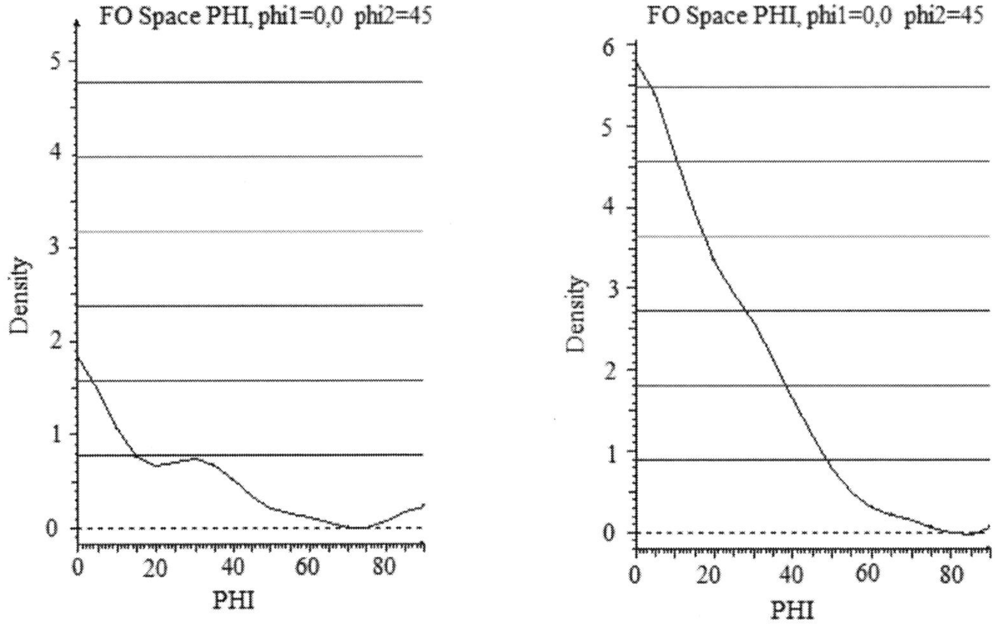

alpha fiber, No. 1.6 (φ=1.435) alpha fiber, No. 1.10 (φ=2.667)

Figure.13. Measured and recalculated pole figures and alpha fiber of the rolled No. 1.6 (φ=1.435) and No. 1.10 (φ=2.667) samples.

References

[1] Gallifa Terricabras A at al 2019 *Assessment of the Mechanical Properties of Ultra-High Purity Niobium after Cold Work and Heat Treatment with the HL-LHC Crab Cavities as Benchmark*, Proc SRF2019. 19th International Conference on RF Superconductivity, Dresden, Germany, 27. Jun. 2019. p 7

[2] Neugebauer at al 2002 *J. Mater. Process. Tech.* **125–126** 856–862

[3] Li X and Al-Samman T 2011 *Mater. Des.* **32** 4385–4393

[4] Wronski S at al 2013 *Mater. Charact.* **77** 116–126

[5] Huh M Y at al 2011 *Mat. Sci. Eng. A* **A315** 35–46

[6] Tang W at al 2015 *J. Mater. Process. Tech.* **215** 320–326

[7] Davim J P 2015 *Modern Manufacturing Engineering* (Switzerland: Springer)

[8] Fukui S and Kudo H *Rep. Inst.Sci. Tech. Univ. Tokyo* **4** 33

[9] Tucker G E G 1961 *Acta Metall.* **9** 275–286

[10] Engler O at al 2011 *J. Mater. Process. Tech.* **211** 1278–1284

[11] Kocks U Fat al 1998 *Texture and Anisotropy, Preferred Orientations in Polycrystals and their Effect on Materials Properties* (UK: Cambridge University Press)

[12] Engler O and Randle V 2010 *Introduction to Texture Analysis, macrotexture, Microtexture, and Orientation Mapping* Second Edition (USA: CRC Press)

IOP Conf. Series: Materials Science and Engineering **903** (2020) 012020 doi:10.1088/1757-899X/903/1/012020

Enhancing the Firework Algorithm ecosystem for the reconstruction of the HTC function

Z Fried[1], Felde I[2] and Szénási S[2]

[1,2]John von Neumann Faculty of Informatics Óbuda University Budapest, Hungary

E-mail: [1]fried.zoltan@stud.uni-obuda.hu

Abstract. The heat transfer coefficients reconstruction could be solved using several ways and configurations. The Inverse Heat Transfer Problems (IHCP) collects all kind of algorithms which try to solve the Heat Transfer Problems. These algorithms try to calculate the heat transfer coefficients inversely. There is some solution for solving this numerical method, but the bio-inspired solutions are better here even though they have some limitations because of this is a kind of genetic one. Using the Fireworks Algorithm (FWA), it is possible to introduce new methods, and modify the original FWA controlling system and parameters in an easy to adapt to this problem. Extensions are suggested with the major limitation of the genetic methods which solved by the proposed one, and with the faster computation of the new modified method.

1. Introduction

The heat transfer phenomena have been described by the principles of the thermal boundary conditions. These are taken place during the heat treatment process. The Heat Transfer Coefficient (HTC) function has been in time by the heat exchange between the examined surface and the surrounding medium. To solve the one-dimensional IHCP in the workpiece, we are using the time-varying HTC only, and the numerical solutions could approximate it. Subtracting an ambient temperature over the boundary temperature as a dimensionless temperature is a possible solution for a system which shifts the function procedure without any integral transforms. Determination of HTC will become a typical Inverse Heat Conduction Problem if using the boundary conditions [1–5]. The IHCP systems are using the thermal collected samples to predict the HTC functions of the measured data of the rod by simulation at the location of the rod. The IHCP solutions usually are using various optimization methods, where the objective function has been defined as the minimum of the differences between the recorded and calculated temperature data [6, 7]. The genetic algorithms are using successfully for solving many types of IHCP cases for example but most of the cases, the genetic algorithms can approximate one or more looked for results but are not able to calculate the expected result [8–10]. For this reason, hybridized solutions need to use to get the expected result. The classification of the result depends on the expectation. The Fireworks Algorithm (FWA) became popular in the last 10+ years due to its power versatility and flexibility. It makes a perfect balance between speed, convergence, cooperation and diversity. Many articles have been published to analyse the FWA technique in inverse heat conduction problems. They have been shown that FWA could reduce the time of processing, reach an expected result by an easy algorithms adaptation process for the task opposite the similar classical methods. In this article an IHCP analysis of the time-dependent and one-dimensional HTC(t) has been calculated [11–15]. The cooling curves were measured in the centerline of the cylinder, and it has been applied to the algorithms to get the inverse heat transfer computation. The objective function has been defined as the quadratic residual between the measurements and the calculated temperatures, and it should be minimized at all. The optimization technique has been implemented in CPU and into GPU architecture for parallelized computation for better performance.

2. The inverse heat transfer problem

The temperatures of a cylindrical workpiece surface have been measured during the heat transfer process at several p points inside the boundary. The variations of the temperature of the thermal boundary conditions determine the solution of the IHCP [6, 7, 12]. Calling the T_k^m as the measured and T_k^c as the calculated temperature at the measure points the objective function can be followed in Eq. (1). The inverse heat conduction problem solution is received by minimizing the objective function.

$$S = \sum_{k=1}^{p} (T_k^m - T_k^c)^2 \rightarrow min \tag{1}$$

According to Eq. (1), the IHCP analysis has been transformed into a solvable numerical optimization problem.

3. Thermal field calculation

The axisymmetric heat conduction model has been considered due to cylindrical workpiece to estimate the temperature distribution. The heat capacity, density and thermal conductivity were assumed as a function of the temperature respectively $C_p(T)$, $\rho(T)$ and $k(T)$. The Eq. (2) describes the mathematical formulation of the nonlinear transient heat conduction problem.

$$\frac{\partial}{\partial r}(k\frac{\partial T}{\partial r}) + \frac{k}{r}\frac{\partial T}{\partial r} + \frac{\partial}{\partial z}(k\frac{\partial T}{\partial z}) + q_v = \rho \cdot C_p\frac{\partial T}{\partial t} \tag{2}$$

using the following boundary conditions Eq. (3) and Eq. (4):

$$T(r, z, t = 0) = T_0 \tag{3}$$

and

$$k\frac{\partial T}{\partial r}|_{r=R} + k\frac{\partial T}{\partial z}|_{0 \leq z \leq Z} = HTC(z, t)[T_q - T(r, z, t)] \tag{4}$$

where

- R is the radius of the cylinder
- L is the length of the cylinder
- HTC(t, z) is the heat transfer coefficient
- k is the thermal conductivity
- C_p is the heat capacity
- r and z are local coordinates
- t is the time
- ρ is the density of the object
- T_0 is the initial temperature of the work-piece
- T_q is the temperature of the cooling medium

and the z is constant if the calculation is one-dimensional.

In this article the material phase transformation does not calculated during the measuring process and the latent heat is not generated ($q_v = 0$).

4. The Fireworks algorithm

The FWA model [16], by definition, contains a firework and several calculating points (sparks) around the firework. The process of the solution is based on the movement rules of the sparks and firework. In our implementation, the firework is a spark also. Each spark is moving around in the searching space, which is implemented in $\mathbb{R}^n$ space. The [15] article is used many times in this article. This article introduces a new statistical solution to determinate a direction from the actual position of the spark, which brings a higher possibility to decrease the objective function.

The number of dimensions represents the number of points of the HTC function. Each spark represents a point in the searching space like $X_i = (x_{i1}, x_{i2}, \ldots, x_{iN})$ which can be transferred to an HTC function as well, and every spark could be a potential solution of the IHCP problem. The possible lowest value of the objective function is determining the solution. Every spark has a limited history for storing earlier calculated objective functions and the selected HTC function as the $\vec{x}_k$.

The next position of every spark is defined by

$$\vec{d} = \vec{A} \odot \vec{r} \tag{5}$$

and

$$\vec{X}_{new} = \vec{X} + \vec{d} \tag{6}$$

where $\vec{r}$ unit vector is the direction of the movement and it depends on the type of the spar, $\vec{A}$ is the amplitude vector (each dimension of the displacement vector has an amplitude value), $\vec{X}$ is the point of the spark in the searching space (vector), $\vec{d}$ is the displacement vector (by Hadamard product).

Every type of sparks have:

- a moving base which is defining the movement base position
- collect own history as storing all own information which was calculated the latest 2 interactions earlier. This information will only be used by a new type of spark in the 4.10 section and the amplitude correction coefficient in the 4.2 section
- the number of sparks in an iteration
- two moving profiles. These moving profiles determine the moving rules of the spark:
  - init state moving profile: this profile is only working in the init state and put the spark to the init position.
  - next state moving profile: this profile is using after the init state all-time

In this article, we have modified the ecosystem of standard FWA model in some cases. For example, in the standard algorithm, there are two types of spark [17, 18]. One of them is the 'explosion spark', and the other is the 'gaussian spark'. We are using both of them, but the movement rules have been changed, and we are introducing three new types of sparks as 'firework spark', 'best spark' and 'blind spark'.

4.1. Firewok population count

In the standard FWA model, there are some firework populations running in the searching space parallelly. This is necessary to avoid the calculation to stuck in local minima and rerun the calculation many times. Moreover, the original definition describes that these populations can contact to each other. In our implementation, the count of the population has been calculated by Eq. (7).

$$ln(F_w^{count}) = log_{10} n \tag{7}$$

where F^{count}_w is the firework population count, the n is the dimension.

4.2. Amplitude coefficient

This coefficient is calculated for controlling the amplitude value due to direction of fitness changes between the actual and last iteration. The controlling mechanism is based on the Fibonacci numbers. The Fibonacci numbers have been defined as

$$F_0 = 0, F_1 = 1, F_q = F_{q-1} + F_{q-2} \tag{8}$$

Define the o counter in $\mathbb{N}$ which has minimum −20 and a maximum 20 value. This o counter determines the q index from Fibonacci numbers Eq. (8) as q = |o| where o counter steps by +1 or −1 forward in each iteration depends on the type of spark as follows

- firework spark: the o counter is starting from 1 to 20. Every iteration the o is stepping by +1 towards and if the o > 20 the o starting from 1 again.
- random spark: as same as the firework spark.
- gaussian spark: there is not any usage due to 4.7 section
- best spark: for selecting the next o number, the calculation is based on the historical statistical calculation [15]. The o is increasing by 1 if the fitness decreasing. If fitness does not change over 3 steps the o counter is increasing by 1, otherwise the o counter does not change. If the o > 20 or o < 1 the o set to 1.
- blind spark: in this case the o counter can be a negative number too. The o is increasing by 1 if the fitness decreasing and the o is decreasing by 1 if the fitness increasing. If fitness does not change over 3 steps the o counter is increasing by 1, otherwise the o counter does not change. If the o > 20 the o = 20 or if the o < −20 the o = −20.

The o is starting from 1 in the initialisation state of the algorithm. Based on the above definitions the A_{coef} is calculating by the Eq. (9) in every iteration

$$A_{coef} = \begin{cases} F_o + F_{o-1} & \text{if } o > 0 \\ \frac{1}{F_{|o|} + F_{|o|-1}} & \text{if } o < 0 \\ skip & \text{if } o = 0 \end{cases} \tag{9}$$

4.3. Amplitude maximum and minimum value calculation

The $\tilde{A}_{max}$ has been defined by Eq. (10) as the maximum value of the A_{max} value Eq. (11). In our terminology the A_{max} value is changing every iteration and there is a maximum value which can be depend on the size of searching space and the number of firework populations running in the same time at the same iteration. It is a theoretically maximum value of the A amplitude.

$$\tilde{A}_{max} = \frac{X_{max} - X_{min}}{2 \cdot F_w^{count}} \tag{10}$$

$$A_{max} = \frac{\tilde{A}_{max}}{A_{coef}} \cdot \frac{log_2(F)}{log_2(F_{max})} \tag{11}$$

where the F is the actual fitness value of the spark and the F_{max} is the maximum fitness value of the same spark in the whole calculation iteration. The A_{min} always has been calculated by A_{max} in Eq. (12)

$$A_{min} = A_{max} \cdot \frac{3 - \sqrt{5}}{2} \tag{12}$$

IOP Conf. Series: Materials Science and Engineering **903** (2020) 012020 doi:10.1088/1757-899X/903/1/012020

4.4. Amplitude boundary calculation

The Amplitude is always ≥ 0. If the new position run out of the searching space in the k dimension, the amplitude in that dimension will be reduced with an uniform distributed random value between 0 and 1, where 1 is exactly the new position when it is reaching the boundary as the following condition

$$A_{new}^k = \begin{cases} A_u^k & \text{if } X_{max}^k > X_{new}^k \\ A_l^k & \text{if } X_{new}^k < X_{min}^k \\ A_{new}^k & \text{if } X_{min}^k \leqslant X_{new}^k \leqslant X_{max}^k \end{cases} \quad (13)$$

where A^k_u is

$$A_u^k = rand(0,1) \cdot (\vec{X}_{max}^k - \vec{X}_{new}^k) \quad (14)$$

A^k_l is

$$A_l^k = rand(0,1) \cdot (\vec{X}_{new}^k - \vec{X}_{min}^k) \quad (15)$$

and k = 1...n dimension index and the X^k_{min} and the X^k_{max} is the searching space boundary minimum and maximum values in each dimension.

4.5. Iteration to live (ITL)

This value was developed to controlling iteration count between the borning and dying event. The main theory is if a spark can decrease the fitness value many times they need to set a high ITL credit and if a spark cannot decrease the fitness value they need to set a low ITL credit. For this reason, if a spark has been born, they have 10 ITL credits. The rule of the ITL credit has been described by the Eq. (16) If the ITL credit reaches the zero, the spark will have died. This calculation is based on the [15] article. The ITL credit is changing as follows

$$ITL_{i+1} = \begin{cases} ITL_i + 1 & \text{if } f(x_{i+1}) - f(x_i) < 0 \\ ITL_i - 1 & \text{if } f(x_{i+1}) - f(x_i) > 0 \\ ITL_i - \frac{1}{2} & \text{if } f(x_{i+1}) - f(x_i) = 0 \end{cases} \quad (16)$$

where i is the iteration number and the $f(x_i)$ is the fitness value of a spark in the ith iteration.

4.6. Explosion spark

This type of spark position calculation is based on a uniformly distributed random number which can be −1 or 1 as a direction in each dimension and a uniformly distributed random number as the distance from firework spark position between the A^k_{min} and A^k_{max} in each dimension also. The A^k_{max} and A^k_{min} are based on the firework spark (see the 4.9 section) amplitude calculation.

4.7. Gaussian spark

This type of spark position calculation is based on a normal distributed random number as a distance from the actual position where the mean is 1, and stddev is 0.293. The direction is the position of the best spark or a uniformly distributed random selected another spark in the population.

4.8. Best spark

The best position means the lowest value of the objective function through the all process. This position has been held by this spark. This spark in every iteration is moving towards the position of spark, which has the best value of the objective function. Although the A_{max} is not used in this type of spark, the value needs to calculate because the firework spark uses it.

4.9. Firework spark

Based on the [15] article the firework spark should be placed to close the Best spark in a random point. The A_{max} calculation is modified as the 4.3 section describes as follow

$$A_{max} = A_{min}^{Best} \cdot \frac{3-\sqrt{5}}{2} \tag{17}$$

where the A^{Best}_{min} is the A_{min} of the best spark value.

4.10. Blind spark

This kind of spark has a complex displacement strategy and rules which are based on the [15] article. This is a kind of hill climbing attempt to minimize the objective function. The spark name comes from the way that the blind people walk from the top of the hill to valley. This spark uses its walk history to calculate the next direction of the way. This spark has been started from the parent $\vec{x}_p|_{k+1}$ position, and the spark inherits its parent spark history too. Any spark in the new position there are two possibilities between the last $\vec{x}_{k-1}$ and the new $\vec{x}_k$ position. If

$$f_p(x_k) < f_p(x_{k-1}) \tag{18}$$

then the spark stay in the new position and the o counter for A_{coef} is increasing by 1. If

$$f_p(x_k) \geqslant f_p(x_{k-1}) \tag{19}$$

then the spark go back to the previous position in the next iteration and the o counter is decreasing by 1. The spark go forward if fitness speed [15] is decreasing but the o counter is decreasing also by 1. The $f_{p(x_k)}$ is the parent spark $\vec{x}_k$ position of kth iteration of objective function.

In the first step of the blind spark uses its parent spark history. The first step direction and distance are as same as its parent spark moved towards in the latest iteration by

$$\vec{x}_b|_{k+1} = \vec{x}_p^H|_k - \vec{x}_p^H|_{k-1} \tag{20}$$

where $\vec{x}_b|_{k+1}$ is the blind spark position in the $(k+1)^{th}$ iteration, $\vec{x}_p^H|_{k-1}$ is its parent spark previous $(k-1)^{th}$ position. To calculate the next position of blind spark in every further iteration are based on the following Eqs. which are using the Eq. (18) and Eq. (19):

$$\vec{x}_{k+1} = \vec{x}_k - \vec{x}_{k-1} \text{ and } -\vec{x}_{k+1} = \vec{x}_{k-1} - \vec{x}_k \tag{21}$$

$$\vec{x}_b|_{k+1} = \begin{cases} \vec{r}(\vec{x}_{k+1} \pm (\frac{45}{2})^\circ) & \text{if } f(x_{k-1}) > f(x_k) \\ -\vec{x}_{k+1} & \text{if } f(x_{k-1}) \leq f(x_k) \end{cases} \tag{22}$$

where the $\vec{r}$ is an unified random vector around $\vec{x}_{k+1}$ in $\pm\frac{45}{2}^\circ$.

To avoid creating lots of or irrelevant new blind sparks it will not born if:

- the possible parent spark is on top of the hill while the best spark is in the valley

$$\frac{f(x_p^k)}{f(x_{\text{best}}^k)} > 10 \tag{23}$$

and

$$\frac{f(x_p^k)}{T_N} - \frac{f(x_{\text{best}}^k)}{T_N} > 1 \tag{24}$$

- there is not free slot to store the new blind spark. The maximum numbers of blind sparks are defined as $2 \cdot (N_e^{max} + N_g^{max})$ where N_e^{max} is the maximum number of explosion sparks and N_g^{max} is the maximum number of gaussian sparks.

where the T_N is the number of calculated temperature points on the workpiece, $f(x_k^p)$ is the possible parent spark fitness value at stage k and the $f(x_k^{best})$ is the best spark fitness value at stage k.

5. Comparing the proposed FWA and the original FWA

To compare the above proposed FWA and the standard FWA, we define a theoretical HTC* function with 50 points in 60 sec in the time axes (Fig. 1). The workpiece radius was 6.25cm. The new type of FWA used 50 explosion sparks and 20 gaussian sparks. The maximum number of blind sparks were 140. The original FWA used 150 explosion sparks and 60 gaussian sparks. The iterations were stopped at 1000 in case of proposed FWA and 10000 in case of standard FWA. The simulation was run ten times. The shape of the calculated HTC has been checked at the end of each run, and that ran was selected where the shape was the best. Therefore Fig. 2 shows the calculated HTCs. Fig. 3 shows the convergence speeds. This figure shows how the fitness value decreasing depends on the iteration. The '+' line shows the proposed FWA, the 'o' line shows the standard one. The Fig. 4 shows a zoom into the Fig. 3 near the (0,0) corner. The '-' line is the theoretical HTC*, the '+' line shows calculated HTC based on the proposed FWA and the 'o' line shows calculated HTC based on the standard FWA. We think that the standard FWA could produce as same as HTC shape like the proposed one, but it takes much more time and computation effort.

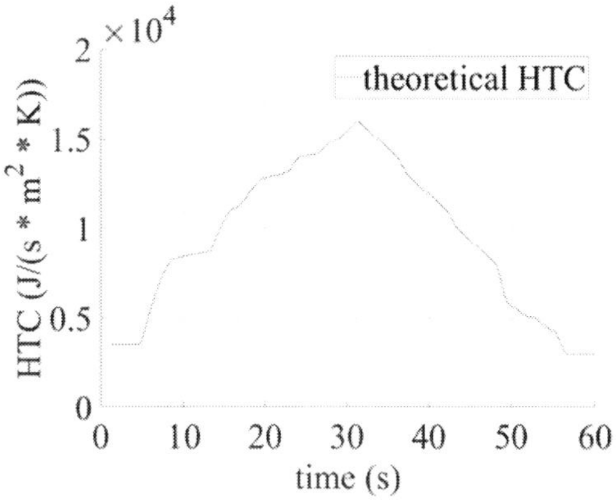

Figure 1. Theoretical HTC function

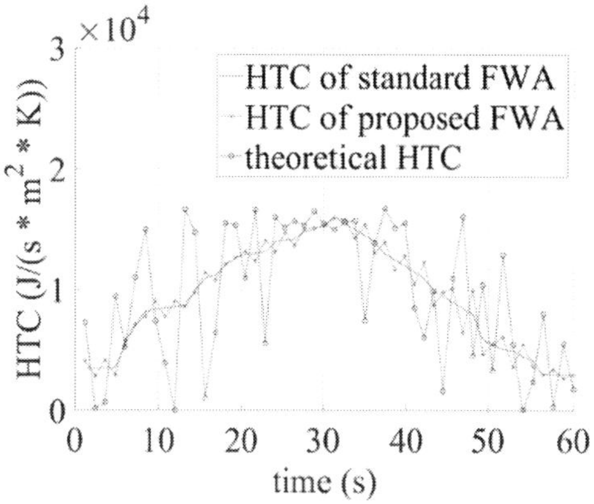

Figure 2. The theoretical HTC and Calculated HTC functions by standard and proposed FWA

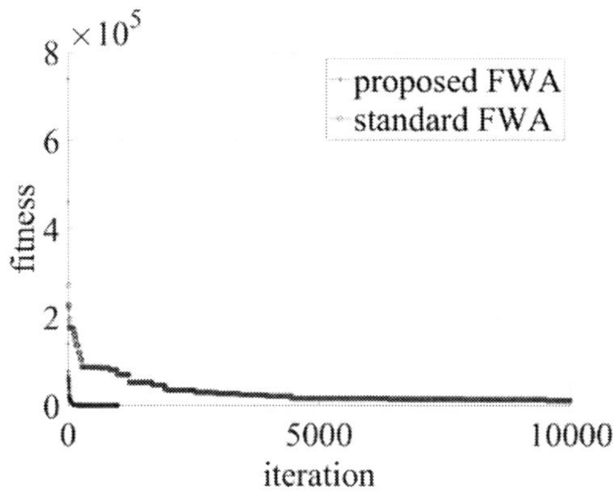

Figure 3. Fitness values by standard and proposed FWA

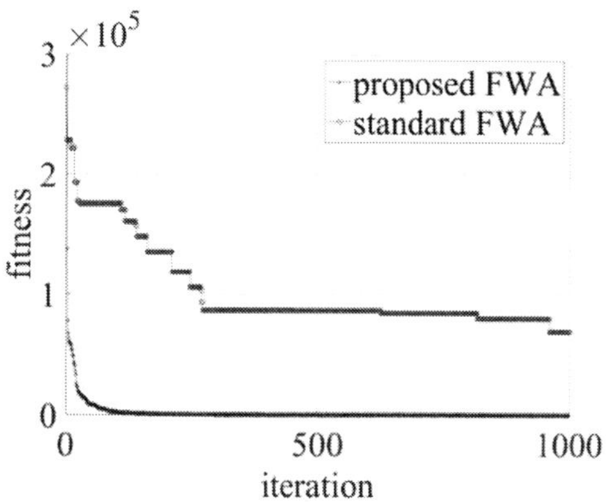

Figure 4. Fitness values by standard and proposed FWA (zoomed version)

6. Conclusions

The Fireworks Algorithm for estimating the Heat Transfer Coefficient was prepared. After analysis of the Fireworks Algorithm has turned out, there is possible to integrate a new type of sparks or modify the exists one. The measurement of the prepared FWA shows that the inverse thermal analysis by a numerical process speed could be fast, and the accuracy could be precise. Comparing the prepared and the original FWA shows that the proposed FWA is more accurate than the original. We need to continue the investigation to make the algorithm more precise and faster.

Acknowledgements

We acknowledge the financial support of this work by the Hungarian State and the European Union under the EFOP-3.6.1-16-2016-00010 project and Hungarian-Japanese bilateral Scientific and Technological (TÉT 16-1-2016-0190) project. The authors of this work also acknowledge the financial support of the Hungarian State for supported by UNKP-19-3-I-OE-35 new national Excellence program of the ministry for innovation and technology.

References

[1] Heinrich B and Hofmann B B J V B B S C C R j 1987 *ZAMM - Journal of Applied Mathematics and Mechanics / Zeitschrift für Angewandte Mathematik und Mechanik* **67** 212–213
[2] Ivanov V V and Salomatov V V 1965 **9** 63–64 ISSN 0022-0841
[3] Ivanov V V and Salomatov V V 1966 **11** 151–152 ISSN 0022-0841
[4] Kozlov V N 1970 **18** 100–104 ISSN 0022-0841
[5] Postol'nik Y S 1970 **18** 233–238 ISSN 0022-0841
[6] Alifanov O 1994 *Inverse heat transfer problems* International series in heat and mass transfer (Springer-Verlag) ISBN 9783540536796
[7] Özisik M N and Orlande H R B 2000 *Inverse Heat Transfer: Fundamentals and Applications* (Taylor &Francis)
[8] Ozisik M *2007 International Journal of Heat and Mass Transfer* **50** 1706–1714 ISSN 0017-9310
[9] Kim K W and Baek S W 2004 *Numerical Heat Transfer, Part A: Applications* **46** 367–381
[10] Felde I and Szénási S *2016 Int. J. Microstructure and Materials Properties* **11** 288–300
[11] Chiba R 2018 *Mathematical Problems in Engineering* **2018** 1–11
[12] Chantasiriwan S 1999 **42** 4275–4285 ISSN 0017-9310
[13] Tian N, Zhu L and Lai C H 2014 **8** 233–248 ISSN 1748-3018
[14] Fried Z, Szénási S and Felde I 2018 (IEEE) pp 000099–000104 ISBN 978-1-7281-1118-6 ISSN 2380-8586
[15] Fried Z, Szénási S and Felde I 2019 (IEEE) pp 000305–000308 ISBN 978-1-7281-0685-4
[16] Tan Y, Yu C, Zheng S and Ding K 2013 **4** 39–70 ISSN 1947-9263
[17] Zheng S, Janecek A, Tan Y, 2013 Enhanced Fireworks Algorithm (IEEE) pp 002069–002077 ISBN 978-1-4799-0454-9
[18] Tan Y 2015 Fireworks algorithm ISBN 978-3-662-46352-9

The effect of microstructure on the dynamic mechanical properties of carbon fiber and carbon nanotube reinforced multi-scale composites with a polyamide 6 matrix

R Petrény[1] and L Mészáros[1, 2]

[1]Department of Polymer Engineering, Faculty of Mechanical Engineering, Budapest University of Technology and Economics, Budapest, Hungary
[2]MTA–BME Research Group for Composite Science and Technology, Budapest, Hungary

E-mail: meszaros@pt.bme.hu

Abstract. We investigated the microstructure and the related dynamic mechanical properties of carbon fiber and carbon nanotube (CNT) reinforced, polyamide 6 (PA6) matrix multi-scale composites. Differential scanning calorimetry (DSC) showed that in composites reinforced with only carbon nanotubes, the aggregated nanotubes have a moderate crystal nucleating effect. However, in the hybrids, the carbon fibers and the uniformly dispersed CNTs have a notable crystal nucleating effect. It means that a crystalline interphase is formed around both the carbon fibers and the carbon nanotubes. As the interphase has higher modulus than the bulk matrix, it has an additional reinforcing effect and transfers the load better to the microfibers. We confirmed this theory by dynamic mechanical analysis.

1. Introduction

The development of multi-scale composites with a thermoplastic polymer matrix is one of the most popular research areas today. However, the widespread use of these materials in practice is limited by the fact that their strength and their dynamic mechanical properties often differ from what is expected from conventional fiber-reinforced composites. This is because nanostructured composites have a different microstructure than conventional fiber-reinforced composites, which may result in different reinforcing mechanisms [1, 2].

Currently, the designing of nanocomposites is based on conventional composite mechanics, which ignores these differences in material structure, thus providing an inaccurate estimate of the macroscopic mechanical properties of nanocomposites. Since the structural and mechanical properties of nanocomposites are poorly understood, designing them accurately is a great challenge and so their use in industrial practice remains modest [3, 4].

In composites, as a result of interfacial interactions between the matrix and reinforcement interface, a three-dimensional interphase layer is formed, whose properties are different from those of the matrix and reinforcing material itself. As nanoparticles have a large surface/volume ratio, the matrix-nanoparticle interface is large, thus the volume ratio of this interphase layer is significant in nano- and hybrid composites, and moreover, it influences the mechanical behavior of the composite. Therefore, accurate designing of such composites requires precise knowledge of the characteristics of this interphase [5-8].

There is a more and more widely accepted approach that, in semi-crystalline thermoplastic polymers, three different phases can be distinguished: a crystalline fraction, a rigid amorphous fraction and mobile amorphous fractions that surround them. Several publications reported that nano-sized and micro-sized reinforcing materials can have a crystal nucleating effect in a thermoplastic polymer matrix, so the interphase around the reinforcement should be mostly crystalline. However, due to the interactions between the matrix and the reinforcement, the movement of the polymer chains is hampered around the reinforcement, which hinders crystallization and results in a rigid amorphous interphase. Moreover, around the crystallites, a similar rigid amorphous interphase is formed. As the rigid amorphous phase needs more energy for segmental movement, the heat capacity step at the glass transition is lower if the rigid amorphous fraction is larger. This means that the volume fractions of the rigid and the mobile amorphous parts are calculable. The structure of the above-mentioned three phases can be analyzed by differential scanning calorimetry, as they have different thermal characteristics [8, 9].

Recently many publications deal with the microstructure of the interphases; however, the effect of the interphase structure on the mechanical properties is less investigated. If a crystalline or a rigid amorphous interphase surrounds the reinforcing material, the mobility of the polymer molecules is lower in either of them, compared to the bulk matrix. This immobilized interphase layer helps transfer the stress from the matrix to the reinforcement. Macroscopically, it results in a higher modulus and lower dynamic dampening, which can be detected by dynamic mechanical analysis [9-14].

The aim of this research was to analyze the microstructure, and particularly, the interphases of carbon fiber and carbon nanotube reinforced multi-scale composites and investigate the effect of the microstructure on mechanical properties.

2. Materials and methods

We used polyamide 6 (PA6) homopolymer (SHULAMID 6 MV 13 F) manufactured by A. Schulman GmbH as matrix material for the composites. It had a density of 1.13 g/cm^3 (at room temperature) and a melt flow index of 14.7 g/10 min (2.16 kg, 230 °C). The multi-wall carbon nanotubes used as reinforcing material were produced by Nanocyl s.a. The NANOCYL NC7000 carbon nanotube (CNT) had an average diameter of 9.5 nm and a length of 1.5 μm. As fiber reinforcement, we used Panex 35 Chopped Pellet 95 type carbon fiber (CF) manufactured by Zoltek Zrt. The fibers had a mean initial length of 6 mm and a diameter of 8.3 μm. The density of the fibers was 1.81 g/cm^3.

The PA6 granules were dried for 4 hours at 80 °C, after which the nanotubes and carbon fibers were mixed first by mechanical mixing in a closed vessel and then with a Labtech LTE 26-44 twin screw extruder. The zone temperatures were 230 °C, 230 °C, 235 °C, 235 °C, 240 °C and 240 °C towards the die, and screw rotational speed was 25 rpm. After granulation and re-drying for 4 hours at 80 °C, MSZ EN ISO 527-2-1A type specimens were produced with an Arburg Allrounder 370 S 700-290 injection molding machine.

We made PA6 matrix nanocomposites with 0, 0.25, 0.5, 0.75 and 1 wt% carbon nanotube content (PA6, PA6+0.25CNT, PA6+0.5CNT, PA6+0.75CNT, PA6+1CNT), and composites reinforced with 30 wt% carbon fiber, and 0, 0.25, 0.5, 0.75 and 1 wt% carbon nanotube content (PA6+30CF, PA6+30CF +0.25CNT, PA6+30CF +0.5CNT, PA6+30CF +0.75CNT, PA6+30CF +1CNT).

For morphological tests, we used a TA Instruments Q2000 type DSC in modulated DSC (MDSC) mode. The test temperature range was 0-250 °C with a heating and cooling rate of 5 °C/min, the amplitude of modulation was 1 °C, and its period was 60 s. The mass of the test samples ranged from 10 to 15 mg. The volume ratio of rigid amorphous phase (ϕ_{RAF}) in the matrix is [9]:

$$\phi_{RAF} = 1 - \phi_{MAF} - \phi_C,$$

where ϕ_{MAF} is the volume ratio of the mobile amorphous fraction, and ϕ_{C} is the crystalline fraction. For this, the volume ratio of the mobile amorphous fraction is calculated as follows [9]:

$$\phi_{\mathrm{MAF}} = \frac{\Delta C_{\mathrm{p}}}{\Delta C_{\mathrm{p,0}}},$$

where ΔC_{p} is the specific heat increment of the semi-crystalline polymer or composite, $\Delta C_{\mathrm{p,0}}$ is the specific heat increment of the fully amorphous polymer in the glass transition temperature range. In the case of polyamide 6, $\Delta C_{p,0} = 0{,}475 \frac{J}{g°C}$, according to the literature. Crystallinity can be calculated with the following equation [9]:

$$\phi_C = \frac{\Delta H}{\Delta H_0 (1 - \varphi_R)}$$

The related ΔH is the crystalline melting enthalpy as determined by the DSC test, $\Delta H_0 = 240\ J/g$ is the melting enthalpy of 100% crystalline polyamide 6, and φ_R is the weight ratio of the reinforcing material [9].

The fracture surfaces of the tensile tested specimens were investigated with a Jeol 6380 LA-type scanning electron microscope (SEM) after they were sputtered with gold.

Dynamic mechanical properties were measured with a TA Instruments Q800 dynamic mechanical analyzer (DMA) in 3-point bending mode. The temperature range was 0-210 °C with a heating speed of 3 °C/min. The amplitude of deformation excitation was 0.02% and the frequency was 1 Hz.

3. Results and discussion

3.1. The microstructure of the nano-and hybrid composites

In the composites reinforced with only carbon nanotubes, the crystalline fraction slightly increased with increasing carbon nanotube content, which indicates the crystal nucleating effect of the nanoparticles (Figure 1.a). The crystal nucleating effect of carbon nanotubes in the PA6 matrix was investigated in one of our former articles [10]. However, the rigid amorphous fraction showed a slight decrease with increasing carbon nanotube content. This can be attributed to the fact that the rigid amorphous fraction can be found in both the interphase around the crystalline fraction and the interphase around the nanoparticles, so the volume fraction of the rigid amorphous phase is significantly influenced by the size of the surface of the nanotubes and the crystallites . If the carbon nanotubes form large aggregates as it can be seen on the SEM images (Figure 1.b), the crystallites formed around them are larger, which results in a smaller surface/volume ratio of the crystallites and a smaller rigid amorphous fraction despite the higher crystallinity.

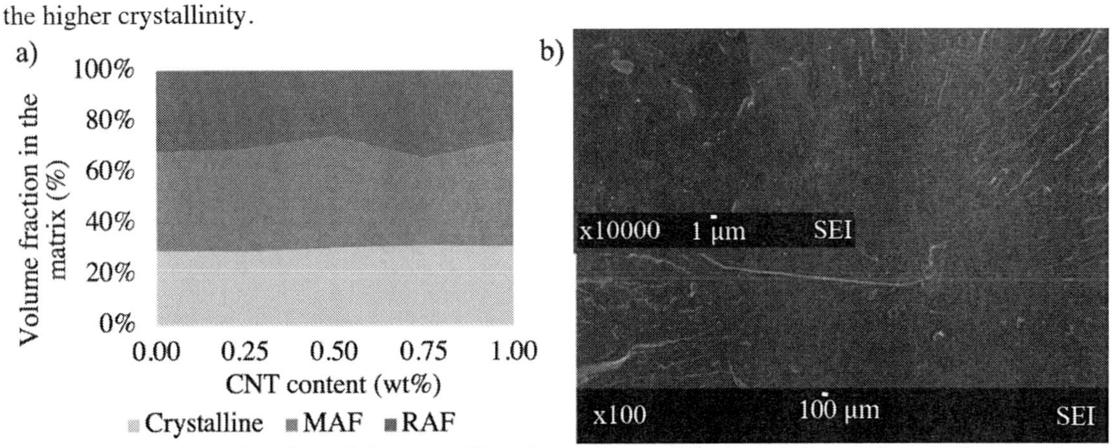

Figure 1. a) Volume fraction of the crystalline, the rigid amorphous (RAF) and the mobile amorphous (MAF) fractions in the matrix of the nanocomposites
b) SEM images of a CNT aggregate on the fracture surface of the PA+0.5CNT nanocomposite

To prove this theory, we investigated the cooling curves to understand the crystallization process in the nanocomposites. The crystallization peaks became wider as nanotube content increased, which was related to the size distribution of the crystallites, i.e. with higher nanotube content, the scatter of crystallite size increased (Figure 2). The crystallization peaks also shifted to higher temperatures with increasing nanotube content, which shows that the nanotubes had a crystal nucleation effect [11].

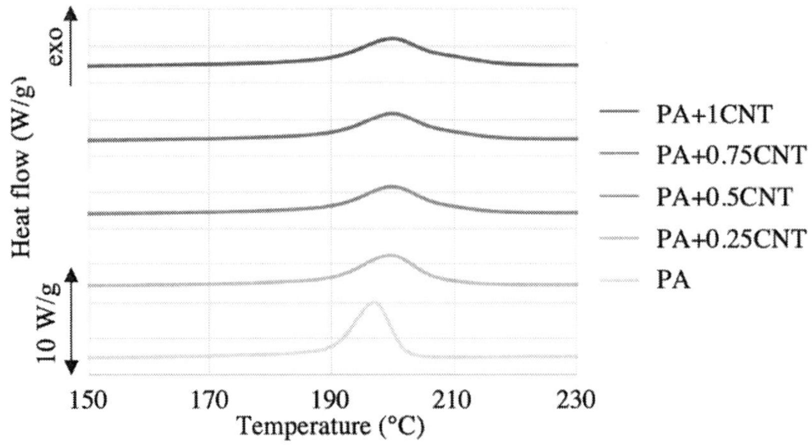

Figure 2. Cooling DSC curves of the nanocomposites

Table 1 shows that the width at half maximum of the peaks became wider with increasing carbon nanotube content, which clearly indicates the wider distribution of crystallite size [11]. The explanation for this is that during processing, the carbon nanotubes remained mainly in aggregated form, and the size distribution of these aggregates was wide [10]. As the starting points of crystal nucleation were these aggregates, the size distribution of the crystallites should also be wide. The crystallization peaks shifted to higher temperatures with increasing nanotube content, which proves the crystal nucleating effect of the nanotubes. The higher crystallization peak temperature means that in the nanocomposites, the crystallites became larger, which resulted in a smaller surface/volume ratio of the crystallites. This explains the smaller volume fraction of the surrounding rigid amorphous interphase.

Table 1. Peak temperature (T_p) and full width at half maximum of the peaks (T_{fwhm}) of the cooling curves of the nanocomposite

Sample	T_{fwhm} (°C)	T_p (°C)
$PA6$	6.9	197.3
$PA6 + 0.25CNT$	10.7	199.9
$PA6 + 0.5CNT$	11.7	200.2
$PA6 + 0.75CNT$	12.0	200.2
$PA6 + 1CNT$	12.6	200.3

In the carbon fiber reinforced composites, the crystalline fraction slightly, and the rigid amorphous fraction significantly increased compared to the pure PA6; adding carbon nanotubes to fiber reinforced composites increased the volume fraction of the above-mentioned fractions even further (Figure 3.a). The increase in the rigid amorphous fraction can be related to both the increase of the crystalline fraction and the decrease of crystallite size. This effect may be explained by that the more uniformly dispersed carbon nanotubes influences the crystallization process differently, than the aggregated ones (Figure 3.b).

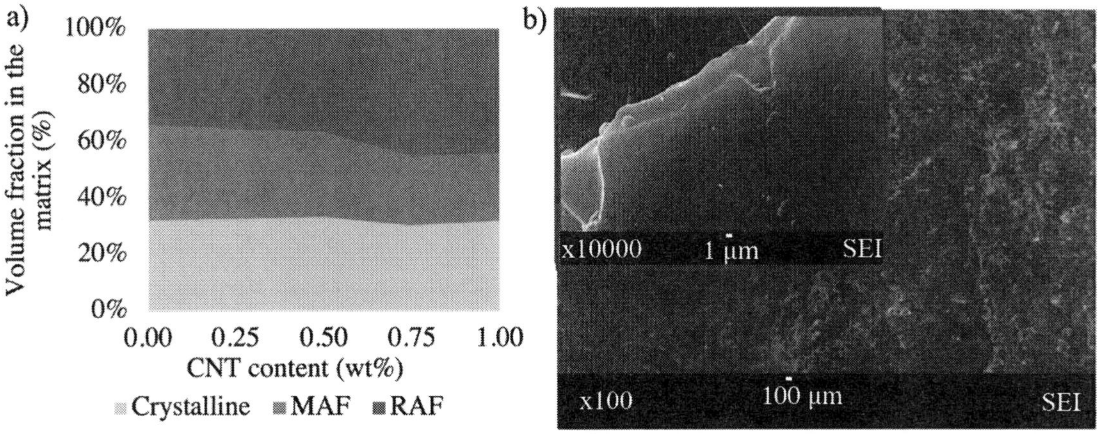

Figure 3. a) Volume fraction of the crystalline, the rigid amorphous (RAF) and the mobile amorphous (MAF) fractions in the matrix of the carbon fiber reinforced composites
b) SEM images of the fracture surface of the PA+30CF+0.5CNT nanocomposite. No aggregates visible, uniformly dispersed nanotubes (white spots) on the higher magnification image

When carbon fiber is added to polyamide 6, a moderate crystal nucleating effect can be observed, which did not significantly change the size and the size distribution of the crystallites compared to the pure PA6. However, crystallinity slightly increased. On the cooling curves of the hybrid composites, an additional peak can be observed around 210 °C, which became more significant with higher nanotube content (Figure 4).

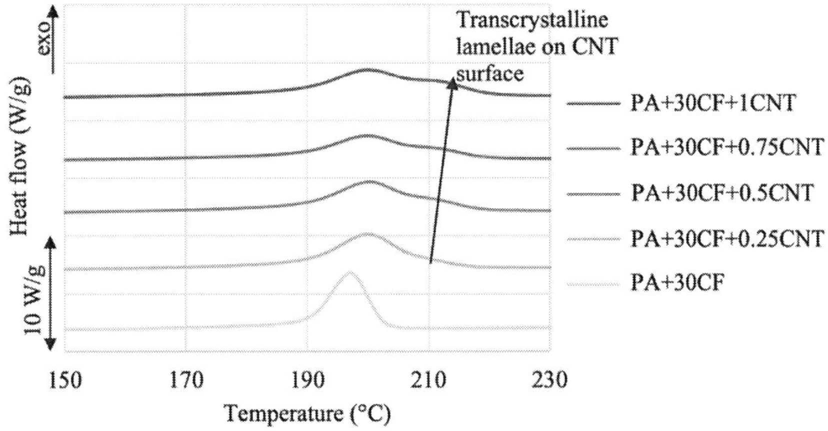

Figure 4. Cooling DSC curves of the composites containing carbon fiber

Brosse et al. [15] reported that the second crystallization peak in carbon nanotube reinforced PA6 indicates trans-crystalline lamellae that grow perpendicularly on the surface of the well dispersed nanotubes. In our hybrid composites, this second peak also appeared and became more significant and shifted towards higher temperatures with the increasing nanotube content which indicates that a significant amount of well dispersed nanotubes are present in the composite (Table 2). If these trans-crystalline lamellae grew on the surface of the dispersed nanotubes, crystallite size should be smaller than in the nanocomposites, where the crystals grew on large nanotube aggregates. Therefore, besides higher crystallinity, the smaller average crystallite size may be the explanation for the increase of the rigid amorphous phase in the hybrid composites.

IOP Conf. Series: Materials Science and Engineering **903** (2020) 012021 doi:10.1088/1757-899X/903/1/012021

Table 2. Peak temperatures (T_{p1} and T_{p2}) of crystallization of the composites containing carbon fiber

Sample	T_{p1} (°C)	T_{p2} (°C)
$PA6 + 30CF$	197.2	-
$PA6 + 30CF + 0.25CNT$	200.1	209.12
$PA6 + 30CF + 0.5CNT$	199.9	204.17
$PA6 + 30CF + 0.75CNT$	200.4	206.13
$PA6 + 30CF + 1CNT$	200.0	207.83

The DSC tests showed that in the nanocomposites, the carbon nanotubes had a moderate crystal nucleating effect. This means that the carbon nanotube aggregates were surrounded by a crystalline interphase, which was surrounded by a rigid amorphous interphase (Figure 5.a). Similarly, in the hybrid composites, both the carbon fibers and the well-dispersed carbon nanotubes were surrounded by a two-layer interphase; the inner layer was the crystalline and the outer one was the rigid amorphous phase (Figure 5.b).

a) b)

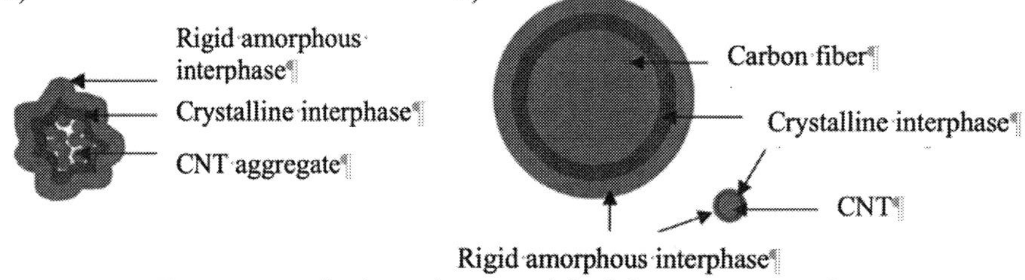

Figure 5. a) the interphase model of the nanocomposites, b) the interphase model of the hybrid composites

3.2. Dynamic mechanical behavior of the nano- and hybrid composites

The effect of the two-layer interphase on the viscoelastic behavior was perceptible on the increase of the storage modulus (Figure 6.a), which indicates that the nanocomposites behaved more elastic compared to the pure PA6. This has an essential role in the case of the long-term mechanical properties, especially the creep resistance of the materials. The loss factor ($tan\delta$) curves (Figure 6.b) showed the changes in the mechanical properties more spectacularly, in the case of the nanocomposites, the first peak was typically lower than that of the pure PA6. This means that the magnitude of the drop in the storage modulus at the glass transition temperature was lower by adding nanotubes to the composite. The position of this first peak is shifted towards higher temperatures in the nanocomposites, which means that the drop in the storage modulus is not only lower compared to the pure PA6, but also the temperature is higher, where it happens. As this first peak shows the glass transition process, where the micro-Brown movements of the amorphous parts of the polymer chains start, it can be stated that this improvement of the mechanical properties is not only due to the nanotubes themselves, but the change in the microstructure of the matrix. This may be explained by that even the aggregated carbon nanotubes effectively blocked the movement of the polymer chain segments, which hampered the glass transition process. However, in the carbon nanotube filled PA6, there was no clear correlation between the position of the glass transition peak and nanotube content, due to the stochastic size and size distribution of the nanotube aggregates.

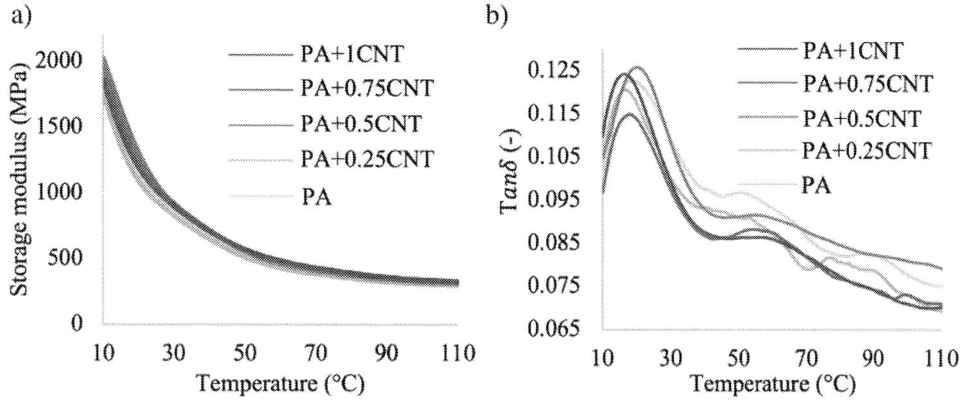

Figure 6. a) Storage modulus and b) loss factor curves of the nanocomposites

Besides the previously mentioned glass transition peak, an additional peak can be observed between 40 and 60 °C, which indicates the drop in the storage modulus caused by the segment motions in the rigid amorphous fractions, i.e. the outer layer of the interphase. By adding carbon nanotubes to the PA6, this peak slightly decreased, which indicates that the drop in the storage modulus slightly decreased by the presence of this rigid amorphous interphase layer. The size and the position of this peak does not show any correlation with the nanotube content, which is in good agreement with the relationship between the nanotube content and the volume fraction of the rigid amorphous interphase.

In the case of the composites containing carbon fibers, the composite reinforced with only carbon fibers had the lowest storage modulus and the highest loss factor (Figure 7). Adding carbon nanotubes to the fiber reinforced composites further increased the storage modulus with increasing nanotube content, which indicates that the nanotubes dispersed well in the composite and connected strongly to the matrix.

The peaks of the loss factor curves decreased with increasing nanotube content and shifted to higher temperatures, which means that the storage modulus drop at the glass transition temperature was significantly lower compared to either the PA6 reinforced with only carbon fibers or the pure PA6 (Figure 7.b). This is because the mobile amorphous fractions represented a smaller volume fraction and were also less mobile in the nanocomposites than in the pure PA6. The size of the second peak, which depends the modulus drop caused by the motion of the segments in the rigid amorphous phase, also increased with higher nanotube content. This is caused by the higher volume fraction of the rigid amorphous parts. As a result, the previously mentioned two-layer interphase that formed an immobilized polymer layer around the reinforcing materials had an additional reinforcing effect in addition to the carbon fibers and the carbon nanotubes.

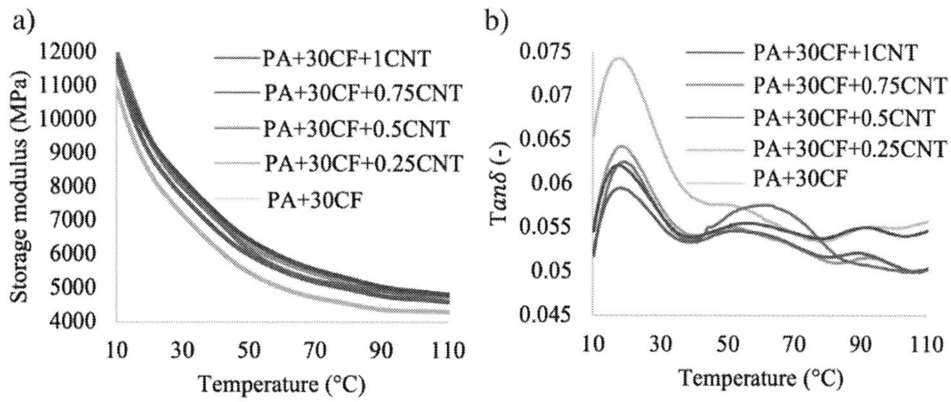

Figure 7 a) storage modulus and b) loss factor curves of the composites containing carbon fiber

4. Conclusion

In this project, we investigated the microstructure and dynamic mechanical properties of carbon fiber and carbon nanotube reinforced nano- and hybrid composites. The DSC tests showed that in the nanocomposites, a two-layer interphase was formed around the CNT bundles, where the inner layer is a crystalline phase and the outer layer is a rigid amorphous phase. However, the volume fraction of the crystalline and the rigid amorphous fractions decreased with the presence of carbon nanotubes compared to the pure PA6. In the hybrid composites, both the carbon fibers and the carbon nanotubes had a crystal nucleating effect; similarly to the nanocomposites, a two-layer interphase was formed with a trans-crystalline inner, and a rigid amorphous outer layer. In the hybrid composites, the nano- and the micro-sized reinforcement had a synergistic effect as the volume fraction of both the crystalline and the rigid amorphous fractions increased compared to the pure or the PA6 reinforced with only carbon fibers.

The dynamic mechanical tests showed that in the nanocomposites, the lower volume fraction of the crystalline interphase and the rigid amorphous interphase around the carbon nanotubes slightly decreased the loss factor peaks and the storage modulus. In the hybrid composites, where the carbon fibers helped disperse the carbon nanotubes uniformly, the storage modulus drop at the glass transition temperature was significantly lower compared to that of either the PA6 reinforced with only carbon fibers or the pure PA6. The two-layer interphase around the reinforcing materials not only helped transfer the stress from the matrix to the reinforcement but acted as an additional reinforcing component as well.

Acknowledgements

This research was realized in the framework of TÁMOP 4.2.4.A/1-11-1-2012-0001 "National Excellence Program – Elaborating and operating an inland student and researcher personal support system". The project was subsidized by the European Union and co-financed by the European Social Fund. This work was also supported by the Higher Education Excellence Program of the Ministry of Human Capacities in the framework of the Nanotechnology research area of the Budapest University of Technology and Economics (BME FIKP-NAT).

References

[1] Eitan A et al 2006 *Compos. Sci. Technol.* **66** 1162-73
[2] Gojny F H et al 2005 *Compos. Sci. Technol.* **65** 2300-13
[3] Odegard G M et al 2005 *Polymer.* **46** 553-62
[4] Ansari R and Hassanzadeh-Aghdam M K 2016 *Int. J. Mech. Sci.* **115-116** 45-55
[5] Bershtein et al 2019 *Express Polym. Lett.* **13** 656-672
[6] Patel R K et al 2008 *Mech. Res. Comm.* **35** 115-25
[7] Prusty R K et al 2017 *Adv. Colloid Interfac.* **240** 77-106
[8] Lipatov Y S 1977 *Physical Chemistry.* (Berlin, Heidelberg: Springer Berlin Heidelberg)
[9] Chen H and Cebe P 2007 *J. Therm. Anal. Calorim.* **89** 417-25
[10] Szakács J et al (2019) *J. Therm. Anal. Calorim.* **137** 43–53
[11] Wang X D et al (2008) *Polym. J.* **40** 450-454
[12] Cruz et al 2019 *Express Polym. Lett.* **9** 825-834
[13] Hajba S and Tábi T 2019 *Period. Polytech. Mech. Eng.* **63** 270-277
[14] Logakis et al 2009 *J. Polym. Sci. Pol. Phys.* **47** 764-774
[15] Brosse et al 2008 *Polymer.* **49** 4680-4686

IOP Conf. Series: Materials Science and Engineering **903** (2020) 012022 doi:10.1088/1757-899X/903/1/012022

Modelling of segregation during injection moulding

Dorottya Réka Fris, Ferenc Szabó

Department of Polymer Engineering, Faculty of Mechanical Engineering, Budapest University of Technology and Economics, H-1111 Budapest, Műegyetem rkp. 3., Hungary

E-mail: szabof@pt.bme.hu

Abstract. The goal of this study is to establish a proper simulation model of segregation, a phenomenon which causes inhomogeneity in filled polymer products. We did preliminary research, in which we calcinated injection moulded specimens of 80 mm x 80 mm x 2 mm, with 75 μm, 125 μm and 250 μm glass beads with 10 m%, 25 m% and 40 m% filler rates. According to preliminary simulations, the boundary conditions were set and an EDEM model was created.

1. Introduction

Since the second half of the 20th-century, polymer products have been conquering the industry. Injection moulding is one of the most productive and most widely used polymer processing technologies, which allows engineers to design virtually any kind of product while guaranteeing fast and cost-effective manufacturing. The versatility of injection moulding is also used to create complex manufacturing procedures, like when combined with 3D printing technologies [1]. Even though primarily injection moulding is a technology of polymer manufacturing, processing metal [2] and foam [3] are also possible, demonstrating the endless potential of the subject.

Homogeneity is essential to produce products which meet the needs of the market. Unfortunately, homogeneity might be difficult to accomplish, especially when fillers are added to the product. In this case, more fillers will be located at the end of the path of the melt flow and closer to the surface of the mould, in the so-called frozen layer. This phenomenon is called segregation. Perhaps the most common example to this is using masterbatch colouring, for in this case, the segregation of the colouring pellets causes unevenly coloured lines across the surface [4]. Depending on the type of the filler, the effect of segregation also differs. According to some investigation, the phenomenon at the end of the flow path is nearly neglectable in case of using reinforcing fibres, while observing spherical fillers like beads, the segregation across the flow path is significant [5]. Further research confirmed, that regarding beads, different sizes induce different degree of segregation, and in this case shrinkage is also a problem that must be considered [6]. Another important issue when it comes to fibre fillers is their orientation, which is a widely researched subject [7].

In the design phase, FEM simulation software is used by engineers to predict and prevent errors that may occur during the actual process of injection moulding. This way, more complex tasks can be optimised, which saves precious time [8]. However, simulating segregation is difficult. Despite this fact, quite few research was made in this field. Even though continuum modelling is often used in polymer manufacturing simulation software [9, 10], this technique cannot model segregation. A discrete element method can analyse the behaviour of the particles one by one, therefore it may be better for this purpose than the finite element method [11, 12].

According to the literature review, segregation occurring during injection moulding can significantly alter the distribution of the fillers, resulting in degrading properties. Since DEM can be used to simulate the segregation of distinct particles, the aim of this study was to investigate, whether FEM-DEM simulations are capable of modelling segregation during injection moulding.

2. Materials, machinery and methods

We injection moulded glass bead-filled specimens. The matrix material was TIPPLEN H 145 F polypropylene, as it has remarkable processability. Based on the literature, we used three different sizes (<75 µm, 70–125 µm, 150–250 µm) of glass beads. First we sieved the Cerablast G120, G100 and G50 glass beads to sort them according to size. Then we compounded the beads into the polypropylene with an LTE 25-30/C simple screw extruder, creating 6 mm long pellets, and injection moulded the 80 mm x 80 mm x 2 mm specimens using an Arburg Allrounder 370 S 700 290 machine. It is a servo-hydraulic machine with a maximum clamping force of 70 tons. Its maximum injection pressure is above 2000 bars and it is equipped with a 30 mm diameter position-regulated screw, so the settings during procedure remain unchanged and the reproducibility of the products is assured. We determined the glass bead content of the specimen by calcination, using a Denkal 6B furnace.

3. Experiments

We performed several tests on the specimens, then compared the data to the segregation simulation results.

3.1. Preliminary experiments

Before creating the specimens, we ran simulations in Moldflow 2018, in order to determine the ideal injection speed. We found that injection speed affects the frozen layer, which is a determining factor in segregation. The simulations showed that injection speeds between 20 cm^3/s and 80 cm^3/s result in a significantly thinner frozen layer. After the simulations, we injection moulded the specimens (Figure 1).

3.2. Segregation

In order to observe segregation across the flow path, we divided a specimen into four sections (Figure 1), in which we examined the distribution of the glass beads by calcination. According to the ISO 3451-1 standard, calcination lasted for four hours, at 600 °C. After all the polypropylene matrix material burnt, the mass of the leftover material was weighed. In order to see how the size and quantity of the beads affect segregation, we performed the tests using 75 µm, 125 µm and 250 µm fillers with 10 m%, 25 m% and 40 m% glass bead contents each. There was no significant change in density at 10 m% from the mass differences in the case of smaller beads. However, the density of 250 µm fillers increased by 2–5% at the end of the flow path. We found, that the larger the filler, the more inhomogeneous the system is, since at the end of the flow path the glass beads tend to pile up. With 40% content of the 125 µm glass beads, the maximum difference in distribution was 8%, while with 250 µm this number was over 12%. This phenomenon may be a result of the drag force, since the larger surface is affected by greater forces in the melt flow. Moreover, the larger beads are also more likely to collide with each other.

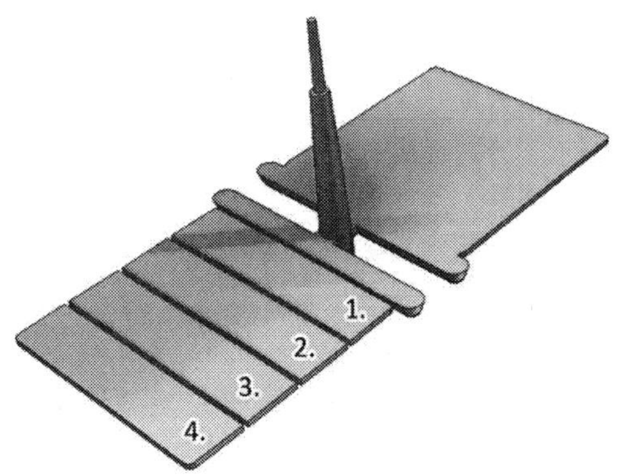

Figure 1. The specimen and the four different sections for the examination of segregation

3.3. Preliminary simulations

We ran preliminary simulations in Autodesk CFD 2018 and Moldflow 2018 to set the boundary conditions of the behaviour of the polypropylene and glass beads. These boundary conditions are necessary for the discrete element method.

3.3.1. The torque limit required to tear out a glass bead. The torque limit above which a glass bead is torn out is a crucial boundary condition. As the mechanical model suggests, this torque is related to how much the filler is embedded into the frozen layer. Whether the bead stays still or is dragged out by the flow depends on the balance of the torques applied to point "A" (Figure 2). The moments balancing each other are derived from the flowing material and the vacuum between the frozen layer and the embedded bead. There is vacuum in the system due to the fact, that when a bead is torn out, there is a small gap between the frozen layer and the bead, which the viscous melt cannot flow into immediately. In the simulations, we examined different rates of embedment—half, a quarter and an eighth of the bead. Preliminary studies suggest that examining a higher rate of embedment than half is unnecessary, since in that case the bead stays in place as it is self-locking by shape. The cross-section of the flow was set to be five times the diameter of a bead and a p=0 Pa boundary condition was set at the end of the flow path (Figure 3).

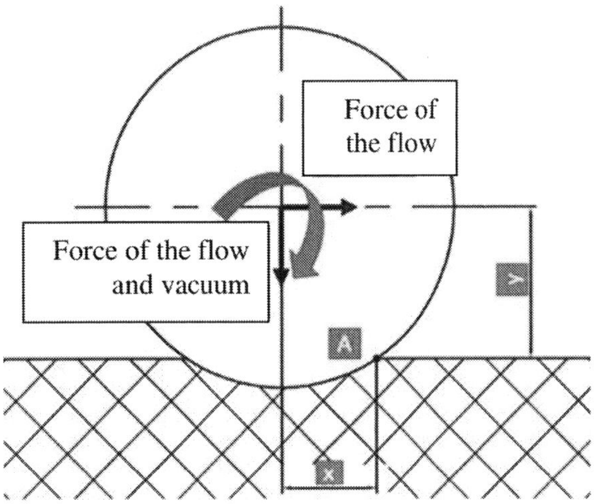

Figure 2. Mechanical model of the embedment of a glass bead

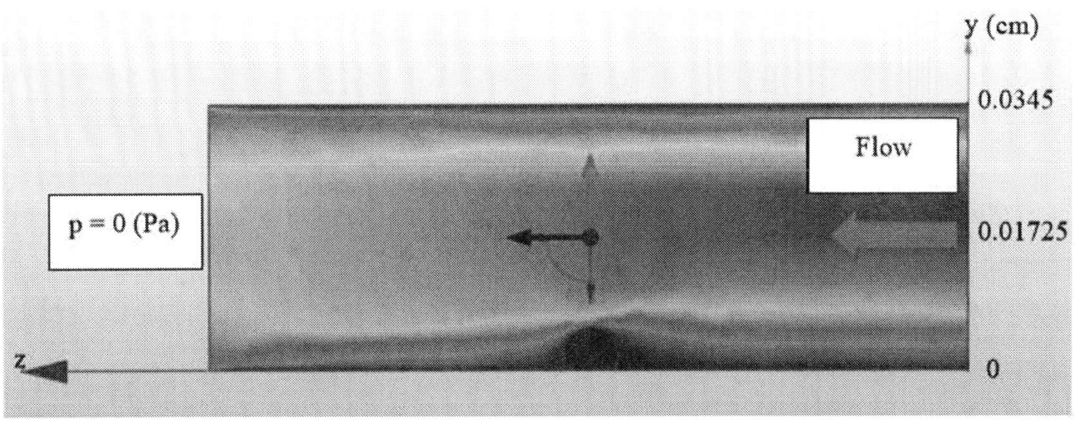

Figure 3. CFD simulation determining the force of the flow

We found that with an injection speed of 5 cm^3/s, the 250 µm glass beads are torn out when the contact surface of the filler and the melt is below 70%, while in the case 75 µm beads this is 75. Higher injection speeds decrease this limit, but it never went below 50%. Thus, it is probably true that if a bead is embedded at least 50%, it will not be torn out by the flow. However, a bead may stay immobile even if far less than half of it is embedded, especially in the case of smaller beads and slower injection speeds.

3.3.2. Determining injection moulding parameters for EDEM 2018.2 simulations. The two essential parameters are filling time and the velocity profile in the cavities. For the three sizes of the fillers, we used three different, reduced-sized models in the simulations. The Moldflow simulations showed that the velocity of the melt in the frozen layer is 0 and it increases towards the middle of the section, where it reaches its maximum. Along the flow path, melt velocity decreases, and reaches its minimum at the end of the flow path (Figure 4). Naturally, these phenomena cannot be neglected in EDEM simulations.

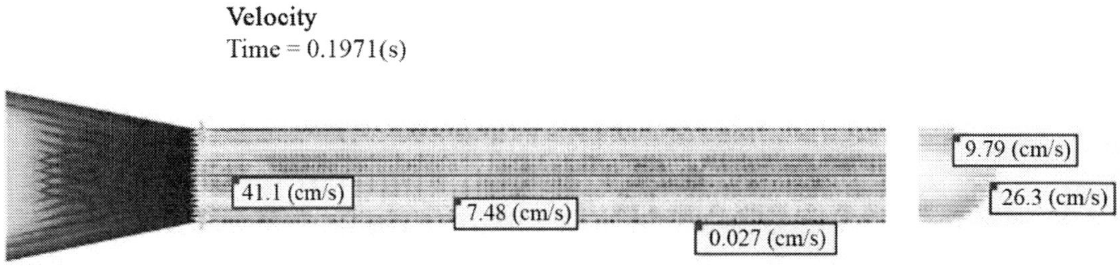

Figure 4. The velocity profile in Moldflow 2018

3.4. Modelling segregation in EDEM 2018.2 software

Modelling segregation is a rather complex problem, for several factors must be considered. For this reason, EDEM simulation would seem to be ideal to perform this task. Software packages using a discrete element method do not consider complex, multi-element systems as continuums. This enables the observation of the behaviour of the glass beads one by one. Considering the velocity and the viscosity results of the Moldflow simulations, drag force model was used. The particles' collision with each other or the wall was handled as a linear spring contact model. The data of the system gathered from the previous preliminary simulations were handled with the use of C++. Establishing such a complex simulation did pose some difficulties. In the end, we ran 2D simulations based on the velocity profile according to the values in the middle line, disregarding the effect the mould's surface might have on the velocity towards it (Figure 4). These profiles were chosen according to the right time step from Moldflow. Since we did the procedure manually, we considered only 10 time steps in a simulation. The established model consisted of four parts (Figure 5):

- the factory, which emits the glass beads modelling particles according to filling time, m%, and the size of the beads,
- the shear flow, which describes the characteristics before the melt front, according to the velocity profile,
- the overall dimension of a 2 mm x 10 mm x 80 mm geometry
- and the fountain flow which describes the behaviour of the melt front itself.

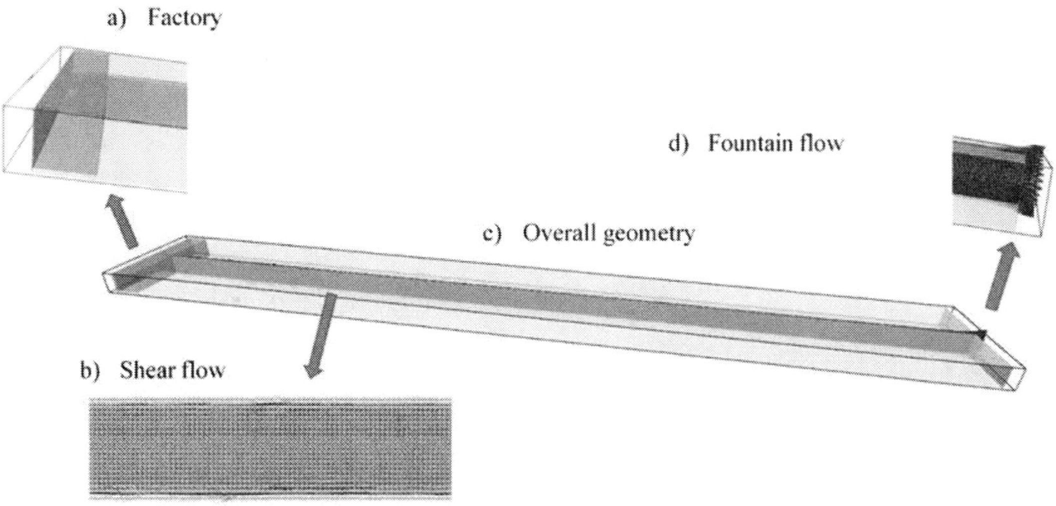

Figure 5. The parts of the EDEM model

4. Results

In this section we present the results of the simulations.

4.1. The results of the simulations

The first obvious phenomenon of the model is the increase of density of the beads across the x axis in the frozen layer (Figure 6). This is due to the characteristics of discrete element modelling, for the generated ten velocity profiles of the ten time steps mean ten differently characterised fountain flows. Since there are velocity components in the y axis, the beads located in the melt front gain velocity in the y direction, getting stuck in the frozen layer, and having their speed reduced to 0. Regarding the frozen

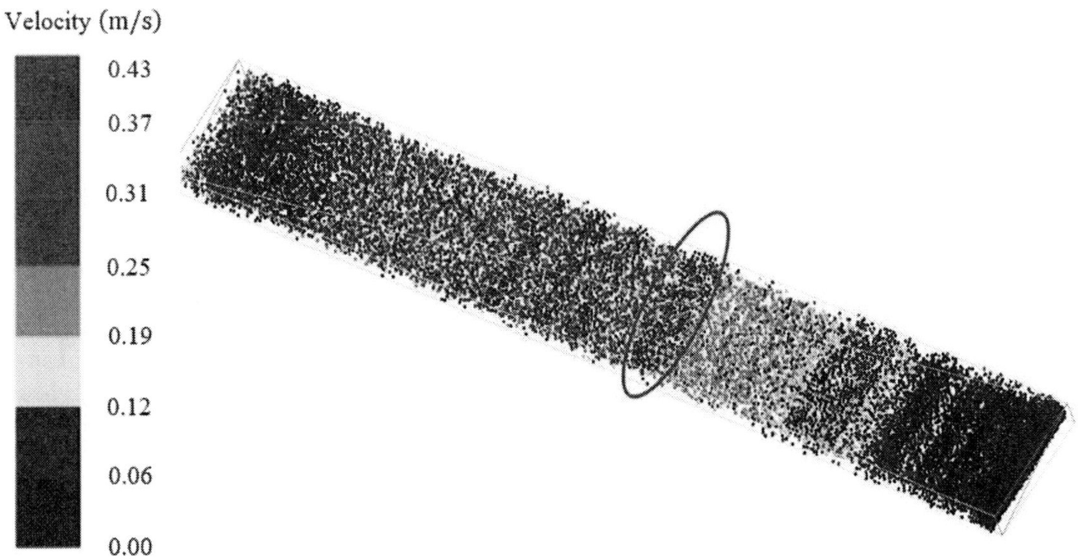

Figure 6. The increase of density across the x axis in the frozen layer

layer, the results are satisfying, as the effect of injection speed can be observed quite well, according to the expected tendency (Figure 7). However, a great error of the model is that it does not consider the frozen layer as a physical wall, rather a layer whose speed is 0. This results in the anomaly that a glass bead can be embedded into it, regardless of the fact that it is already in the solid state (Figure 8). Another rather unrealistic anomaly is that at the beginning of the flow path at the top and the bottom of the geometry, beads are stuck in the frozen layer (Figure 9), even though it is not even formed at that time. This increase in density is in contrast to results in the literature and what happens in practice, as the beginning of the path is the very last place where the melt solidifies. The anomaly is somewhat connected to the other error demonstrated in Figure 8 and probably occurs due to the fact that several beads are located at the beginning of the flow path, and the possibility of collision is very high, therefore more beads are pushed into and embedded in the frozen layer.

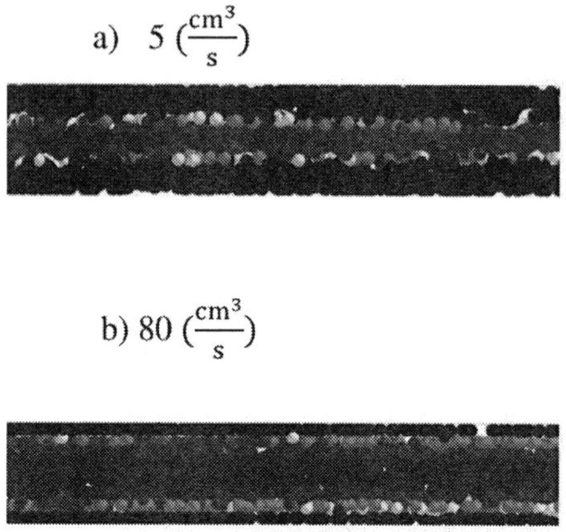

Figure 7. The effect of injection speed on the frozen layer

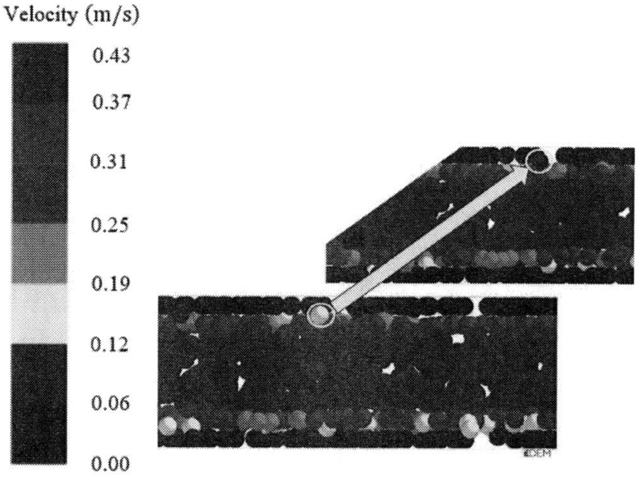

Figure 8. The anomaly of the filler particles in the model

Figure 9. Increase in density at the beginning of the flow path

4.2. The distribution of the glass beads

The numbers of beads according to the EDEM simulation were placed into histograms with 2 mm-wide domains. These histograms were converted to a diagram (Figure 10). The anomaly of increasing density at the beginning of the flow path can be observed in general, but more significantly in the case of the injection speed of 5 cm^3/s. Even though the distribution is quite uniform, at the end of the flow path, there is an expected peak of density. In this respect, the EDEM model worked well, but it is not precise enough, therefore it needs further research and development.

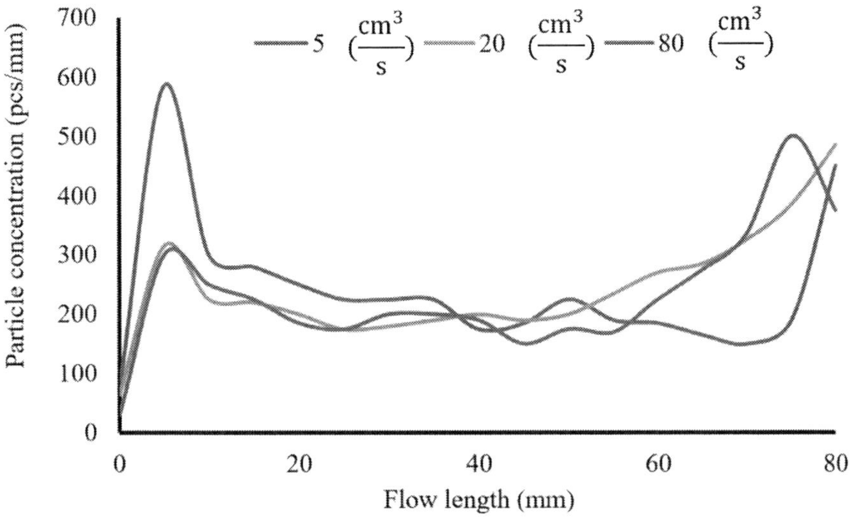

Figure 10. The distribution of glass beads as a function of injection speed (250 μm and 25 m%)

5. Conclusion

In this section we summarise the results of the research and provide some ideas for further development.

5.1. Summary of the results

Comparing the results of the EDEM simulations to the preliminary experiments, it can be stated that the model is promising regarding the frozen layer, as it shows similar tendencies in the thickness of the frozen layer when the injection speed is altered. The simulations could also indicate that the glass beads pile up at the end of the flow path. However, several rather unrealistic anomalies occurred, which need to be corrected. Considering these, using EDEM simulation is promising and the model could be able to simulate segregation after improvement.

5.2. Ideas for further development

As already mentioned before, the boundary condition of 50% embedment may be too much on the safe side, thus finding a more accurate limit might be beneficial. The increase of density across the x axis could be reduced with the use of more than ten time steps, since the error of numerical instability can be eliminated this way. To solve the problem of the 0-speed frozen layer, mobile walls could be used in the model, which would function as a physical obstacle to the filler particles. However, these walls must enable the beads to be freely embedded and perhaps get dragged out by the flow itself or get pushed out by other beads. These walls should also be able to simulate the thickening of the frozen layer. Another way to reduce error is to use a dynamic factory, which adapts to the continuously altering frozen layer. Even though only mono-disperse systems were examined, it might also be beneficial to observe poly-disperse systems, for they are more common in practice.

Acknowledgements

The project is funded by the (NKFIH) National Research, Development and Innovation Office (NVKP_16-1-2016-0038, 2017-2.3.7-TÉT-IN-2017-00049, 2018-1.3.1-VKE-2018-00001) and by the National Research, Development and Innovation Fund (TUDFO/51757/2019-ITM, Thematic Excellence Program). The authors would like to thank Arburg Hungaria Ltd. for the injection moulding machine, Lenzkes GmbH for the clamping tool system, and Piovan Hungary Kft. and Tool-Temp Hungária Kft. for their support.

References

[1] Boros R, Kannan Rajamani P and Kovács J G 2019 Combination of 3D printing and injection molding: Overmolding and overprinting *Express Polymer Letters* **13** 889-897

[2] Ammosova L, Mönkönnen K and Suvanto M 2020 Precise fabrication of microtextured stainless steel surfaces using metal injection moulding *Precision Engineering* **62** 89-94

[3] Wu H, Zhao G, Wang J, Wang G and Zhang M 2019 A kind of special weld lines with high specific strength and elongation obtained by core-back chemical foam injection molding *Express Polymer Letters* **13** 1041-1056

[4] Zsíros L and Kovács J G 2018 Surface homogeneity of injection molded parts *Periodica Polytechnica Mechanical Engineering* **62** 284-291

[5] Hegler R P and Mennig G 1985 Phase separation effects in processing of glass-bead- and glass-fiber-filled thermoplastics by injection molding *Polymer Engineering and* Science **25** 395–405

[6] Kovács J G 2011 Shrinkage alteration induced by segregation of glass beads in injection molded PA6 Experimental analysis and modeling *Polymer Engineering & Science* **51** 2517–2525

[7] Albrecht K, Baur E, Endres H-J, Gente R, Graupner N, Koch M and Müssig J 2016 Measuring fibre orientation in sisal fibre-reinforced, injection moulded polypropylene – Pros and cons of the experimental methods to validate injection moulding simulation *Composites Part A: Applied Science and Manufacturing* **95** 54–64

[8] Wang M-V, Fatahul A and Vu V-H 2019 The Study of Optimal Molding of a LED Lens with Grey Relational Analysis and Molding Simulation *Periodica Polytechnica Mechanical Engineering* **63** 278-294

[9] Ingber M S, Graham A L, Mondy L A and Fang Z 2009 An improved constitutive model for concentrated suspensions accounting for shear-induced particle migration rate dependence on particle radius *International Journal of Multiphase Flow* **35** 270–276

[10] Nott P R and Brady J F 1994 Pressure-driven flow of suspensions: simulation and theory *Journal of Fluid Mechanics* **275** doi:10.1017/s0022112094002326

[11] Liu G, Yu F, Wang S, Liao P, Zhang W, Han B and Lu H 2017 Investigation of interstitial fluid effect on the hydrodynamics of granular in liquid-solid fluidized beds with CFD-DEM *Powder Technology* **322** 353-368

[12] Yu H, Cheng W, Wu L, Wang H and Xie Y 2017 Mechanisms of dust diffuse pollution under forced-exhaust ventilation in fully-mechanized excavation faces by CFD-DEM *Powder Technology* **317** 31-47

IOP Conf. Series: Materials Science and Engineering **903** (2020) 012023 doi:10.1088/1757-899X/903/1/012023

Characterisation of Inhomogeneous Plastic Deformation of AlMg Sheet Metals During Tensile Tests

Sz Szalai[1], D Harangozó[1], I Czinege[1]

[1]Széchenyi University, 1. Egyetem tér. H-9026 Győr, Hungary.

E-mail: harangozo.dora@sze.hu

Abstract. The Portrevin-Le-Chatelier (PLC) effect was analysed quantitatively in the selected AlMg alloys with Mg-content between 2,8-4,6%. The propagation and characteristics of PLC bands were detected by digital image correlation (DIC) technique and the parameters of the bands were evaluated from strain distributions of tensile test specimens. In parallel, stress-strain curves were analysed evaluating stress serrations. The numerically defined parameters of changes in stress and strain were used for characterising the PLC effect on the behaviour of different sheet metals. General statements of literature were numerically proven for the tested materials and the effect of magnesium content on stress amplitude was analysed in detail.

1. Introduction

It is well known from the literature and experiments that stress oscillations – serrations – can be observed during tensile tests in AlMg alloys. Serrations are defined as sudden stress drops or strain jumps in the stress-strain curves of certain metals. The deformation mechanism in these alloys is known as the Portevin-Le-Chatelier (PLC) effect [1] and it is derived from dynamic strain aging (DSA) which is the repeated action of diffusing-solutes which pin the dislocations. As the stress increases, dislocations break free from this pinning [2-5]. On the macroscopic scale, this phenomenon is observed as strain localization in the form of narrow bands of intense shear [6]. The deformation bands leave undesirable traces on the surface of the sheet product during forming operations, thus restricting the application of AlMg alloys for car outer body panels. In addition, the PLC effect is harmful for formability because the PLC bands can lead to a premature onset of necking and fracture [7].

According to the literature, serrations are usually divided into three types: Type A, B and C [3,8]. The main characteristics of serrations can be defined from tensile test diagrams. The first serration can be observed at ε_c critical strain after the onset of plastic deformation. The amplitude of serrations can be characterised by $\Delta\sigma$. The frequency spectrum of serrations also provides useful information for characterising the PLC effect which can be evaluated from normalized stress amplitude diagrams by fast Fourier transformation (FFT) [9, 10]. The serration parameters are influenced by the tensile strain (ε), the strain rate ($\dot{\varepsilon}$), the testing temperature (T), the grain size of the alloy (d) and the dimensions of the specimen (aspect ratio and thickness). The effects of the parameters are the following:

- The increase of **strain** during tensile tests increases $\Delta\sigma$ stress amplitude, modifies stress drops from types A to B and C. This indicates that DSA is becoming more significant, as the strain increases [8].

- As function of the **strain rate,** the critical strain first shows a decreasing tendency, then reaches a minimum value and finally increases. The serration amplitude decreases as the strain rate increases [8, 11, 12]
- As the **temperature** increases – similarly to the strain effect – the stress amplitude increases and the stress drop type changes from A to B and later to C [8].
- Increasing the **grain size** decreases $\Delta\sigma$ in case of AlMg alloys.
- The stress-drop size was found to slightly decrease with increasing the **specimen thickness**. The cross-section form factor of the specimen also influences $\Delta\sigma$, from a circle to a rectangular shape, $\Delta\sigma$ slightly increases [10].

It is obvious that serrated yielding in AlMg alloys is influenced by the Mg-content and heat treatment as presented in [13]. Comparing AA5052 and AA5182 alloys higher stress drops occurred on the stress-strain curve of AA5182 alloy and serrations showed a higher rate than in case of AA5052 alloy.

Further analysis of the PLC effect is based on techniques which show the local strains of specimens during tensile tests. Namely, these are the digital image correlation (DIC) methods [4, 6, 10, 15, 16, 17] and laser speckle techniques [14]. Using the graphic representation of the strain field or its numeric results, all characteristics of PLC bands can be detected and evaluated such as the band strain, the width and the velocity of the band front edge.

However, it should be mentioned that the exact definition of PLC band parameters is missing from the literature, although many publications deal with them. For example, the PLC band strain (ε_{PLC}) is an extent of the incremental strain according to [4] or the strain difference between the internal and external band strain [6]. Some authors suppose that the base strain is zero which can be found in strain curves in publications [14, 16, 17]. In [17] the width of the PLC band is defined from the strain-length function at half-height and ε_{PLC} is given as the maximum strain of the peak.

2. Experimental procedure

The experiments were focusing primarily on AlMg3 sheets but some tests were carried out on AlMg4,5 alloys for comparison. Table 1 shows the thickness and chemical composition of the sheets.

Table 1. Basic data of the tested sheets

	Thickness (mm)	Mg (wt.%)	Fe (wt.%)	Si (wt.%)	Mn (wt.%)
A1	1,5	2,82	0,23	0,15	0,34
A2	2,5	3,30	0,11	0,05	0,33
A3	2,5	3,41	0,23	0,17	0,16
A4	2,5	3,52	0,27	0,13	0,16
A5	1,25	4,57	0,28	0,13	0,22
A6	1,5	4,69	0,11	0,09	0,25

A2 and A3 sheets were tested both in annealed (O) and cold rolled (H22) condition. The received sheets were cold rolled and subsequent annealing was carried out at 320 °C/2 hours. The sheets were tested in rolling (RD), transversal (TD) and diagonal (DD) directions.

2.1. Evaluation of the DIC measurements

The time series of the strain fields of the tensile test specimens were recorded by GOM ARAMIS [18] hardware-software system. Graphic representation of the strain distribution was available in colored images with scale bars. The GOM Inspect software offered digital output of measured strain values along straight lines parallel with or perpendicular to the tensile axis. Evaluation of strain fields was carried out from 2 to 15% global strain by 1...2% strain increments. From one strain matrix the following parameters were calculated: the average strain ε_{AVG} (which is equal to the global strain measured by extensometer); the standard deviation of strains ε_{STD}; the maximum and minimum values of the strain matrix ε_{max} and ε_{min}; the strain distribution (histogram) characterized graphically and approached by a four-parameter Gaussian function.

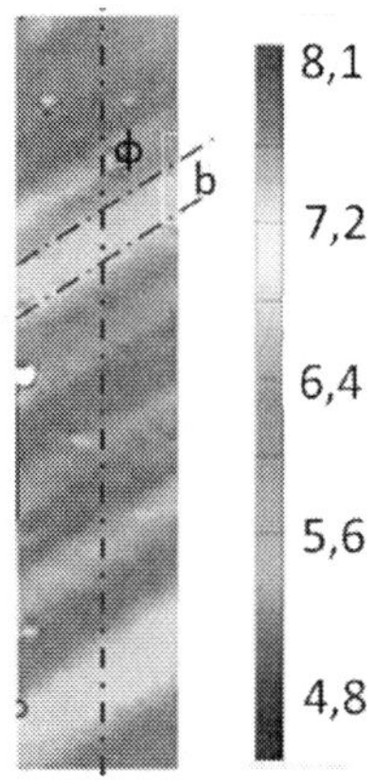

Figure 1. Strain distribution

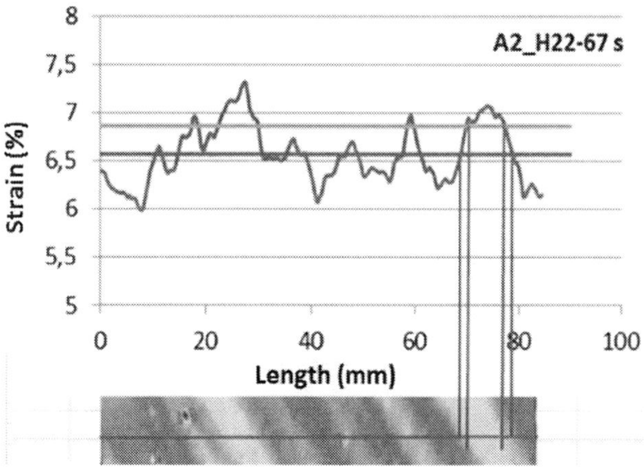

Figure 2. Digitized and plotted strain values and evaluation of PLC band parameters.

Figure 1 shows the graphic evaluation of a PLC band. The band width (b) is calculated from the distance of dashed lines and the angle between the tensile axis and the border line is given by ϕ. The absolute band strain is estimated from the color scale as 7,2% but according to the former definition, PLC strain is the increment of the internal and the external strain. The external strain is marked by green which might be ~6,5% (this is similar to the global strain measured by the extensometer) – so ε_{PLC} is calculated to 0,7% which is ≈0,007 in logarithmic strain. This estimation shows that a better definition can be given for the PLC strain if the average (or global) strain along the test line is the basis and the increment is related to that value.

Digitized strain values collected from the specimen center line can be seen in Figure 2 together with a colored strain map. The first impression is, as more PLC bands are on the specimen, their width and maximum strain strongly differs from each other, as they are in different phases of band propagation. This is why a uniform definition of band parameters is not possible. Two alternatives can be offered for evaluation: (i) one characteristic band is selected for evaluation, or (ii) considering that the strain is a random variable, some statistical measurements should be implemented for characterization of band parameters.

Observing the band on the right side of the specimen (Figure 2) the first question is band width. It can be defined as the intersection of the strain graph with the average strain line (marked with red) or with the line above it (marked with green) – for example at half height of the band. It is suggested by the authors of this paper that band width should be defined as the intersection with the average strain line. The second question concerns band height, which might be the strain maximum or a characteristic average strain. The actual band height is 7,07% and the average strain is 6,55% – resulting that ε_{PLC} is 0,52%. Regarding the peak on the left side (at L=27,6 mm), the strain maximum is 7,25% and the PLC strain is 0,7%, which is the same as ε_{PLC} evaluated from the graphic image in Figure 1.

The referenced authors suggest a global parameter for the characterisation of the average PLC strain. This might be the average of the strain values above the $\varepsilon_{AVG}+\varepsilon_{STD}$ line (in green). This is a bit lower than the maximum strains (7,01%) but more balanced than the value of one individual peak. For example, this average can be compared with the average of the five highest peaks. From that it follows that the values are very similar – in the first case ε_{PLC} =7,01-6,55=0,46% while the average of the five peaks is 0,48%.

The average band width can be calculated from the sum of intersections where the strain is equal to the average – now it is 36,8 mm and the number of bands is six so b_{AVE}=6,13 mm.

A possible characterisation of inhomogeneous local deformations during tensile tests is suggested by [19] which is independent from the PLC effect and valid for any kind of inhomogeneity and allows to compare the strain fields of different materials. The so-called local inhomogeneity factor, Λ can be calculated with the following equation:

$$\Lambda = \frac{\varepsilon_{max} - \varepsilon_{min}}{\varepsilon_{AVG}} \quad (1)$$

This parameter is dependent of global strain, Λ is the function of ε_{AVG}.

2.2. Evaluation of serrations from tensile test diagram

A typical true stress-time diagram is illustrated in Figure 3. Fitting a four-parameter Voce-equation on the measured data stress amplitudes can be evaluated as the differences of the measured data and the calculated curve points. Figure 4 shows that type A serrations can be identified between 35 s and 45 s and later high-frequency B-type serrations occur with C-type stress jumps. The strain amplitude is increasing versus time.

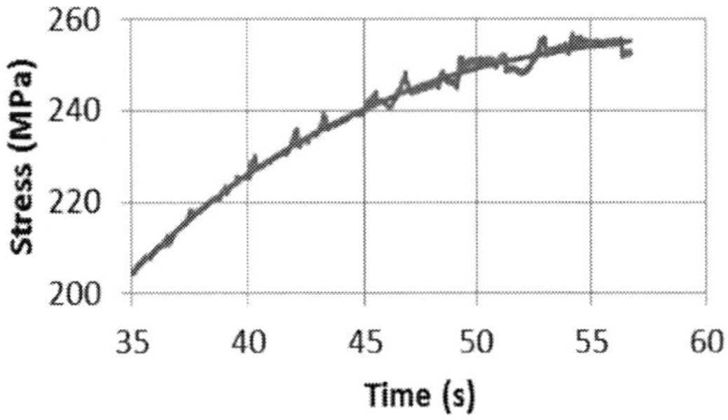

Figure 3. Stress-time diagram, approximated by Voce equation.

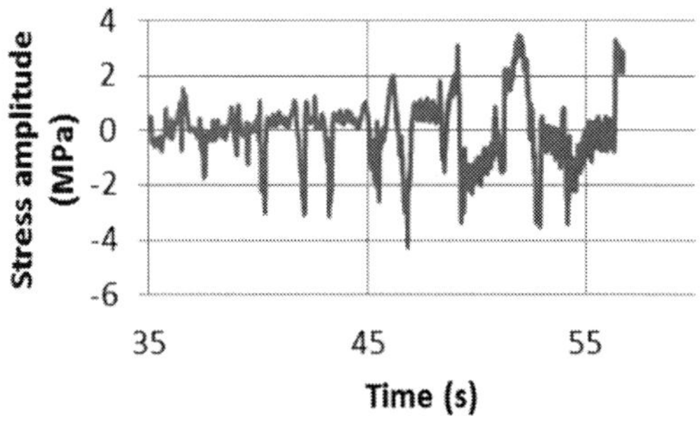

Figure 4. Stress amplitudes versus time ($\dot{\varepsilon}$=0,0022 1/s).

As strain amplitude is a random variable, statistical evaluation is needed for calculating the serration parameters. For one approach, the referenced authors suggest to follow the evaluation method similar to surface roughness – that means to calculate average stress amplitude analogue to R_a and maximum amplitude R_z. The new parameters will be designated as $\Delta\sigma_a$ and $\Delta\sigma_z$ respectively and they can be calculated according to equation (2).

$$\Delta\sigma_a = \frac{(\sum_{i=1}^{n} abs(\Delta\sigma_i))}{n} \; and \; \Delta\sigma_z = \left[\left(\sum_{i=1}^{5}\Delta\sigma_{i\,max}^{+}\right) - \left(\sum_{i=1}^{5}\Delta\sigma_{i\,max}^{-}\right)\right]/5 \quad (2)$$

Another approach is the distribution analysis of the serrations. The absolute values of the stress amplitudes are illustrated in a relative density histogram (Figure 5).

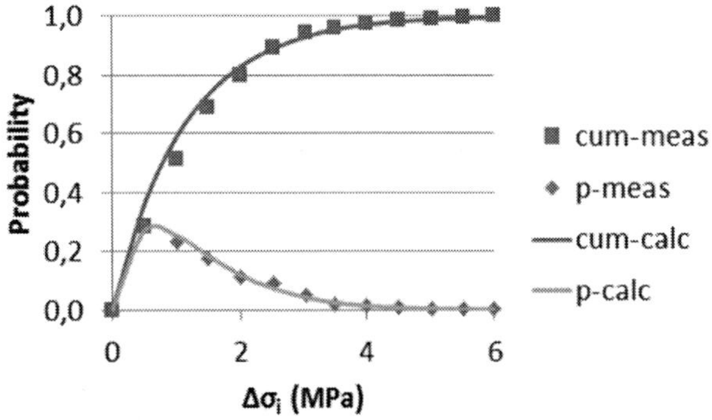

Figure 5. The Weibull distribution of the serrations. The points refer to measured probability values while the lines illustrate the calculated probability density and cumulative distribution functions.

The shape of the point arrangement refers to the Weibull distribution, which was validated by regression parameters. Dozens of calculations have proven that R^2 is near to 1 and always over 0,96 – so the hypothesis is acceptable. Using these Weibull functions the stress amplitudes can be calculated for different probabilities. Three values were tested and evaluated: p=0,97; 0,98 and 0,99. It was concluded that $\Delta\sigma_p$ is stable at p=0,97 and this parameter correlates well with $\Delta\sigma_z$.

3. Results and discussion

3.1. Results of DIC measurements

The strains defined in point 2.1. are illustrated in Figure 6 and 7. Figure 6 shows the comparison of maximum strain values, the calculated average strain plus twice the standard deviation ($\varepsilon_{AVG}+2\varepsilon_{STD}$) and the PLC peak strain, which is the average of the strain values above $\varepsilon_{AVG}+\varepsilon_{STD}$. The horizontal axis contains the global or average strain, its points were detected at selected moments of the tensile test.

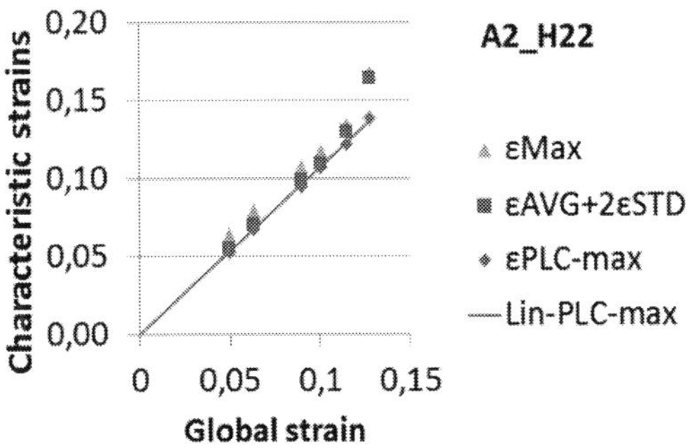

Figure 6. Characteristic strains against the global strain.

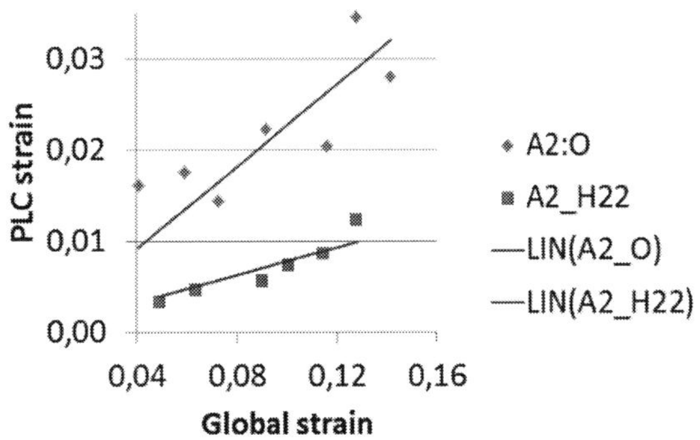

Figure 7. PLC strains as function of the global strain

It follows from Figure 6 that for the first approach the maximum strain values (ε_{MAX}) and $\varepsilon_{AVG}+2\varepsilon_{STD}$ are very similar to each other and do not differ significantly from the evaluated average PLC peak strains. For easier evaluation of strain maps maximum values can be used also as the characteristic strain for evaluation of the PLC strain increment: $\varepsilon_{PLC}\approx\varepsilon_{MAX} - \varepsilon_{AVG}$. Figure 7 illustrates the PLC strain, which is calculated as the difference of the PLC peak strain minus the average strain – by definition it is the strain increment caused by the PLC effect related to its environment. The measured values show an increase as the global strain increases [4, 6, 8]. A further conclusion is that the annealed sheet exhibits higher PLC strain than the cold rolled one.

The local inhomogeneity factor (Λ) can also be calculated from the points of maximum and minimum strain values as the function of the global strain. The slope of linear fit to maximum and minimum strain values (s_{MAX} and s_{MIN}) have to be evaluated, and their difference gives Λ. Similarly plotting the $\varepsilon_{PLC}\approx\varepsilon_{MAX} - \varepsilon_{AVG}$ points versus the average strain (Figure 7) the slope of the fit line characterises the material behaviour as a global parameter of the PLC strain increment.

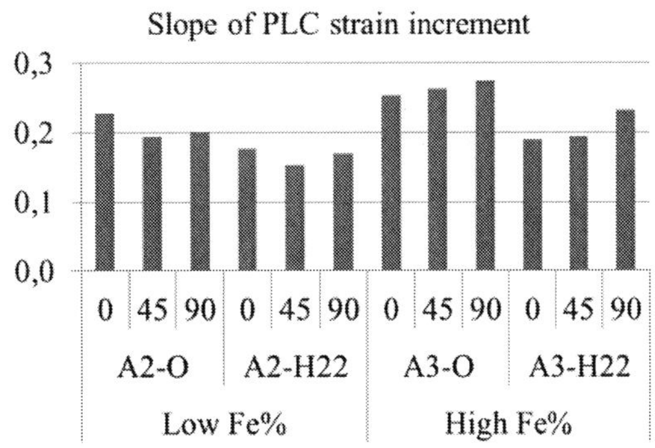

Figure 8. Comparison of sheets by their PLC strain increment

The comparison of the two different AlMg3 alloys is illustrated in Figure 8. As can be seen from Table 1, the A2 sheet contains 0,11% Fe while the A3 has 0,23%. Therefore the volume of intermetallic particles in alloys are respectively 0,53% and 1,35% which means that the formability of the A2 sheet is better than that of the A3 alloy. The slopes of PLC strain increments show this effect very clearly as values are lower in case of the A2 sheet. Similar results can be found in [6]. Another important observation is that inhomogeneity parameters of cold rolled H22 sheets are lower than annealed ones (see Figure 7).

Band velocity can be measured easily using DIC technique. Figure 9(a) shows a typical A-type band which develops continuously from one end of the tensile specimen to other. Knowing the geometric dimensions, the band displacement can be estimated from the strain maps as the function of time.

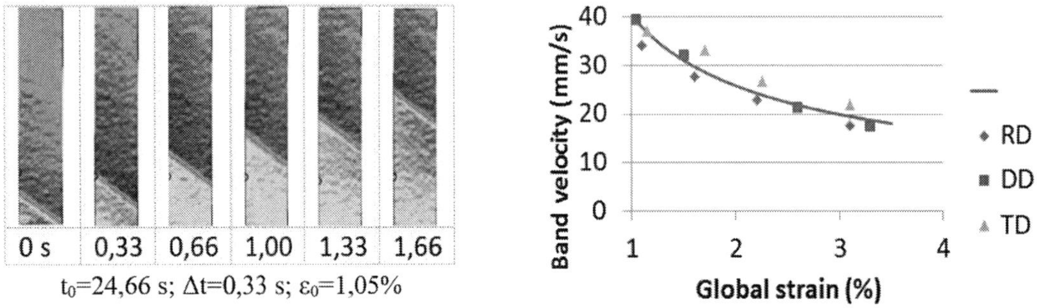

(a) (b)

Figure 9. Map of the PLC band (a) and the band velocity versus the global strain (b)

Figure 9(b) contains the detected points of band velocity in rolling, diagonal and transverse directions and the power function which was fitted to all points. Both the shape of the function and the measured band velocities are in accordance with [5].

The average band width was calculated two ways, (i) from the sum of intersections where strain is equal to the average (ε_{AVG}) or (ii) from intersections with the $\varepsilon_{AVG}+\varepsilon_{STD}$ line, which was formerly used for calculation of the average band strain as well. Band widths do not show regular tendency as the function of the global strain at the tested strain rate ($2,2*10^{-3}$ 1/s), therefore their average was calculated between 5-15% strain range in case of the A2_H22 sheet. The band width evaluated from the intersections of ε_{AVG} line is 7,18 mm while evaluated from the $\varepsilon_{AVG}+\varepsilon_{STD}$ line the width is smaller, only 3,88 mm. Having studied the strain distributions equation (i) is more realistic, as it is in accordance with the basic value of the PLC strain increment while using equation (ii) approximately half of the bands were not intersected with the higher line. At a certain moment bands along the specimen are in different phases of propagation and this explains why their band strain and width is different. Another factor is that the type of serrations changes as global strain increases.

3.2. Results of serration analysis

Figure 3 and 4 show typical stress-time functions and stress amplitudes. Average and maximum amplitudes can be obtained by evaluating the stress amplitudes in 5 s time intervals according to equations (2). These are illustrated in Figure 10 for A2_H22 sheet. Both parameters are increasing versus time and obviously versus strain.

It is also clear that the maximum stress amplitude ($\Delta\sigma_z$) varies more significantly than the average ($\Delta\sigma_a$), this is why the former parameter will be used during the following evaluations, together with ($\Delta\sigma_w$) which is calculated from the Weibull-distribution of the stress amplitudes. Figure 11 illustrates the plot of amplitudes derived from the Weibull-function at p=0,97 and from 5-5 positive and negative stress peaks.

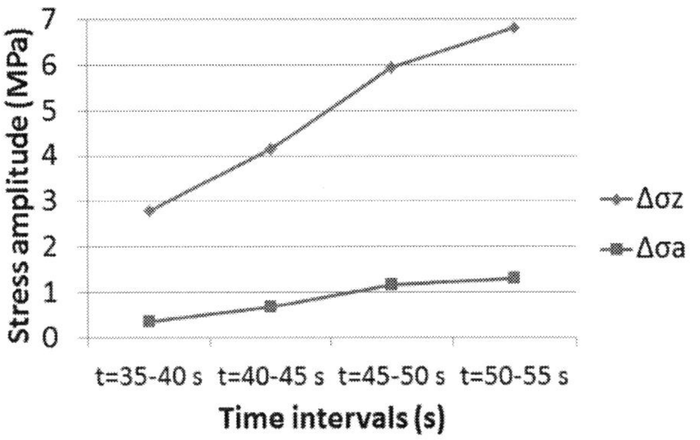

Figure 10. Stress amplitude versus time and global strain

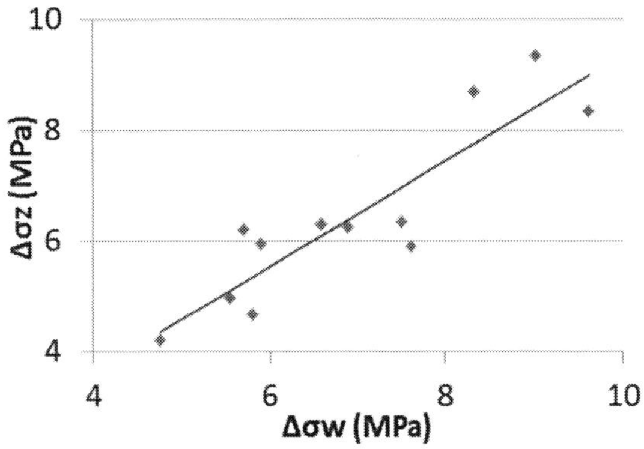

Figure 11. Correlation between stress amplitudes

Considering the common character of Figure 7 and 10 – namely both the PLC band strain and the stress amplitude increases as the function of the global strain – it follows that close correlation exists between the PLC band strain and the stress amplitudes. It is illustrated in Figure 12, where the PLC band strains were calculated at the given global strain points while the stress amplitudes were evaluated from the neighbouring stress-strain values. The mechanical background of this correlation is explained by the sudden PLC strain increments which cause stress drops. The amplitude of serrations is influenced by the magnitude of strain increments and by the global stiffness of the testing machine-specimen system.

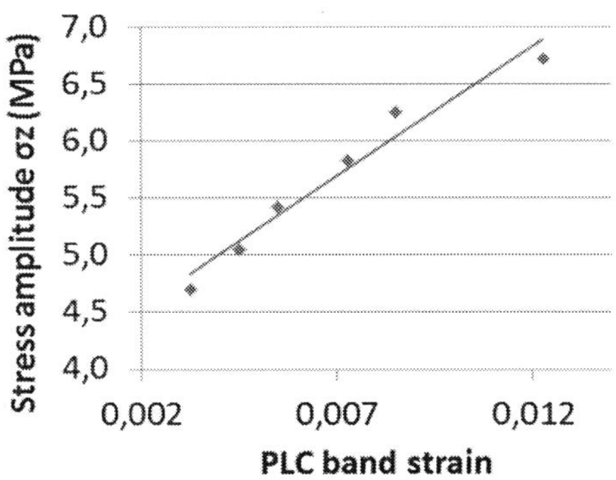

Figure 12. Correlation between the PLC band strain and the stress amplitude (A2_H22)

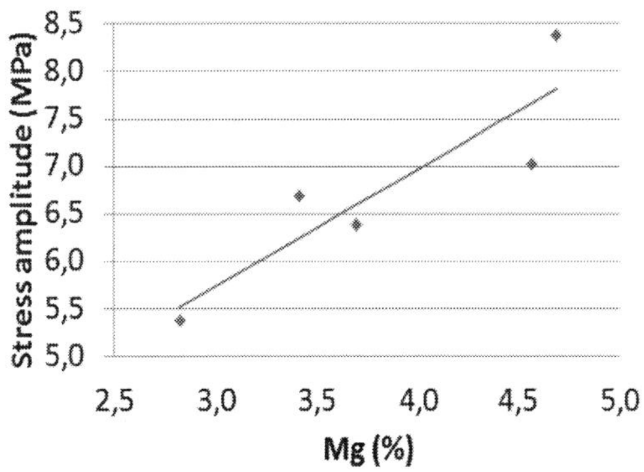

Figure 13. Correlation between the stress amplitude and the magnesium content

For characterising the serration behaviour of one sheet metal authors suggest to evaluate the formerly defined stress amplitude parameters between 8% and 12% global strain – similarly to plastic anisotropy. This strain domain represents the steady-state process of serrations and its position is approximately in the middle of the uniform elongation area. Having collected the stress amplitude parameters of the tested materials in rolling, diagonal and transversal directions the average of the measured points are displayed in Figure 13 as the function of the magnesium content (Mg%). As the graph shows a relative close correlation exists between the stress amplitudes and the Mg-content, as it is published in [13] for AA5052 and AA5182 alloys. This can be explained by the interaction of Mg atoms and dislocations – the higher the magnesium content, the more obstacles occur and DSA becomes more effective. Increased DSA activity causes stronger band propagation and increases the differences between stress peaks and drops.

4. Summary and conclusions

The PLC band attributes and the stress serrations were numerically characterised using newly defined parameters for strain inhomogeneity and stress oscillations. In both cases a statistical approach was used for better evaluation of random variables. The PLC band parameters were measured and gained from DIC measurements. The strain distributions were characterised by their average, standard deviation and minimum-maximum values.

The global characterisation of a selected sheet metal is possible using the graphs of global strain and PLC strain increment. The slope of the fitting line is a numerical parameter of strain inhomogeneity caused by the PLC effect. It was concluded from the measurements that the formability of sheets can be characterised by this parameter, the lower the slope, the better the formability. It can also be declared that cold forming lowers strain inhomogeneity.

The evaluation of serrations was carried out applying similar statistical methods. The absolute values of stress amplitude distribution over a given strain interval was approached by a two-parameters Weibull distribution and 97% probability value was used for the definition of the stress amplitude. It has been proven that strong correlation exist between the Mg-content and the stress amplitude. Another important observation is that the PLC band strain and the stress amplitude increase together.

Acknowledgment

The research presented in this paper was carried out as part of the EFOP-3.6.2-16-2017-00016 project in the framework of the New Széchenyi Plan. The completion of this project is funded by the European Union and co-financed by the European Social Fund.

References

[1] Portevin A and Le Chatelier F 1923 *Compt. Rend. Acad. Sci.* Paris **176** pp 507-510
[2] McCormick P G 1988 *Acta metall* **36** pp 3061-3067
[3] Jiang H, Zhang Q, Chen X, Chen Z, Jiang Z, Wu X and Fan J 2007 *Acta Mater* **55** pp 2219-2228
[4] Halim H, Wilkinson D S and Niewczas M 2007 *Acta Mater* **55** pp 4151-4160
[5] Benallal A, Berstad T, Børvik T, Hopperstad O S, Koutiri I and Nogueira de Codes R 2008 *Int. J. Plast.* **24** pp 1916-1945
[6] Kang J, Wilkinson D S, Embury J D, Jain M and Beaudoin A J 2005 *Scr. Mater.* **53** pp 499–503
[7] Yuzbekova D, Mogucheva A, Zhemchuzhnikova D, Lebedkina T, Lebyodkin M and Kaibyshev R 2017 *Int. J. Plast.* **96** pp 210-226
[8] Zhang Y, Liu J P, Chen S Y, Xie X, Liaw P K, Dahmen K A, Qiao J W and Wang Y L 2017 *Prog. Mater. Sci.* **90** pp 358–460
[9] Yuan Z, Li F and He M 2011 *Mater. Sci. Eng:A* **530** pp 389–395
[10] Zdunek J, Spychalski W L, Mizera J and Kurzydłowski K J 2007 *Mater. Charact.* **58** pp 46-50
[11] Mazičre M and Dierke H 2012 *Comput. Mater. Sci.* **52** pp 68-72
[12] Casarotto L, Dierke H, Tutscha R and Neuhauser H 2009 *Mater. Sci. Eng:A* **527** pp 132–140
[13] Wen W, Zhao Y and Morris J G 2005 *Mater. Sci. Eng:A* **392** pp 136–144
[14] Shabadi R, Kumar S, Roven H J and Dwarakadasa E S 2004 *Mater. Sci. Eng:A* **364** pp 140–150
[15] Nguyen V T, Kwon S J, Kwon O H and Kim Y S 2017 *Procedia Eng.* **184** pp 381-389
[16] Tong W and Zhang N 2007 *J. Eng. Mater. Technol.* **129** pp 332-341
[17] Sene N A, Balland P and Bouabdallah K 2018 *Arch. Civ. Mech. Eng.* **18** pp 94-102
[18] https://www.gom.com/metrology-systems/aramis.html
[19] Tisza M, Gál G, Kiss A, Kovács P Z and Lukács Zs 2014 *Multidiszciplináris Tudományok* **4** pp 39-48

IOP Conf. Series: Materials Science and Engineering **903** (2020) 012024 doi:10.1088/1757-899X/903/1/012024

Modeling the deformation behavior of polymer sandwich structures with inhomogeneous core

Laszlo Takacs[1,2], **Ferenc Szabó**[1]

[1] Department of Polymer Engineering, Faculty of Mechanical Engineering, Budapest University of Technology and Economics, H-1111 Budapest, Műegyetem rkp. 3., Hungary
[2] eCon Engineering Kft., H-1116 Budapest, Kondorosi út 3., Hungary

E-mail: szabof@pt.bme.hu

Abstract. Polymer sandwich structures become widely used in the transportation industry due to their high bending stiffness and strength combined with low weight. In the conceptual design phase, it is essential to model the mechanical behavior of the sandwich properly in full-vehicle scale in order to analyze different design variants effectively. In this paper, a finite element modeling method is shown. The method is introduced on a sandwich structure with glass fiber reinforced, vinyl-ester matrix composite face-sheets and a PET foam as core material with an inhomogeneous structure. To model the sandwich panel with layered shells, where the core material is a single layer, equivalent stiffness constants of the inhomogeneous core are needed. To determine these constants, a detailed finite element model was created and virtual tensile and shear tests were performed. On the other hand, an analytical method was also shown. By applying the Voigt- and Reuss-rule on the inhomogeneous core, the needed stiffness constants can also be determined properly. Results of the two methods were compared and they showed a good correlation. Validation of the model was performed via comparing the results of the 4-point bending experimental tests and the simulation results.

1. Introduction

Composite materials are increasingly present in the industry [1]. Polymer sandwich structures with composite face-sheets and foam core are widely used not only in aerospace but in automotive, autobus, marine and construction engineering as well. The main reasons are their high bending stiffness and strength combined with low weight. Therefore, extensive studies have been conducted on sandwich structures due to their significance in industrial applications. Although a comprehensive review of current trends in research and applications of sandwich structures was accomplished by Birman and Kardomateas [2] in 2018, here, some of the papers in this field are reviewed briefly. Different case studies have been published from automobile industry with the aim of weight reduction by using sandwich panels in the whole body, the floor panel or the luggage panel [3-5]. However, polymer sandwich structures are present in the autobus, marine and construction engineering as well [6-8]. Almost all the applications use thermoset polymers but beside the dominance of thermosets, thermoplastic core materials are strongly developed to take the advantages of thermoforming and recycling. The aim of their technology process development is cost efficiency [9].

To make the design process of sandwich structures effective it is essential to model the mechanical behavior accurately. Many researches have been published using numerical modeling and also experimental tests [10,11]. The main micromechanical methods of the modeling of sandwich structures are classified on the basis of whether they are modeled as an equivalent single layer, or as multiple layers. A more detailed modeling method is to use 3D finite elements. Ivanez et al. [12] analyzed the dynamic flexural behavior of a sandwich plate with a three-point bending test. They haven't used a layered shell model but a 3D-model instead with a proper damage-model of the core material as well. It was concluded that the compressive strength of the core material affects the failure of the sandwich more than the tensile strength of the composite face-sheets. Manalo et al. [13] investigated strength and failure modes of a polymer composite sandwich beam with 4-point bending in edgewise and flatwise position. Here, also a 3D model was used in the FEM simulation. The edgewise position showed stiffer behavior as expected, the failure mode was the progressive failure of the face-sheet, while in flatwise position, it was the shear failure and debonding of the core. Awad et al. [14] tested a newly developed GFRP sandwich panel with point load. A crushable foam model and Hashin failure criteria were used in the finite element analysis. Polymer sandwich panels with composite face-sheets and foam cores were investigated under impact loading by Long et al. [15]. Drop-tests were performed with different impact energy. Foam density and stacking sequence of the composite skins were varied. A new user-defined material was developed in Abaqus in order to model the sandwich failure more effective. Delamination, fracture and foam crush regions were analyzed to better understanding of the strength of such structures.

Our goal is to develop a finite element modeling method of sandwich panels with inhomogeneous core material that can be effectively used to model the deformation behavior of a structure even in full-vehicle scale, where the sandwich is modeled as layered shells. If the core is modeled as a single layer, equivalent stiffness constants are to be determined. Different methods are shown to extract these constants. Our results hold the promise of the development of a cost-effective sandwich modeling method that can be effectively used in the product development phase in the industry.

2. Materials and Experimental Tests

The material chosen to analyze the polymer composite sandwich structure is a glass fiber reinforced composite with vinylester matrix. This type of composite is typically used in the transportation industry. The fiber reinforcement is a multidirectional fabric with a stacking sequence of 0°/45°/90°/-45°. The commercial name of the product is QE fabric as it is a so-called quadraxial fabric (Saertex GmbH, Germany). The specific weight is 1232 g/m^2. The face-sheet of the tested sandwich structure has 3 layers of this fabric with the same orientation and with a symmetric lay-up. The thickness of the face-sheet is 2.5 mm. The matrix material of the composite is a vinylester, its commercial name is Distitron VE220 (Novia Kft., Hungary). It is recommended for resin-transfer molding or vacuum injection technology. The specimens were manufactured with vacuum injection technology, Butanox-M50 (methyl ethyl ketone peroxide, solution in dimethyl phthalate) was used as initiator and 0.2 wt% cobalt solution as activator. The curing time was 24 h in room temperature and then 3 h at 100 °C.

The core material of the sandwich panel is a PET foam with the commercial name of Airex T90. It is a closed-cell foam with a density of 110 kg/m^3. The type of the foam is named FlexiCut. It has a thickness of 25 mm and 1.2 mm thick cuts every 30 mm. Both sides have cuts, on one side the 85 % of the thickness is cut, on the other side the 20 % of the thickness. These cuts help the manufacturing, they transfer the resin and this structure allows the full impregnability of the sandwich. The structure of the foam is shown in Figure 1. These foams are likely to be used in the industry, they have good recyclability and flame-retardant properties can be also enhanced by different additives [16].

The mechanical tests were performed on a Zwick Z020 uniaxial testing machine on room temperature and with a relative humidity of 46±2 %. The Young's moduli and Poisson-ratios of the composite face-sheets were carried out with a displacement controlled tensile test following EN ISO 527-4 standard with the Type 3 specimen. The test speed was 2 mm/min. The test was performed until failure. The strain components were measured with two unidirectional strain gauges perpendicular to each other on one side of the specimen.

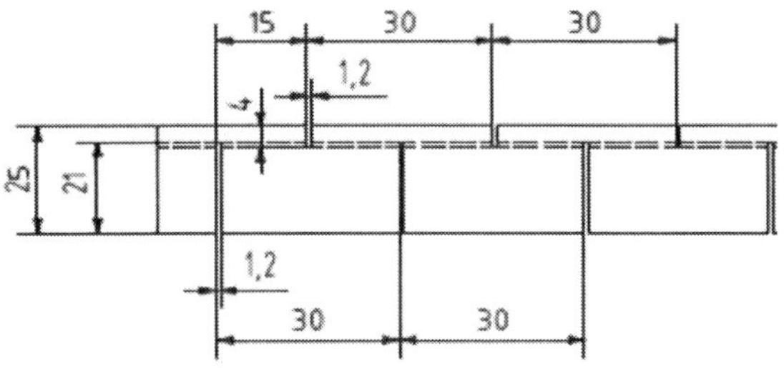

(a)

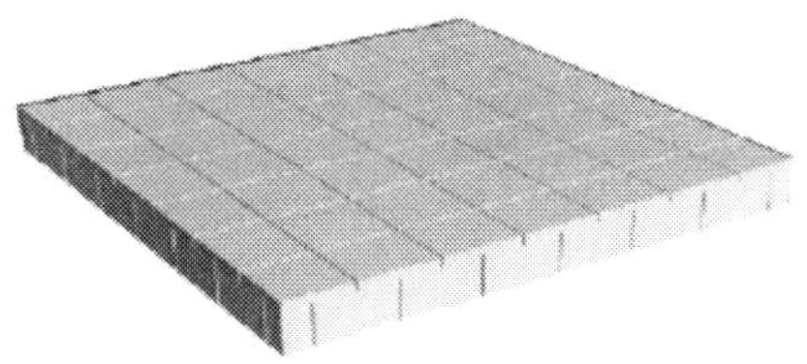

(b)

Figure 1. The structure of the FlexiCut foam: (a) sketch (b) 3D-view

The in-plane shear modulus was carried out with Iosipescu-test of a V-notched specimen following the standard ASTM-D5379 [17]. The test speed was 2 mm/min. To apply a clear shear loading on the specimen a special fixture is needed. During the test the specimen must remain plain and the edges parallel to each other. The sketch of the fixture can be seen on Figure 2 (a). To measure the shear deformation, strain gauges were used with a placement that can be seen on Figure 2 (b).

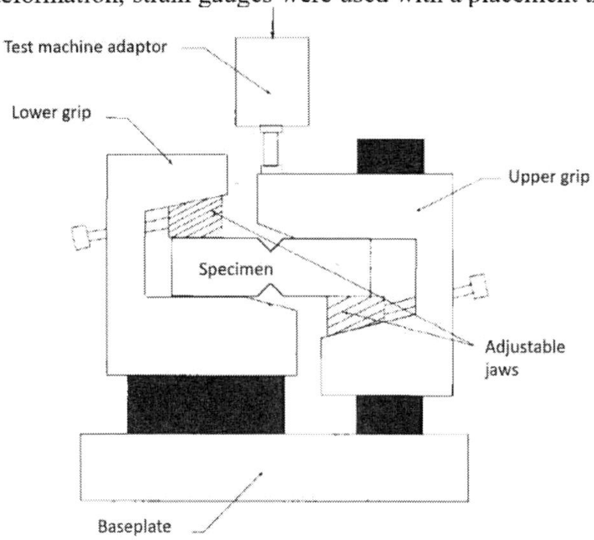

(a)

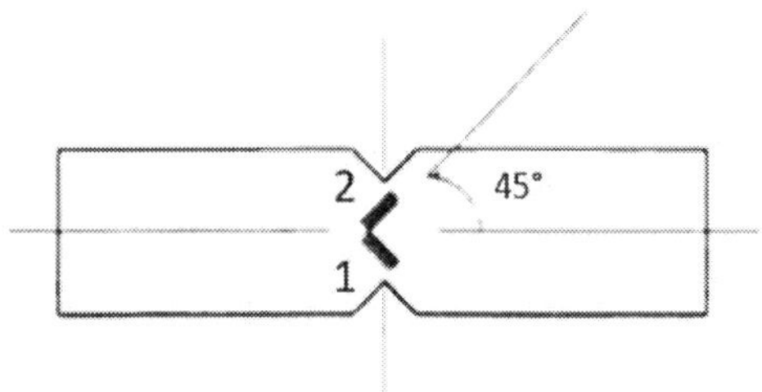

(b)

Figure 2. (a) The test fixture of the shear-test (b) Placement of the strain gauges

The shear deformation can be calculated as [17]:

$$\gamma_{xy} = \varepsilon_1 - \varepsilon_2 \tag{1}$$

Figure 3. 4-point bending of the sandwich beam

The stiffness parameters of the face-sheets can be determined with these tests but the core material cannot be directly tested. During manufacturing, the resin flows into the cuts of the used FlexiCut foam and that can significantly influence the stiffness of the core. This effect can be examined by the test of the complete sandwich. For the sandwich beams, a 4-point bending was performed following the standard ASTM-D7249. The test speed was 6 mm/min. The measured sandwich beam can be seen on Figure 3.

The aim of the study is to develop a method with which the deformation behavior of the sandwich structure can be effectively modeled in full structure scale in finite element analyses. When modeling a complete structure, e.g. vehicle body, layered shell elements are the most widely used instead of using solid elements. Solids allow deeper understanding of stress-state or even failure modes but the modeling and the calculation as well are more time-consuming so they are used generally in specimen scale.

In this study, first, we derived the stiffness parameters of the composite face-sheets from the tensile- and shear-tests, then we investigated the effect of the resin-walls in the core material on stiffness. For that, we used a mixed model using layered shells for face-sheets, solids for the core and shells for the resin-walls in the core. After that a method is introduced to derive the equivalent stiffness parameters of such a core material with the help of the Voigt- and Reuss-rules and the numerical shear-test of the core with its resin-walls. When having the equivalent stiffness constants, the modeling of the whole sandwich structure with a layered shell is possible. Results are validated with the mixed solid-shell models and the experimental tests.

3. Modeling Method

Modeling composites with layered shells is a widely used modeling technique. With this method, the layers are taken as homogeneous, anisotropic ones and the stiffness of the whole laminate is calculated with the classical laminate theory (CLT). Various kinds of commercial finite elements software are using this method as well. Quadrax fabrics are made by sewing unidirectional reinforcing layers together using thin polyester fibers, so the quadrax layer can be taken as an asymmetrical sub-laminate with four unidirectional reinforcing layers rotated relative to one another. We used an orthotropic material model and based on the structure of the reinforcing layer, moduli E_1 and E_2 are equal. These tensile moduli and the Poisson-ratio can be determined from the tensile test, shear-modulus can be expressed from the Iosipescu-test results the following way:

$$G_{12} = \frac{\tau_{12}}{\gamma_{12}} = \frac{F_{shear}}{A\,\gamma_{12}}. \quad (2)$$

where F_{shear} is the shear force, A is the cross-sectional area and γ_{12} is the shear strain.

The deformation behavior of the composite face-sheet can be properly modeled using orthotropic material model with the above constants, but the modeling method of the core-material depends strongly on its structure. Vehicle chassis structures are commonly modeled with shell elements due to their shell-like geometrical build-up. Furthermore, modeling with shell elements requires less work capacity and means less computational time compared to using solid elements. Our aim is to characterize the mechanical behavior of the core-material and present a method with which the complete sandwich panel can be modeled as a shell.

The investigated core material is closed-cell PET foam which has thin grooves. These serve that the foam can be better formed in a 3D-shape mold and also helps the impregnation process as the resin passes through the grooves during injection. After manufacturing, thin resin walls are formed in the core which affect the mechanical behavior of the sandwich panel. In order to investigate the effect of this wall structure a solid-shell finite element model was built based on the ASTM D7249 measurement layout and specimen geometry. The average element size is 2 mm, face-sheets are modeled using shell elements with orthotropic material model, the foam with solid elements with isotropic material model and the resin walls with isotropic shells having common nodes with the solids of the foam. The elastic modulus of the foam is 110 MPa and the resin is 3412 MPa. These were determined via tensile tests. The structure of the model is shown in Figure 4.

We simulated the same four-point bending as the experiment. The contact surfaces of the supports and the load introductions were considered infinitely rigid and frictionless contact definitions were used. As a vertical displacement, 7 mm was applied.

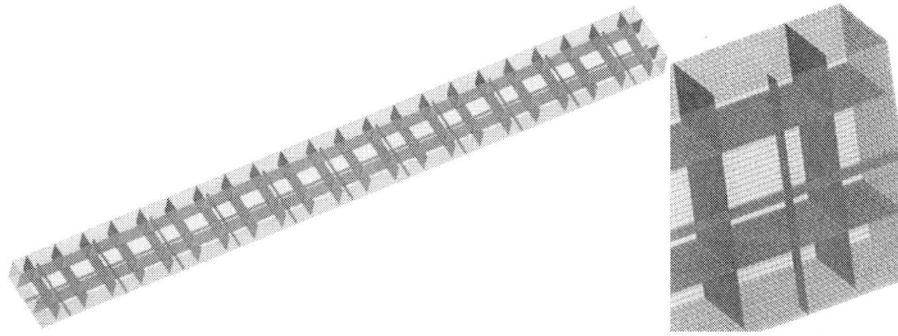

Figure 4. Finite element model of the sandwich core – resin walls with shell elements

A method is required to determine an equivalent modulus that can be used to characterize the resin impregnated foam. To determine this, we used the relationships of the micromechanics, respectably the rules of mixture. When applying the rules of mixture, the premise is that there are no cavities or foreign material in the composite and there is perfect adhesion between the components.

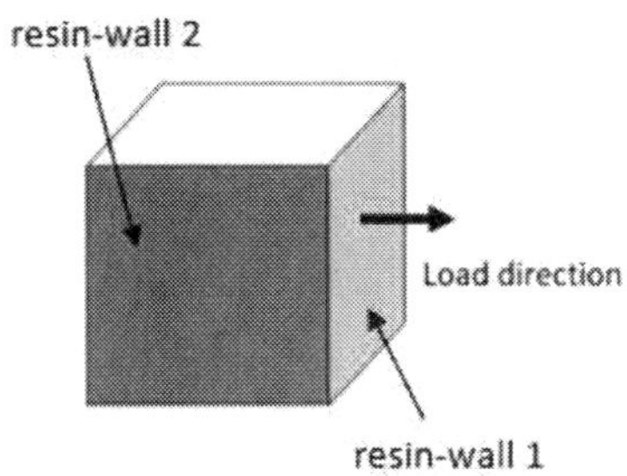

Figure 5. A "primitive cell" of the foam including resin walls

The equivalent modulus is deducted of a primitive cell of the core containing two resin walls, which is shown in Figure 5. The normal of resin-wall 1 is parallel to the load direction while the normal of resin-wall 2 is perpendicular to the load-direction. The direction of the resin walls is important in determining the equivalent modulus, but since a primitive cell is examined and the volume of the resin walls are the same, the equivalent modulus can be considered as a direction independent parameter.

If we first consider only the foam with resin-wall 2, the Voigt-rule [18] can be used to determine the E_{e1} equivalent modulus as follows:

$$E_{e1} = \Phi_1 \, E_r + (1 - \Phi_1) \, E_f, \tag{3}$$

where E_r is the elastic modulus of the resin, E_f is the elastic modulus of the foam and Φ_1 is the resin content of the primitive cell without the volume of resin-wall 1. Φ_1 can be derived from the volumetric ratio as:

$$\Phi_1 = \frac{V_{rw1}}{\left(V_{primitive\ cell} - V_{rw2}\right)}, \tag{4}$$

where V_{rw1} is the volume of resin-wall 1 of the figure, V_{rw2} is the volume of resin-wall 2 of the figure, and $V_{primitive\ cell}$ is the volume of the entire primitive cell.

If we now consider resin-wall 2 also, then the Reuss-rule [19] can be used to determine the equivalent modulus of the complete primitive cell. The expression is as follows:

$$E_e = \left(\frac{\Phi_2}{E_r} + \frac{1 - \Phi_2}{E_{e1}}\right)^{-1} = \frac{E_r E_{e1}}{\Phi_2 E_{e1} + (1 - \Phi_2) E_r}. \tag{5}$$

where Φ_2 can be interpreted with the volumetric ratio as:

$$\Phi_2 = \frac{V_{rw2}}{V_{primitive\ cell}}, \tag{6}$$

Based on these, using the two simple relationships described above, the equivalent modulus of the heterogeneous sandwich core material can be determined.

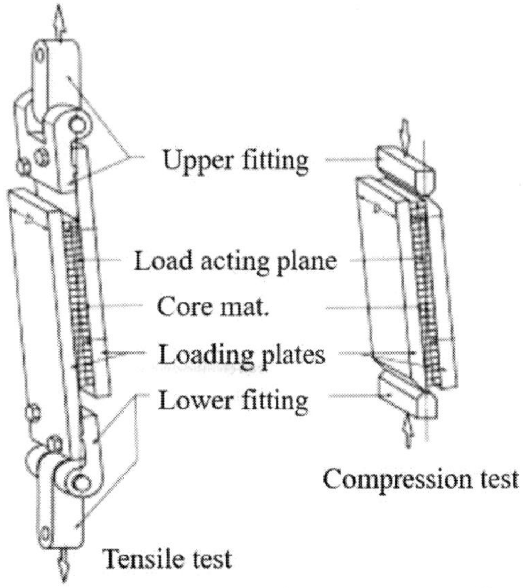

Figure 6. Shear test of sandwich core according to ASTM-C273

The sandwich panels are predominantly subjected to bending load, the bigger part of tension and compression is taken by the face-sheets, while the foam is subjected to considerable shear. Thus, an equivalent shear modulus is also required to model the heterogeneous core material of sandwich structures. Recommendations for this are given in ASTM C273, Figure 6 shows the standard measurement layout.

From the reaction force and the displacement, the shear modulus can be calculated in the following way:

$$G_e = \frac{S\,t}{L\,b} \quad (7)$$

where t is the thickness of the core, L and b are the length and width of the specimen. S can be calculated as follows:

$$S = \frac{F_{reaction}}{\Delta x} \quad (8)$$

where $F_{reaction}$ is the evaluated reaction force and Δx is the crosshead displacement, whereas in the virtual test it is the defined displacement.

4. Results and Discussion

Table 1 summarizes the orthotropic elastic constants of the quadrax fiber reinforced composite face-sheet obtained from the tensile- and the shear-test. Both the tensile and the shear tests were performed with 5 specimens. Average values and relative scatter are shown.

Table 1. Stiffness parameters of the glass fiber reinforced vinylester matrix composite face-sheet

Stiffness parameters of composite face-sheet	
E1 (MPa)	18051 ± 15%
E2 (MPa)	18051 ± 15%
ν12 (-)	0.298 ± 29%
G12 (MPa)	7035 ± 9%

After having the elastic constants of the face-sheet, the sandwich panel was analyzed. The 4-point bending test results of the detailed model show that neglecting the resin walls in the core, the stiffness would be underestimated by approximately 16%, as the reaction forces at 7 mm vertical displacement are 1181.5 N with and 1586.5 N without resin walls.

As the wall structure of the investigated foam is more complex in reality, we applied the procedure of rules of mixtures to the entire heterogeneous structure of the specimen. With a wall thickness of 1.2 mm, the calculation gives an equivalent modulus of 242.2 MPa

To test the accuracy and sensitivity of the procedure, a virtual tensile test was performed on the heterogeneous core material, without face-sheets, by fixing the nodes at one end and moving the nodes at the other end 10 mm longitudinally. With the calculations we investigated the effect of the thickness of the resin wall. By evaluating the longitudinal relative elongation (ε_y) and the reaction force ($F_{reaction}$), the equivalent elastic modulus can be determined as follows:

$$E_{e\,FEM} = \frac{\frac{F_{reaction}}{A}}{\varepsilon_y}. \quad (9)$$

The finite element simulations were performed at 0.7 mm, 1.2 mm and 1.5 mm resin wall thicknesses. The equivalent elastic modulus values derived from the simulations were then compared to the ones determined with the rule of mixtures.

The comparison is shown in Figure 7. The results obtained by the two methods are in good agreement with each other. With 0.7mm wall thickness, the match is perfect, for bigger wall thicknesses, the method minimally overestimates the modulus, for 1.5mm resin wall thickness the difference is 4%.

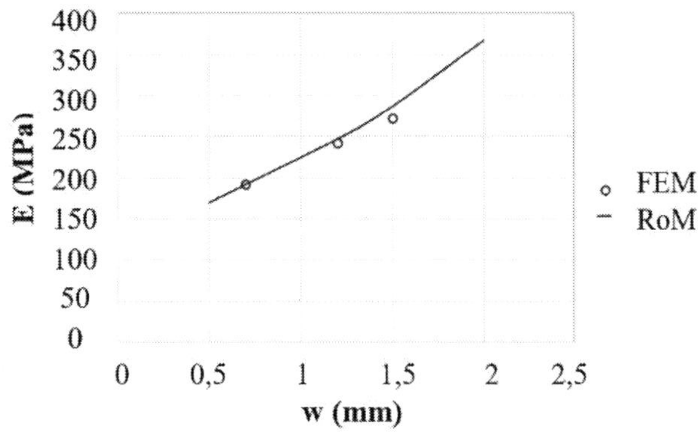

Figure 7. Equivalent elastic moduli carried out from FEM simulations and Rule of Mixture at different wall thicknesses

We performed a virtual test with the shear-measurement lay-up using the solid-shell model of the panel. The nodes of the upper sheet were fixed and the nodes of the lower sheet were moved by 20 mm.

The obtained equivalent shear modulus of the heterogeneous core is 74.6 MPa.

Beside the moduli, the Poisson-ratio is also to be mentioned. The bending behavior of the sandwich is hardly sensitive to the Poisson-ratio of the core. We investigate this with the virtual bending of a sandwich specimen modeled with shell elements. The core had an orthotropic material model with the moduli values obtained above. The Poisson-ratio was varied from 0.1 to 0.4 with a 0.05 step size. The evaluated reaction forces showed less than 0.5% difference between the two extreme cases. Based on this for further simulations we have used the Poisson-ratio of 0.35 which comes from the data sheet of the raw material.

Finally, the comparison of the 4-point bending results is summarized in Table 2.

Table 2. Comparison of the 4-point bending results

	$F_{reaction}$ (N) @ 7 mm vertical deflection	difference to experiment	simulation time (s)
experimental test	1811	-	-
FEM model – detailed	1823	3.2%	17446
FEM model – layered shell	1881	3.7%	317

By having the equivalent engineering constants of the heterogeneous core, it is possible to model the complete sandwich panel with layered shell elements. We performed a standard 4-point bending with both the detailed model and the layered shell model. The test lay-up follows the standard ASTM-D7249.

The model with layered shells has the middle layer with a thickness of 25 mm and an orthotropic material model with the engineering constants obtained with the method described above. The average element size was 5 mm. The prescribed motion was 7 mm for both models. The reaction force was 1881.5 N for the layered shells and 1822.9 N for the solid-shell model. To compare, the average reaction force value of the tested specimens at 7 mm displacement was 1811.4 N. This means a 3.2% difference in comparison to the detailed model and 3.7% difference in comparison to the experiments. In terms of simulation time, we got a huge difference. The simulations were performed on a computer with an Intel Xeon CPU and 128 GB RAM. This means only a slight difference taking into account that using this method is much less time-consuming regarding the modeling complexity and also the simulation time. So, the method can be effectively used to model sandwich panels with heterogeneous core materials.

5. Conclusions

The aim of the study is to develop a method with which the deformation behavior of the sandwich structure can be effectively modeled with shell elements in finite element analyses. In this study, first, we derived the stiffness parameters of the composite face-sheets from the tensile- and shear-tests, then investigated the effect of the resin-walls in an inhomogeneous core material on stiffness. The 4-point bending tests of a detailed model show that neglecting the effect of these resin walls would underestimate the stiffness with almost 16%. Voigt- and Reuss-rules can be effectively used to determine the equivalent tensile moduli of such core materials. When having the equivalent stiffness constants, the modeling of the whole sandwich structure with a layered shell is possible. Results are validated with the mixed solid-shell models and the experimental tests. They show that a difference of less than 4% in comparison to the experimental test can speed up the simulation time with two orders of magnitude.

Acknowledgement

The project is funded by the National Research, Development and Innovation (NKFIH) Fund, Project title: "Production of polymer products by a short cycle time, automatized production technology for automotive applications, with exceptional focus on the complexity and recyclability of the composite parts"; The application ID number: NVKP_16-1-2016-0046. The developers are grateful for the support.

References

[1] Mochane M.J. et al 2019 *Express Polymer Letters* **13** 159-198
[2] Birman V.; Kardomateas, G.A. 2018 *Composites Part B: Engineering* **142** 221-240
[3] Velea M.N. et al 2014 *Composite Structures* **111** 75-84
[4] Hara, D.; Özgen, G.O. 2016 *Transportation Research Procedia* **14** 1013-1020
[5] Alok Raj et al 2018 *IOP Conference Series: Materials Science and Engineering* **422** 012004
[6] Wu H.C. et al 2003 *Composites Part B: Engineering* **34** 51-58.
[7] Shen W. et al 2017 *Ocean Engineering* **144** 78-89
[8] Kulpa M.; Siwowski T. 2019 *Composites Part B: Engineering* **167** 207-220
[9] Sewell J. et al 2016 *Reinforced Plastics* **60** 146-150
[10] Kovács L.; Romhány G. 2018 *Periodica Polytechnica Mechanical Engineering* **62** 158–164
[11] Asadi A. et al 2018 *Express Polymer Letters* **12** 781-789
[12] Ivañez I. et al 2010 *Composite Structures* **92** 2285-2291
[13] Manalo A.C. et al 2010 *Composite Structures* **92** 984-995
[14] Awad Z.K. et al 2012 *Engineering Structures* **41** 126-135
[15] Long S. et al 2018 *Composite Structures* **197** 10-20
[16] Szabó V.A.; Dogossy G. 2020 *Periodica Polytechnica Mechanical Engineering* **64** 81–87
[17] ASTM-D5379 2019 *ASTM International*
[18] Voigt W. 1889 *Wiedemann's Annalen der Physik und Chemie* **38** 573-587
[19] Reuss A. 1929 *Zeitschrift für Angewandte Mathematik und Mechanik* **9** 49-58

IOP Conf. Series: Materials Science and Engineering **903** (2020) 012025 doi:10.1088/1757-899X/903/1/012025

Development of high power femtosecond laser microstructures on automotive stainless steel

Ferenc Tajti[1], Miklós Berczeli[1]

[1] Department of Materials Technology, GAMF Faculty of Engineering and Computer Science, John Von Neumann University, Hungary

E-mail: cbfferi@gmail.com

Abstract. It is indispensable for the industry to introduce the most accurate manufacturing technologies. Therefore, laser beam machining centres have been gaining popularity in machining equipment over the last few years. We can perform quick and highly accurate machining with the latest generation of laser equipment which operates in femtosecond pulse mode. The purpose of this study is to find the best femtosecond laser surface machining technique on stainless steels by changing technological parameters. We have compiled a general table that lists the parameters, with variable values in bands and columns to select the most accurate surface treatment with minimal heat affected zone.

1. Introduction

Industrial participants consider it essential the precise and accurate adjustment of the processing, treatment technologies. It is not enough to buy a production equipment from a distributor, the company's engineers are expected to program it according to their expertise for any process. This enables fast and complete product production on the machine.

Laser beam machining centers have been growing in popularity for decades, but over the past few years, they have seen a surge in popularity. Most of the equipments used by the industry are high performance continuous wave lasers. However, laser source manufacturers now have being able to build ultra-short-pulse laser equipment with above average performance. In short pulse laser beam machining, at about 300 fs, the laser peak power can reach a nuclear power plant's power [1], this results in a whole new kind of laser beam-material interaction [2]. Thus, it is extremely important to know and explore the processes that occur when using these devices.

The absorption of high-power electromagnetic waves in metals causes hyper-sublimation, because of this, the metal changes from a solid state to a direct plasma state [3]. This phenomenon has minimal thermal effects and makes micromachining possible. With such applications, vascular stents, microfluidic channels, surface microstructures, channels can be formed in metals, depending on the appropriate parameters [4, 5]. Furthermore, with this technology we can make different surface engineering on materials without damaging the base material [6, 7].

Our aim was to investigate the effect of the different laser impulses on stainless steel using various laser frequency and impulse burst number. According to the different parameters the paper should reveal the difference between the material removals methods. The key goal was to find the best sharpening edges' grooves, micro structured by the femtosecond laser impulses. This application can be used as a preparation of different joining technologies, micro structuring working surface of polymer and metal injection molding tool and change the wettability behaviors on the surface.

2. Materials and Methods

In our study we used stainless steel (AISI 304), one of the most common type of corrosion resistant steels. Before the laser machining, the surface of the steel was cleaned with methanol.

The laser treating process was performed using a Coherent Monaco Femtosecond Laser with wavelength of 1035 nm. The laser impulses had 277 fs of width and the average power was set between 2 W and 40 W (ie. from 5% to 100% power). The beam was focused on the surface with a diameter of 80 μm.

Due to the pulses of the laser, the lines were constructed from points. First, it was necessary to determine the optimum distance between the points, where the resulting pattern resembles a single line. The laser beam diameter was 80 μm, therefore, we had to choose shorter distances. We set the distances, 60 μm, 50 μm, 40 μm, 30 μm, 20 μm, 10 μm, 5 μm. After that we examined with microscope which distance was the ideal to create a solid line, as shown in Figure 1.

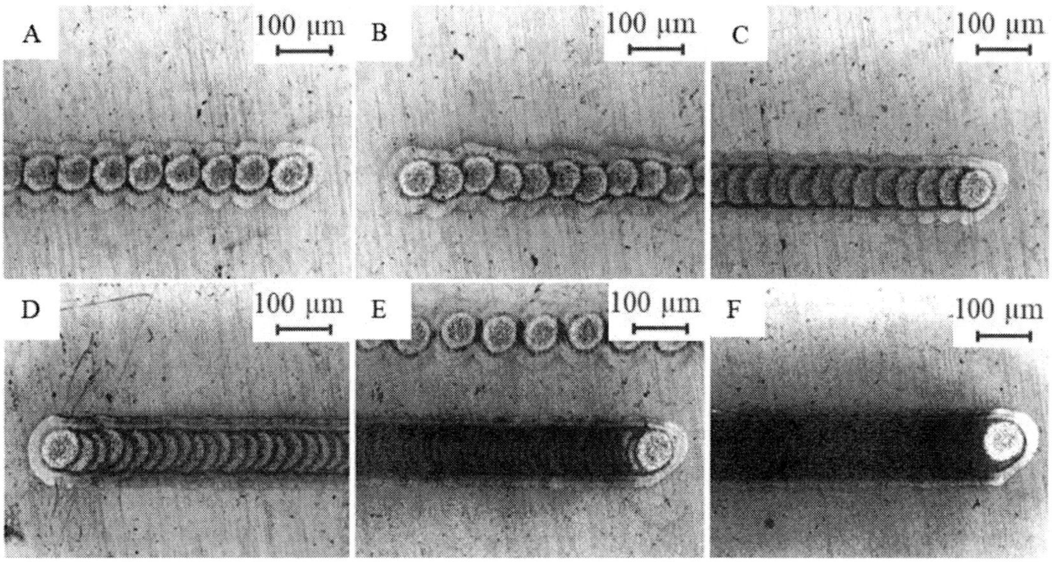

Figure 1. Distances between the laser spots A)50 μm, B)40 μm, C)30 μm, D)20 μm, E)10 μm, F)5 μm)

After examining the results, we found that the optimal distance is 10 μm.

After that we created a parameter table showing the power values and the number of pulses fired at one point. We draw four lines with each parameter, but repeated different times. First line wasn't repeated, second was repeated 10 times, third was repeated 100 times, fourth was repeated 1000 times. The test was performed on both 188 kHz and 50 MHz. The final table is shown in Figure 2.

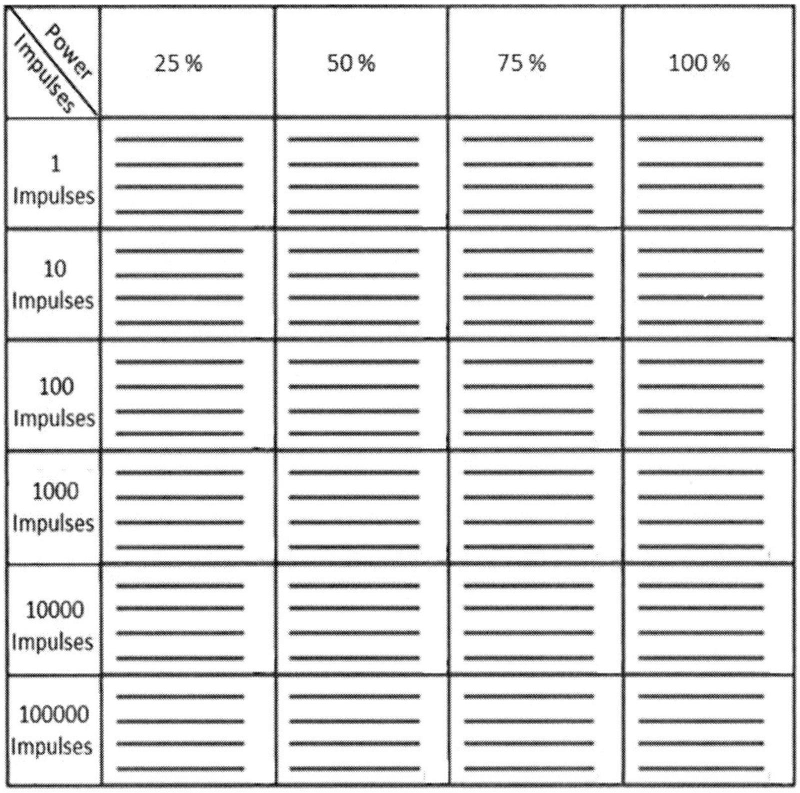

Figure 2. Parameter table: in all rectangles the first line wasn't repeated, second was repeated 10 times, third was repeated 100 times, fourth was repeated 1000 times.

After completing the table with the different parameters, we used microscope to examine the different lines to determine which parameters makes the optimal results.

3. Results

If we look at the best results between Figure 4 and Figure 5, we get different results.

At 50 MHz, good lines are created at higher power, and pulse numbers can be counted between 1000 and 10000 and in general 100 repetitions proved to be appropriate. While at 188 kHz, line quality was the best around 25% average power. In terms of pulse numbers, it was clear that there were mainly 1, up to 10 pulse numbers that characterized good images. With this parameter 100 repetitions proved to be the optimal setting.

Examining the two tables, it is striking to see what frequency was characteristic. At 50 MHz, with little pulse and power, almost no material removal occurs. We only got the first test value after 100 pulses. This is due to the fact that at such high frequencies the laser is able to emit less power. At 188 kHz, it has a lower frequency, allowing it to emit more pulses per cycle, and the laser operates at a much higher average power. This is also evident in the heat input, it removes much more material at a lower frequency and thus the heat affected zone is larger.

In our research, we examined all the results for the thickness of the machined lines and the heat affected zone. Figure 3 illustrates which parameters and widths we specifically examined.

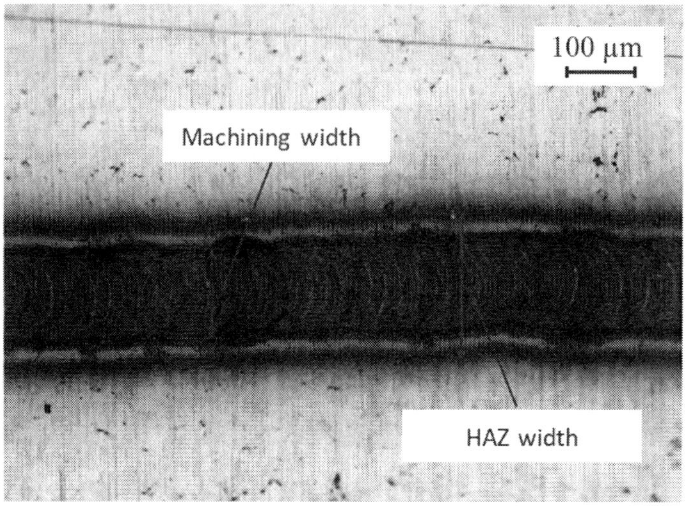

Figure 3. Examined widths

These values have been tabulated for both frequencies. First, we examined the machining width. The values of the two frequencies are plotted in a separate diagram.

At 50 MHz, it is clear that the width of the machined structure is also likely to increase with pulses which shown in Figure 4. This is around 80 μm, but if the pulse rate is higher than 10,000, this width will increase by leaps and bounds with higher power.

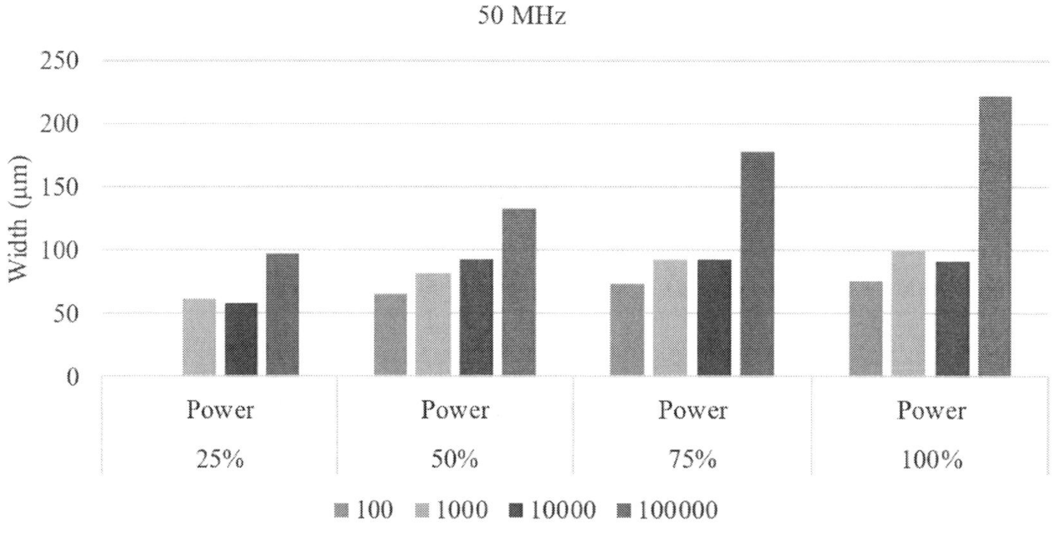

Figure 4. 50 MHz machining width as a function of pulses

Similar results can be observed at 188 kHz as shown in Figure 5. At low power, the width increases proportionally, but at 50% power, 1000 pulses are enough to make the 80 μm spot diameter works 3x wider.

IOP Conf. Series: Materials Science and Engineering **903** (2020) 012025 doi:10.1088/1757-899X/903/1/012025

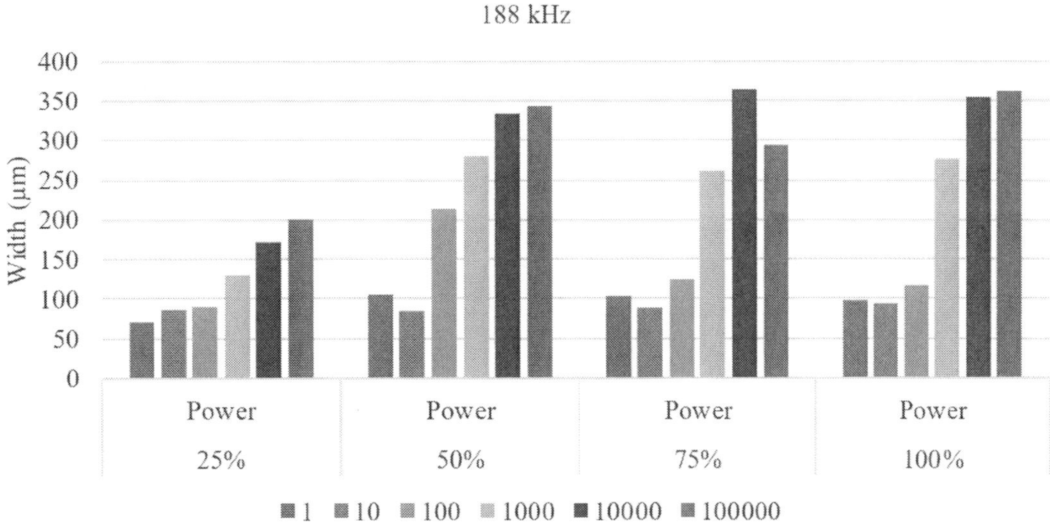

Figure 5. 188 kHz machining width as a function of pulses

After analyzing the working widths, we examined the heat affected zone. In the same way, we used to evaluate widths in diagrams. In most cases it was measurable, but at higher impulses and power, the heat effect was sometimes so large that we could only estimate the width.

Also starting at 50 MHz, it can be seen that above 50% the amount of heat input increases drastically at 100000 pulses, but below that it changes steadily between acceptable values as shown in Figure 6.

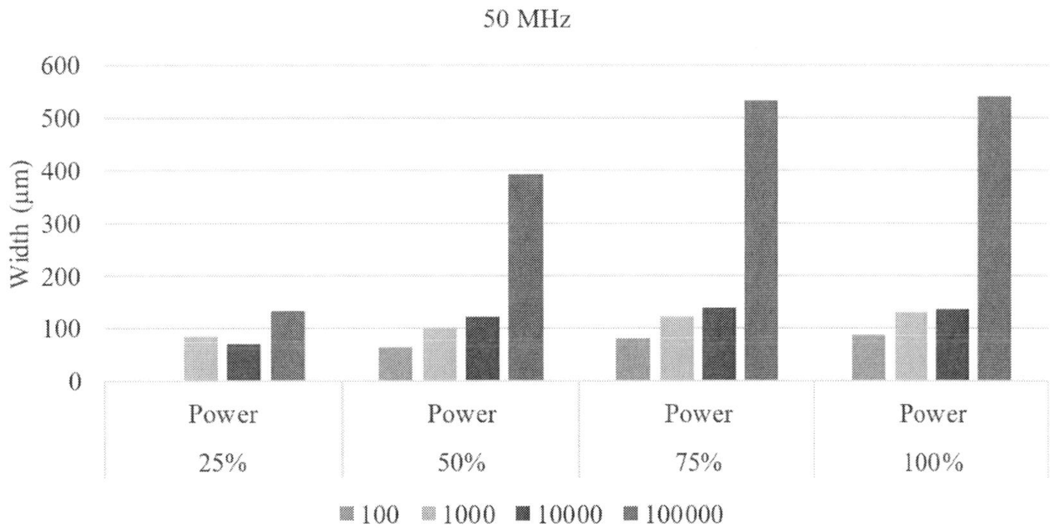

Figure 6. 50 MHz HAZ width as a function of pulses

The same phenomenon can be observed at 188 kHz, but here we were able to examine several pulse ranges, so the diagram shows the changes better as shown in Figure 7. Also, over 50 % average laser power, the heat affected zone increases after 100 pulses and can even reach width over 1 mm at higher power.

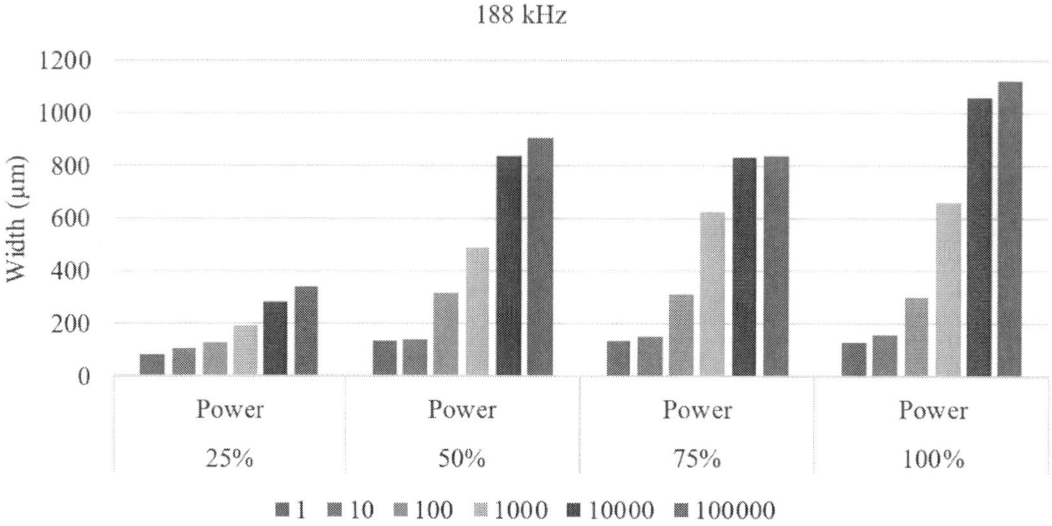

Figure 7. 188 kHz HAZ width as a function of pulses

The best structures were first selected by visual inspection. Compared to other similar patterns, we selected 2 images to show our results. Starting with the tests at 188 kHz, the results achieved there are shown in Figure 8. It is clear that the edges of the machining are sharp, forming a straight line. At this frequency, there was also enough lower power to get the structure right. The machining width is excellent, almost the same as the laser spot diameter, and the heat affected zone is.

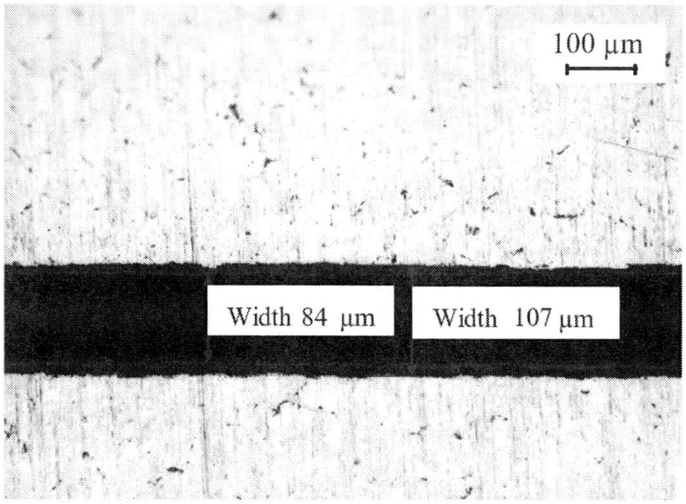

Figure 8. Results reached at 188 kHz

The set parameters are as follows: Frequency is set to 188 kHz, power is set to 25%. The number of pulses is not significant, only 10, and the number of repetitions is also 10.

After that we also checked at 50 MHz, what settings were needed to achieve similar results. Figure 9 shows the successful test at 50 MHz.

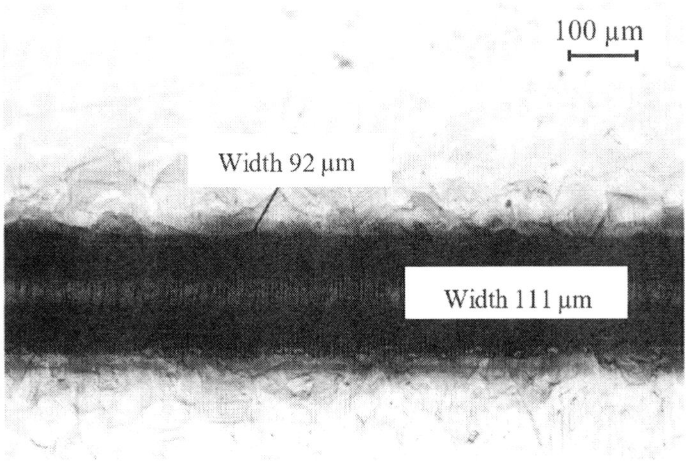

Figure 9. Results reached at 50 MHz

The image has sharp borders, which also give a straight line along the machining. The heat effect is seen to be minimal here, as with the previous frequency. The exact parameters at 50 MHz are as follows: Machining at 100% power. Pulse count of 1000, which was recorded with 100 repetitions.

4. Conclusion

Laser beam technology is one of the best machining tools available today. Its industrial development is progressing in enormous pace, and developments are almost impossible to follow. It is gaining ground in every area, which is understandable as it allows them to do work that would be difficult to do by other means. More and more material removal and machining tools are being replaced by laser units, as developments give this vision. They are more economical, faster, more accurate, less energy intensive than their traditional counterparts. These devices are most suitable for developing microstructures. During our research we were able to create the appropriate surface structures.

Summarizing the results of our research:

- Laser parameters greatly influence the surface structure of stainless steel,
- 188 kHz and 50 MHz also have parameters that achieve the desired surface structure,
- 188 kHz, 25% average laser power, 10 pulses and 10 repetitions caused 84,2 μm width of the lines,
- 50 MHz, 100% average laser power, 1000 impulses and 100 repetitions caused 92,4 μm width of the lines,
- very precise machining can be achieved with both frequencies with minimal heat effect.

Acknowledgments

This research was supported by EFOP-3.6.1-16-2016-000014. The Project is supported by the Hungarian Government and co-financed by the European Social Fund.

References

[1] Jean-Claude D and Wolfgang R 2006 Ultrashort laser pulse phenomena *Fundamentals, Techniques, and Applications* (Amsterdam)
[2] Jiwang Y. and Takayama N 2019 Micro and nanoscale laser processing of hard brittle materials *Elsevier Science* (Amsterdam)
[3] Youqiang X. 2017 High friction and low wear properties of laser-textured ceramic surface under dry friction *Optics & Laser Technology* **93** 24-32
[4] Yongchao Y. et al 2018 Ultra-short pulsed laser manufacturing and surface processing of microdevices *Engineering* **4** 779-786

[5] Orbulov I N and Májlinger K 2013 Microstructural aspects of ceramic hollow microspheres reinforced metal matrix composites *International Journal of Materials Research* **104** 903-911
[6] Weltsch Z 2016 The relation between the surface tension and the surface texture formation*Gradus* **3** 38-44
[7] Weltsch Z and Lovas A. 2013 The relation between the surface tension and the bulk oroperties of diluted silver based melts *Materials Science Forum* **729** 19-24

IOP Conf. Series: Materials Science and Engineering **903** (2020) 012026 doi:10.1088/1757-899X/903/1/012026

Optimization of material removal parameters by femtosecond laser pulses

Miklós Berczeli[1], Ferenc Tajti[1]

[1] Derpatment of Materials Technology, GAMF Faculty of Engineering and Computer Science, John Von Neumann University, Hungary

E-mail: berczeli.miklos@gamf.uni-neumann.hu

Abstract. Lasers are widely used and have become indispensable in many areas. Their application is very advantageous in many cases, the technology is very fast and high-precision machining can be achieved. Lasers are increasingly used in the automotive, aerospace, electronics, and medical industries. With this technology, we can perform almost any kind of material processing whether drilling, cutting, welding or even heat treatment. The laser can be well integrated into various machining systems and well automated. With the advancement of technology, we are capable of producing femtosecond pulsed lasers which machines are capable of high peak performance. Because of the high peak power and short pulses, other physical phenomena occur when the laser interacts with the material, so the method of material removal is different from longer laser impulses. Their application is expected to provide a much more controlled, more precise material removal, with very good surface quality, free of burrs and other dirt and deposits. One of the most important benefits expected from the use of technology is the minimal or no heat affected zone. In our research we are looking for a parameter that can achieve the most accurate material removal with a smaller heat affected zone.

1. Introduction

Lasers have evolved a lot since their launch and many new types of lasers have been developed over the last 50 years. The increase in performance, continuous improvement in laser beam quality and the development of laser beam conduction offers new opportunities for expanding applications. Lasers are widely used and have become indispensable in many areas. Their use is very advantageous in many cases, the technology provides high precision and fast machining. Lasers are increasingly used in the automotive, aerospace, electronics and medical industries [1]. With this technology, we can perform almost any kind of material processing, stripping, drilling, cutting, welding heat treatment or even surface treatment [2-4]. Laser beam technology can be well automated and integrated into machining processes [5].

Although laser technology is widespread, but femtosecond lasers have not spread yet because the equipment usually had very low power. Nowadays also the equipment with sufficiently high power is available. Although less widespread in the industry, science is paying ever greater attention to technology, and more research and studies are being conducted on the subject. Femtosecond lasers deliver tremendous peak power. Because of the high peak power and short pulses, other physical phenomena occur when the laser interacts with the material, so the method of material removal is different from general lasers [6-7].

IOP Conf. Series: Materials Science and Engineering **903** (2020) 012026 doi:10.1088/1757-899X/903/1/012026

Compared to machining with nanosecond or longer pulses, drilling or cutting with femtosecond pulses does not result in melting. In the case of long pulses or continuous operation, we are essentially talking about thermal machining, which only works for materials that absorb the specific wavelength at which the laser operates. The process of laser ablation is related to the relaxation time of electrons in the case of metallic materials. The relaxation time can be between 1-10 ps for metallic materials. For longer pulses, the material completely melts in the irradiated area. This results in an uncontrolled flow of molten material, leading to the formation of irregular patterns.

However, it should be added that this phenomenon can be easily influenced and used to create patterns. If the duration of the pulses is less than the relaxation time that exists for femtosecond pulses, the material does not melt. Femtosecond laser radiation is accompanied by several phenomena, which depend on the working conditions, laser parameters and material properties.

Experience has shown that the ablation threshold can be reduced by as much as an order of magnitude for femtosecond pulses compared to nanosecond pulses, which allows for much more precisely controlled material separation, and the thermal impact zone can be minimized. This is why ultrashort pulses can be used to machine materials that conduct heat well. Experiments have also shown that the interaction of material and ultrashort laser pulses is a very complex phenomenon, which is also strongly influenced by the properties of the laser and the material. Furthermore, unlike nanosecond pulses, the thermal conduction of electrons can play an important role in the process of dissipating absorbed energy. However, at high power densities (above 1015 W / cm2) the benefits of ultrashort pulses may be lost. In such cases, a significant amount of melt can already be observed, and a significant temperature difference can cause either deformation arms or cracks [8].

When machining with ultrashort pulses, we can expect higher efficiencies at higher frequencies, since several pulses reach the surface in the same time interval. For example, if 1000 pulses are generated in 1 second, the frequency is 1 kHz, 50000 pulses in the same time is 50 kHz. However, the increased frequency also means smaller intervals between each pulse, resulting in two phenomena. This is one that less time is available for the propagation and distribution of heat until the arrival of the next pulse, which clearly influences the phenomenon of heat accumulation. The heat absorbed by the workpiece does not contribute to the material removal, but causes a splash around the hole, something re-solidified on the wall of the hole, which has a very negative effect on the quality of the hole. Another problem is that the removed material, which contains both liquid and gas phases, and the resulting plasma have less time to leave the treated area, which can also result in the accumulation of the removed material at the bottom of the created cavity. This can result in deflection and absorption of the laser beam, energy does not reach the bottom of the hole, the hole does not deepen further. From this it can be concluded that the repetition frequency significantly influences the process of micromachining, its result.

If the frequency is increased, exceeding the frequency of a few hundred kHz, the delay time between each pulse is shorter than the thermal relaxation time and therefore heat accumulation develops at the machined point, resulting in an increase in temperature. Increased temperature, in turn, results in a lower ablation threshold.

The burr along the edges of the hole is due to the accumulation of thin layers of molten material which adheres to it under the action of vapor pressure. From this it can be concluded that if this accumulation can be reduced, if the energy density is reduced, until it just exceeds the ablation limit. The disadvantage of this is that it greatly reduces the ablation rate. Furthermore, only very shallow shapes can be created at energy densities just above the ablation threshold. The minimum required energy density increases strongly with the depth of the hole. Experience has shown that spiral drilling, originally developed for nanosecond pulses, can be used in the femtosecond range as well. In the case of spiral drilling, the melt can also move sideways, which is later removed by a subsequent pulse, so it is not necessary to move axially and travel the relatively long way to the hole entrance to exit the hole, unlike impact drilling technology.

With these lasers, we have the ability to make machinations that have been expensive or complicated before. We have the ability to machining materials very accurate and precise. We can also work with special materials such as stainless steel, aluminum, copper. In our study we investigated drilling with femtosecond laser on stainless steel [9].

2. Materials and Methods

In this paper we used stainless steel (AISI 304), one of the most common type of stainless steels. Machining stainless materials by other methods would be difficult, but very good results can be achieved with femtosecond laser. Before the laser machining, the surface of the steel was cleaned with methanol.

The laser drilling process was performed using a Coherent Monaco Femtosecond Laser with wavelength of 1035 nm. The laser impulses had 277 fs of width and the maximum average power was 40 W. The beam was focused on the surface with a diameter of 80 μm. There are properties that are given, we cannot change them. Such as the wavelength and modus of the laser beam. However, there are parameters that we can change within certain limits. Such parameters include frequency or pulse width. Another variable is the time of interaction between the material and the laser beam. There are several ways to control this: you can set how long the process lasts or how many laser pulses are received at a given point. The goal is to determine the parameters that can deliver the finest, highest quality material removal. We tried to find a setting where the material removal is as regular as possible, the heat affected zone is as small as possible.

Two setup strategies are used for the tests. In the first case the frequency was the fixed parameter (188 kHz and 50 MHz). These parameters are the extreme values of the frequency. The variable parameters were the laser average power (50 %- 100 %) and the number of pulses fired at one point (1 – 60000).

In the second case, the repetition number was set to a fixed value (10000) and the variable value was examined at 10 adjustable frequencies (188 kHz, 250 kHz, 330 kHz, 500 kHz, 750 kHz, 1 MHz, 2 MHz, 4 MHz, 10 MHz and 50 MHz) at different power levels.

For evaluation, the laser-drilled holes were examined under a microscope. The results were compared based on the hole and the heat affected zone diameter, as shown in the Figure 1.

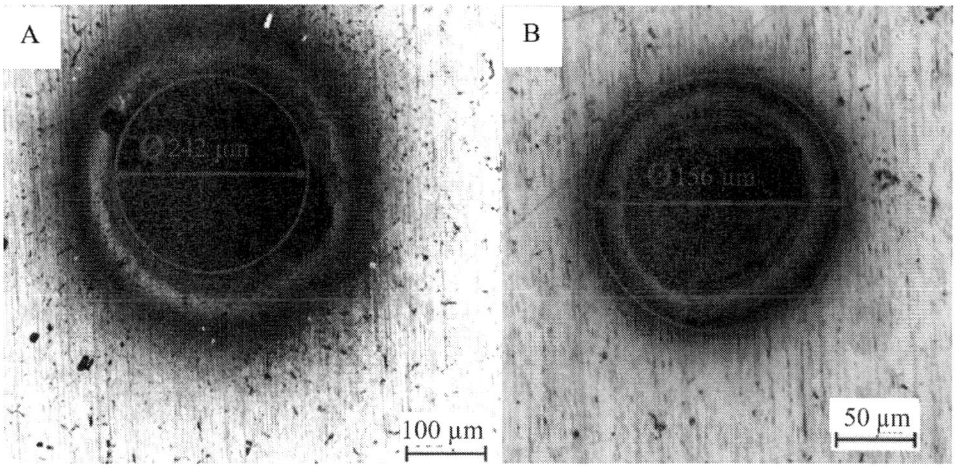

Figure 1. A) hole diameter measurement, B) heat affected zone diameter measurement

3. Results

In the first case as expected, the amount of material removed increased as the number of pulses applied to the surface. Also, the diameter of the hole formed gradually increased as the pulse rate increased. The diameter of the craters also increased gradually as the pulse rate increased.

3.1. Results at 188 kHz

Along with these parameters, changes in the surface of the material were visible after a single pulse. As the number of pulses increases, more marked changes can be seen in the surface of the material. Already after 4 pulses, it is noticeable that material has been removed in the treated area, and a crater has formed which is gradually increasing in depth. Figure 2 shows the hole diameter as a function of the number of pulses delivered to the surface at two power levels.

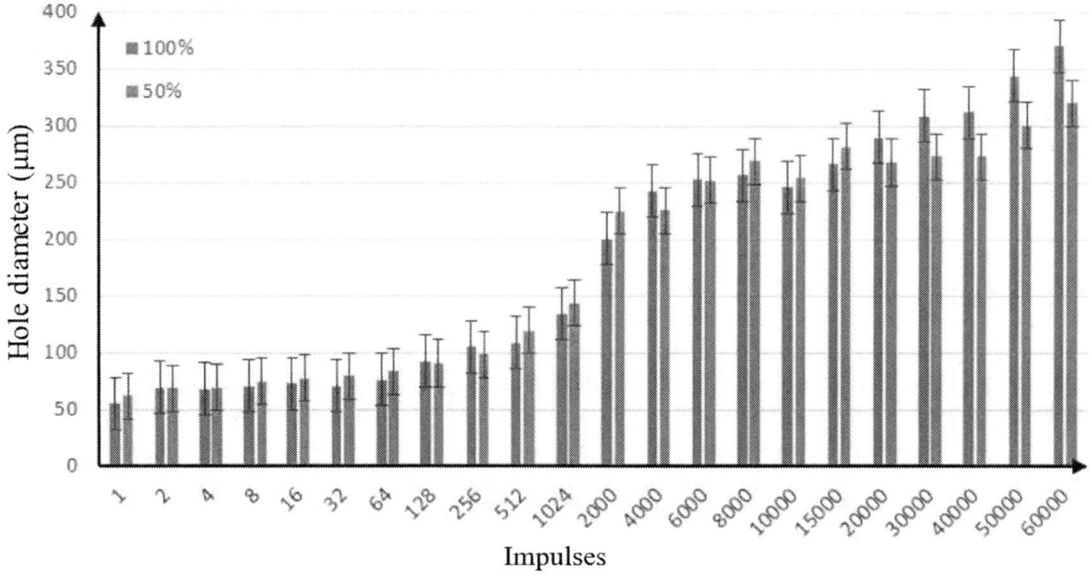

Figure 2. Hole diameter as a function of pulse rate at two power levels at 188 kHz

Overall, the hole diameters showing an upward trend. It can be observed that between 2 and 64 pulses the central diameter of the crater remains almost unchanged, typically around 70 μm, but from the recordings it appears that its depth increases slightly. Above the 2000 repetition number, the depth and geometry of the hole is such that we cannot see it with the help of a microscope. Between 4,000 and 15,000 iterations, stagnation is observed, with neither the hole nor the heat affected zone increasing significantly. Above this range, both diameter and heat affected zone show further increases. Figure 3 shows the evolution of the heat affected zone, which is similar to the change in hole diameter.

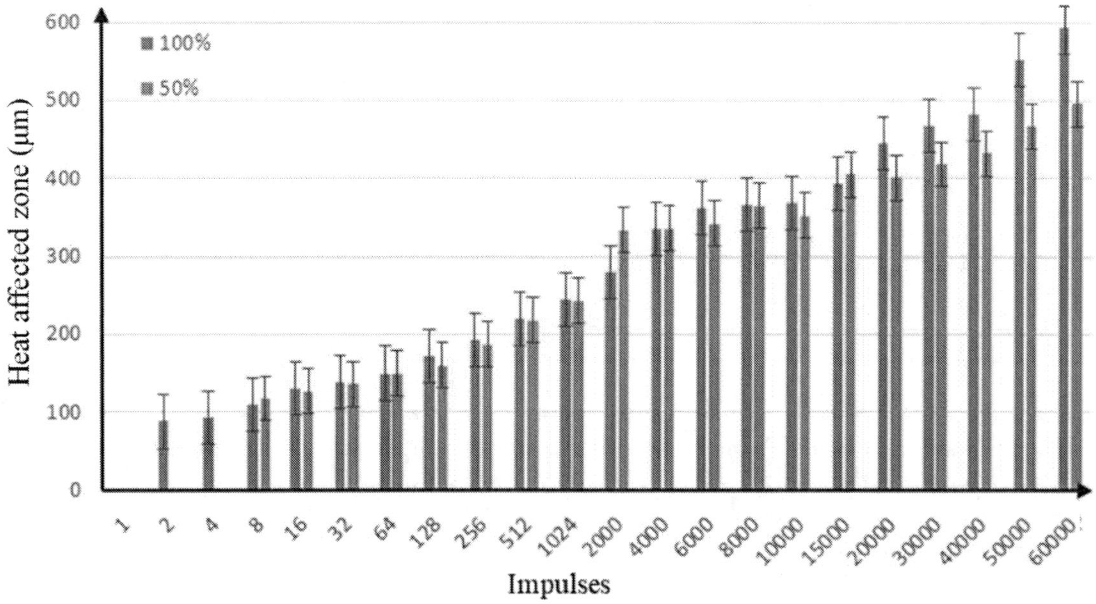

Figure 3. Changes in the size of the heat affected zone as a function of frequency at two power levels at 188 kHz

The diagrams show that there is little difference between the diameters of the holes created by the two laser parameters, but the difference in the heat affected zone is larger. From this it can be concluded that with less power more accurate material removal can be achieved.

3.2. Results at 50 MHz

Now examining material removal with the highest adjustable frequency. Because the maximum average power is given, in our case 40 watts, two important factors are significantly different from experiments at 188 kHz. One is the time elapsed between the two pulses, which is obviously much smaller at 50 MHz, and the other is the pulse energy, which is also significantly below the 188 kHz values. The two factors affect the properties of the hole in different ways. According to the literature, a smaller time interval between each pulse may result in greater heat accumulation, which may not only improve the efficiency of the process but also reduce the quality. In contrast, lower pulse energy can result in a more controllable process. Figure 4 below shows the evolution of hole diameter at 50 MHz at two different power levels.

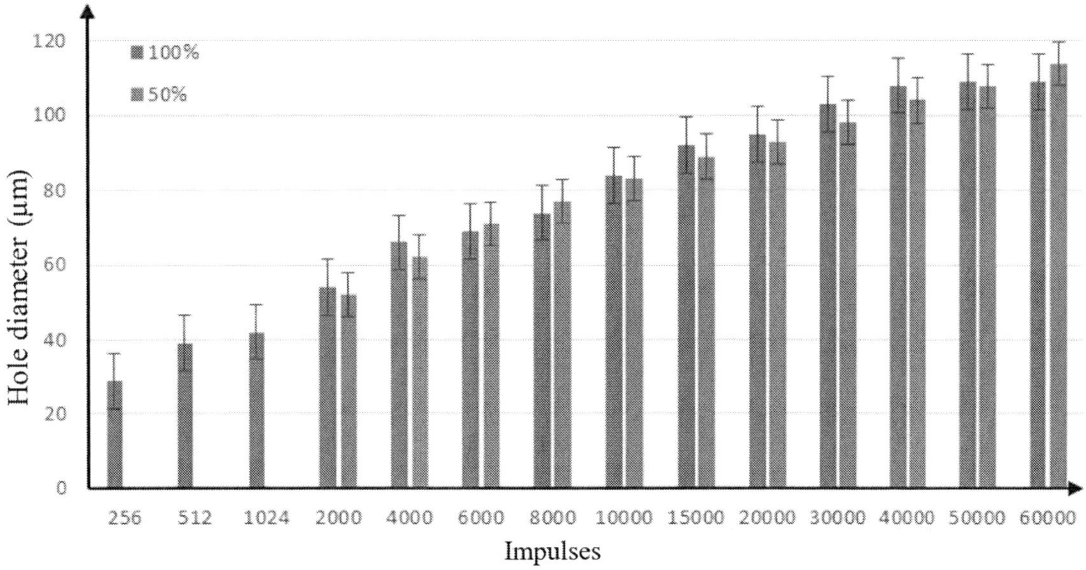

Figure 4. Development of hole diameter as a function of pulse rate at two power levels at 50 MHz

If you look at the hole's diameter, you can see that they are much smaller than the 188 kHz frequency. It is striking that after 256 pulses the laser has a noticeable effect, which is even smaller than at 188 kHz per pulse. The tendency is the same at the lower frequency, the hole and the heat affected zone increase with the increase of the pulse rate, but the size is much smaller than at the lower frequency. Figure 5 shows the evolution of the heat affected zone at the frequency of 50 MHz.

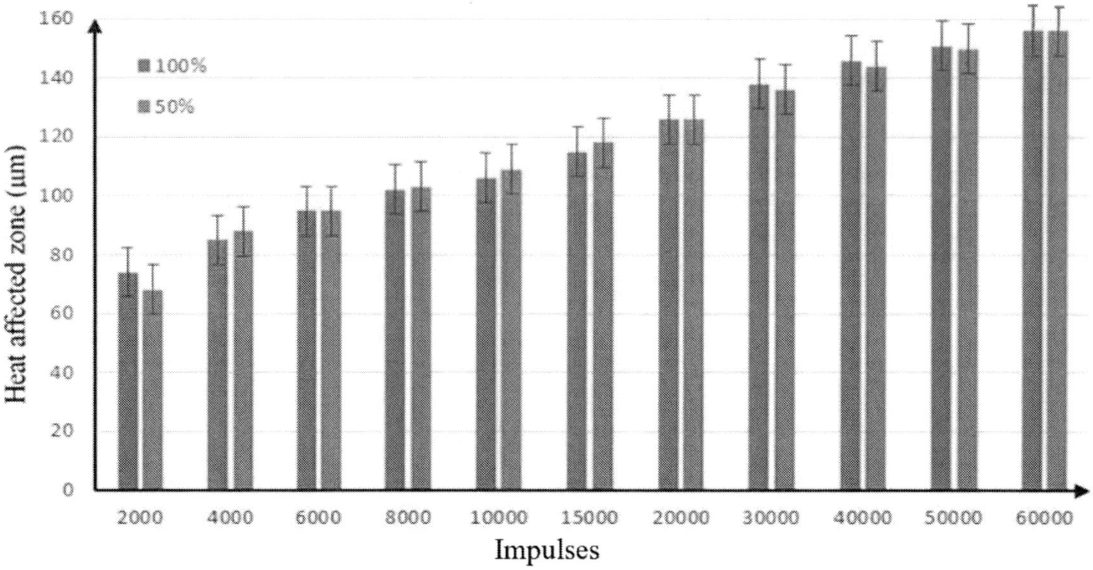

Figure 5. Changes in the size of the heat affected zone at two power levels at 50 MHz

Although the quality of machining has improved, the heat affected zone is still significant. The results at 50 % power are almost identical to those at 100 % power. Apparently, there is no significant difference in quality or geometric properties. From the data it can be seen that with the two power levels, similar results were obtained, the diameter of the holes increased evenly throughout the range.

The above shows that the frequency has a great influence on the diameter and quality of the holes. The results clearly reflect the significant difference between pulse energies, at 188 kHz, even a single pulse left a trace larger than 256 pulses at 50 MHz.

3.3. Second treatment strategy

In the second phase, the effect of frequency was examined in more detail. In this case, the pulse number was not changed, it was 10,000 each time. The variable setting was power and frequency. Figure 6 shows the evolution of hole diameter as a function of frequency at three power levels.

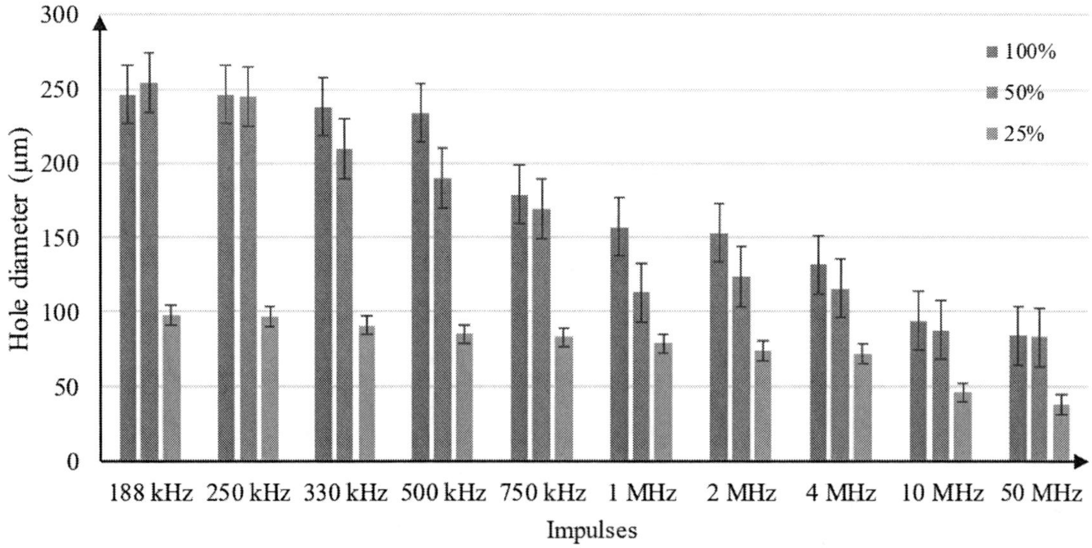

Figure 6. The evolution of hole diameter as a function of frequency at three power levels

As expected, as the frequency increases, hole diameters typically decrease. When looking at 100 % performance, it can be seen that the changes between 188 kHz and 500 kHz are almost negligible, and the hole diameter decreases very slightly. Then, at 500 kHz, the diameter begins to decrease sharply. At the same average power, peak power should theoretically increase with frequency decrease. That is, at 188 kHz, theoretically, much higher peak power is achieved, along with pulse energy.

If we examine the heat affected zone, similar processes can be observed. Figure 7 shows the evolution of the heat affected zone size as a function of frequency at three different power levels.

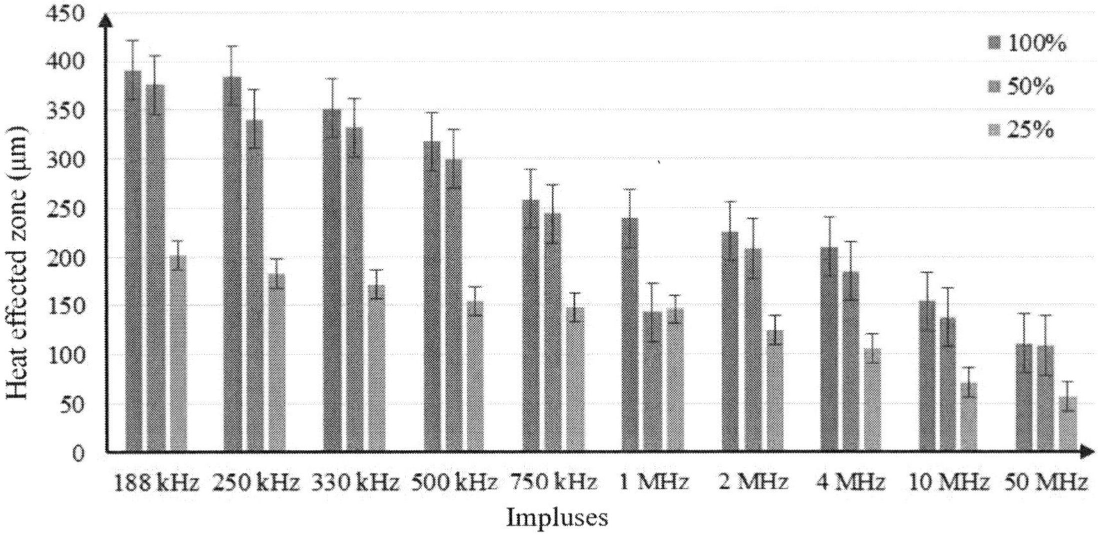

Figure 7. Size of the heat affected zone as a function of frequency at three power levels

In contrast to the hole diameter, which was constant between 188 kHz and 500 kHz at 100 % power, the heat affected zone decreases with increasing frequency. This contradicts the expectation of the literature that an increase in frequency can cause heat congestion, which can increase the heat affected zone. The size of the heat affected zone alone carries little information without the size of the hole. More important is the size of the heat affected zone relative to the size of the hole. Figure 8 shows the size of the heat affected zone relative to the hole size.

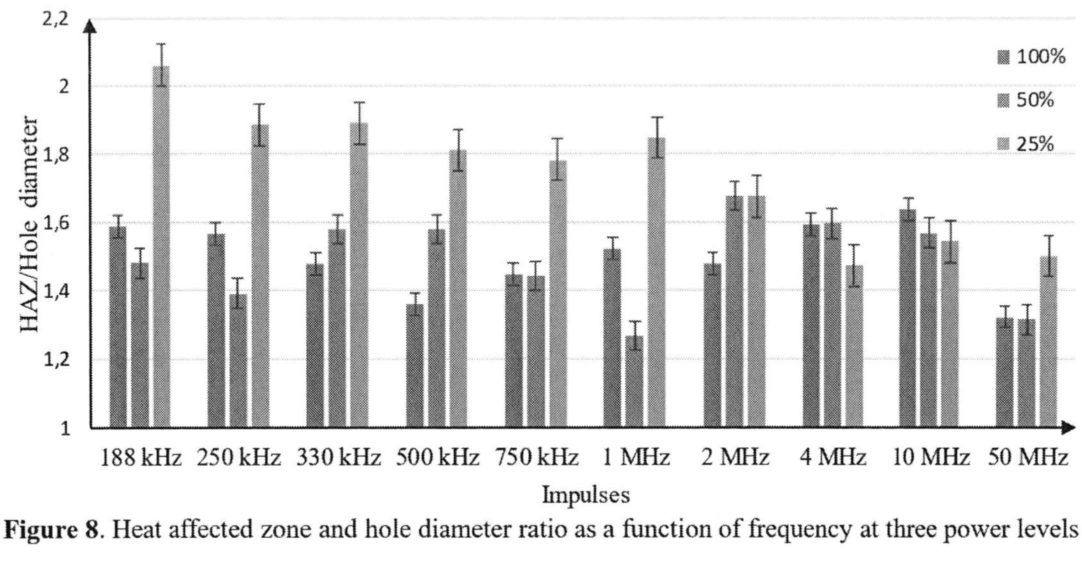

Figure 8. Heat affected zone and hole diameter ratio as a function of frequency at three power levels

In this case, it is more difficult to talk about clear trends, the relative size of the heat affected zone seems to be less dependent on frequency. In addition to the different power levels, there was a minimum and maximum heat affected zone at different frequencies. In terms of proportions, the smallest thermal effect occurred at 50% power and 1 MHz, and the highest at 25% and 188 kHz. Performance affects the relative size of the heat affected zone. At almost all frequencies, 25% power produced the highest rate of heat effect, and in most cases, the deviation was very significant. The lowest rate of heat effect is typically 50%. The proportion of the laser beam's intensity distribution may also explain the higher proportion of heat effects. At lower power, only the central part of the beam has enough intensity to remove material, although the outer parts are subject to material changes, but material removal is not significant.

However, it is interesting that the decreasing power results in a larger heat affected zone, since the impulse energy decreasing due to the frequency typically results in a better quality end result and a smaller heat affected zone.

4. Conclusion

During our experiments it became clear that pulse energy plays a very important role in the process of material removal. Although in our case the frequency was the primary variable, but since each series was made with a constant average power, the change in frequency obviously brought about a change in the impulse energy.

The significant difference between the results of experiments at different frequencies is actually due to the pulse energy, but the effect of the frequency is not so significant. However, in the future, the effect of frequency alone should be investigated with constant pulse energy. The often mentioned advantage of femtosecond lasers is that no or only minimal heat effects are produced. With lower pulse energy, a clearly smaller heat affected zone was produced, and we think it is possible that even with a significant decrease in frequency we can achieve even better results.

In our research, we found that high pulse energy results in poor quality, but low pulse energy provides better control and more precise machining. We can produce a very high quality hole with 1MHz frequency, 50% power, 10,000 pulses, which has small heat affected zone.

Acknowledgments

This research was supported by EFOP-3.6.1-16-2016-000014. The Project is supported by the Hungarian Government and co-financed by the European Social Fund.

References

[1] Antar M et al 2016 High speed EDM and laser drilling of aerospace alloys *Procedia* **42** 526–531

[2] Weltsch Z and Lovas A. 2013 The relation between the surface tension and the bulk oroperties of diluted silver based melts *Materials Science Forum* **729** 19-24

[3] Weltsch Z 2016 The relation between the surface tension and the surface texture formation *Gradus* **3** 38-44

[4] Orbulov I N and Májlinger K 2013 Microstructural aspects of ceramic hollow microspheres reinforced metal matrix composites *International Journal of Materials Research* **104** 903-911

[5] Mishra S and Yadava V 2015 Laser beam micromachining *Optics and Lasers in Engineering* **73** 89–122

[6] Pattanayak S and Panda S 2018 Laser beam micro drilling *Lasers in Manufacturing and Materials Processing* **5** 366–394

[7] Huang H et al 2014 Optical metrology of a large deformable aspherical mirror using software configurable optical test system *Optical Engineering* **53** 0091-3286

[8] Nedialkov N N et al 2004 Ablation of metals byultrashort laser pulses *Applied Physics* **37** 638 – 643

[9] Kamlage G et al 2003 Deep drilling of metals by femtosecond laser pulses *Applied Physics* **77** 307–310

IOP Conf. Series: Materials Science and Engineering **903** (2020) 012027 doi:10.1088/1757-899X/903/1/012027

Producing compressive stresses in the surface layer of spheroidal graphite cast iron by laser hardening

Anna Maloveczky[1*], Ádám Filep[2], Gábor Buza[1], Ádám Vida[1]

[1]Bay Zoltán Nonprofit Ltd. for Applied Research, Kondorfa Street 1, Budapest, 1116
[2]University of Miskolc, Miskolc Campus, Miskolc, 3515

E-mail: anna.maloveczky@bayzoltan.hu

Abstract. The purpose of the examined laser beam induced surface modification technology was to produce compressive stresses in the surface layer of the workpiece. The presence of residual compressive stresses increases the fatigue limit. The compressive stresses generated by investigated laser surface treatment were compared to shot peening technology results, and has been found to be equivalent.

1. Introduction

Compressive stress works contrary to the forces that occur during fatigue loading, thus increasing the fatigue limit. During shot peening the crystal lattice is distorted, and the lattice constant decreases due to the impact of the particles, thus creating residual compressive stresses in a material. The goal was the same during laser hardening, but on a different principle. The advantage of laser beam technology is that it is precisely controllable, more localized than shot peening, and can also handle hard to reach areas [1-5].

Transformations during laser hardening are the followings: At room temperature in cast iron body-centered cubic lattice structured ferrite presents, which heated above A_{C3} transforms to face-centered cubic lattice structured austenite. Ferrite can only dissolve a few carbon atoms, and austenite can the most, compared to other forms of iron. Above A_{C3}, carbon diffuses from graphite into austenite. With a sufficiently high cooling rate and carbon content, austenite will not transform back to ferrite, but martensite. Martensite (body-centered tetragonal) lattice is a distorted version of ferrite lattice, because it has an intersticial carbon atom in the structure. High cooling rate is needed to prevent the diffusion of carbon. So martensite has a lattice structure that is supersaturated with carbon, and has larger unit cell than both the ferrite and the austenite. (see Figure 1).

This is how the compression stress is created, but also the direction of the heat transport plays an important role in generating residual stress: During bulk hardening, the workpiece is cooled from the case to the core by the applied cooling medium, so that the martensitic transformation (i.e., the compressive stress of its formation) resulting in a higher specific volume also occurs from the case to the core.

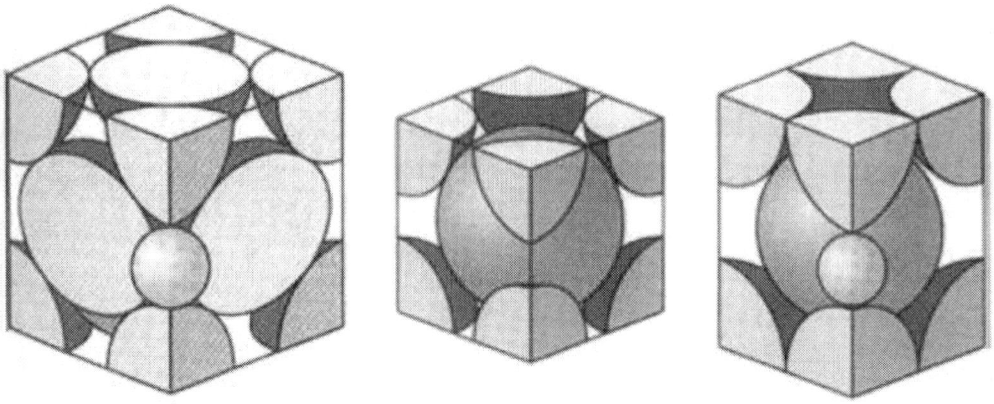

Figure 1. Unit cell of austenite, ferrite, martensite (from left to the right)

As a result, tensile stress is created near the surface and compression stress is developed far from the surface, in the core of the workpiece. This process is reverse and the results are the opposite when performing laser hardening: because the whole piece does not heat over the AC3, only the surface layer, so the workpiece itself functions as a cooling medium, with significant efficiency

Therefore, residual compressive stresses can be produced in the surface by laser beam hardening but cannot be done by bulk hardening [6-8].

2. Experimental investigations

2.1. Test material

The investigated material was GGG50 (material no. 0.7050) according to DIN EN 1563, a ferritic-pearlitic spheroidal cast iron. The chemical composition is shown in Table 1. The metallographic image of the base material is shown on Figure 2.

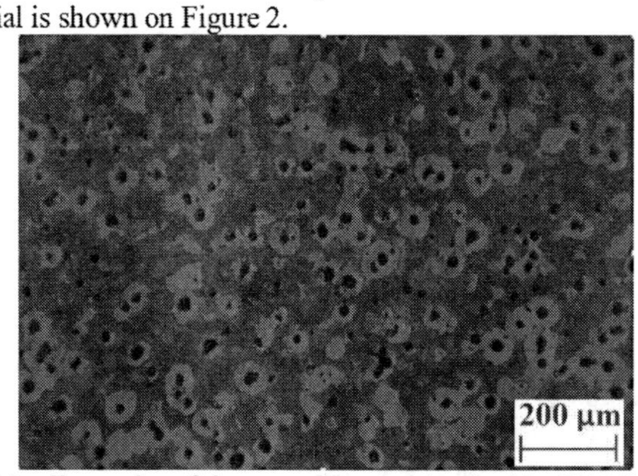

Figure 2. Metallography of the test substance (etched in 5% Nital solution)

The base material consists of three microstructural elements: ferrite (white), pearlite (light brown) and graphite (black). The graphite morphology is spheroidal and has an average diameter of 30 μm.

Table 1. Chemical composition of our test substances compared to the standard

	Chemical components (wt%)				
	C	**Si**	**Mn %**	**Cu %**	**Mg %**
Test substances	3.55	2.4	0.28	0.48	0.041
DIN EN 1563	3.5 – 3.7	2.2 – 2.8	max. 0.4	0.4 – 0.6	0.035 – 0.055

2.2. Laser equipment and parameters

A Trumpf TLC 105 high-frequency CO_2 laser device with maximum 5 kW output power was used for the experiments. The laser beam was deflected by a 5 axis CNC machine tool. One-axis scanning head with a vibrating mirror was used for laser hardening.

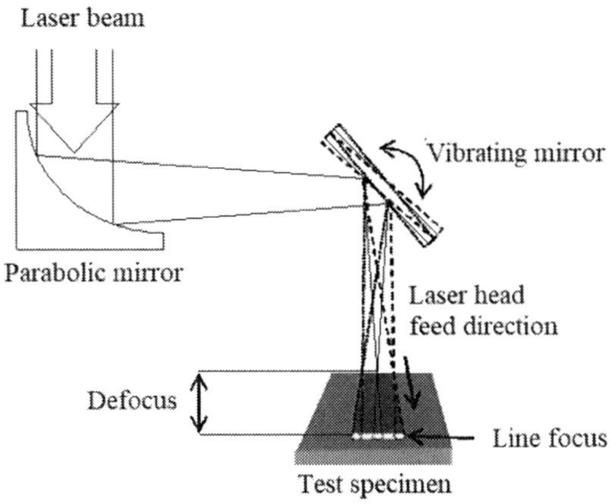

Figure 3. Sketch of the vibrating mirror and the line focus design

The vibrating mirror causes virtual line focus on the surface. Harmonic oscillation was used at 100 Hz frequency with an angular deviation of ± 4°. (as the sketch shows in Figure 3). This scan head is able to produce a 20 mm wide treated band, which helps to prevent the overlapping. A graphite spray was used to improve the absorption of the laser beam. Defocus was 84 mm, except for sample 12 and sample 14.

The applied laser parameters for the hardening experiments can be seen in Table 2.

Table 2. Laser hardened test specimens (with sample numbers 11-18), and the experimental parameters

Feed speed (mm/min)	**Laser power (kW)**			
	2.8	**3.4**	**3.8**	**4.5**
200	18	17		
250			16	15
300				11, 12*, 14**
400				13

* 130 mm defocus, ** 100 mm defocus

3. Results and discussion

3.1. Metallography

In the following, two representative laser hardened specimens will be introduced (with sample number 11 and 17). The metallographic images were taken by a Keyence VHX2000 digital optical microscope (Figures 4, 5).

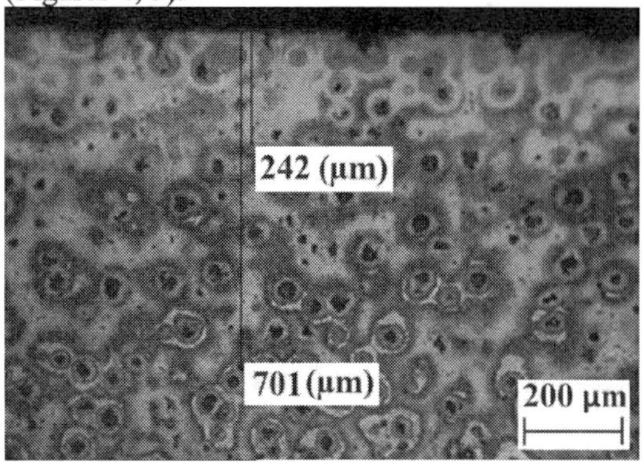

Figure 4. Metallographic image of sample 11 (etched in 5% Nital solution)

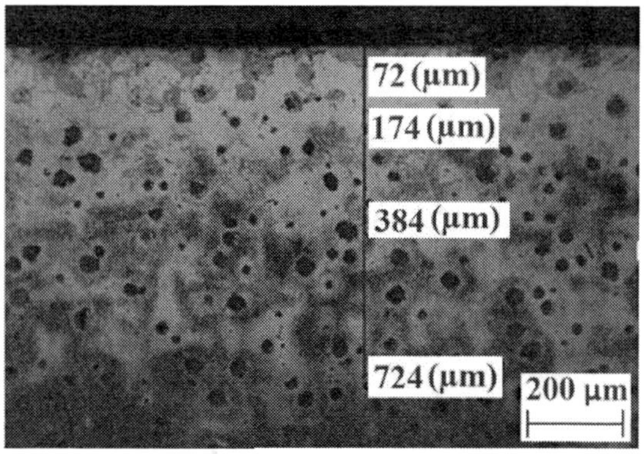

Figure 5. Metallographic image of sample 17 (etched in 5% Nital solution)

The amount of martensite were significant in both samples, and both the dark and light coloured matrix surrounding the microstructural elements was martensitic, as shown in Figure 6. The surface of sample 17 was melted. For a more exact investigation, sample 17 was over-etched to give a better view of the microstructural structure (Figure 6).

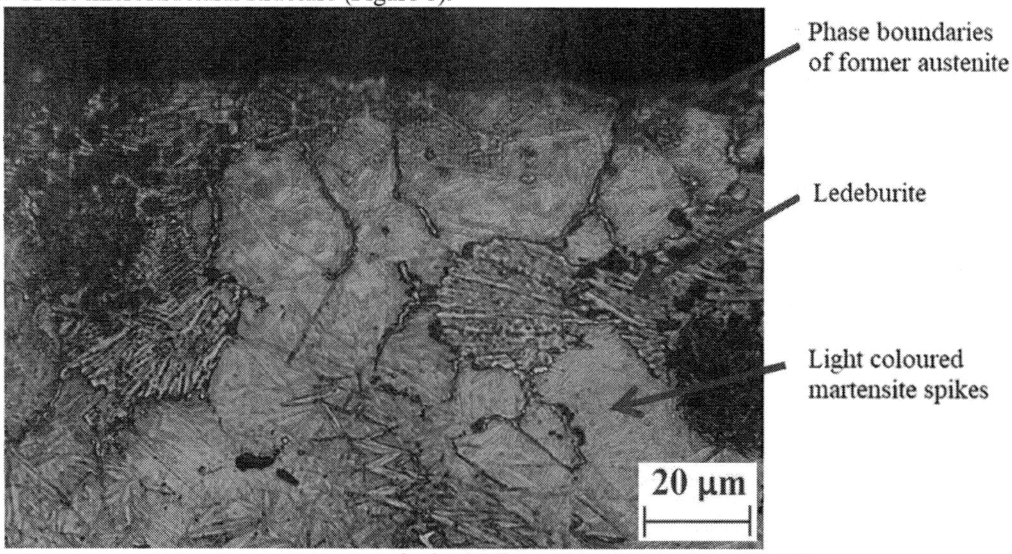

Figure 6. Over-etched metallographic image of sample 17 (etched in 5% Nital solution)

The over-etched image shows that the light colour matrix was indeed (spiky) martensitic, which was more difficult to etch due to its high carbon content. This produces a higher compressive stress than the dark spiked martensite because of its higher carbon content. In some parts near the surface the sample was melted, for example along the former austenite phase boundaries, and ledeburite crystallized from the melt. (Ledeburite is the eutecticum of carbon and iron. It consists of cementite and pearlite microstructural elements.)

3.2. Residual stress analysis by X-ray diffraction

The principle of diffraction analysis is that residual stress - from a crystalline point of view - results in the change of the lattice parameter. Residual stress can be calculated by measuring the changes in the distance of the lattice plates in the material.

The test was performed by a Cr X-ray tubed Stresstech Xstress 3000 G3R X-ray diffractometer and an XTronic software. The values were calculated from the shift of ferrite {211} reflection. The measurements were carried out in 3-3 tilt position (45°-45°) of the goniometer. The exposure time was 10 seconds and the beam spot diameter was 3 mm.

Residual stress values were measured outward from the centre of the treated band (Figure 7, Table 3).

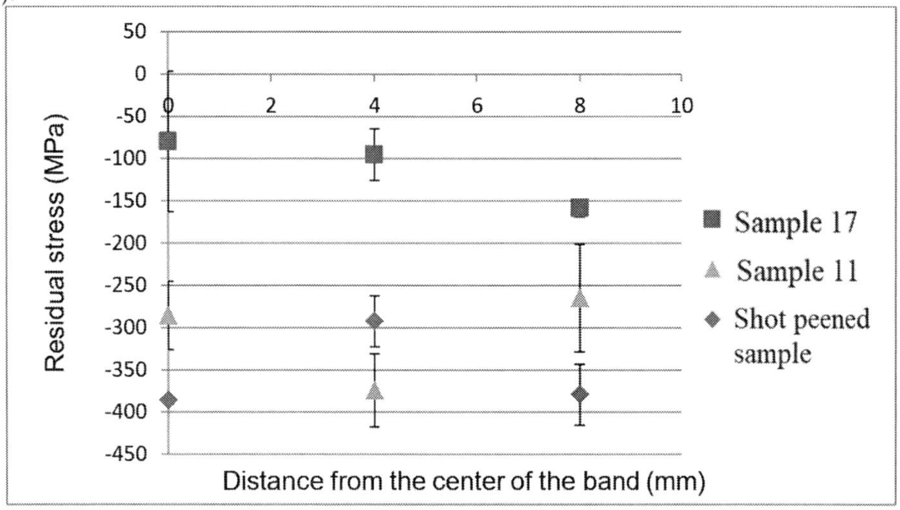

Figure 7. Average residual stresses of the samples (+ represents tensile, - represents compressive stress)

Table 3. Residual stress values of the samples

	Distance from the centre of the band (mm)		
Samples	**0**	**4**	**8**
Shot peened sample	-386.1	-292.8	-379.1
Sample 11	-285.7	-374.3	-265.1
Sample 17	-79.9	-95.6	-159.0

The character of the stress distribution is not comparable in the case of two LASER hardened samples due to the surface melting of sample 17. On the other hand, taking into account that sample 11 was not melted by the beam, the measured values can be compared to that shot peened one. Based on the measured values, in sample 11, the compressive stress is as high as obtained by shot peening technology (~ 300-400 MPa).

4. Conclusion

According to the metallography, there were no cracks in the laser beam hardened samples, and a significant amount of martensite phase was formed in both samples.

The effective compressive stress for increasing the fatigue limit is 300-400 MPa. In sample 11, a similarly high compressive stress (300-380 MPa) was achieved.

Although metallographic examinations showed that sample 17 contained the more high-carbon content martensite (which was expected to produce higher compressive stress), it still has much lower compression stress than sample 11. This is explained by the fact that sample 17 had too high energy input during laser beam treatment, which resulted in melting. There were more ledeburite grains in the surface layer and the former austenite phase boundaries were visible. The melt produced tensile stress when cooled, which is contrary to the desired effect.

Therefore, it is advisable to select laser beam parameters for the optimum: producing the highest energy input while avoiding surface melting.

Our study shows that laser hardening can be a good alternative of the conventional technologies - like shot peening and cold rolling - used for increasing the fatigue limit. Laser technology has the advantage of being highly productive and controllable, and also enables to treat hard to reach areas. We plan to test our theory further by submitting the samples to fatigue tests.

References

[1] Krawitz A D 2001 *Introduction to diffraction in materials science and engineering* (USA: John Wiley & Sons INC)

[2] 2002 *Handbook of Residual Stress and Deformation of Steel* ASM International (USA: ASM International)

[3] Martin U, Altenberger I, Scholtes B, Kremmer K and Oettel H 1997 *Cyclic deformation and near surface microstructures of normalized shot peened steel* (Material Science and Engineering A vol 246) pp 69-80

[4] Wang C, Jiang C, Zhaoa Y, Chen M and Ji V 2017 *Surface mechanical property and residual stress of peened nickel-aluminum bronze determined by in-situ X-ray diffraction* (Applied Surface Science vol 1) p 420

[5] Dalaeia K, Karlssona B and Svenssona L E 2010 *Stability of shot peening induced residual stresses and their influence on fatigue lifetime* (Procedia Engineering vol 2) pp 1008-15

[6] Kattoura M, Mannava S R, Qian D and Vasudevan V K 2017 *Effect of laser shock peening on elevated temperature residual stress, microstructure and fatigue behavior of ATI 718Plus alloy* (International Journal of Fatigue vol 1) p 104

[7] Mertinger V Filep Á and Benke Márton 2015 (in Hungarian) *A lézersugaras megmunkálások hatására kialakuló maradó feszültségállapotok* (Gépgyártás vol LV) p 1

[8] Buza G Kálazi Z Kálmán E and Sólyom J 2004 (in Hungarian) *Öntöttvas felületi lézersugaras átolvasztásának néhány fémtani kérdése* (Bányászati és kohászati lapok vol 2) p 137

Application of Rietveld refinement in orientated material structures

E Nagy [1], F Kristály[2], V Mertinger[3]

[1] Researcher, MTA-ME Materials Science Research Group, H-3515 Miskolc-Egyetemvaros
[2] Researcher, Institute of Mineralogy and Geology, University of Miskolc, H-3515 Miskolc-Egyetemvaros
[3] Professor, Institute of Physical Metallurgy, Metalforming and Nanotechnology, University of Miskolc, H-3515 Miskolc-Egyetemvaros

E-mail: femzsofi@uni-miskolc.hu

Abstract. The phase analysis based on the X-ray diffraction method is a well-known procedure for determining the phases crystalline in the sample. The phase identification is simple in case of the isotropic powder samples, however the standard investigation protocol cannot always be used, e.g. in case if the presence of anisotropy like preferred orientation (PO) is observed. In case of steels, the diffracted intensity of a given phases is influenced by the directional phase development having a needle character caused by the martensitic transformation; some reflections can even disappear from the diffractograms as a consequence of the strong PO. One impacting consequence of PO is over quantitative evaluation, percentages close to actual values are obtained only if diffracted intensity is integrated for the whole measured 2θ range for each phase and anomalous intensities caused by PO are corrected.
Incorrect quantitative values were obtained by using the full-profile analysis method in the course of our earlier investigations, because no correction method had been available for the recognition and evaluation of the orientated structures in phase-identification of metals. By using the Rietveld refinement method for whole measured 2θ range, both the PO and the size-strain broadening effects are accounted for. Strain results from stress developing in the lattice and can be taken into consideration in the calculated model and results are corrected to accurately describe the real structure. In the course of our research work, orientated, complex structures developing in the Cu-base alloys were investigated by using the Rietveld refinement method. The quantity of phases was determined in the different samples in which clear signs of the used technological processes could be observed in the structures. The character of orientation appearing in the structure was observed, evaluated and the necessary corrections were performed.

1. Introduction

Complex structures showing clearly anisotropic character (size, strain and crystallite arrangement) often develop in the course of the solidification or transforming during cooling in the metallic alloys [1]. The crystallite arrangement anisotropy – lack of randomness – results in the preferred orientation (PO) of phases. PO and the stress of lattice-structure are not taken into consideration by the majority of methods suitable for the exact determination of phases quantity during the identification of phases developed in the microstructure.

In case of alloys, the production as well as the applied technological processes leave clear signs in the microstructure. An oriented structure can develop under the influence of heat extraction in the cast structure while the uniaxial- or multiaxial deformation of the material occurs by the forming processes e.g. by rolling or drawing where the main orientation is determined by the direction of deformation [2,3]. The PO can appear in metals as a result of the different heat treatment methods as well, e.g. following the crystal-oriented phase transformation [4]. Moreover, it is very important to take into consideration the dissolution of the individual alloying elements in the crystal lattice and the change of unit cell building up the crystal lattice in case of the alloy-families [5].

The phase analysis based on the X-ray diffraction method is a basic procedure for determining the phases existing in the sample. It is a universally recognized method in case of isotropic materials consisting of some components (mainly in case of powder samples) however it is probable that the standard investigation protocol can rarely be used owing to the presence of anisotropy. The diffracted intensity received from the measured phases is influenced by the presence of PO; certain reflections can even be missing from the diffractogram. Quantitative results are strongly altered by presence of PO, since phase percentages are calculated from measured intensity. The presence of PO leaves a clear sign on the diffractogram, and it is necessary to identify these signs and correct their effect [6,7]. E.g. the strong PO for phase x on given $d_{(001)}$ peak results in increased measured diffracted intensity, which needs to be corrected in relation to all the measured peaks to obtain the theoretical intensity ration, a.k.a. relative intensity.

All the reflections coming from the material are taken into consideration in case of the phase analysis performed by using the full pattern analysis applied mostly in case of metallic materials. In an ideal case, the integrated intensity of all the possible reflections are taken into consideration for each phase existing in the material, however merely the reflection appearing in the diffractogram can be taken into consideration in case of orientated structures. Correction methods are available in this procedure, by which the presence of anisotropy can be taken into consideration and correct for the orientation effects, like March-Dollase or spherical harmonics [8]. A realistic structure can be described more exactly by using the Rietveld refinement method [9,10] which relies on minimizing the differences of the measured pattern in relation to a calculated pattern by minimizing the sum of least squares. The method is convolution based, instrumental profile is defined to calculate the individual diffraction pattern of the identified phases, while background and baseline are modelled by polynomial functions. Peak broadening is modelled by crystallite size and strain contribution. In case of the phase analysis performed by the Rietveld refinement method, it is possible to take into consideration and to correct both the PO and the strain accumulated in the lattice, while crystallite sizes are calculated [9]. During this procedure, each phase is fitted individually and the unique PO characterizing the phase and the lattice strain are corrected.

In the course of our experiments, the usability of Rietveld refinement method was investigated in the complex structures developing as a consequence of heat treatment and plastic deformation in the Cu-based alloys. The microstructure of samples was investigated by optical and electron microscopy method, the phases were identified by X-ray diffraction method. The phase amounts were determined by Rietveld refinement method.

2. Experimental

The investigations were performed on Cu-Zn alloy of Standard No. C35000 medium leaded brass. The composition of alloy can be seen in Table 1.

Table 1. Composition of C35000 alloy

Chemical composition (wt%)				
Cu	**Zn**	**Pb**	**Mn**	**Si**
59.3	36	1	2.2	1.5

IOP Conf. Series: Materials Science and Engineering **903** (2020) 012028 doi:10.1088/1757-899X/903/1/012028

Samples of complex structures used for performing the investigations were obtained by heat treatment and plastic deformation. The heat-treated sample was homogenized at a temperature of 800°C for half an hour, it was tempered at 550°C for 4 hours and it was cooled on air. The plastic deformation of the hot-formed sample was performed by forging at a temperature of 750°C. The structure of samples was investigated by using optical- and scanning electron microscope (SEM); the chemical composition of phases was analysed by energy dispersive X-ray (EDX) microprobe. The phases developing in the microstructure were identified by X-ray diffraction (XRD) and they were matched with the microscopy results. Grinded and polished samples were used for the investigations, after etching with 5% hydrochloric acid and ferric-chloride solution. A Zeiss Imager m1M type light microscope and a Zeiss EVO MA 10 type scanning electron microscope were used for the microscopic investigations performed in the Institute of Physical Metallurgy, Metalforming and Nanotechnology of University of Miskolc. The XRD investigations were performed on a Bruker D8 Discovery XRD equipment, in Bragg-Brentano geometry with Cu Kα source (40 kV and 40 mA generator settings) in a range of 20-120 °(2θ), an increment of 0.01 °(2θ) and a counting time of 1s using a LynxEye X-ET energy dispersive detector with linear 2° opening. Evaluation was done in Bruker EVA 5.0 software and a PDF2 database were used for the identification of phases and Bruker TOPAS 6.0 software was used for Rietveld refinement. The instrument is aligned with SRM NIST 1976b corundum standard and calibrated with SRM NIST 640d silicon powder standard. Instrumental profile is calculated with the Fundamental Parameters Approach (FPA), baseline is modelled with 1st order Chebyshev polynomial. PO is corrected with March-Dollase function [8] or spherical harmonic functions [10], where required. Peak broadening was resolved by simultaneous size and strain calculation by size(nm)=FWHM(2θ)*cos(θ)/λ and ε_0=FWHM(2θ)/(4tanθ), a dimensionless parameter which can be related to distortion of lattice parameters. The two parameters were refined until the best fit, i.e. the smoothest difference curve was obtained. In cases when size-strain modelling did not resolve the measured pattern, a second structure of the same phase was implemented to distinguish between fractions defined by size range and/or strain amount. Measurements were made in the 3DLab Fine Structure Analysis laboratory of the University of Miskolc.

3. Results and Discussion

On the basis of the metallographic investigations, it was stated that oriented, long-shaped beta'-CuZn grains with protracted shape developed in the alpha-CuZn solid-solution matrix in the heat-treated structure created during the equilibrium cooling. A structure having a martensitic character was formed in the hot-forged sample as a consequence of the faster cooling process. In the microstructure, the alpha-CuZn solid solution forms well defined grains besides the beta'-CuZn acicular phase. The Pb droplets added to the alloy for the sake of the easier processing are distributed uniformly. The long-shaped phase, i.e. the Mn_5Si_3 compound-phase which is of blue colour (Fig 1) in etched state and can be recognised in the structure, develops during casting.

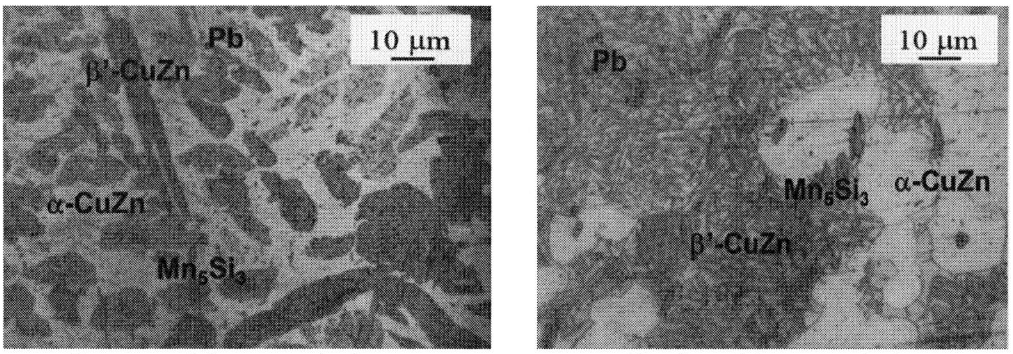

Heat treated (cross section) Hot forged (cross section)

Figure 1. Optical microscopic images about the microstructure of samples

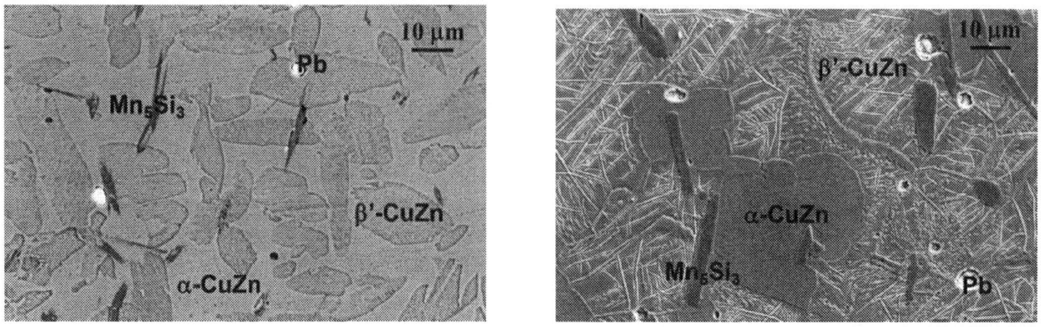

Heat treated (cross section) Hot forged (cross section)

Figure 2. Scanning electron microscopic images about the microstructure of samples

On the basis of EDX measurements and X-ray diffraction phase analysis, it can be stated that an alpha-CuZn solid solution (face centered cubic unit cell, space group Fm-3m (225), a: 3.697Å) and a beta'-CuZn (face centered cubic unit cell, space group Pm-3m (221), a: 2.949Å) phase as well as Mn_5Si_3 intermetallic phase (mavlyanovite, hexagonal unit cell, space group P63/mcm (193), a: 6.910Å, c: 4.8168182Å) and Pb (face centered cubic unit cell, space group Fm-3m(225), a: 3.704Å) can be identified in the alloy.

The exact quantity of phases can be obtained by Rietveld refinement in case if the phases contained in the alloy are known. In this paper, we show a step by step application of this method for oriented structures. In the course of Rietveld refinement, the measured curve obtained from the sample is fitted with the cumulative calculated curve obtained from the single curves calculated from identified phase's crystal structures, comprising all the corrections (PO) and broadening modelling (size and strain) that we apply. We aim at minimizing the difference between the measured and calculated curve. Conclusions concerning the phenomena (e.g. orientation) causing the difference can be drawn from the shape of difference curve and from the absolute value of residual peaks on the difference curve.

After identification of phases, the calculated patterns are fitted to the measured diffractogram (Fig. 3) without any PO correction, theoretical unit cell parameters, peak broadening modelled with crystallite size calculation.

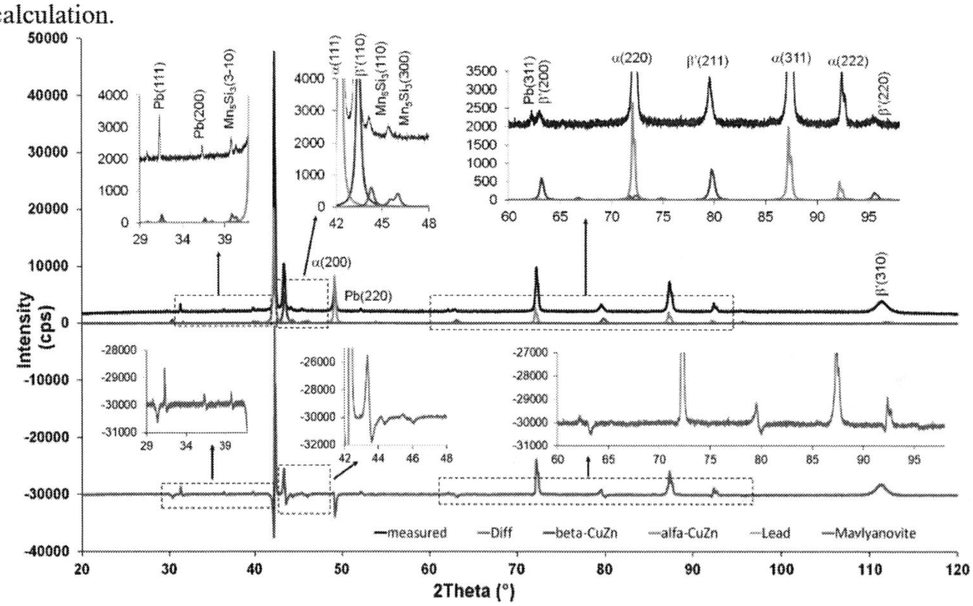

Figure 3. Fitting the diffractogram of heat-treated sample, without PO correction, theoretical unit cell sizes

Since unit cell parameters reflect the presence of alloying elements in the phases, this is the first parameter to be fitted, after which the peaks presenting PO and strain broadening can be identified. In Fig 3. and Fig 4. dotted line box marks the enlarged regions in which the differences after each step can be traced, beyond the more visible high intensity peaks. The residual peak intensities on the difference curve are significantly reduced by the fitting of cell parameter (Fig. 4).
The following step requires to perform the PO correction to the *(hkl)* peaks marked in Fig 4. With the March-Dollase function we can correct for two *(hkl)* series – defined by the user – of each phase. Correcting for one or two *(hkl)* sets can be decided based on the peaks on the difference curve, as well as the corresponding values of *(hkl)*, which may require several trials until the necessary correction is found.
In case of the alpha-CuZn solid solution, residual intensity value of the third peak *(200)* at 49°(2θ) on the difference curve decreases significantly as well, simultaneously with the PO correction for (*111*) at 43°(2θ) and (*220*) at 72°(2θ) peaks (Fig. 5A). Thereafter the difference curve was further reduced by modelling the lattice strain contribution to peak broadening. For the alpha-CuZn solid solution, the modelling of lattice strain contribution resulted in improvement for (*111*) and (*022*) peaks (Fig. 5B).

Presence of lattice strain is reasonable in case of an orientated structure obtained by thermal or mechanical processing. Intensity distribution on the difference curve as well as on the measured pattern (the reflection has got a shoulder) indicates the presence of second fraction for both α and β'-CuZn phases. This phenomenon is easily visualized by modelling crystallite size on calculated pattern, since the model applies lognormal distribution, thus any bimodal size distribution in the size broadening (after strain is taken into account!) is leaving peaks on the difference curve. Therefore, the introduction of a further phase in the model with identical crystal-structure but lower size and smaller lattice parameters becomes necessary and is supported by the minimization of the difference curve (Fig. 6A).

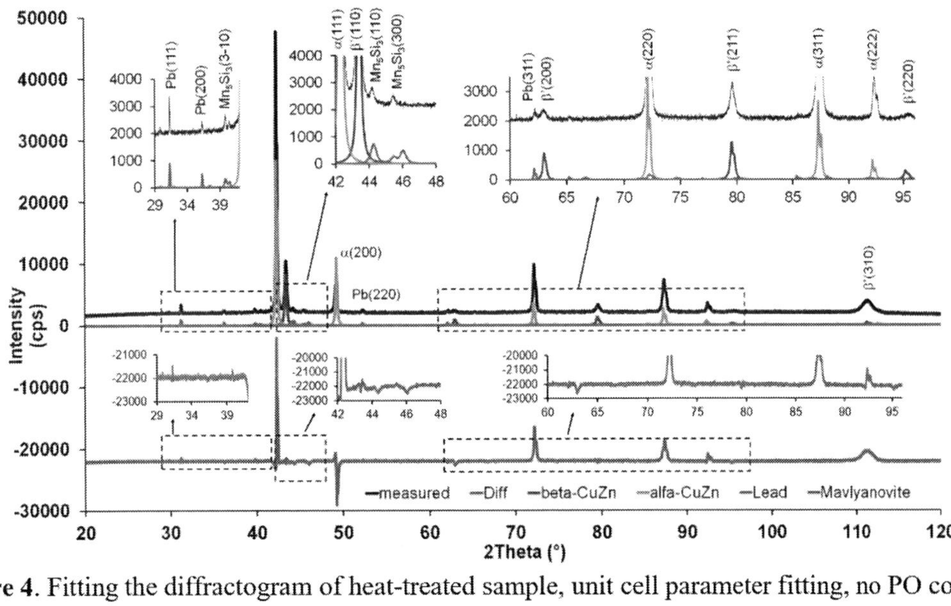

Figure 4. Fitting the diffractogram of heat-treated sample, unit cell parameter fitting, no PO correction

It seems reasonable to perform the PO correction even for these second phases introduced into the model. In case of alpha2-CuZn phase, this step means the orientation correction of (*311*) and (*400*) peaks while in case of beta'2-CuZn phase, it means the orientation correction of (*031*) peak and by this action the difference curve is further minimized (Fig. 6A).
It is very important to note that this fraction having a lower crystallite size shows a clearly different orientation in the complex structure, both in case of alpha2-CuZn and beta'2-CuZn phases. In case of the alpha-CuZn phase, the introduction of a 3rd fraction arose as the difference curve, especially the peak at 114 °(2θ) was not solved under the influence of any fitting possibility.

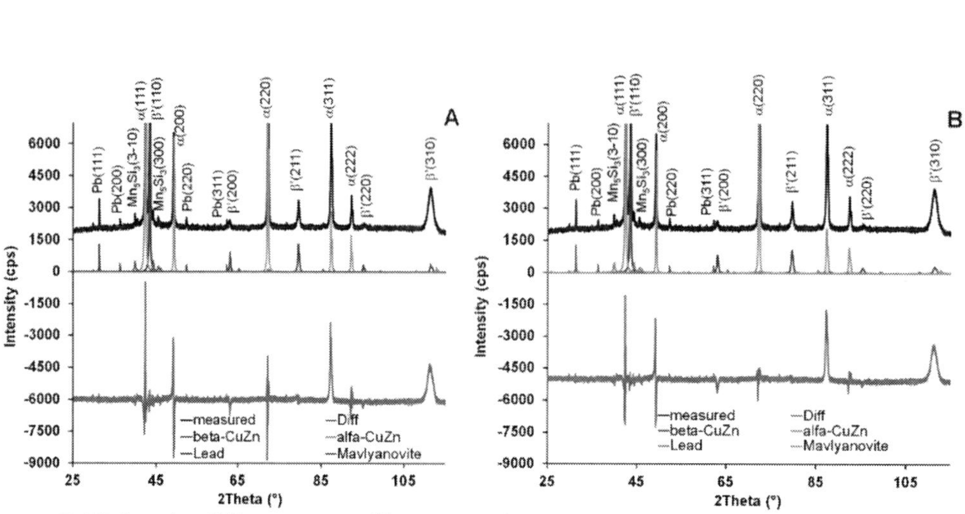

Figure 5. Fitting the diffractogram of heat-treated sample A: correction of PO, B: lattice strain correction

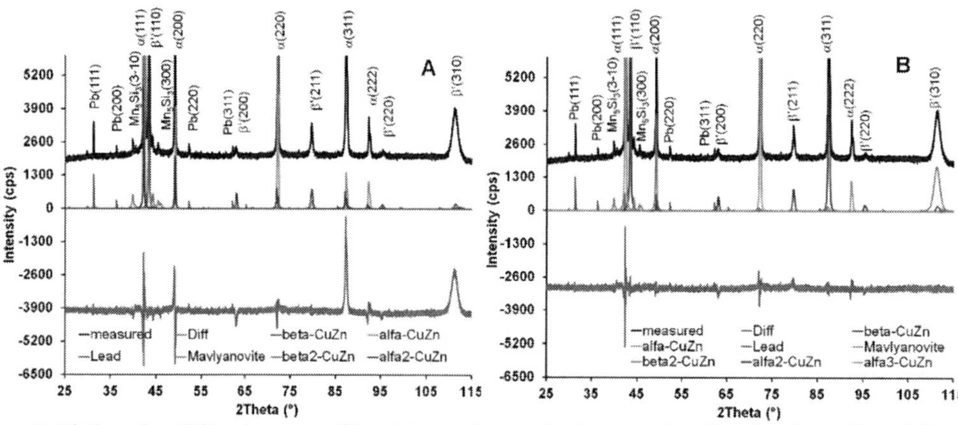

Figure 6. Fitting the diffractogram of heat-treated sample A: new size fraction introduced for alpha- and beta-CuZn, unit cell fitted and PO corrected; B: new fraction introducing

This 3rd phase with strong PO on *(002)* peak did solve the last major peak on the difference curve (Fig. 6B), thus its presence is reasonable.

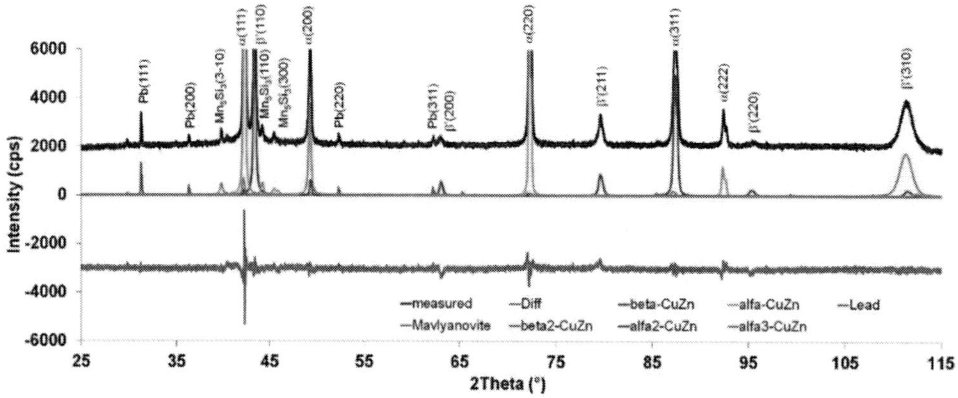

Figure 7. Final fitting result with all corrections and modelling, heat treated sample (upper part omitted)

At last the refinement procedure is completed by fitting the Debye-Waller thermal parameters (Fig. 7). As observed on the fitted pattern, only minor changes are obtained, while phase percentages (Table 2)

IOP Conf. Series: Materials Science and Engineering **903** (2020) 012028 doi:10.1088/1757-899X/903/1/012028

are not affected significantly. The phase quantities obtained during the different fitting steps are shown in Table 2. During the first steps, the quantity of alpha-CuZn is approximately 60wt%, the sum of the two phases corresponds to this value when introducing the alpha2-CuZn phase as well. The quantity of beta'-CuZn phase is around 20-25wt%. As soon as the third alpha-CuZn fraction is introduced, the quantity of sum alpha decreases to about 30wt% while the quantity of beta'-CuZn reaches a value of more than 50wt%. It can clearly be observed from the change of phase quantities that the results obtained without performing the corrections does not reflect the real phase composition.

Table 2. Amounts of phases in the different steps of Rietveld refinement

Refinement step	**Phases (wt%)**						
	α	β'	Pb	Mn_5Si_3	α2	β'2	α3
Without correction	64.72	23.20	0.90	11.18	-	-	-
Cell parameter fitting	69.54	19.48	0.67	10.30	-	-	-
Orientation correction	64.95	25.22	0.63	9.21	-	-	-
New fraction introduced, cell parameter fitting, orientation correction	45.68	16.90	0.58	10.45	16.07	10.33	-
New fraction introduced, cell parameter fitting, orientation correction	7.56	44.91	1.36	16.37	9.90	7.64	12.26
Thermal parameter fitting	7.61	45.50	1.38	16.18	9.50	7.37	12.47
Hot-forged sample*	31.41	11.01	0.51	6.81	6.22	25.52	11.75

* also contains 7.03 wt% CuZn(MnAl) martensite

In case of the forged samples, the changes caused by the solidification and transformation as well as by the plastic deformation were taken into consideration when fitting the complex structure. Based on the residual peaks of difference curves, a martensite phase of possible CuZn(Mn) composition was identified (Fig. 8).

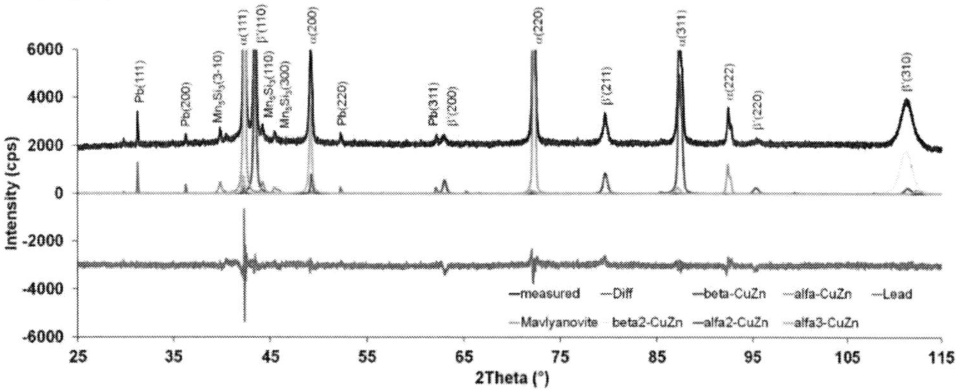

Figure 8. Diffractogram of hot-forged sample, complete refinement (upper part omitted)

As a discussion of limitations and possibilities of our calculations several remarks must be highlighted. Most important is related to quantitative phase analysis, which is improved towards realistic sample values by the corrections but is limited only to that volume of sample which is penetrated by the X-rays. Thus, such a measurement for metal or alloy block sample is problematic to carry out, since the distribution of phases is highly inhomogeneous. Even with all the correcting and measuring techniques an average (bulk) composition for block samples cannot be obtained with X-ray diffraction, it will always reflect the local composition of the illuminated area, depending on X-ray wavelength, instrumental configuration, sample stage and sample composition.

IOP Conf. Series: Materials Science and Engineering **903** (2020) 012028 doi:10.1088/1757-899X/903/1/012028

Average phase quantitative composition can be obtained only on proper powder samples. Another issue is size distribution of phases. In the XRD evaluation phases are generally regarded as one single structure with the according size, not a realistic model for most materials. Hot-forging and heat treatment inevitably produces new size fractions of existing phases through recrystallization. These fractions can be evidenced by Rietveld refinement of XRD patterns, if all corrections and modelling are done properly.

4. Summary

Complex microstructures of alloys represent a problem in XRD analysis, due to preferred orientation, multimodal crystallite size distribution and lattice strain. These properties can be adequately modelled and corrected by Rietveld refinement to obtain reasonable quantitative phase composition and characterization of individual phases. The unit cell size, after fitting to the measured pattern gives information on alloying ratio, while strain present in the phase is also modelled. After PO corrections, the existence of multiple size +/- strain +/- orientation fractions are revealed for the modelled phases. All the properties are in connection with processing of analyzed sample. In the case of heat treated CuZn material, the alpha-CuZn phase develops 3 fractions, while the sample was obtained in three steps: homogenization at 800 °C, tempered at 550 °C and cooled in air. The Pb content, theoretical in the 0.8-2.0 wt% range, is well reflected in the refinement. Also, phase transformation effect of forging is evidenced, probably left unrecognised without all the corrections and modelling. The 3rd fraction of alpha-CuZn represents a phase which is not influenced by heat treatment, neither forging, like relict grains of the original, unprocessed CuZn cast. Whatsoever, the numbering and development sequence of the different phase fractions is not readily linked to the processing steps, since it would require a step by step processing and Rietveld refinement investigation.

Acknowledgement
Supported by the ÚNKP-19-4 New National Excellence Program of the Ministry for Innovation and Technology.

References

[1] Fredrikson H, Akerlin U. 2012 *Solidification and Crystallization Processing in Metals and Alloys* (John Wiley & Sons Publication, Chichester) pp 267
[2] Verhoeven J.D 1975 *Fundamentals of Physical Metallurgy* (John Wiley & Sons Publication, New York)
[3] Van Vlack L.H. (1982) Materials forEngineering Concept and applications (Addison-Weslwy Publishing, Masssachusetts) pp. 71
[4] Brooks C.R 1982 *Heat Tratment, Structure and Properties of Nonferrous Alloys* (ASM, USA) pp 20.
[5] Askeland D.R. (1989) *The Science and Engineering of Materials* (PWS Publishing Company, Boston) pp.175
[6] Cullity B.D. 1956 *Elements of X-ray Diffraction* (Addison-Wesley, Massachusetts)
[7] Krawitz A. 2001 *Introduction to Diffraction in Material Science* (John Wiley & Sons Publication, New York) pp. 255
[8] Percharsky V.K, Zavalij P.Y 2003 *Fundamentals of Powder Diffraction and Structural Characterization of Materials* (Kluwer, Dordrecht)
[9] Young R.A ed. 1993) *The Rietveld Method* (Oxford University Press, Oxford)
[10] Popa N.C 2008 *Microstructural properties: texture and macrostress effects*, In Dinnebier R.E , Billinge S.J.L ed. 2008 Powder Diffraction:Theory and Practice (RoyalSociety of Chemistry, Cambridge)
[11] Velterop L et al 2000 *J.Appl.Cryst.* **33** 296-306
[12] Bergmann J et al 2001 *J.Appl.Cryst.* **34** 16-19

The effects of the steel's surface quality on the properties of anti-corrosion coatings

B. Herbath[1] and K. Kovacs[2]

[1]BPW-Hungaria Ltd. Szombathely

[2] University of Pannonia, Institute of Materials and Mechanical Engineering, Veszprem

E-mail: HerbathB@bpw.hu

Abstract: The requirements of painted vehicle construction in terms of corrosion resistance apply not only to cars but also to heavy duty lorries, trucks and agricultural machinery. Phosphating is the most commonly used method for the pre-treatment before painting of ferrous metals in the vehicle manufacturing industry. The task of the phosphate layer between the metal surface and the paint is to protect the metal from corrosion under the paint film and promotes the adhesion of the paint film to the metal substrates. In this study the zinc phosphate conversion layer was deposited onto the steel surfaces from a bi-cationic (nickel-free) phosphating bath by dipping using identical technological parameters. The developed crystal structure and morphology was examined in terms of surface roughness and blasting quality of the metal. The surface quality of the metal was tested by digital light microscopy (LM), and the structure of phosphate coating was investigated by scanning electron microscopy (SEM).

1. Introduction

Surface preparation of steel parts is one of the main factors influencing the effectivity and longevity of anticorrosion coatings. Improperly prepared surfaces prevent the formation of an even conversion layer, meaning insufficient paint layer adhesion, leading to corrosion under the protective layer and causing continuous peeling-off of the coating [1,3,4,5].

In addition to cleaning the surfaces also must be modified prior to applying the anti-corrosion layers. A zinc phosphate layer is added onto metal surfaces, to ensure good adhesion of the additional layers to the raw metal surface and protects the base metal from water, carbon dioxide, sulfur oxides, ozone, aggressive ions and other harmful substances getting through the damaged paint coatings [1,4,5]. The performance of zinc phosphate layer on steel depends on the fraction of the total covered surface area. This coverage fraction is effected by process parameters and chemical composition of phosphating bath and surface morphology of the deposit [6].

Besides chemical surface cleaning, the blasting is used for cleaning, surface preparation, and surface treatment as well. This process reduces the susceptibility to corrosion, seal porous surfaces. A blasted surface can be a very clean surface providing excellent mechanical adhesion. Media selection plays an important role in effective blasting. Shoot blasting can be widely used to remove rust, oxides, oils and mill scales, even the welding beads, welding silicate layer from the surface and welding seams of steel parts, structures and even complex assemblies. Its main scope is to provide surfaces free of adhesion preventing materials. Also, increased surface roughness provides good adhesion for the first layer of the coating for the interfacial reactions. [2,4,5].

Quality of the phosphate layer is influenced by the material quality [2], surface roughness, method of surface preparation and parameters of the phosphating process (concentration of ingredients, temperature, time, etc.) [1,4,5], structural defects and presence of contaminants in the crystallizing method [7]. The purpose of this work is to investigates the properties of the zinc phosphate conversion layer on blasted steel surfaces with different surface roughness after blasting under the same technological parameters.

2. Materials and Methods

2.1 Metal substrates preparation

As a substrate for zinc-phosphate coatings deposition both sides of a S420MC steel sheet (70 mm × 150 mm × 10 mm) and for comparison a cold rolled steel Q-Panel (102 × 152 × 0,81 mm) were used.

Chemical composition of **Q-Panel** by SAE Material Designation 1008/1010 is as follows: max. 0.5 % manganese, max. 0.15 % carbon, max. 0.030 % phosphorus, max. 0.035 % sulfur.

Chemical composition of **S420MC** steel panel by EN 10149-2 (High-strength steels for cold-forming, thermomechanically-rolled, normalized): max. 1.6 % manganese, max. 0.12 % carbon, max. 0.50 % silicon, max. 0.025 % phosphorus, max. 0.015 % sulfur. This steel is used for cold-formed components, easy to cut and bend, suitable for welding, and can be easily machined. Due to the favorable mechanical properties this is a commonly used sheet material in different sheet thicknesses in BPW axle and chassis manufacturing. Table 1 lists the chemical composition of samples.

Table 1. Chemical composition of samples

Steel		wt %									
		Fe	C	Si	Mn	P	S	Cr	Mo	Ni	Al
S420MC	1.	98.8	0.0750	0.0118	0.960	<0.004	0.0048	0.0287	<0.004	0.0068	0.0405
	2.	98.8	0.0764	0.0152	0.969	<0.004	0.0052	0.0289	<0.004	0.0100	0.0413
	3.	98.8	0.0759	0.0083	0.977	<0.004	0.0054	0.0288	<0.004	0.0053	0.0393
	mean	98.8	0.0758	0.0118	0.969	<0.004	0.0051	0.0288	<0.004	0.0074	0.0404
Q-Panel	1.	99.2	0.0734	0.0146	0.355	0.0067	0.0095	0.0837	0.0076	0.0245	0.0394
	2.	99.2	0.0756	0.0125	0.359	0.0058	0.0102	0.0830	0.0081	0.0246	0.0397
	3.	99.2	0.0769	0.0129	0.361	0.0069	0.0098	0.0843	0.0081	0.0236	0.0394
	mean	99.2	0.0753	0.0133	0.359	0.0065	0.0098	0.0836	0.0080	0.0242	0.0395

To analyze the chemical compositions was use Metal analyzer PMI-MASTER Pro2 spark excited optical emission spectrometer (OES) from Oxford Instruments.

Different surface structure was visible on micrographs of the Q-panel and S420MC panel.

Surface roughness of S420MC steel panel before blasting Rz: 6.04 µm

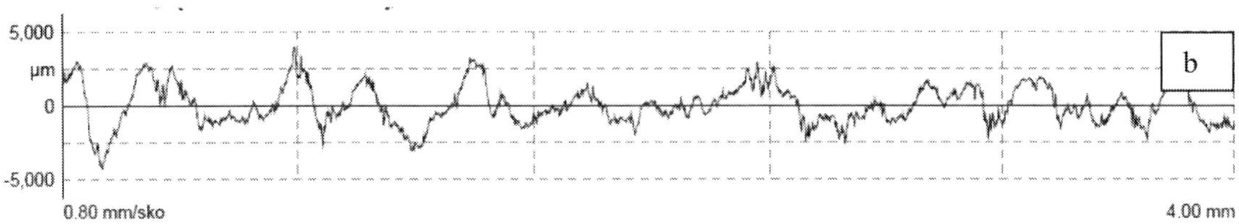

The Q-Panel did not receive mechanical surface pretreatment (generally accepted industry specimen plate for corrosion testing). The surface roughness Rz: 4.50 µm

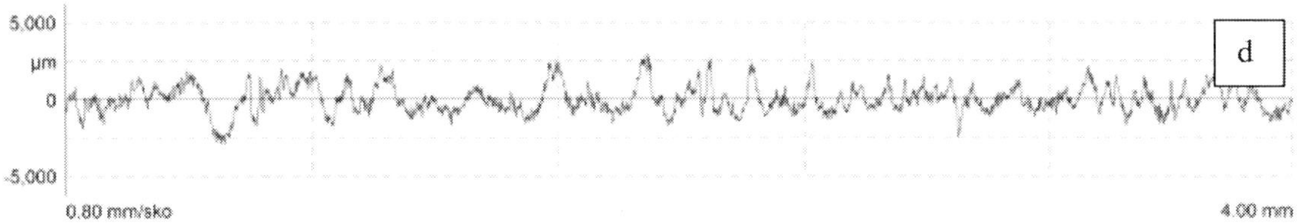

Figure 1. Surface measurements of steel surfaces: a) Micrograph of S420MC steel panel, b) Surface roughness of S420MC steel panel before blasting, c) Micrograph of Q-Panel, d) Surface roughness of Q-Panel

Prior to deposition of zinc-phosphate coatings, the S420MC test panel was blasted with steel shots in an automated equipment. One side of the panel was properly blasted ("good side"), while the other wasn't ("not good side"). The two different sides under the same conditions received the surface pretreatment.

2.2 Pretreatment and zinc phosphating process

The tests were done in day-to-day production conditions on a working production line of the company and within the limits of process capability. Free acid and total acid were determined by titrating of 10 ml bath sample against 0. 1 N NaOH and total alkalinity was determined against 0.1 N H_2SO_4 using indicators. The classical volumetric titration methods were performed with Schilling burettes, in Erlenmeyer flasks.

A SenTix® pH electrode with IDS (Intelligent Digital Sensors) was utilized to measure and adjust the pH values of solutions. This electrode was normally calibrated every day by using buffer solutions of pH values 4.11, 7.0 and 10.0. The conductivity of solutions was measure with TetraCon® Conductivity cell. This electrode was normally calibrated every week by using 1.413 μS/cm (25°C; 0,01 mol/l; KCl) calibration solution.

After shot blasting (Q-Panel without shot blasting) the test panels were pre-degreased at 60 °C temperature for 5 min. in 20 g/l of Gardoclean S5165 and 2 g/l Gardobond Additiv H7375 alkaline degreasing dip bath (pH = 12.7; total alkalinity = 21.5 points). Degreasing was done at 60 °C temperature for 5 min. in 40 g/l of Gardoclean S5165 and 3 g/l Gardobond Additiv H7375 alkaline degreasing dip bath (pH = 11.7; total alkalinity = 23.8 points) and dip rinsed with tap water in three successive cascade flushing baths. Treatment time was 1minute in each bath by stirring the rinsing water, at ambient temperature. The conductivity of the rinsing baths during the progress of the technological process were: 1871 μS/cm and 875 μS/cm and 717 μS/cm.

Zinc-phosphate coatings were formed chemically after dip activation with 2 g/l Gardolene V6599 bath (pH = 9,17), by immersion in phosphating bath (pH = 2,84 ; free acidity = 2,0 points; total acidity = 23.4 points; accelerator = 5.5 points) for 4 min, at same temperatures of the nickel-free double cation Zn/Mn phosphating bath (58 °C). It should be noted here that nickel has long been known to significantly improve paint adhesion and corrosion protection. However, nickel compounds are noxious and closely regulated in the effluent stream. Nickel-free processes, therefore, are desirable to satisfy health and environmental demands. $NaNO_2$ – acting as an accelerator – was added to the phosphating bath (5.5 points).

Finally, all plates were dip rinsed with tap water (pH > 4,5), three times in a successive cascade flushing bath. The conductivity of the rinsing baths during the progress of the technological process were < 300 μS/cm and < 150 μS/cm and < 50 μS/cm) and finally they were also rinsed by dipping and by spraying in deionized water (conductivity < 20 μS/cm). Treatment time was 1 minute in each bath by stirring the rinsing water, the spraying was done with fresh deionized water at ambient temperature. Total acidity/alkalinity and free acidity were determined with acid – base titration.

All solutions were prepared with chemicals of Chemetall Ltd, according to the technical data sheets, with deionized or tap water. The pretreatments are applied in a dipping process, all technological parameters were set to the specified mean value within the allowable tolerances.

Examination of appearance after the pretreatment process: parts of different tones visible on the surface on "not good" side (Figure 2 a). On other side the coating is consistent and whole-colored (Figure 2 b)

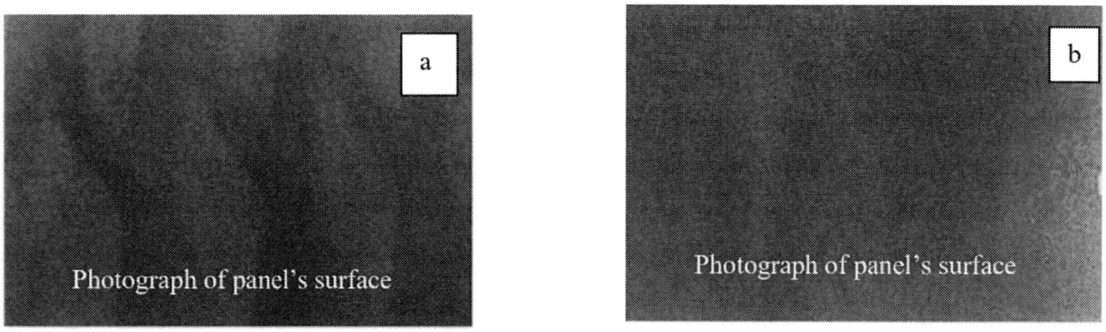

Figure 2. Surface measurements of S420MC steel surfaces after blasting and zinc phosphating: a) photograph of „not good" side, b) photograph of „good" side

IOP Conf. Series: Materials Science and Engineering **903** (2020) 012029 doi:10.1088/1757-899X/903/1/012029

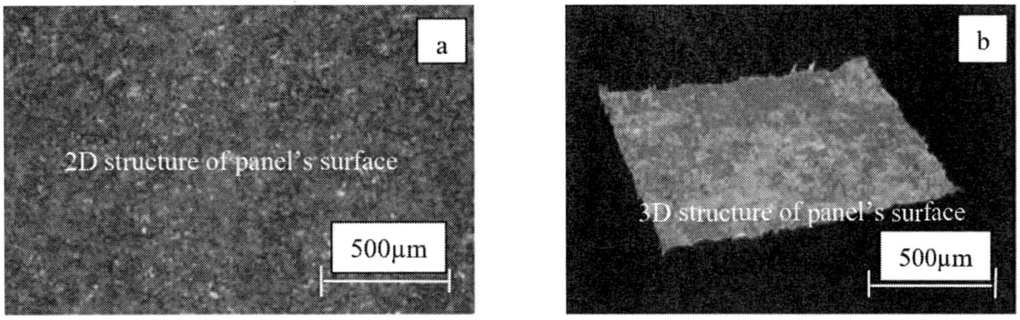

Surface roughness on "not good" side: Rz = 21.62 μm; Rz_{max} = 28.35 μm

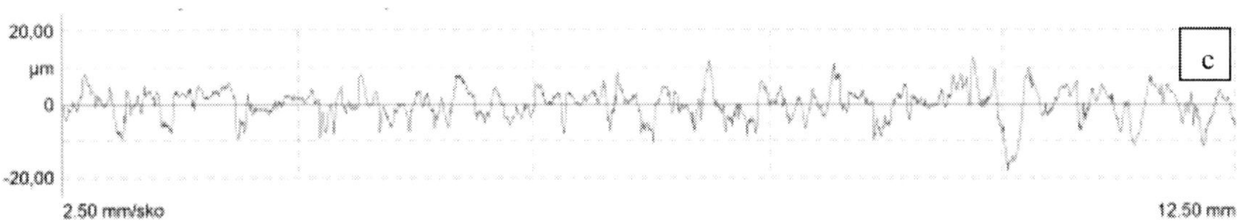

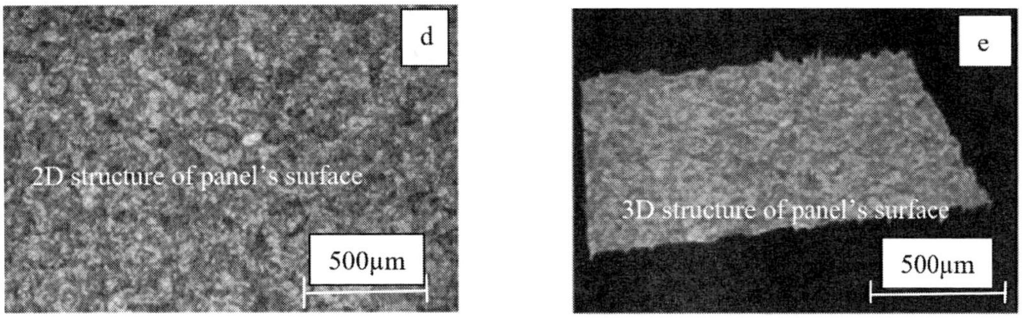

Surface roughness on "good" side: Rz = 60.13 μm; Rz_{max} = 70.15 μm

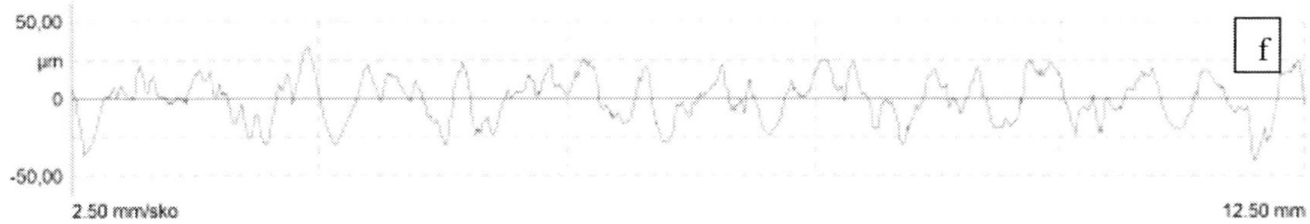

Figure 3. Surface measurements of steel surfaces after blasting and zinc phosphating: a) 2D micrograph of S420MC steel panel on „not good" side, b) 3D micrograph of S420MC steel panel on „not good" side, c) Surface roughness of S420MC steel panel on "not good" side, d) 2D micrograph of S420MC steel panel on „good" side, e) 3D micrograph of S420MC steel panel on „good" side, f) Surface roughness of S420MC steel panel on "good" side,

There is a separating boundary section between properly blasted and not properly blasted surface. Scanning electron micrographs show irregular crystal growth at the border (see Figure 4).

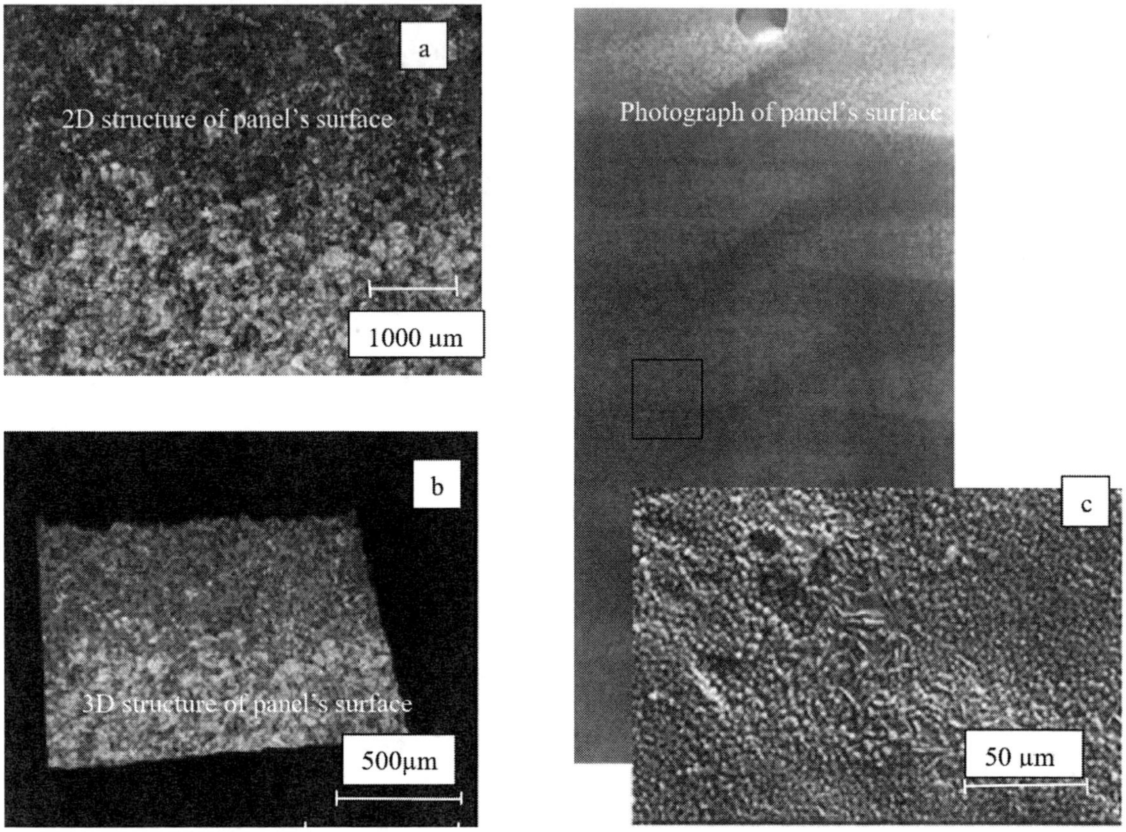

Figure 4. a), b), Digital light micrograph and 4.c) SEM micrograph of irregular crystal growth at the border line

2.3 Surface and surface coverage measurements

The weight per unit area in g/m^2 of the phosphate layers coatings was determined by chemical dissolution of the layers in a solution of 4 % TEA + 12 % Na_2EDTA x $2H_2O$ + 9 % NaOH + 75 % water at 70 °C degree. The phosphate conversion coatings weight of the Q-Panel was 2.65 g/m^2, which corresponds to the regulatory mean value of line.

Surface roughness (Rz and Rz_{max}) according to DIN EN ISO 4287 was measured by a MAHR/MarTalk Drive Unit GD 25 / tactile: MFW-250:1 .

2D and 3D structure of surfaces was visualized by a KEYENCE digital light microscope Model WHX J20T. Morphology, crystal shape and size, as well as orientation was measured by a ThermoFisher/FEI Apreo S scanning electron microscope (see Figure 5).

Results

A: Properly blasted base material **B:** Incompletely blasted base material **C:** Q-Panel surface

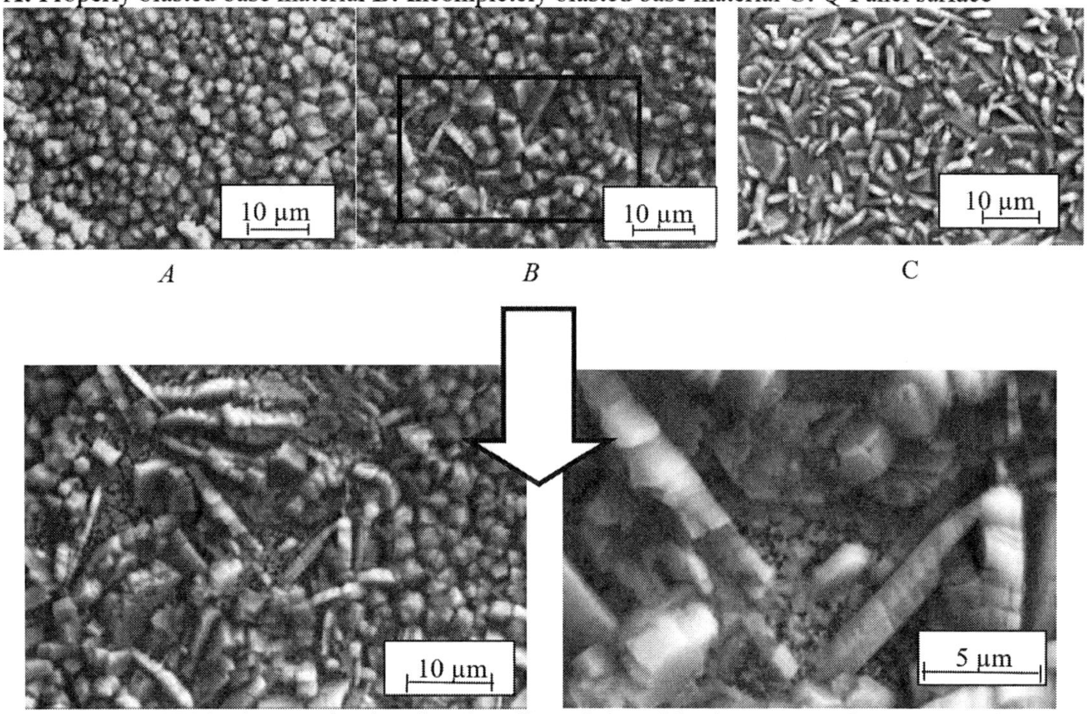

Figure 5. SEM micrographs showing the effect of shoot blasting on crystal structure and coverage
A: Properly blasted base material B: Incompletely blasted base material C: Q-Panel surface

Structure of the metal surface, the micro- and macrostructure is just as important as the material composition. As the surface roughness increases, so does the calculated surface area per coating weight, and better coverage is achieved on steel surfaces due to reactions taking place on the boundary layer, corrosion resistance in case of damaged top coating is improved. Increased surface roughness correlates with finer structure of the coating, as smooth surfaces have weaker reaction to phosphating. If a surface contains lots of indentations and microcracks, then the acid corrosion effect is amplified during phosphating, leading to stronger layer adhesion but also changing the layer's microstructure.

3. Conclusions

The crystalline size and crystal type were found to be surface roughness after the blasting dependent. Surfaces properly prepared with shoot blasting have a regular and evenly covered microstructure (Figure 5A). Improper or incomplete shoot blasting causes irregular grain growth in the corrosion protection layer as shown in Fig. 5B and decrease the fraction of the total covered surface area by the deposit. On the sample plates representing the "ideal" surface to be phosphate layer, the structure of the zinc phosphate does not show the same structure as the phosphate layer structure used in the vehicle manufacturing steels exhibiting different surface roughness and quality. Irregular crystal growth for reason of improper shot blasting, on work pieces resulting from defective mechanical surface preparation causes subsequent corrosion problems because the base metal surface is not completely covered by the zinc phosphate crystals.

A completely different zinc phosphate crystal structure was deposited onto the surface of Q-Panel (Figure 5C). When evaluating SEM micrographs of zinc-phosphate conversion coatings on steel surfaces in surface pre-treatments plants, it should be taken into account, that the SEM image of the zinc phosphate conversion layer formed on industrial sample plates (e.g. Q-Panel) should not be used for qualifying technology parameters.

References

[1] Rausch W 2005 Die Phosphatierung von Metallen (Eugen G. Leuze Verlag D-88348 BAD SAULGAU ISBN 3-87480-197-7)

[2] Lohmeyer S 1993 Phosphatierfehler aus Zink-und Alkaliphosphatierungen (Eugen G. Leuze Verlag D-7968 SAULGAU/WÜRTT ISBN 3-87480-086-5)

[3] Fettis G 1995 Automotive Paints and Coatings (VCH Verlagsgesellschaft mbH, D-69451 Weinheim ISBN 3-527-28637-3)

[4] Narayanan S 2005 Surface Pretreatment by Phosphate Conversion Coating – *A rewiew Rev.Adv.Mater.Sci.*Vol.9 pp.130-177

[5] Debnath N 2013 Importance of Surface Preparation for Corrosion Protection of Automobiles *Journal of Surface Engineered and Advanced Technology*, 2013, 3, 94-105

[6] Abdalla K, Rahmat A, Azizan A 2013 The effect of pH zinc phosphate coating morphology and its corrosion resistance on mild steel *Advanced Materials Research* Vol. 626 (2013) pp.569-574

[7] Narayanan S and Subbaiyan M1993 Effect of Surfactants on the Growth and Crystal Habit of Zinc Phosphate Coating *Trans.Inst.Metal Finishing*, 1993, 71(1), pp.37-40

Microstructure examination of additive manufactured 316L steel

Dávid Miklós KEMÉNY[1], Dávid PAMMER[1], Dorina KOVÁCS[1]

[1]Budapest University of Technology and Economics, Faculty of Mechanical Engineering, Department of Materials Science and Engineering, Address: 1111, Hungary, Budapest, Műegyetem rakpart 3.; Phone / Fax: +36 1 463 1079

Email: kemenyd@eik.bme.hu

Abstract. The additive manufacturing (AM) is a new field in research. In our research, the samples were made by direct metal laser sintering (DMLS). During the production of 316L material, the piece was formed layer by layer, so the existing layer was re-heat treated when the new layer was formed. Due to the multiple heat treatment, the post-heat treatment of the product is necessary. Microstructure changes were investigated after two different heat treatments. The first was at 450 °C for 4.5 hours, and the second was at 1100 °C for one hour, followed by annealing in each case. The heat-treated sample at 450 °C showed no significant difference. In contrast, the sample heat-treated at 1100 °C the particles were better separated, and segregation in grain boundary was observed, which showed high phosphorus concentration.

1. Introduction

There are many ways to design a product. Which option was chosen depends among other things, for example, the available technology, the required quality, and the cost of production. Recently, additive manufacturing (AM) is gaining ground, and more and more technology is available. AM technology can also provide a geometry that is not or only very difficult to produce with traditional manufacturing technologies [1,2]. With AM techniques are easy to produce complex shaped functional metallic components such as tubes, gear wheels, or toolmaking, e.g. blow moulding, extrusion, die casting and sheet metal forming.

1.1. AM technology

During the research, direct metal laser sintering (DMLS) was used to produce the test specimens. The DMLS does not melt the powder of the base material but heats it to a temperature close to the melting point, which results in a chemical bond between the powder particles, i.e. the sintering process [3]. With this technology cannot use pure metal powder, only a mixture of multiple powders [4], but DMLS is still the most common additive technology.

The equipment consists of 4 main components: a metal powder feeder, a continuous carbon dioxide (CO_2) laser, a powder spreader (Figure 1), and a computer system that controls and monitors the process [5]. For the manufacturing of the product, a 3D model is required. It can be converted to an STL file format for the AM device [6].

The STL file has a layer thickness of approximately 50 μm, which depends on the technology [5,7]. During layer by layer formation [8], a new layer reheats the existing layer(s). As a result, a re-heat treatment was made layer by layer, which can result in a characteristic microstructure. Special care must be taken to the microstructure of the additively manufactured piece because it can influence the mechanical properties. If the application area only allows a slight change in properties, it is recommended a post-heat treatment after the product is finished [7].

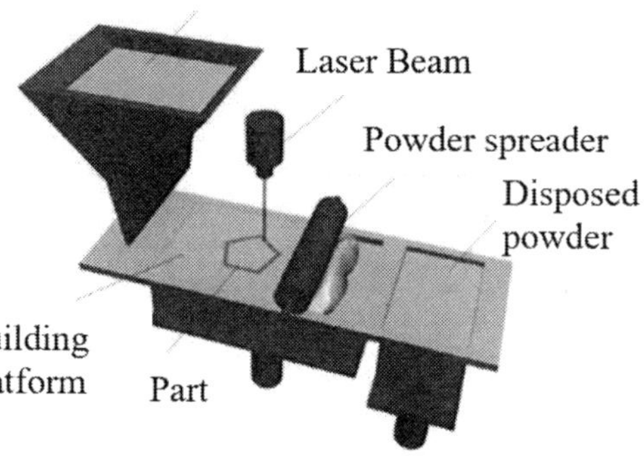

Figure 1. Schematic diagram of DMLS [8]

1.2. Heat treatment

The purpose of heat treatment is to achieve structural changes in the products or to eliminate the residual stress due to technology [9].

According to Katherine et al. [10], controlling residual stresses and microstructure was investigated for 3D printed 316L material. Two different heat treatments were carried out, one at 490 °C for 5 hours, and the second product was heat-treated at 1200 °C for 1 hour, followed by annealing in each case. The heat treatment has been prepared according to Chen et al. [11] whose stated that mechanical and microstructural properties were improved in a positive direction when the heat treatment was made at 490 °C. The aim was a regenerative, restorative heat treatment at 1200 °C, which was sharpened to the sigma phase, but neither recrystallisation nor grain growth was observed. During mechanical tests, heat treatment at 490 °C resulted in a reduction in the hardness of the specimens, while at 1200 °C, the hardness of the pieces increased. After the tensile test, the heat treatment at 490 °C gave substantially different tensile curves similar to the reference (untreated) sample, which broke after nearly the same elongation. After the heat treatment at 1200 °C, the value of tensile strength decreased, but the elongation increased [10].

The aim of this research is to compare the effects of heat treatment on two different test specimens relative to the initial untreated state.

2. Materials and methods

The specimens were made from 316L stainless steel with EOS M100 equipment. The composition, mechanical and physical properties of 316L material are presented in Tables 1-2 [12]. The first test specimen was heat-treated at 450 °C for 4.5 hours (later marking: HT1 – heat treated 1) and annealed. The second test specimen was heat-treated at 1100 °C for 1 hour (later marking: HT2 – heat treated 2) and annealed. The heat treatment was carried out in the air for both samples. After the heat treatment, the samples were mechanically ground with 60 to 2500-grit SiC and polished with 1 μm diamond suspension. After the polishing, the samples were etched in Kalling's No. 2 reagent. The cleaning fluid was acetone, and it was dried with hot air. The images were taken with Olympus SZX16 stereomicroscope, Olympus PMG 3 optical microscopic, and Zeiss EvoMa 10 scanning electron microscope (SEM). The composition analyses were performed with energy-dispersive spectrometry (EDS). The hardness was measured with Vickers microhardness tester (Buehler IndentaMet 110/220).

Table 1. The 316L material chemical composition in wt% [12]

Chemical composition									
C	**Mn**	**Si**	**P**	**S**	**Cr**	**Mo**	**Ni**	**N**	**Fe**
max 0.03	max 2.0	max 0.75	max 0.045	max 0.03	16-18	2-3	10-14	max 0.1	balance

Table 2. 316L material mechanical properties [12]

Ultimate Tensile Strength (MPa)	Yield Strength 0.2% Offset (MPa)	Modulus of Elasticity in Tension (GPa)	Hardness (HV1)
min 485	min 170	200	223

3. Results and discussion

The microstructure of 3D products was different from traditionally manufactured products. Figure 2 shows the microstructure of the 316L specimen produced by additive and traditional manufacturing technology. In Figure 2 (a) can be observed that the grain boundary cannot be distinguishable clearly. Figure 2 (b) shows that the grain boundary can be distinguishable, and the twins plane of the austenitic structure can be seen.

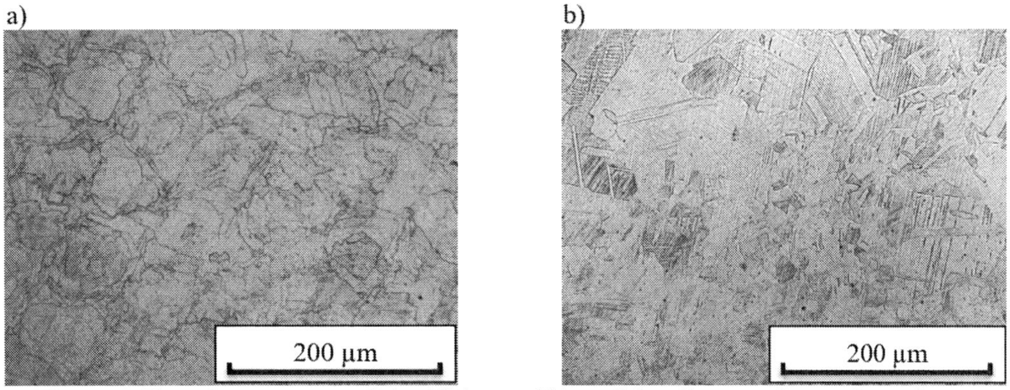

Figure 2 The microstructure of a) additive and b) traditionally manufactured specimens

As previously mentioned, the product was re-heat-treated layer-by-layer, which can produce unpredictable properties. The finished product was polished before heat treatment to provide a basis for comparison with the heat-treated samples.

3.1. The original state

Stereomicroscope images were made from the unpolished surface of the samples, which are seen in Figure 3. At lower magnification, the surface was like drum-pressed, and at higher magnification, the surface was more like caterpillar seam.

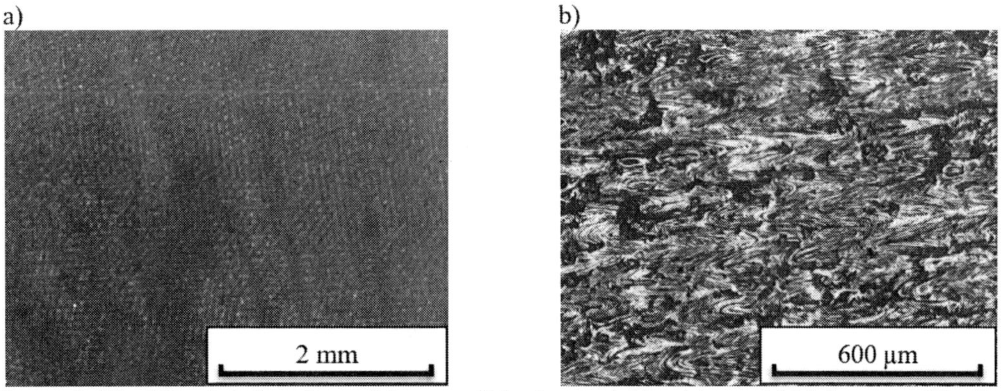

Figure 3. The 3D printed surface of the unpolished material in a) lower b) higher magnification

Before the heat treatment, the surface of the product was prepared and etched with Kalling's No. 2 reagent. Figure 4. shows the microstructures. At higher magnification, the forming lines were visible, which was also comparable to a caterpillar seam. Due to the technology, it resulted inhomogeneous appearance of austenitic grains with precipitations and/or inclusions.

Figure 4. The 3D printed surface of the polished and etched material. From left to right, the magnification is increasing.

3.2. Heat-treated sample at 450 °C temperature – HT1

The microstructure of HT1 is seen in Figure 5. After etching for less time in lower magnification, the surface showed caterpillar structure and the reflecting of the laser beam movement [13], but after 2 minutes etching [14], it can be seen in higher magnification that there were enrichments, precipitation or inclusions along the grain boundary and no twin plane was observed.

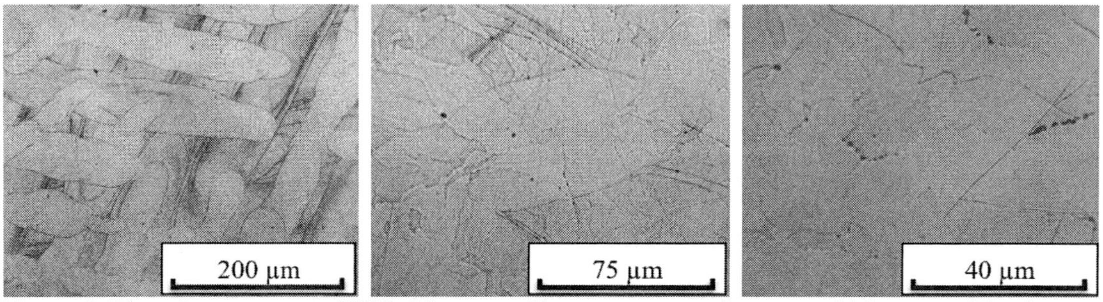

Figure 5. The microstructure of HT1. The caterpillar structure can be seen at higher magnification as well.

3.3. The heat-treated sample at 1100 °C temperature – HT2

The microstructure images of HT2 are seen in Figure 6. In this case, a much more spectacular change has occurred, since the grain can be distinguishable, and they were more closely resembled austenite microstructure in lower magnification. In higher magnification, Figure 6 shows the twin plane of austenite and a large number of tiny inclusions.

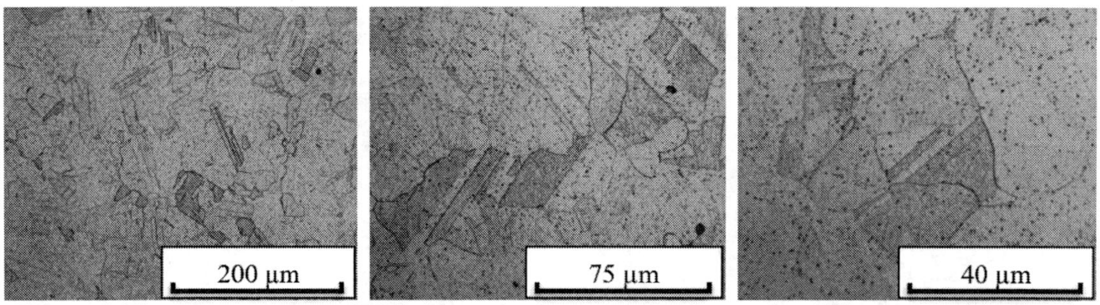

Figure 6. The microstructure of HT2. The twin planes can be observed at different magnifications.

3.4. Hardness test

Table 3 shows the results of the hardness tests. The hardness was measured with a load of 1 kilogram for 11 seconds. As we can see, the hardness of the HT2 was higher, but in these two procedures, the hardness was higher than the base material according to the metals catalogue [12,15].

Table 3. The hardness of the heat-treated products

Number of measurements	Hardness of HT1 (HV1)	Hardness of HT2 (HV1)
1	161	220
2	154	221
3	154	217
4	153	222
5	155	210
Average	**155**	**218**

3.5. SEM and EDS results

After the stereo and optical microscope measurements, the samples were examined with SEM. The main aim was to specify precipitations origin. The surface of the samples can be seen in Figure 7. The twin planes of HT2 can be observed, which were not visible at the HT1.

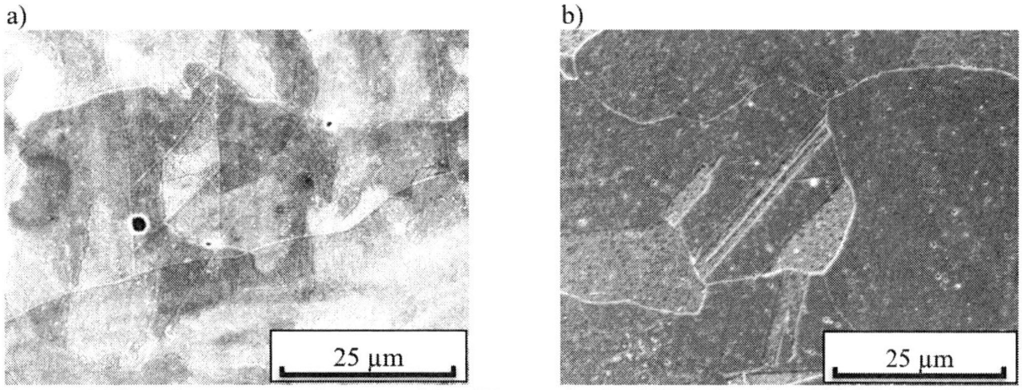

Figure 7. The microstructure images of the 3D samples by SEM after a) at 450 °C b) at 1100 °C heat treatment

The identification of the precipitations can be seen in Figure 8. In the two images show the microstructure and the measured area was marked with the red circle. The EDS examination results of Figure 8 (a) was: Si: 0.29%; P: 0%; Mo: 2.52%; Cr: 20.21%; Mn: 1.36%; Fe: 62.78%; Ni: 12.84% in wt%. The EDS examination result of Figure 8 (b) was: Si: 0.39%; P: 0.22%; Mo: 3.01%; Cr: 19.98%; Mn: 1.35%; Fe: 62.13%; Ni: 12.91% in wt%. The EDS results did not show differences in the composition of the base material.

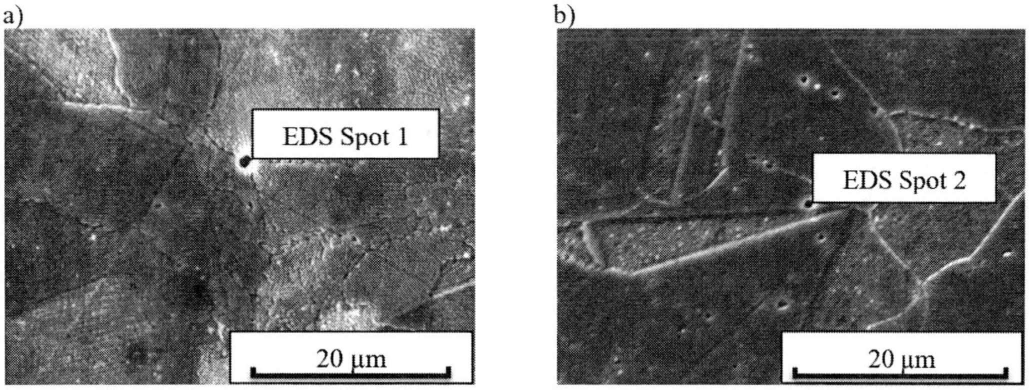

Figure 8. The EDS analysis results and the measured are of the samples by SEM after a) at 450 °C b) at 1100 °C heat treatment

According to the EDS results, it can be stated that these formations at grain boundary are porosity, not precipitation. By the HT2, the EDS showed five times more phosphorus in the material like the allowed value. Similar values were measured for several different measurement points as well.

4. Conclusion

It can be stated that the microstructure of the 3D product was showed austenitic characteristic. Homogenisation was observed after the post-heat treatment, which resulted in a more ordered and larger grain size. The austenitic structure was observed by the heat-treated sample at 1100 °C, because of the twin plane. After heat treatment, porosity appeared at 450 °C along the grain boundary, and it can be visible at 1100 °C both in the grain and also their boundary. The hardness increased compared to the base material in both cases, but the hardness of HT2 was more significant approximately 60 HV. One possible reason for the difference in hardness value is the grain coarsening. The higher temperature of the heat treatment has a greater chance to coarse of the grains, which remains coarse-grained even after cooling. This can even affect hardness.

The HT2 indicated the microstructure similar to that traditionally manufactured microstructure. The high phosphorus concentration was measured and was not favourable as it can negatively affect the mechanical properties. Phosphorus segregation was formed due to the heat treatment, but further investigations are needed to determine this phenomenon. The effect of this transformation on mechanical properties requires further investigations. More examinations are required to minimise the number of porosity in the procedure and/or in heat treatment as well.

Acknowledgement

Our research was supported by the ÚNKP-19-3 New National Excellence Program of the Ministry for Innovation and Technology. The authors are grateful for Budapest University of Technology and Economics to providing our research.

References

[1] Colditz P 2020 *Proc. Man.* **47** 1164-1169
[2] Gebhardt A 2011 *Carl Hanser Verlag*
[3] Keresztes Z 2019 *Periodica Pol. Mech. Eng.* **63** 195-200
[4] DebRoy T 2018 *Progress in Mater. Sci.* **92** 112-224
[5] Bertol L S 2010 *Materials & Design* **31** 3982-3988
[6] Kemény D M 2018 *Acta Mater. Transyl.* **1/2** 81–84
[7] Simchi A 2003 *Journal of Mater. Proc. Tech.* **141** 319-328
[8] Santos V M R 2020 *Additive Manufacturing* **32** 100964
[9] Sisson R D Jr 2001 *Encyc. of Mater. – Sci. and Tech.* 3736-3741
[10] Adelman K 2018 *Mater. Eng.* pp. 1-16
[11] Chen X 2018 *Mater. Sci. &Amp, Eng. A* **715** 307-314
[12] Michler T 2001 *Encyc. of Mater. – Sci. and Tech.* 406-410
[13] Avanzini A 2019 *Procedia Struc. Int.* **18** 119-128
[14] Varbai B 2019 *Measurement* **147** 106832
[15] Orbulov I N 2009 *Periodica Pol. Mech. Eng.* **53/2** 93-99

IOP Conf. Series: Materials Science and Engineering **903** (2020) 012031 doi:10.1088/1757-899X/903/1/012031

Characterization of short fiber-reinforced polylactic acid composites produced with Fused Filament Fabrication (FFF)

Csenge Tóth and Krisztián Norbert Kovács

Department of Polymer Engineering, Faculty of Mechanical Engineering, Budapest University of Technology and Economics, H-1111 Budapest, Műegyetem rkp. 3., Hungary

E-mail: kovacsn@pt.bme.hu

Abstract. The interdisciplinary expansion of additive manufacturing (AM) technologies resulted in an increasing demand for high-performance materials for engineering applications. In order to enhance the mechanical properties of printed products, short fiber composites are of great interest recently. In this study, the development of short fiber-reinforced composite materials is performed along with the mechanical characterization of the samples produced by fused filament fabrication (FFF) technology.

1. Introduction

Fused filament fabrication (FFF) is one of the most popular AM technologies due to its cost efficiency and low waste production, and as the technology attracted attention in various industries, the demand for high-performance materials rose as well. A commonly used material for FFF technology is polylactic acid (PLA) due to its good printing properties, and as a result of the recent increasing interest in sustainable solutions, PLA became a desirable alternative to oil-based polymers. While the good tensile strength (60 MPa) and competitive price (2 Euro/kg) make PLA a standout among biopolymers, it still has some disadvantages that need to be addressed [1,2].

A way to improve the mechanical properties of printed PLA products is the optimization of manufacturing parameters such as the build orientation, feed rate, and layer thickness [3], however, these technology-dependent solutions might not be applicable for all geometries. Another way to enhance the mechanical properties of PLA is the use of blends. For instance, Hajba et al [4] reported that toughness can be improved with natural rubber particles. Fiber reinforcement could also be a viable solution especially in case of FFF technology since short fiber filaments can be processed with most any FFF printer, although the fiber content is reported to be limited to 30 w% [5,6]. In addition to the limitation in fiber content, short fibers seem to bring several further challenges. First, selecting the most suitable fiber type for PLA is essential as it affects the overall mechanical properties greatly. Sang et al [7] reported that basalt fiber composites showed better tensile and flexural properties compared to carbon fiber PLA composites due to rheological differences. Ferreira et al [8] also found the tensile strength of PLA/CF composites to be unsatisfactory which was caused by poor fiber-matrix adhesion.

Reaching the critical fiber length is also a challenging task for short fiber FFF technology, as the shear stress of the filament manufacturing process results in fiber breakage [9]. Moreover, the increase of fiber content results in further fiber breakage as more fiber-fiber interactions are present [6]. However, a promising feature of the FFF composite technology is the fiber orientation. It is demonstrated in studies that during printing, a highly oriented material is gained due to the small diameter of the nozzle and the draw speed [6,10]. Since the orientation is parallel to the printing direction, it is feasible to produce composite parts with unique reinforcement patterns with FFF technology. This aspect of the technology could be beneficial in producing final parts and additionally, FFF printing has a good potential in complementing other thermoplastic manufacturing technologies as well [11]. The strength of the technology lies in its design freedom since the production of complex geometries with traditional composite technologies could face numerous challenges [12]. Also, the high degree of automation of AM technologies makes it ideal for integration into industry 4.0 systems as well.

To the best of our knowledge, most of the studies investigated short carbon fiber reinforced materials regarding FFF processes, therefore this paper aims to expand the range of materials. In addition to carbon and glass fibers, basalt fibers are also investigated. Their excellent mechanical properties and the lower energy consumption during production can make basalt fibers a favourable alternative to glass fibers [7,13]. The printing orientation is also considered in this study, for which samples were made with two types of printing orientation. Mechanical tests were carried out as well as optical experiments to gain knowledge on the adhesion conditions and the degree of fiber fracturing during FFF technology.

2. Experimental methods

2.1. Raw materials

The PLA (Ingeo 3100HP, Melt Flow Rate 24 g/10 min (210 °C/2,16 kg)) used in this study as matrix material was obtained from NatureWorks. The self-made composite filaments contained chopped Zoltek PX35 carbon fibers (CF), Camelyaf glass fibers (GF), and Kamenny vek basalt fibers (BF) respectively. The average fiber length is 6 mm. Mixtures of 5 weight % fiber content were prepared for each and a neat 3100HP PLA filament was made for control. The commercially available Proto-Pasta PLA/15 w% CF composite filament was also examined and the unfilled PLA (NatureWorks Ingeo 4043D, Melt Flow Rate 6 g/10 min (210 °C/2,16 kg)) material was used as a reference. All composite filaments are summarized in table 1.

2.2. Preparation of composite filaments

First, the mixtures were fed to a twin-screw extruder (Labtech LTE 26-44) in a dry state (80 °C/4h). Preforms were made and granulated of 6 mm length, then extruded again using a cylindrical die. The two-step manufacturing process was necessary as the first extrusion resulted in an uneven filament diameter, which led to irregular material flow during printing.

Table 1. Composite filaments.

Matrix	Reinforcement	Weight fraction (weight %)	Manufacturer
	BF	5	self-made
3100HP PLA	CF	5	self-made
	GF	5	self-made
4043D PLA	CF	15	Proto-Pasta

2.3. 3D printing

Samples were prepared by a desktop FFF printer (CraftBot Plus). Table 2. show the process parameters. The intention was to maintain the same process parameters for each material, however, the extrusion of the 15 w% CF composite filament required a higher nozzle temperature as the flowability of the material decreased by the rigid carbon fibers [7]. The samples were fabricated with two types of unidirectional printing orientations (see figure 1.) and the infill was set to 100%. One contour layer of 0,4 mm was applied.

Table 2. Printing parameters.

Nozzle temperature 5 w% (°C)	215
Nozzle temperature 15 w% (°C)	230
Bed temperature (°C)	60
Nozzle diameter (mm)	0,4
Speed (mm/s)	60
Infill (%)	100
Layer height (mm)	0,2

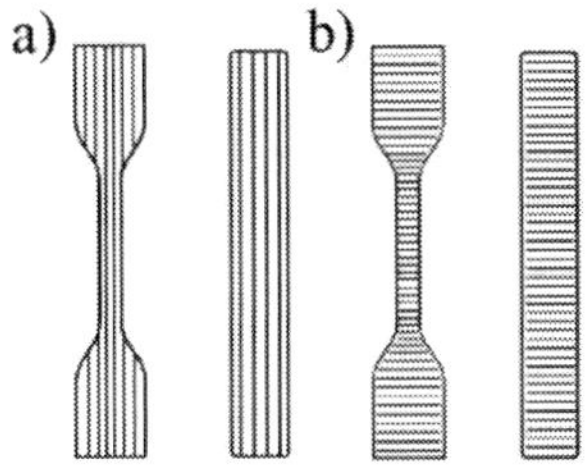

Figure 1. Printing orientations of the specimens: a) parallel (0°) and b) perpendicular to the length (90°).

2.4. Mechanical properties

Tensile tests were conducted on a Zwick Z005 testing equipment according to the ISO 527-2 [14] standard, with a crosshead speed of 5 mm/min. Flexural properties were determined by 3-point bending tests in accordance with the ISO 178 [15] standard. During these experiments, 5 specimens per type were examined. Finally, impact strength was measured on notched specimens using a Charpy impact machine (Ceast Resil Impactor Junior equipped with a pendulum of 2 J) according to the ISO 179 [16] standard. Notch radius was 0,25±0,05 mm. 10 specimens per type were used to measure the impact strength. All tests were performed at room temperature.

Fiber length distributions were measured to determine the extent of fiber breakage during the material extrusion and FFF printing. For this purpose, samples were obtained from the composite filaments and the FFF processed materials both, then the samples were heated up to 500 °C for 3 hours to remove the PLA matrix. Heating was conducted in a laboratory furnace under air atmosphere (type Denkal 6B). The extracted fibers were transferred onto glass plates, then the images of the fibers were taken with a Keyence VHX 5000 digital microscope. Fiber length distributions were obtained from the pictures using ImageJ software and about 200 fibers were measured each.

SEM micrographs of cryogenic fracture surfaces were taken with a JEOL JSM 6380LA equipment at an acceleration voltage of 10 kV. The surfaces of the samples were gold plated in vacuum. The internal morphology was also observed via non-destructive micro-computed tomography (Nikon XT H 225 ST). The sample for the μ-CT analysis was prepared with 0° printing orientation with 80 x 10 x 4 mm dimensions. The high-resolution technique resulted in a complete 3D visualization of the composite specimen.

3. Results and discussion

3.1. Fiber length distribution

The fiber length distribution (FLD) in the composite filaments and the FFF processed materials are shown in figure 2. All filaments were extruded twice, and all materials have the same initial fiber length of 6 mm. Significant fiber breakage occurred during the filament extrusion due to the high shear forces generated by the twin-screw extruder. However, it can be stated that the FFF process causes further fiber breakage only to a slight extent. The carbon and glass fibers longer than 300-400 μm decreased more while the basalt fibers showed no significant change in the FLDs. In summary, all fiber types followed a similar trend, and the material extrusion through the nozzle during FFF printing did not alter the average fiber lengths drastically.

Theoretical critical fiber lengths for polylactic matrix and basalt, carbon, and glass fibers were calculated based on literature data [17,18,19] and this resulted in the ranges of 600-1000 μm, 1000-1200 μm and 800-1200 μm, respectively. For carbon and glass fibers this theoretical length is yet to be achieved, however, the FLD of the basalt fibers shows a promising amount of fibers in the desired range. The remaining fiber lengths could be improved by applying a gentler extrusion process, or the theoretical critical fiber length could be decreased by optimizing the fiber-matrix adhesion thus increasing the interfacial shear strength.

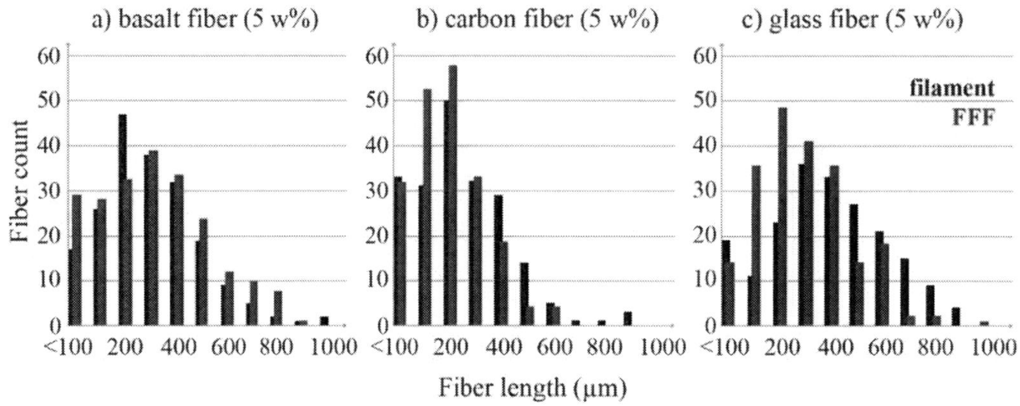

Figure 2. Fiber length distributions obtained from filaments and FFF processed materials: a) PLA/5 w% BF, b) PLA/5 w% CF, c) PLA/5 w% GF.

3.2. *Mechanical characterization*

In the following chapters, the mechanical properties are presented as percentage deviations from the corresponding references. The values of the matrix materials are shown in Table 3.

Table 3. Reference materials.

Properties	3100HP 0°		3100HP 90°		4043D 0°		4043D 90°	
	Avg.	Dev.	Avg.	Dev.	Avg.	Dev.	Avg.	Dev.
Tensile strength (MPa)	45	1,33	41	1,30	52	2,88	49	2,16
Tensile modulus (GPa)	2,4	0,05	2,5	0,04	0,7	0,11	0,4	0,11
Flexural strength (MPa)	88	5,88	74	3,97	91	1,50	84	5,18
Flexural modulus (GPa)	2,4	0,14	2,3	0,06	2,4	0,04	2,2	0,23
Impact strength (kJ/m^2)	3,2	0,55	3,0	0,45	5,2	0,89	3,6	0,48

3.2.1. Tensile properties. The effects of the printing orientation and the different types of fiber reinforcements on the tensile strength and Young's modulus are shown in figure 3. and figure 4., respectively. An increase in tensile strength can be observed in all 5 w% samples where the printing direction is longitudinal (0°), however, among the 90° printed samples only the BF material showed a significant rise in strength. Contrary to expectation, the increased fiber content did not result in bigger tensile strengths (as the 0° printed 15 w% CF specimens showed less than 5% development), moreover, the strength of the 15 w% CF 90° samples decreased. Literature also states that PLA/CF composites tend to have a bigger dependency on infill orientation, which is caused by relatively weak interlayer adhesion [7, 20].

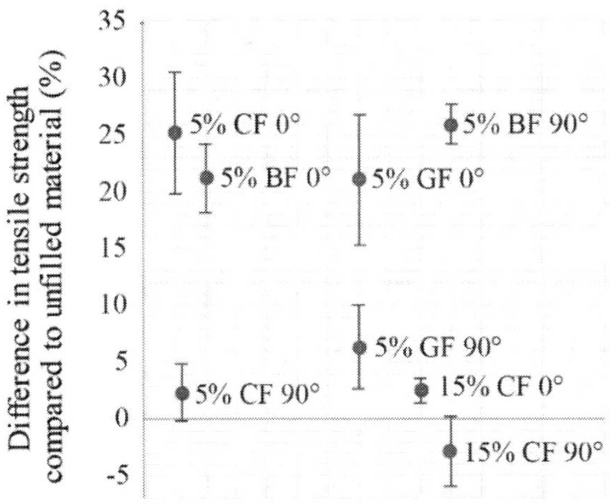

Figure 3. Percentage deviation of tensile strengths compared to the matrix PLA.

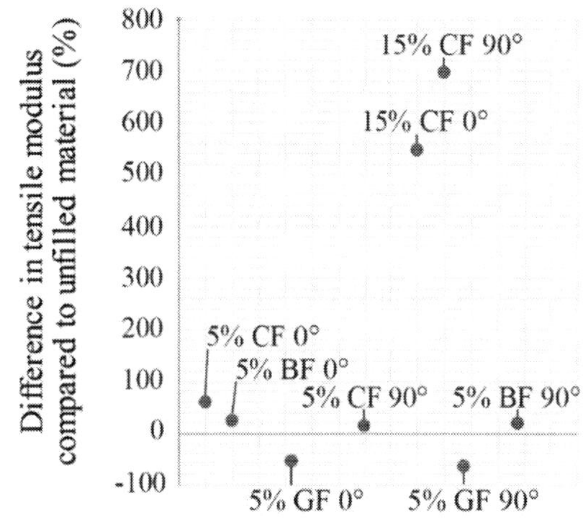

Figure 4. Percentage deviation of tensile moduli compared to the matrix PLA.

3.2.2. Flexural properties. A comparison of flexural properties (strength and modulus) between composites and neat materials are shown in figure 5. and figure 6. Only the glass fibers could improve the flexural strength, all the other types of reinforcements resulted in a decrease. On the other hand, the flexural modulus values mostly improved. In comparison with the tensile properties, it can be stated that the flexural properties have a bigger dependency on printing orientation.

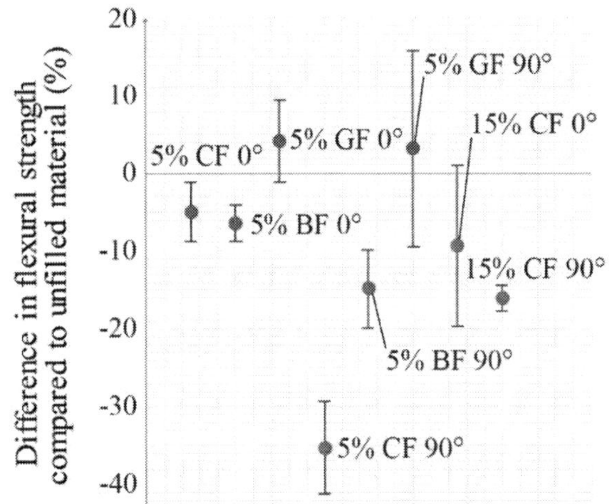

Figure 5. Percentage deviation of flexural strengths compared to the matrix PLA.

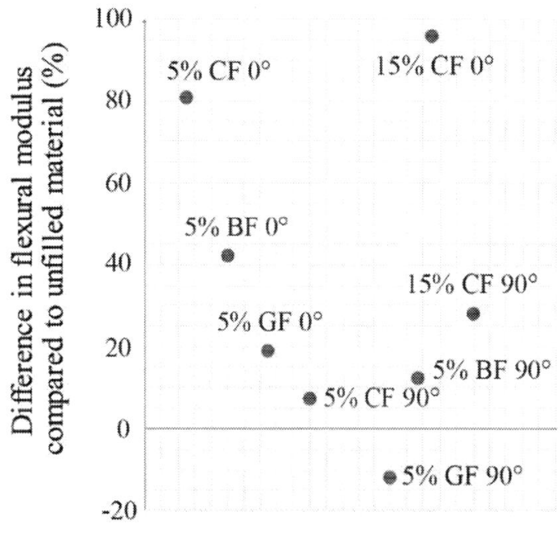

Figure 6. Percentage deviation of flexural moduli compared to the matrix PLA.

3.2.3. Impact strength and ductility. The effects of fibers and infill orientation on impact strength are shown in figure 7. It can be observed that the 5 w% and 15 w% fiber content mostly decreased the impact strengths, except for the 5 w% carbon and glass composites with parallel infill. The results presented may be due to low average fiber length or poor interfacial adhesion. Furthermore, it can be stated that with lower fiber content, it is statistically less likely for fibers to prevent crack propagation. figure 8. shows the differences in ductility between composites and neat plastic specimens. The ductility index decreased as fibers were added which means the composites became less rigid than the matrix PLA.

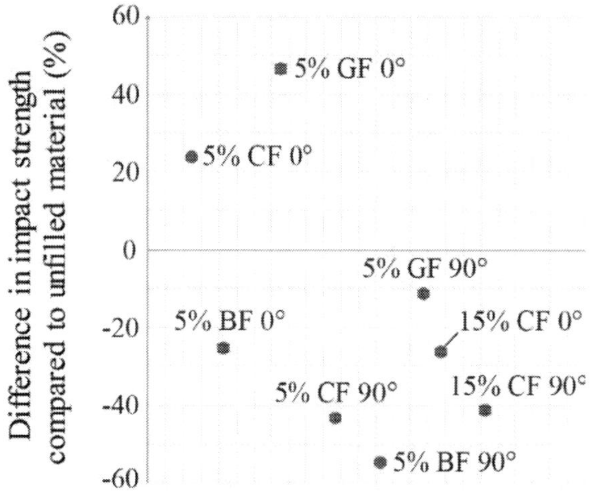

Figure 7. Percentage deviation of impact strengths compared to the matrix PLA.

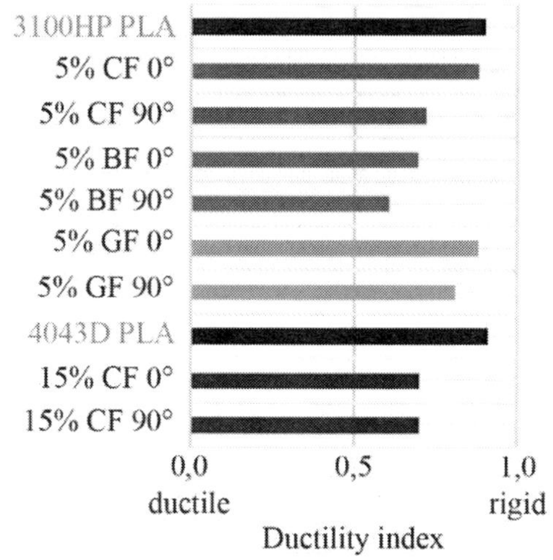

Figure 8. Ductility index of the composites and the unfilled PLA materials.

3.3. Microstructure

3.3.1. Void formation. SEM images of the cryogenic fracture surfaces of 5 w% basalt and carbon samples and 15 w% carbon sample are shown in figure 9. The printing orientation of the specimens is perpendicular to the fracture surface. Significant porosity can be observed on the 15 w% carbon sample made from the commercially available filament. The voids are of similar size and they are evenly dispersed through the surface. In contrast, the samples produced with self-made filaments show an unevenly dispersed porosity and the voids also differ in size. This is likely to be caused by the uneven diameter of the self-made filaments which potentially generates irregular material flow during printing. The pore formation can also be observed on the microtomographic images shown in figure 10. The complete 3D visualization confirmed that the samples present an all-over uneven nature in porosity which can be traced back to the fluctuating polymer flow.

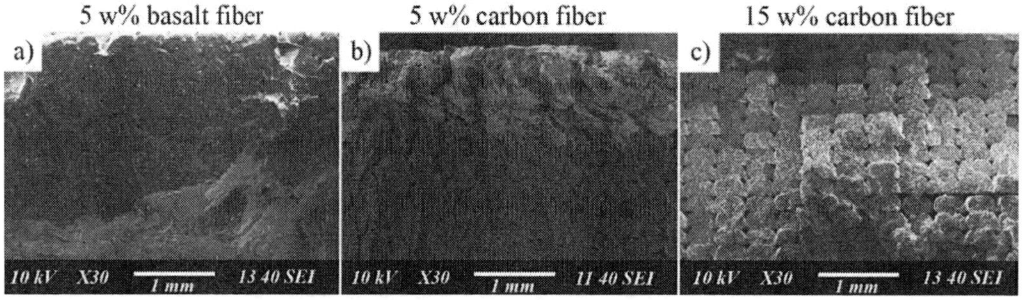

Figure 9. SEM images of cryogenic fracture surfaces of samples with 0° printing orientation: a) PLA/5 w% BF b) PLA/5 w% CF and c) PLA/15 w% CF specimens.

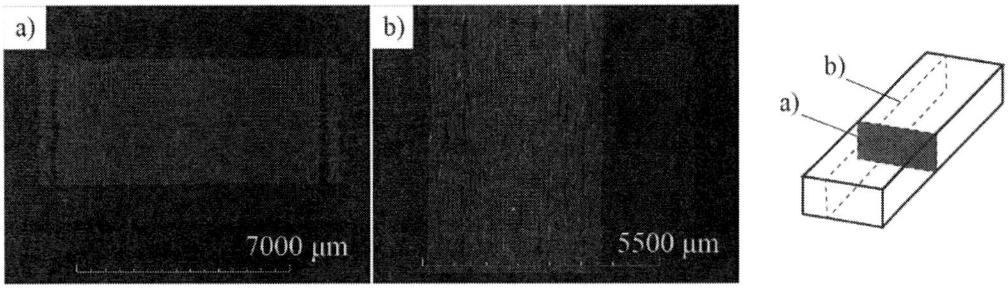

Figure 10. Microtomographic images of PLA/5 w% BF FFF printed sample with 0° printing orientation. Cross-sections shown are a) perpendicular and b) parallel to the printing direction.

3.3.2. Fiber orientation. Figure 11. shows the 3D microtomographic image of a basalt fiber-reinforced specimen with 5 w% fiber content. It can be observed that most of the fibers are oriented in the print direction in agreement with the literature [7]. The slight sinking of the upper layers can also be detected which is presumably caused by the relatively high degree of porosity underneath.

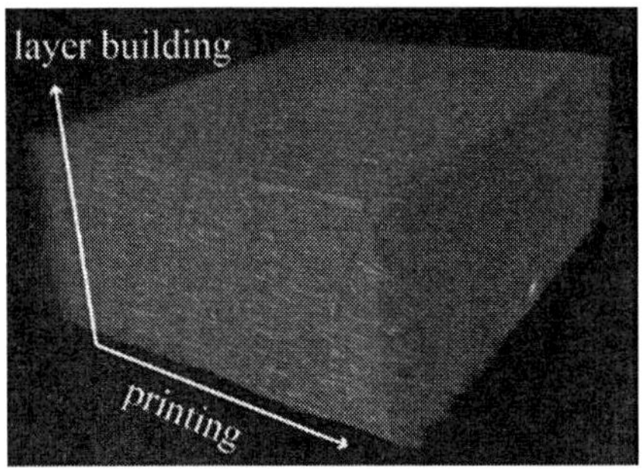

Figure 11. 3D microtomographic image of PLA/5 w% BF FDM printed sample with 0° printing orientation.

3.3.3. Fiber-matrix adhesion. SEM images of basalt and carbon fiber-reinforced samples are shown in figure 12. It can be seen that protruding carbon fibers are mostly clear from PLA, indicating poor fiber-matrix adhesion. On the other hand, the basalt fibers are covered with the matrix material to a moderately bigger extent. When examining the adhesion, it should be borne in mind that the basalt fibers were treated with a silane-based coupling agent, and it is recommended by the manufacturer for epoxy resin matrices. The surface treatment of carbon fibers used for the self-made filament is optimal for thermoplastic matrix materials.

3100HP PLA + silane treated basalt fiber | 3100H PLA + carbon fiber treated for thermoplastics | 4043D PLA + carbon fiber

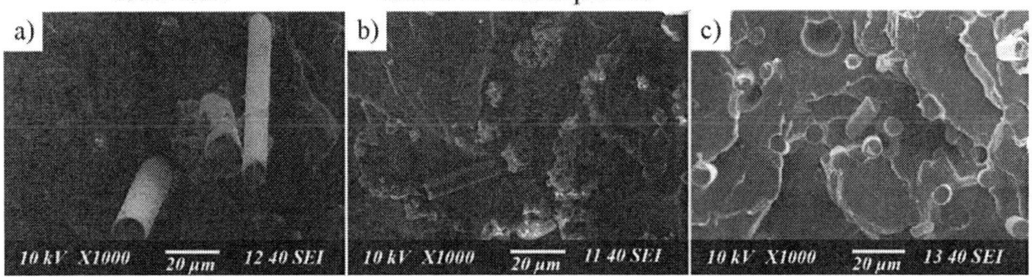

Figure 12. SEM images of FDM-printed samples: a) PLA/5 w% BF and b) PLA/5 w% CF are prepared with self-made composite filaments and c) PLA/15 w% CF is from commercially available material.

4. Conclusion

In this research, carbon, glass and basalt fiber-reinforced composite filaments are successfully fabricated and examined along with a commercially available PLA/CF filament. The fiber content is considered as the self-made composite filaments are prepared with 5 w% fiber content each and the PLA/CF material is loaded with 15 w% of fibers. Specimens were fabricated with FFF technology and two types of printing orientation were applied. Mechanical and optical characterizations were conducted and fiber length distributions were measured to determine the extent of fiber breakage during the filament extrusion and the FFF process.

The results show that the average fiber length decreased significantly during the filament extrusion process due to high shear rates, however, the FFF printing does not cause further significant breakage. The 5 w% fiber content with the parallel infill resulted in an average of 20% increase in tensile strength values, but contrary to expectation, the 15 w% carbon fiber content resulted in a slight increase only. The flexural properties and the impact strengths of the composite materials showed no significant development compared to the unfilled matrices. Based on the results presented, the mechanical properties of the CF composites are the most dependent on the printing orientation. Scanning electron microscopy showed a better fiber-matrix adhesion for basalt fibers, which can be further improved by applying a more suitable surface treatment, specifically for the PLA matrix. Finally, microtomography showed a relatively high porosity in the specimens which is likely caused by irregular material flow during printing. The porosity might be reduced by increasing the feed rate or decreasing the draw speed.

With FFF technology it is possible to produce complex geometries with custom fiber orientation, therefore it could be an excellent alternative for thermoplastic composite manufacturing. However, further research and development are still necessary for engineers to be able to use the technology in its full potential. The results of this paper contribute to a deeper understanding of the relationships between processing parameters and mechanical properties. Optimization is required especially in terms of the fiber length, fiber-matrix adhesion, and porosity. Our further research focuses on basalt fiber-reinforced composites due to the good tensile properties and the favourable features of the basalt fibers.

Acknowledgments
The project is funded by the (NKFIH) National Research, Development and Innovation Office (2018 - 1.3.1 – VKE – 2018 - 00001).

This publication was supported by the Italian-Hungarian bilateral agreement of the CNR (Research National Council) and MTA (Hungarian Academy of Sciences).

References

[1] Auras R et al 2001 *Poly(Lactic Acid): Synthesis, Structures, Properties, Processing, and Applications* (John Wiley&Sons Inc., Hoboken, New Jersey)
[2] D Garlotta 2001 *Journal of Polymers and the Environment* **9(2)** 63-84
[3] J M Chacón et al 2017 *Materials&Design* **124** 143-157
[4] Hajba S and Tábi T 2019 *Periodica Polytechnica Ser. Mech. Eng.* **63(4)** 270-277
[5] Hull et al 2015 *Proceedings of the ASME 2015 Int. Manuf. Sc. and Eng. Conference*
[6] H L Tekinalp et al 2014 *Composites Science and Technology* **105** 144-150
[7] L Sang et al 2019 *Composites Part B: Engineering* **164** 629-639
[8] R T L Ferreira et al 2017 *Composites Part B* **124** 88-100
[9] Dunai A and Macskási L 2003 *Műanyagok fröccsöntése* (Lexica Kft, Budapest)
[10] X Wang et al 2017 *Composites Part B* **110** 442-458
[11] Boros R et al 2019 *Express Polymer Letters* **13** 889-897
[12] Geier N and Pereszlai Cs 2020 *Periodica Polytechnica Ser. Mech. Eng.* **64(1)** 67-80
[13] Czigány T et al 2005 *Periodica Polytechnica Ser. Mech. Eng.* **49** 3-14
[14] MSZ EN ISO 527: Plastics. Determination of tensile properties. (2012)
[15] MSZ EN ISO 178: Plastics. Determination of flexural properties (2011)
[16] MSZ EN ISO 179: Plastics. Determination of Charpy impact properties. (2010)
[17] V Fiore et al 2015 *Composites Part B: Engineering* **74** 74-94
[18] Y Z Wan et al 2001 *Journal of Applied Polymer Science* **82** 150-158
[19] M J Mochane et al 2019 *Express Polymer Letters* **13** 159-198
[20] S Yu 2019 *Composites Science and Technology* **175** 18-27

Simulation of hot rolling by cellular automata

T Hegyes[1] and P Barkóczy[2]

[1] Arconic Köfém Mill Products Hungary Kft., Székesfehérvár, Hungary
[2] Fux Zrt., Miskolc, Hungary

E-mail:tibor.hegyes@arconic.com

Abstract. This development is focusing on one of the most complex and important production steps of flat rolled products that is the simulation of hot rolling. During hot rolling two phenomena, work hardening and the process of regeneration of microstructure have strong influence for physical properties of aluminium alloys. Dynamic recovery has importance in the dynamic softening beside dynamic recrystallization in aluminium alloys. Hot rolled and newly modified grain structure is influenced by these dynamic phenomena. Hot rolled grain structure goes through significant changes under further production steps like cold rolling and heat treatments. But aside from these intermediate production steps the microstructure that we get after hot rolling has significant effect on mechanical and grains structure of the final flat rolled product. Proper cellular automata simulations of recrystallization and recovery can be the basis for good technology planning practices.

1. Introduction

Worldwide, millions of tons of flat rolled aluminium sheets are produced per year for customers. For a technology engineer it is not an easy task to build up a proper technology that fulfils the product requirements. These requirements that we can call material properties like material thickness, grain size, mechanical properties, deep drawing capability etc. are in focus for end users and the rolled material must conform to these specified customer requests [1]. Therefore, a development of a simulation has an importance that could support industrial engineers in their daily practice to find the best optional production route. One possibility is to simulate the whole production process [2], but at first hand the simulation of material behaviour and not the technology itself is useful [3]. For the rolling technology simulation from material workability, rolling passes etc. FEM simulations are more common solutions [4]. But when the calculation of the microstructural parameters is important, e.g. deep drawing, it needs to extend to other calculation or simulation methods [5]. One possible extension is the cellular automata [6]. Cellular automata are different, and they could give better solution for recrystallization simulation [7]. Working with cellular automata is easier and calculation time could be shorter. This is an advantage in the mentioned industrial application where time factor is important and material properties are in focus.

The cellular automata are applied for this simulation task is 3D [8], 2D [9] and 1D [10]. The 3D and 2D simulations give a detailed picture about the development of the microstructure. The large number of the used cells, the resolution of the computation, has a great effect on the computation time. For this aspect the 1D simulations are extremely effective ones, but the mentioned detailed picture is lost. But calculation of the average grain size, and the distribution of grain sizes are easy with these automata.

The automata use arbitrary units during the calculation. Therefor a scaling step is necessary for the evaluation of the calculation [11].

Cellular automata is a discrete dynamic method with a simple calculation scheme. Therefore, the calculated results and parameters have special measures, which derived from the operation of the automata. Such like the time, which measured in automata steps, and the distance which measured by cells. There is no information in the sub-cell region, or no valid states inside a step. These two main measures change the other derived measures of all used parameters. Basically, the automata make the calculation well if the real parameters converted to this specific universe. This conversion called scaling, and method in [11] uses a Nelder-Mead simplex method for the mentioned conversion based on the comparison of measured and calculated results. In this method the time scale, which is the quotient of the real time period and automata steps, is fixed and the kinetic parameters, for example the activation energies of the processes vary by the simplex method.

The [10] automaton is used in the current development. This is a 1D automaton with periodic boundary conditions in both ends of the universe [12]. The periodic boundary condition makes the state of the cells identical at both ends of the universe. This assumption means that the processes take place in the bulk, far from the surface regions. The basic states of the cells are deformed and recrystallized.

The used 1D cellular automaton of recrystallization [7, 10] is used in several study e.g. kinetic analysis of recrystallization of OFHC copper, simulation of crystallographic changes during recrystallization of rolled sheets etc. Only two aspects are missing to apply this automaton to a process simulation: one is the coupling the simulation of recovery to the automaton of recrystallization, and the description of the dynamic recovery and recrystallization in a same way as it used in [7, 10]. This article introduces the solution of these missing calculations to prepare the automaton the process simulation of production of rolled flat sheet products. The key for this is the stored energy as a state variable is used in the description of the states of the cells as in [10]. The different sub-processes changing the value of it. The deformation process increasing the stored energy according to the strain hardening. The recovery and the recrystallization (nucleation and growth) are decreasing the amount of stored energy. These changes are basically described by the kinetics of the processes and ruled by the kinetic constants. The kinetics calculated by the cellular automaton and the activation energies are used to specify the automata as a given material (alloy and state). In this study the behavior of the developed automaton is introduced through a single hot rolling step. The results are discussed and evaluated on the effect of the different kinetic constants. The basic set-up was chosen from prior study of recrystallization of EN AW 3003 [7], based on this calculation were estimated the kinetic constants. It is important to note that these results calculated by estimated constants, not scaled ones.

2. Recovery simulation by 1D cellular automata

The [10] automaton just calculates the changes during a recrystallization process, and it is scaled to OFHC copper [7]. In this case the recovery has no importance but in case of aluminum alloys it is necessary to couple that calculation with a recovery simulation.

The recovery has no significant effect on the deformed microstructure, but the dislocation structure is rearranged. One basic property where it can be evaluated is the stored energy, which is the driving force of the following recovery and the recrystallization. Therefore, the stored energy (E_{st}) is incorporated into the state of the cells. This incorporation is changing the meaning of the basic states. So, the critical value of the stored energy is defined (E_{stc}). The cells which have lower stored energy are recrystallized. In the other case, deformed cells have larger (or equal) stored energy. One other limit is defined, which has a strong connection to the recovery process. The analysis of measured kinetic of softening of aluminum alloys [13] shows that the softening by recovery is asymptotically approximated with a value which depends on the degree of deformation and the alloying. This has a strong connection to the subgrain size and structure. The defined limit of stored energy means that value which can be reached by the recovery process only (E_{str}).

The developed automaton connects 10000 cells in 1D like a chain. Every cell has two other neighbors with periodic boundary condition. It means that the first and the last cells in the lace (universe) are connected to each other. The cell chain becomes unending. During the tests the automaton calculates with 1000 calculation steps. The following basic parameters were set for this capability study of recovery: each cell has deformed state (stored energy: 1000 (J/cell); critical stored energy: 100 (J/cell)). The calculations were made under isotherm circumstances (500 (°C)). It can be demonstrated like a cold rolled material with height dislocation density/stored energy which is annealed. The dislocation movement as well as the recovery are both thermally activated processes. For the calculation of the stored energy of a cell a same stochastic rule is used as in [10]. The activation energy of the recovery (Q_r) is converted to a probability by the Arrhenius function (1). In every step at every updating of the cell states a random number is generated between 0..1. If the generated random number is lower than the calculated probability, the stored energy is changed by the recovery according to (2). This function means an exponential decay of the individual values of the stored energy. K_{st} is another kinetic constant next to the activation energy. It has the same effect to the calculated kinetics of the recovery, so it is constant value in the current study. The T denotes the temperature and R is the gas constant.

$$p = exp\left(\frac{Q_r}{RT}\right) \tag{1}$$

$$E_{st}(t + \Delta t, i) = E_{st}(t, i) - \frac{E_{st}(t, i) - E_{str}}{K_{st}} \tag{2}$$

In this case also true the assumption in the Introduction, that recrystallization kinetics of EN AW 3003 is the basis. But there is no specific calculation till now to the kinetics of recovery alone with 1D automaton. Additionally, the numerical effect to the incorporated kinetic constants are unknown. Therefore, in the following study the automaton's arbitrary units are used. The total stored energy is calculated as a summary of the stored energies of the cells. In the first calculation the changes of this value are examined. Figure 1. presents the stored energy changing in function of automata calculation steps using different activation energy. Results show us the different value of activation energy has different effect on the stored energy. Increasing activation energy slows the recovery. The value of stored energy didn't change efficiently in case of 50 (KJ/cell).

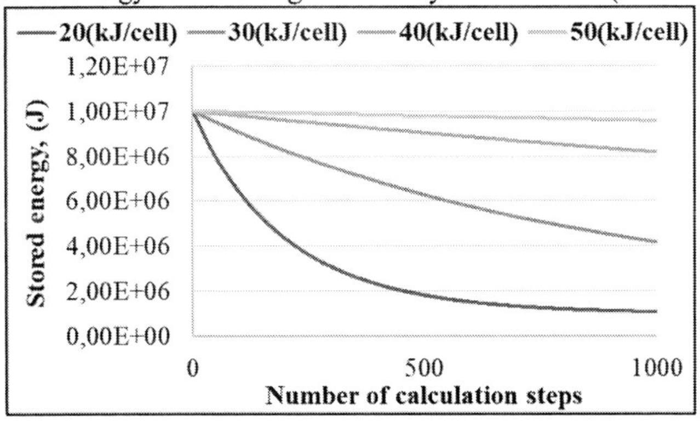

Figure 1. Stored energy changing in function of 1D cellular automata calculation steps with different activation energies

In this case we can say the dislocation density remained high and no significant softening take place. The value of Kst is set to 10 for calculation. Critical stored energy is an important parameter. When a cell reaches this value, the cell becomes recrystallized. Activated cells change their own state in this case from formed state to annealed (recovery) state. Lower value of activation energy is causing drastic drop in stored energy during calculation. The nature of 20 (KJ/cell) activation energy curve is exponential that is typical for recovery mechanism.

The developed automaton is sufficient for calculating the stored energy release during recovery. Additionally, it uses the same type of rule as the basic recrystallization automaton, so the coupling is possible.

3. Coupled simulation of recovery and recrystallization by 1D cellular automata

Recovery and recrystallization are responsible for softening mechanism in aluminium alloys. Our goal was to connect the recovery and recrystallization rules in one automaton that can handle both mechanisms parallel in one calculation.

We used almost the same basic parameters like in the previous chapter for recovery Qr=20 (kJ/cell), Est=1000 (J/cell), E_{stc}=100 (J/cell) and E_{str}=800 (J/cell). For this calculation 1500 calculation steps and different temperatures (iso and non-isothermal) were applied. The activation energy of nucleation was 60 (kJ/cell) while the activation energy of growth was 40 (kJ/cell). With these values the recrystallization related to the recovery is defined.

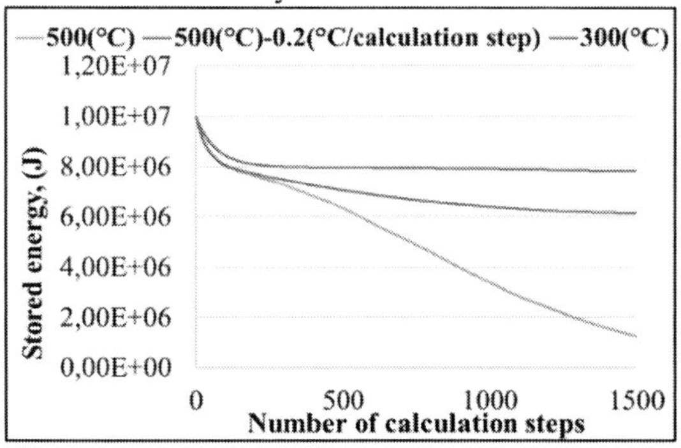

Figure 2. Stored energy changing in function of calculation steps with using different temperature

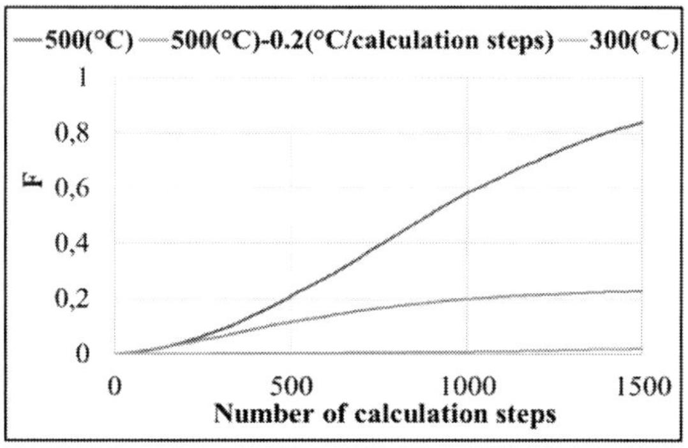

Figure 3. Recrystallized fraction in function of calculation steps with using different temperature

Results were similar until approximately 200 calculation steps (Figure 2-3.), there is an intensive drop in stored energy due to recovery. This is that point where recrystallization become a larger effect in softening than recovery. Over 200 calculation steps curves start to separate from each other and there is big difference in stored energy. Stored energy didn't change significantly at 300 (°C). This was a low temperature for recrystallization, but recovery was responsible for softening. Lowest stored energy was calculated at 500 (°C) and the recrystallized volume fraction (F) is almost 0,9 after calculation. This is a typical curve of Avrami type phase transformation. In the reality there is non-isothermal cooling therefore we simulated this case also. The Orange curve represent this opportunity where starting temperature was 500 (°C) and at every calculation step it was decreased by 0,2 (°C). Recrystallization started but it wasn't finished at the end of calculation. These results demonstrate that the developed automata can differentiate these recovery and recrystallization processes, and the coupling was successful.

4. Hot rolling simulation by 1D cellular automata

In the real factory practice approximately 400-500 (mm) thick semi continuous slabs are hot rolled to 5-8 (mm) final thickness in reverse rolling passes. Number of rolling passes could be different from 15 to 25. Between two hot rolling passes the material lays on roller table and it is cooling. The microstructure is changing during this time not just at hot deformation. Our automata were improved to simulate the material microstructure in case of hot rolling. This capability study focuses on one rolling pass and cooling period which is between two rolling passes. We can call the cooling period holding time. The First part of calculation is the rolling and second part of calculation is the cooling. At this stage the stored energy is an important factor therefore it was the main point in our experiment with recrystallized fraction. These were analyzed with modification of activations energies, recovery limit value and rolling speed. One additional calculation is necessary, the increment of the stored energy during the plastic deformation. For this the rolling operation is also divided by the same calculation steps, which is used by the simulation. Based on the geometry changes of the sheet - which recently calculated by geometric formulas by the rolling speed – the time scale becomes fixed, currently to 0.0001 (s/step). By the change of the thickness of the sheet an increment of the degree of deformation is calculated at every step. The 1D automata cannot hold a detailed information about the position and geometry of the represented material. But here in the simulation of hot deformation is a key question. In this study the universe

represents a small part of the cross section of the sheet. So, as the examined cross section passes through the rolls, with the introduced simple geometric model all cells make the same deformation. It can be refined with other detailed simulation of rolling process of course.
The mentioned increment of deformation per step means an increment of the stored energy. But this increment is depending on the current deformed state of the given cell, which is determined by the actual value of stored energy. To define the dependence between the stored energy and the degree of deformation the formula is used form [10].

$$E_{st}(\Delta h/H) = E_{stmax}(1 - \exp(-\Delta h/H)) \tag{3}$$

E_{stmax} determine the shape of the curve, and 1 (kJ/cell) is used in current calculation. The algorithm of the calculation of stored energy increment is: 1. calculate the equivalent degree of deformation from the stored energy of the cell by (3); 2. increase the degree of deformation by the value of the actual calculation step; 3. calculate the new value of the stored energy by (3). This algorithm can be easily incorporated into the introduced automaton.

Figure 4-5. present the recrystallized fraction and stored energy in function of calculation steps where different nucleation activation energies (Qn) were used in calculations.

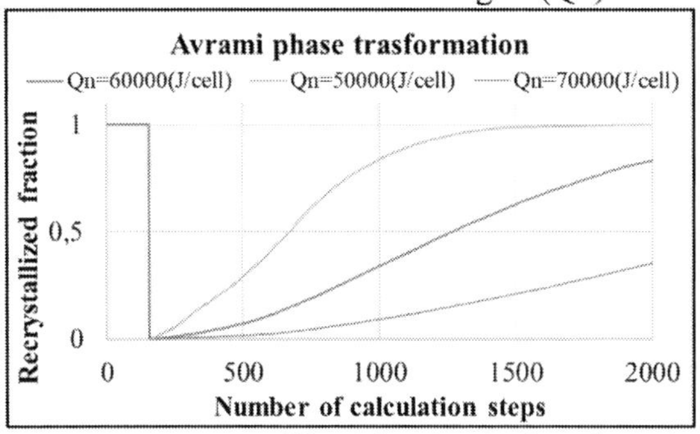

Figure 4. Recrystallized fraction in function of calculation steps with different nucleation activation energy

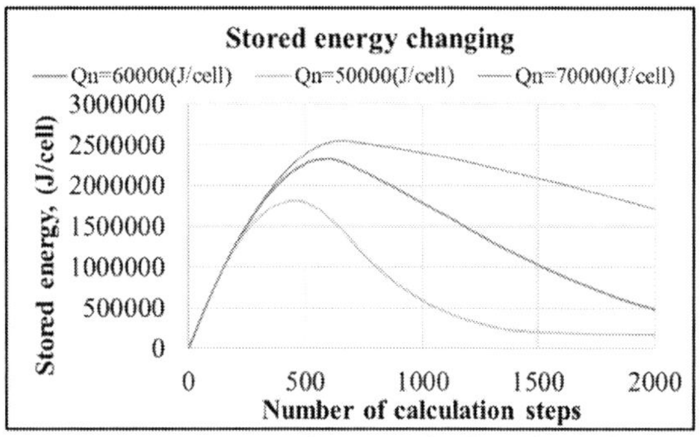

Figure 5. Stored energy in function of calculation steps with different nucleation activation energy

The other parameters were: Q_g=40 (kJ/cell), Q_r=30 (kJ/cell), E_{str}=800 (J/cell), and the same values as presented above. The rolling speed was 1500mm/s. The hot rolling deformation phase was extended to 688 calculation steps and it is followed by a 2 (sec) cooling time which is present for 2000 calculation steps. Increased nucleation activation energy slows down recrystallization. Recrystallization starts at rolling and the smallest activation energy resulted in the highest recrystallized fraction. Recrystallized fraction is app. 5 (%) at the highest activation energy and app. 50 (%) at the smallest activation energy at 688 calculation steps in rolling zone. It can be seen also on figure 5. where curves have local maximum at this point. The maximum value is decreasing with decreasing activation energy that is coming from fast recrystallization and it is getting closer to the starting point of rolling. Recrystallization started sooner at smaller deformation and we got fully recrystallized structure in holding time.

5. Conclusion

The current study focused on the simulation of microstructural processes of hot rolling by 1D cellular automata. Hot rolling is a complex metal forming process where different physical processes are work hardening, softening (recovery and recrystallization). Cellular automata are a good method to simulate the microstructural changes and hence the changes in material properties. The big advantage of the 1D recrystallization automaton is that it calculates complex results next to its simplicity. This simulation was developed elsewhere and is applied for the study of recrystallization of OFHC copper where the recovery has no effect on the softening. In aluminum alloys during the annealing and hot rolling the recovery has a large effect on the mechanical parameters. A simple calculation is developed for recovery and coupled with the 1D recrystallization automata. The most complex step of the aluminum sheet processing technology is the hot rolling. The developed automata are extended with a simple calculation of deformation and the changes of the stored energy and recrystallized volume fraction of the universe of the automaton are examined with different kinetic parameters as activation energies. The simple build-up of the introduced automata shows as complex behavior as the hot rolling step. This article introduced two extensions of the 1D automaton of recrystallization: the calculation of recovery kinetics and coupling it with the recrystallization simulation in that way that the new automata can simulate the kinetics of the mentioned processes during hot rolling and annealing too. This development makes possible to use the new automaton in the process simulation of production of rolled flat sheets. The automata can help a technology engineer to study the different changes of process parameters in the properties of the product. Currently the automata use arbitrary units in the calculation, so the scaling is important.

References

[1] Totten G.E. and MacKenzie D.S. at al 2003 *CRC Press*
[2] Hirsch J. Karhausen K.F and Engler O. et al 2004 *Continuum Scale Simulation of Engineering Material*
[3] Guo Z. and Saunders N. and Schillé J.P and Miodownik A.P et al 2009 Materials Science and Engineering vol 499 pp 7-13
[4] Marthinsen K. Holmedal B. Valle R. and Abtahi S. 2003 Materials Science Forum vol 426-432(5) pp 3777-3782
[5] Gottstein G, Crumbach M., Neumann L. and Kopp R. 2006 Materials Science Forum vol 519-521 pp 93-102
[6] Raabe D. et al1998 Computational materials science
[7] Bubonyi T. and Barkóczy P. 2018 IOP Conf. Ser.: Mater. Sci. Eng.426 012007
[8] Radhakrishnan B., Sarma G.B and Zacharia T. 1998 Acta materialia, Elsevier
[9] Midownik, M. A. et al 2002 Journal of Light Metals vol 2 pp 125-135
[10] Gyöngyösi Sz., Tóth A. and Barkóczy P. 2010 Materials Science Forum vol 659 pp 405-410
[11] Barkóczy P. and Gyöngyösi Sz. 2013 Materials Science Forum vol 729 pp 150-155
[12] Espericueta R. et al 1997 Cellular Automata Dynamics Math Department, Bakersfield College
[13] Farzadi A. et al 2015 Mat-wiss. u. Werstofftech vol 46, no. 12 pp 1218-1225

IOP Conf. Series: Materials Science and Engineering **903** (2020) 012033 doi:10.1088/1757-899X/903/1/012033

Determining the optimum heating time of small sized test specimen made from weldable mild steel

F Tancsics[1] and T Ibriksz[1]

[1]Széchenyi István University AHJK - ATT, Győr, Hungary

E-mail: ibriksz@sze.hu

Abstract. The usage of CHT (Continuous Heating Transformations) diagrams for a given steel or equivalent grade, requires knowledge of heating temperature, average heating rate, and the heating time. Definition of these technological parameters are primarily based on the complex relationship system between the geometric, thermal parameters and the heating device. In our research, we mainly focused on those physical key parameters that can mostly influence the heating and transformation rate. These parameters provide realistic, usable data for analysing the process of thermal diffusion and FEA (Finite Element Analysis) tests. During analysis, an easy-to-use function relationship was determined for approaching the heating rate more precisely. This method allows handling the CHT charts easily, within a selected range, regarding weldable mild steels.

1. Introduction

A steel component of the selected material grade have to be austenitized with a certain average heating rate for heat treatment. The heating rate gives the relationship of the geometry and the heating furnace between the heating temperature and the time required for heating. However, the relationship appears only indirectly on CHT charts for a given material quality. Therefore, the average heating rate is an important technological parameter, its simple definition is important for engineers. In this point of view, the data for heating technology are just relatively well-defined for specific industrial processes and special laboratory procedures [1]. The goal of our research was to work out an easy-to-use and properly accurate approximation method for typical 20 mm of diameter specimens made from weldable mild steels that has an acceptable estimate result for heating rate or the heating time.

2. Applied method and modelling for the task

Our work was fundamentally based on the Reverse Engineering Techniques (RET). Based on practical measurement results and data, we have interpreted the heat transfer processes necessary to use the CHT diagrams and to achieve our goal. During the heating experiments, we used an electric resistance-heated furnace type K-28/1100 with a power of 3.5 kW and a PID (Proportional Integral Derivative) control system, without usage of ventilator.

The specimens were cut out of the columnar crystallized zone of a 100x100 mm billet that originated from continuous casting, in this way ensuring a homogeneous material structure. The material structure was also improved by the rolling ratio (RR is 6) of billet. The grain structure of the specimens consisted of ferrite and perlite, corresponds to the grain size class of ASTM 10-11. The carbon equivalent value (CE according to spec. EN 1011-2/A) is 0.483, so this material grade has a good weldability.

The most important data of the experimental S460N material grade used for the heating experiments compared to standard S460N material grade can be seen in Table 1.

Table 1. Comparison data of experimental and standard steel S460

Composition (wt%)	C	Mn	Ni	Si	Al	Cu	Ti	V
experimental S460N	0.141	1.5	0.62	0.18	0.008	0.2	0.0067	0.186
standard S460N	max. 0.2	1.0-1.7	max. 0.8	max. 0.6	max. 0.02	max. 0.55	max. 0.05	max. 0.2

The geometry of the test specimens is characterized by Ø20x60 mm. Along the centreline of the specimens a cylindrical borehole with a diameter of Ø3 mm and 30 mm deepness was made for the thermocouple. The installation and adjustment of the thermocouple required very careful preparation and several tests [2]. The heating temperature of the specimens were set at 1050 °C and we strove to achieve the fine-grained austenitic microstructure.

The specimens were individually heated to determine the minimum heating time and maximum heating rate. We wanted to characterize the relationship between the thermal conductivity of the small-cross sectioned specimen and the surface heat transfer using the Biot number (1):

$$B_i = \frac{\alpha_{EHTC}}{\kappa_E} \cdot L \tag{1}$$

whereabout:

B_i	the Biot number is a dimensionless quantity (-)
α_{EHTC}	the effective heat transfer coefficient (sum up convection and radiation) (W/m^2·K)
κ_E	the effective thermal conductivity of scaled specimen (W/m·K)
L	the characteristic length (ratio of the volume and the cylindrical surface) (m)

The Biot number is a simple index of the ratio of the heat transfer resistances inside and on the surface of specimen. If this index is smaller than 0.1, the heat flux is determined by the heat transfer ability on the surface. For determination of Biot number, we needed data of time and temperature inside the specimen. Therefore, several heating experiments were performed. A typical result of the experiments is shown in Figure 1.

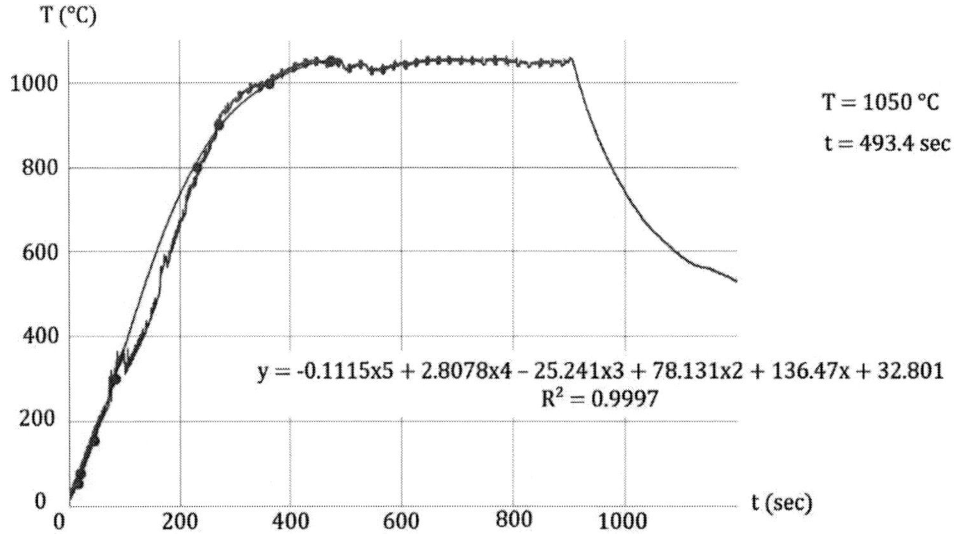

Figure 1. Graph drawn with measured data and a polynomial fitted to it

As shown in Figure 1 we drew a good approximated function onto the heating curve recorded by measurement. The measurement graph contains some measurement noise, which is unnecessary to consider [3]. This solution allowed us to handle thousands of data in the future in a simple way.

In the following step, we determined the effective heat transfer coefficient acting on the surface of the specimen. This factor is the cumulative value of the convective and radiative heat transfer [3,4] (2).

$$\alpha_{EHTC} = \alpha_C + \alpha_R \tag{2}$$

whereabout:

α_C the convective heat transfer coefficient (W/m²K)

α_R the radiation heat transfer coefficient (W/m²K)

In the preheated electric resistance-heating furnace, the heat transfer is basically determined by the thermal radiation. This effect increases with increasing the surface temperature of the specimen (3).

$$\alpha_R = \frac{\sigma \cdot \varepsilon \cdot \left(T_F^4 - T_S^4\right)}{T_F - T_S} \tag{3}$$

whereabout:

σ the Stefan–Boltzmann constant (5.67×10^{-8} W/m²K⁴)

ε dimensionless emissivity constant. (its value is 0.8 because of ceramic walling of furnace) (-)

T_F the air temperature in the furnace (K)

T_S the surface temperature of the specimen (K)

It should consider that in the absence of a ventilator, the airflow near the surface might be a natural convection. In this case, with consideration of all efficiency, for determining convectional heat transfer coefficient, the following mathematical expression could be used which derived from the discontinuous furnace performance:

$$\dot{Q} = \frac{c_P \cdot \dot{m} \cdot (T_S - T_0) \cdot 3.6}{t \cdot 1000} \cdot \eta \;\rightarrow\; q = \frac{\dot{Q}}{A_S} \;\rightarrow\; \alpha_C = \frac{q}{T_F - T_S} \tag{4}$$

whereabout:

$\dot{Q}$ the amount of heat flux per unit time (J/s or W)

q the amount of heat flux per unit time and unit surface (W/m²)

T_0 the initial temperature of surface (298 K or 25 °C)

t the heating time of specimen (measured) (sec)

c_P the amount of specific heat on constant pressure (600 J/kg·K or 600 J/kg·°C)

$\dot{m}$ the performance per hourly of furnace on the actual temperature (kg/h)

A_S the surface of the specimen (m²)

η the technological efficiency of the furnace (80%)

The results of the calculations are summarized in Figure 2. The figure shows two graphs. The upper graph is the sum of heat transfer coefficients, the lower is the heat convection coefficient, as a function of temperature. At a temperature of 500 and 600 Celsius, the heat transfer coefficient was determined by the adjusted polynomial that is visible on Figure 1. From the diagrams, we can conclude that the effective heat transfer coefficient grows parabolically, depending on the heat radiation. The value of the heat convection coefficient decreases below 10 in the range of 300-800 Celsius. The change in this interval is small and its minimum value falls in the range of 500-600 Celsius, which can be related to the powerful scale formation. However, convection heat transfer is not significant. The average of the heat convection coefficient corresponds to the values normally used for natural airflow [4].

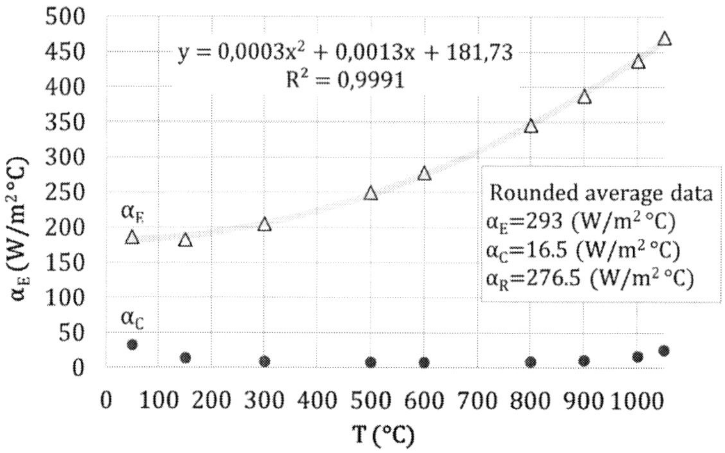

Figure 2. Graph of the effective and the convective heat transfer coefficients as a function of temperature

For structural steels, the thermal conductivity varies linear from room temperature to 800 °C, the thermal conductivity reduces from 55 W/m·°C to 28 W/m·°C. From this point, the thermal conductivity does not change with further temperature increase [4, 5, 6]. During examination, the smallest value of thermal conductivity was considered. For this reason, the average thickness of the scale was considered in this range. Consequently, we had to measure the thickness of the formed scale on the surface after heated to 1050 °C. We considered the thermal conductivity of the scale to be 2.5 W/m·°C, based on the above and related literature references [6, 7]. Based on the law of J. Païdassi the following relationship for pure iron to determine the layer thickness of the scale [8]:

$$x = 24550 \cdot exp - \left(84650 / RT\right) \cdot \sqrt{t} \tag{5}$$

whereabout:

x the scale magnitude at an instant (μm)
R the universal gas constant (8.314 J/K·mol)
T the heating temperature (K)
t the heating time of specimen (measured) (sec)

After cooling of specimen, the scale thickness was measured by HITACHI SN3400 Scanning Electron Microscope. Based on our measurements the relationship (5) was corrected by a simple coefficient. The coefficient takes into account the tendency of manganese alloy mild carbon steels to scaling at a high temperature (6).

$$x = 24550 \cdot exp - \left(84650 / RT\right) \cdot \sqrt{2t} \tag{6}$$

The scale thickness were determined based on the Equation (6). The average scale thickness of the samples heated to 1050 Celsius was calculated of 97.815 microns. This value approximates the average of the measurement results well, which can be seen on Figure 4. The thickness of the scale relative to the cross-section of the specimen is negligible, however, due to its inferior thermal conductivity, we did not ignore it. Therefore, the specimen thermal conductivity was modified to 27.75 W/m·°C.

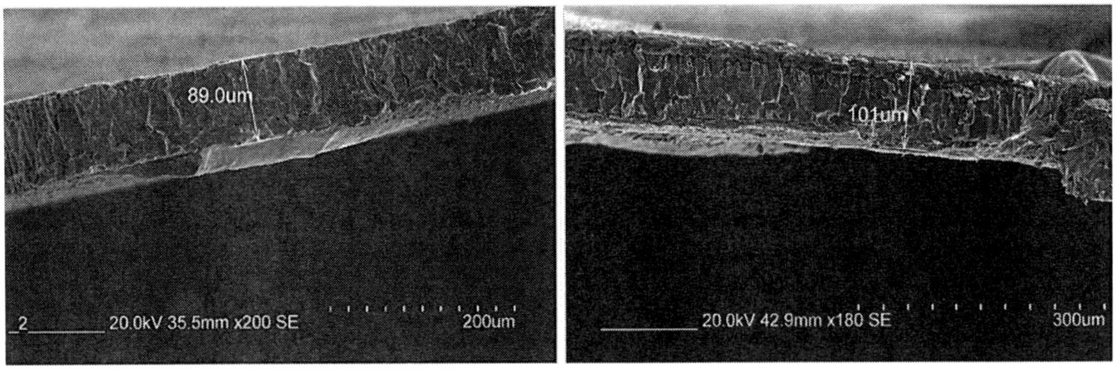

Figure 3. Typical thickness of the scale from different places of specimen

Taking into account the experimental results and calculations, the value of Biot number is 0.053. It can be stated that the value obtained is lower than the threshold value. This means that the heat transfer time only depends on the process of heat transfer on the surface, the heat flux distribution is uniform in the cross-section of material. The test results mentioned above were checked by a finite element method. For finite element studies, we used the Simufact.forming version of 15.0.

The key pre-processing data:

- 3D-FE solver usage, mesh generation by tetrahedron (134), element size is 1.235 mm
- the effective heat transfer coefficient on the surface is 293 W/m2 K
- material grade from softver database, DB_S460M_h
- the heating temperature is 1050 °C
- the heating time is 493.4 sec
- the emission coefficient of steel on temperature of 25 °C is 0.25.

The key post-processing data:

- the exit number of analysis is 3004
- based on the pre-processing data the maximum heating temperature is 1049.85 Celsius.

The difference is only 0.15 °C from 1050 °C. This difference is actually negligible. The result of the analysis can be seen in Figure 4. It confirms that the temperature distribution is uniform over the entire cross section of the workpiece based on the experimentally determined data.

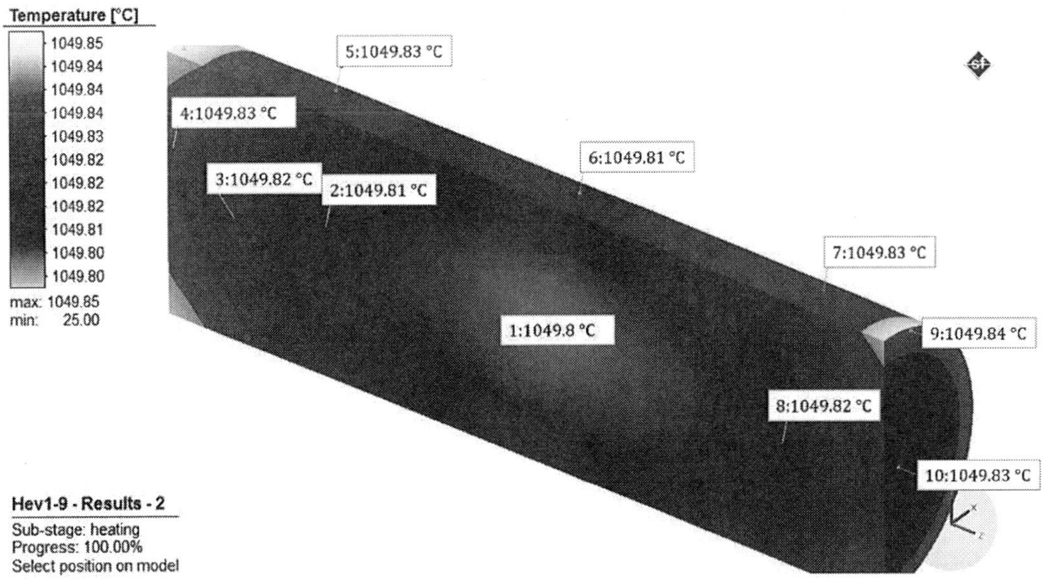

Figure 4. Temperature distribution in the cross section and on the surface of specimen

IOP Conf. Series: Materials Science and Engineering **903** (2020) 012033 doi:10.1088/1757-899X/903/1/012033

3. Examination of heating rate and material structure

In order to produce linear heating rate curves, the temperature-time relationships had to be linearized. For this, we first needed a straight line to fit onto the endpoints of the heating curve. Based on the heating curve in Figure 2, the average heating rate was determined at 2.077 °C/sec, practically rounded to 2.1 °C/sec. Then a simple algorithm was used to determine the temperature - time relationships derived from the furnace power data using this constant heating rate. Certainly, there are simpler mathematical solutions as well, but the furnace technology data cannot be separated from recording the heating curves. Equation (7) gives the linearized time data by furnace power set at a constant heating rate of 2.1 °C/sec.

$$t = \frac{\frac{m}{\dot{m}} \cdot 3600}{\eta} \cdot k_E \qquad (7)$$

whereabout:

t the heating time of specimen (calculated) (sec)
m the mass of the specimen (kg)
$\dot{m}$ the furnace power per hour (kg/h)
η the technological efficiency of the furnace (80%)
k_E the dimensionless cumulative correction factor that takes the additive effects of geometry, shape and furnace type into account (-)

Based on the results, a simple and good approximation function relation can be obtained from the linear equation. The function relation can also provide a good approximation of temperature - time relationship referring to heating rate (8).

$$a)\ t(min) = \frac{T[°C]}{127} \qquad b)\ t = \frac{T}{127} \cdot \frac{v_{(2.1)}}{v_{(3)}} \qquad (8)$$

whereabout:

$\frac{v_{(2.1)}}{v_{(3)}}$ the heating rates ratio in the same dimension (e.g. 2.1(°C/sec) / 3(°C/sec) = 0.7).

The simplified function of formula (8a) (red circle, details below) in Figure 5 fit well to the values calculated from formula (7) (black triangle, details above).

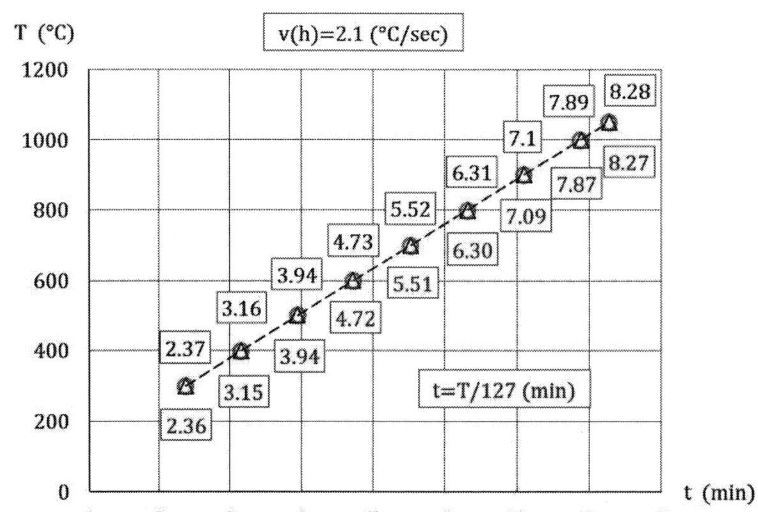

Figure 5. The constant heating rate as a function of temperature - time

Using the equation (7) and the very precise approximation shown in Figure 5, the heating rate functions can be generated in the usual form. In the case of relation (8b), the heating rates, shown in parentheses, we mentioned only as examples. Using this formula the function can be generated at any arbitrary heating rate. It is easy to see, if that heating rate is closer to the determined heating rate, we can obtain results that are more accurate. By plotting the time data of the relationship on a logarithmic scale, we can obtain the well-known heating rate function, which is applicable on the CHT diagram.

The examined structural steel S460 can be considered Mn alloy carbon steel as well. This explains that the scale was stronger than expected, actually can also be detected in the austenitic grade.

The specimens were heated to 1050 degrees Celsius for eight minutes and approximately fourteen seconds for material structure studies. After reaching the specified temperature and time, the specimens were quenched in water at 20 degrees Celsius. As a quenching result, the obtained martensitic structure showed the original austenitic crystal boundaries, which were examinated according to ASTM E112-13 (Heyn Lineal Intercept Procedure, magnification = 50x).

As a result of the examination, we found that the grain size satisfied grade 5-6, closer to grade six. The typical average grain size in the middle of the samples is 51.6 microns. The typical average grain size near the surface cross-sections of the samples 57.3 microns. Observably, the material needs more Ti or Nb alloys for finer grain structure at high temperature [9, 10].

As a result of the microstructural analysis, we found that the material structure does not contain coarsed grain, locally duplex structure. The composition of the material is characterized by a homogeneous, average grain size. It follows that the method presented for determining the heating rate is admissible.

4. Summary, conclusions

In this article, we reported on the partial results of a research process. In our work, we have dealt with a possible method of generating the technological data needed to heat the C-Mn steels with a cross section typical 20 mm of diameter. Based on the experience of the experiment, we drew conclusions that can be found in the relevant chapters.

The results, which are considered important, are summarized as follows:

- the effective heat transfer coefficient grows parabolically, basically depending on the heat radiation
- the scale thickness were well approachable on the Equation (6), although the thickness of the scale relative to the cross-section of the specimen is negligible
- the finite element analysis confirmed that the temperature distribution is uniform over the entire cross section of the workpiece based on the experimentally determined temperature-time data, so the applied method is satisfactorily precise
- the simplified functions of formula (8a) and 8(b) are a satisfactorily precise approximation to the determined heating rate function, so converted to logarithmic scale are applicable on the CHT diagram as well
- the microstructural analysis has confirmed, the heated material of S460 has a homogeneous and average grain sized structure, so the method presented for determining the heating rate is usable

Overall, we can say that our subtask goal was mostly met, and we managed to define a simple method for easier handling of CHT curves.

Acknowledgments

The paper was written with the support of the project titled "Internationalization, initiatives to establish a new source of researchers and graduates and development of knowledge and technological transfer as instruments of intelligent specializations at Széchenyi István University" (project number: EFOP-3.6.1-16-2016-00017).

IOP Conf. Series: Materials Science and Engineering **903** (2020) 012033 doi:10.1088/1757-899X/903/1/012033

We would also like to thank Imre Fekete and Dóra Harangozó for assisting in connection with the theoretical background of this article.

References

[1] Charles M et al 2004 *Heat Treatment Procedure Qualification for Steel Castings* (The Pennsylvania State University)

[2] Meriga V et al 2018 *Leonardo El J Pract Technol* **32** pp 41-54

[3] Lienhard J H 2001 *A Heat Transfer Textbook* (J.H. Lienhard, Cambridge, Massachusetts USA)

[4] Totten G E 2006 *Steel Heat Treatment Handbook* (Taylor & Francis Group, LLC)

[5] Naser M Z 2019 *Constr Build Mater.* **215** pp 192-206

[6] Kawahara T et al 2017 Thermal conductivity measurment of iron scale at high temperature bylaser flash method *Proceedings of the Asian Conference on Thermal Sciences* ACTS-P00463

[7] Li M et al 2017 *ISIJ INT* **57** pp 2097–2106

[8] Suarez L et al 2006 *Defect and Diffus. Forum* **258-260** pp 158-163

[9] Wang J et al 2019 *Materials Sci Appl.* **10** pp 665-675

[10] Wu D et al 2018 *High Temp. Mater. Proc.* **37** pp 899–907

The effect of different grain sizes and heat input on the gas pressure inside artificial resin-bonded sand cores

L Mádi[1], I Budavári[1], L Varga[1]

[1]University of Miskolc, Institute of Foundry, Miskolc-Egyetemváros 3515, HU

E-mail: madi.laura.johanna@gmail.com

Abstract. Sand cores are used to form the cavities of the castings. The quality of castings is directly affected by the quantity of the gases released from the cores. During our research, the effect of different grain sizes and heat input on the pressure of the gases evolved from the cores was investigated. The granulometric properties of the sand used for making the cores have a significant impact on the pressure of the gases. Because of the uneven particle size distribution of multi fraction sands, the gas flow conditions in the cores can be different. The gas pressure in the cores made from sieved sands can be changed between 6 and 9 mbar. According to the recorded cooling curve of the melt, it can be established, that in the case of the presented pouring conditions, a solid metal layer with adequate strength could not be developed before the gas could pass through the core/casting interface. This phenomenon may result in gas blowhole defects on the casting surface.

1. Introduction

The properties of the castings are directly influenced by the amount of gas released from the sand cores, the quantity of precipitated condensate and the deformation behavior of the cores during the casting process. The selection of the binder-system with the expected properties and the optimization of its quantity is also pivotal from an environmental protection point of view [1]. Based on the directive of the European Commission (314/2005. (XII. 25.) Gov. decree) the reduction of the environmental load, moreover, the moderation of the effect on environmental health, the best available technique (BAT) should be applied. The rules for air quality protection, especially the rules for institutes using organic dissolvent are more strict (306/2010. (XII. 23.), 2010/75/EU (5) Gov. decree).

Due to the improvement of casting technology allows the manufacturing of more complex castings with thinner wall thickness. The complex core geometry usually results in a larger core surface, which contacts the liquid metal [2]. Increasing core surface area, coupled with difficulties in venting, significantly raises the pressure inside the cores and increases the tendency to produce gas-related defects in castings. The gas formation and its pressure during the thermal decomposition of the artificial resin binder of the cores at the early stage of solidification could drastically influence the casting properties [3]. If the gas pressure exceeds the metal head pressure, the gas could bubble through the molten alloy. In most of the research reported in the literature, only the amount of gas developed from the core is measured, and the rate of gas formation is used to estimate the pressure peaks [4-10] but this method does not give exact information on the pressure within the core. Moreover, it is possible to measure equal gas quantity with different compositions and heat quantity, but the characteristics of the gas pressure curve could greatly differ.

2. Literature review

Since the start of the application of organic resin binders in mold-, and core making processes, more and more research was conducted to determine the factors that can influence the volume and pressure of gases released from the molds and the cores during the casting process. The relatively simple no-bake process has enabled foundries throughout the world to produce castings more efficiently. In the no-bake process, room temperature curing is involved with the direct addition of a curing agent. In this way, cores and molds are produced from self-setting sand mixtures at the ambient temperature of the foundry. Since the no-bake technology was developed, metal casters throughout the world recognized the productivity and efficiency of making molds and large complex cores at room temperature. The foundry industry has classified the resin-binders used in the sand casting processes into three categories: Cold-box, No-bake, and Heat-Cured binder systems. There are eight types of the no-bake process which are as follows: phenolic urethane, modified phenolic urethanes, ester-cured phenolic, oil urethane, furans, phenolics, silicate-ester cured and magnesium-phosphate based [11]. In this article, we have investigated no-bake furan resin bonded sand cores, so we will only discuss this procedure. The binders contain furan resin systems (based on furfuryl alcohol, not furan as commonly noted), in the no-bake process, acid catalysts are used to promote curing [12], wherein the amount and strength of the acid are used to control the curing rate. The reaction of the catalyst can be triggered by temperature fluctuations and the control of the temperature of the sand can be more easily realized than the continuous adjustment of the amount of catalyst [13]. The examinations were carried out on furan no-bake resin-bonded sand cores. Furan no-bake resin-bonded cores are known for their good high-temperature strength, erosion resistance, and good shakeout properties.

Numerical simulations can be made regarding the gas released from the cores, which can help in the characterization of the gas pressure and rate of gas development within the core. It is highly important to know the exact rate of temperature rise of the core, the mechanism of gas development and the dynamics of gas flow. The base and boundary conditions can be different for all cores, such as the distribution of the void fraction (porosity), shape coefficient and the mean diameter of the sand particles. These parameters can be defined with initial measurements, moreover, they can be used to calculate sand particle size distribution and particle density. The gas venting positions can be exactly defined with the SIMPLE (Semi-Implicit Method for Pressure-Linked Equations) method, so the adequate venting of the cores can be realized [14-17]. (These basic equations are solved by the SIMPLE scheme under the initial and boundary conditions. The velocity, as well as the pressure of the fluid in each cell from which the core domain is made, can be obtained. The SIMPLE method which was proposed by Patankar [18] is one of the well-known solution algorithms of transient fluid flow. The initial conditions are the properties of the sand core, such as the distribution of void fraction, the shape coefficient and the average diameter of the sand particles. These conditions, such as the size distribution of the sand grains and the density of the particles are easily obtained by the results of the preliminary experiments.) The main topic of future developments is to model the gases released from cores with complex geometrical properties and to formulate computer codes, which can be used to predict different core gas pressures caused by the different casting temperatures, so the probability of the gas-related failures can be predicted too [14]. The composition and concentration of the developed gases are needed for more exact models, moreover, the surface heat transfer and the heat conductivity of the sand is also needed to be known [19]. (Future research will deal with the gas evolution when cores are placed in contact with magnesium, aluminum, brass, cast iron and steel at the same temperature to determine if the metal might be affecting the reaction products.)

Based on the literature, the gas pressure curve characteristics are usually similar, on which two maximum peaks can be observed. The first minor pressure increase is caused by the heating and inflating of the air within the pores. The high heat gradient then causes the gas development rate to increase which comes into balance at the maximum pressure value. The produced gas can escape from the core after the pores are filled. The rate of gas formation and escape defines the core pressure [20]. The pressure developed within the core is influenced by many parameters such as casting temperature, the gas permeability of the core, resin content and distance from the core print. The pressure of the gases developed in the core increases with the casting temperature and the viscosity of the developing gases. The gases developed in the core can escape through the gas vents. By increasing the distance from the core print, the maximal gas pressure increases linearly. By increasing the gas permeability the venting rate of the gases increases, which causes an asymptotic gas pressure attenuation [21].

3. Materials and method

In the course of our research work, measurements of the pressure of evolved gas from sand cores during casting were conducted in standard conditions. The experimental conditions were selected based on the literature [21, 22, 23]. For the experiments, core specimens were made with a no-bake furan resin and different core sands. During the research, the effect of the granulometric properties of the sand on the pressure of gas evolved from specimens having dimensions of Ø26.2 x 117 mm, and Ø 50x50 mm was investigated. Specimens were prepared by using core mixtures containing 1.2 w/w% of no-bake furan resin and 50 % of acid hardener relative to the binder content. To prevent the appearance of gas blowholes, the metallostatic pressure was increased according to literature data. Thus, the height of the melt above the core was changed from 30 mm to 50 mm, thereby reducing the hanging length of the core in the mold cavity. The effect of the heat input on the cores with modified geometric parameters (Ø 30x100 mm) was examined. These cores were produced with sand mixtures containing 1.4w/w% of no-bake furan resin and 50 % of acid hardener relative to the binder content. For measuring the evolved gas pressure of the cores, the above-mentioned specimens were made out of different types of no-bake binders. The refractory base sand of the cores was commercially available SH32, SH33, SH34 foundry sands from Sajdikovó. The foundry sand is characterized by slightly variant grain size differences, that can be eliminated with the sieving of the sand. During the experiments, the effect of the application of unsieved foundry sand and sieved sand on the pressure of the evolved gas inside the cores was investigated. The temperature of the cores was recorded. Table 1. shows the granulometric properties of different investigated sand types.

Table 1. Granulometric parameters of the different sands used during the experiments

Granulometric properties	Sand type (mm)					
	SH34	SH33	SH32	0.11-0.22	0.22-0.3	0.3-0.54
Average grain size (mm)	0.19	0.23	0.3	0.19	0.26	0.42
Fineness number	97.73	74.93	53.75	105	55	45
Specific surface (cm^2/g)	140	105	80	155	115	90
Degree of uniformity (%)	34	74	82	100	100	100

4. Experimental process

In order to collect the released gases evolved from the sand core, a steel pipe with a diameter of 6 mm and length of 300 mm was placed 5 mm away from the top of the test bars during the core making process. In the literature [22], the gas pressure was measured at a distance of 0.635 cm (0.25 in.) from the mold/metal interface. The expendable mold was prepared using a green sand mixture with 8 % bentonite and 4 % water based on the silica sand weight. The mold was made from SH33 foundry sand from Sajdikovó.

Approximately 45 kg of the green sand mixture was necessary for each form. The sand mixtures were produced with a Multiserw Morek sand muller mixer. The mold consisted of two parts. The sizes of the drag box were Ø330x200 mm and the sizes of the cope box were Ø330x140 mm. The form filling was carried out simultaneously using a tangential gating channel in a three-cavity form that was made with a cone-shaped wooden pattern (Ø70x150mm, draft: 4°). The different amounts of heat input also influence the solidification time of the castings. During the experiments, the melt temperature was recorded during the casting process 10 mm away from the metal/core interface. For measurements, a K-type thermocouple was applied, that was placed into a 100 mm long metal sheath with a diameter of 3 mm. The released gases passed through a silicone tube to a pressure transducer. Three Baumer type pressure transducers were used for measuring the gas pressure. The data acquisition was carried out with an HBM QuantumX-MX840B universal 8 channel amplifier module, which was connected to a computer with an HBM Catman software. For the experiments, an aluminum alloy EN AC-45500 was poured into the mold cavity with a casting temperature of 720 °C. The produced castings which had different wall thicknesses (20 and 40 mm) were labeled as Q20 and Q40. The steps of the mold-making process can be seen in Figure 1.

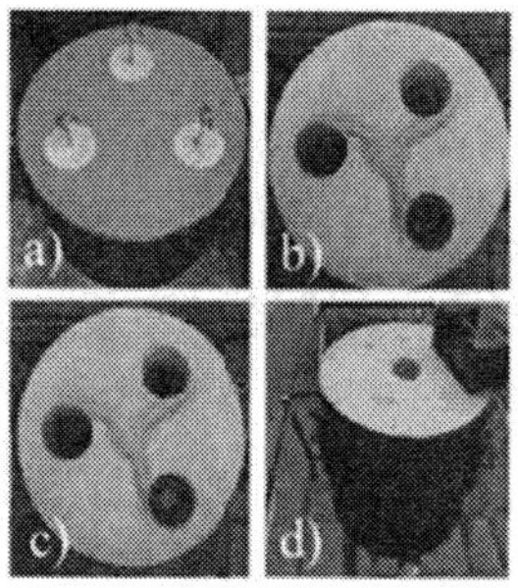

Figure 1. Steps of the expendable mold-making process: a) lower part with the cores, b) tangential access channel without cores and, c) with core, d) lower and upper molding box

5. Experimental results

5.1. The effect of the grain structure on the pressure and temperature of the evolved gases from the cores

During the pouring, the evolved gas pressure and the temperature of cores were measured. Core specimens are characterized by different average grain size and degree of uniformity (see Table 1.). The temperature of the cores was measured at the side of the steel pipe. Figure 2 shows the gas pressure and temperature curves of the released products obtained from cores containing different foundry sands (SH32, SH33, SH34). The bulk density of the cores: δ_{SH32}=1.36 g/cm^3, δ_{SH33}=1.35 g/cm^3, δ_{SH34}= 1.33 g/cm^3.

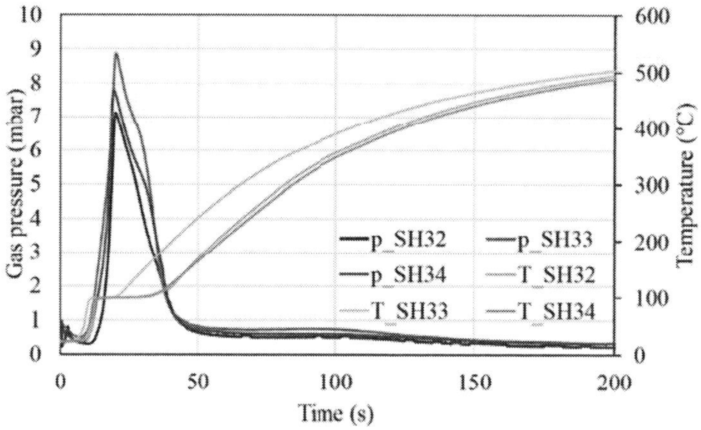

Figure 2. The change of evolved gas pressure and temperature of the cores with time. Cores contained different types of commercially available foundry sands.

As can be seen, there are no significant differences between the gas pressure and temperature curves of the cores containing different types of foundry sands. In order to eliminate differences resulting from the mentioned variables, the measurements were also executed on cores made from sieved sands. Figure 3 shows the results of cores made from sieved sands. The bulk densities of the cores were: δ_{coarse}=1.37 g/cm^3, δ_{medium}=1.35 g/cm^3, δ_{fine}= 1.635 g/cm^3.

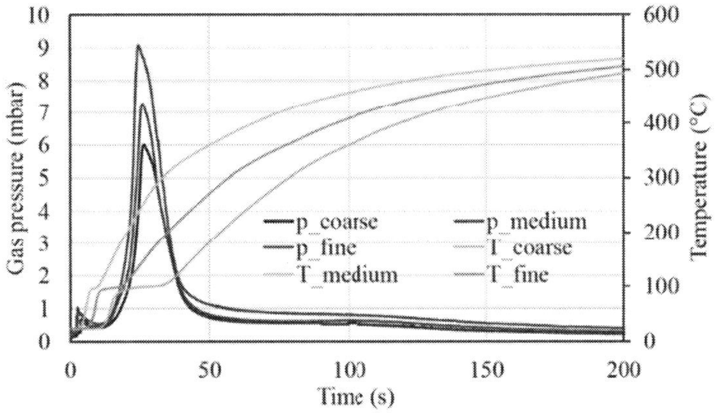

Figure 3. Gas pressure curves and temperature curves of the cores containing different sizes of sieved sands

It can be observed, that all specimens regardless of grain sizes show a peak pressure at approximately the same time of 30 seconds. However, a difference in the peak pressure at this time was observed, with peak pressures ranging from 6 to 9 mbar, depending on the grain size.

Based on the results, the maximum temperature of the core was 500 °C. The heating curve of the cores and their maximum value shows a similar tendency as in the case of the experiments of Bargaoui et al. [24]. There is a breakpoint in the heating curve of the core. This can be observed both in the literature data and in the presented measurements. When the metal initially contacts the mold, there is a huge increase in the temperature of the mold, and the condensed water near the interface vaporizes, expanding in volume over 1000 times [25]. This rapid expansion displaces the air originally present in the mold. The maximum gas pressure is observed at the time the core is heated up to 100 °C, which is explained by the vaporization of volatile components (water, ethanol, propanol, solvents) formed during the polycondensation processes take place during the setting of the binder [26].

The gas permeability of the cores is inversely proportional to the degree of compaction. Gas permeability is primarily dependent on the granulometry and the geometry of the pore structure created during the core shooting [27]. There is a strong correlation between the pressure of the core gas and gas permeability. The applied core specimens used in this research were not proper for the measurement of gas permeability. Because of that, standard Ø 50x50 mm, cylindrical test specimens were produced. These specimens were made from different foundry sands (SH32, SH33, SH34), as well as the sands that were separated into different size fractions (fine, medium, coarse) with the aid of sieving. The bulk density of the specimens was the same (δ=1.60 g/cm^3). An asymptotic relationship can be observed between the measured maximum gas pressure and the gas permeability (Pe), which is the same relationship as the one which is described in the literature (see Figure 4.).

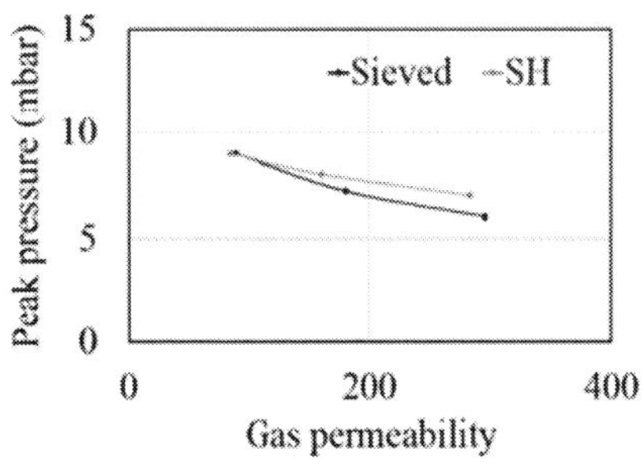

Figure 4. The asymptotic relationship between the measured maximum gas pressure and the gas permeability of the core specimens

It can be seen in Figure 4, that as the gas permeability of the cores increases, the maximum gas pressure is reduced asymptotically. In the case of sieved sands, the effect is more remarkable as the base sands contain only one size fraction of particles.

5.2. The effect of the differences in the amount of heat input

The gas pressure evolving inside the sand cores during the casting process is highly dependent on the amount of heat transferred by the liquid metal. Different melt temperatures, and in this way, non-identical amount of heat input can cause significant differences in the solidification time. During our experiments, the temperature of the liquid metal was measured at a distance of 10 mm from the melt/core interface. Based on the cooling curves, the conditions of the solidification of castings with different geometrical parameters can be evaluated. Table 2. presents the geometrical parameters of the two experimental castings and the sand core, as well as the product of the amount of heat transferred by the liquid metal (calculated based on reference temperature data from ref. [28]) multiplied with the sand/metal volume ratio. The investigated specimens had uniform bulk density ($\delta = 1.5$ g/cm^3), Pe=250.

Table 2. Results of the calculation of the heat input

Markings	Q20	Q40
Mass of the casting (kg)	0.84	2.49
Volume of the casting (m^3)	0.0003	0.0009
Volume of the sand core (m^3)	0.00005	
Quantity of heat, Q (J)	107465	320463
Sand/metal volume ratio	0.16	0.05
Product of heat input and sand/ metal ratio	17194	16023

As can be seen in Table 2, the product of the calculated heat input and the sand/metal ratio are closely equal in both cases. Based on this, for a given sand/metal volume ratio the amount of heat input can be estimated. During the experiments, two sets of samples with different geometries were cast. The evolving core gas pressure and the temperature of the melt above the sand core were measured during the casting and solidification. Figure 5 shows the cross-section of two experimental castings.

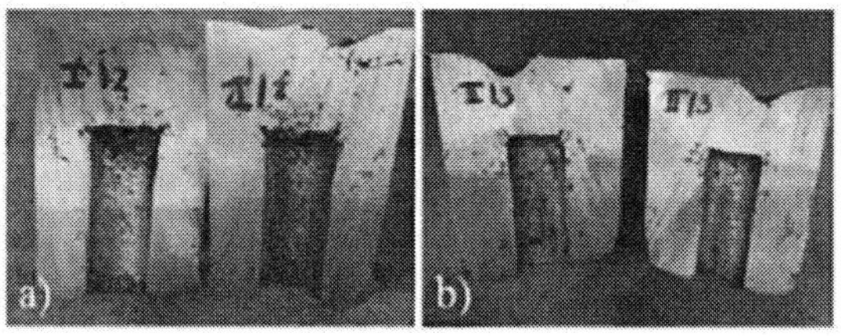

Figure 5. The cross-section of experimental castings with wall thicknesses of a) 20 mm and b) 40 mm

As can be seen, blowhole defects can be observed in the cross-section of the specimens, which influences the accuracy of the gas pressure measurements. This source of error can be evaded during future experiments by the application of different sand mixture composition. Figure 6 shows the results of the gas pressure measurements and the cooling curves of the castings.

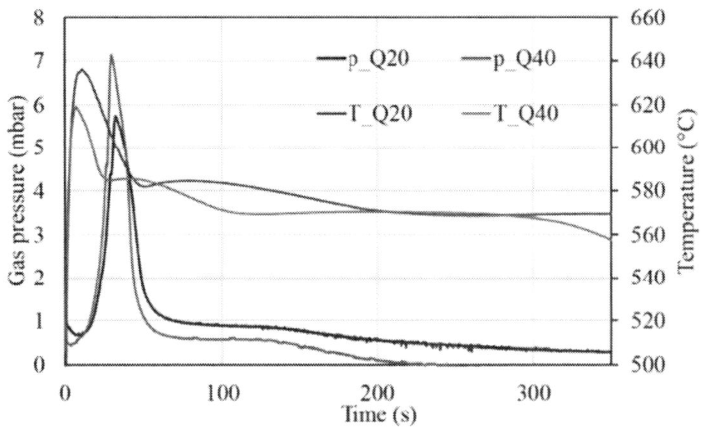

Figure 6. Gas pressure and cooling curves of the castings with different wall thicknesses

Unfortunately, because of the appearance of blowhole defects, clear ascertainments cannot be made based on the pressure curves. However, it can be seen, that the maximum values of the gas pressure were measured 20-30 seconds after casting, which timing was independent of the sand/metal volume ratio. According to our hypothesis, the breaking points on the steep rising section of the gas pressure curves indicate the formation of blowhole defects. Blowhole defects can only occur if the strength of the solidified metal layer around the core is not enough to prevent gas intrusion. The cooling curves noted as T_Q20 and T_Q40 differ from each other to some degree, but the curves follow a similar tendency. In the case of T_Q20, the solidification of the primary phase (α-Al) starts at 40 s, while in the case of T_Q40 this time is 26 s after the pouring. The gas pressure curve of Q40reaches the maximum before the start of the primary solidification which can be the main reason for blowhole defect formation. The eutectic temperature is approximately 570 °C in both cases, while the time of eutectic nucleation is 200 s for T_Q20 and 100 s for T_Q40.

6. Conclusion

A significant amount of casting defects can be traced back to the gases released from the sand cores. There are numerous parameters that can affect the pressure of the gases released from the cores during the casting process. During the course of our research, the effect of different granulometric parameters and the differences in the amount of heat input were investigated. The variation of grain size caused different peak pressures, while the time needed for reaching the maximum pressures has not changed. The cores made from coarse sand had good gas permeability, so the released gases could escape more easily from the core, which resulted in lower gas pressures. On the other hand, the application of coarse sand can cause inadequate surface quality and melt penetration. The selection of the sand mixture composition should be made based on the quality requirements of the casting.

For the investigation of the effect of the differences in the amount of heat transferred to the cores during casting and solidification, castings with different wall thicknesses (20 and 40 mm) were made with inner sand cores. During the experiments, blowhole defects occurred in each case regardless of the sand/metal volume ratio. The reason for this is that short exposure to a melt with high temperature can cause a significant rise in the temperature of the cores, which results in accelerated gas pressure build-up and blowhole defect formation. In these cases, the simultaneous investigation of the gas pressure curve and the size of the gas blisters should be satisfactory. Based on the cooling curves, the solidification time was too long for the development of a solid metal layer with adequate strength, which resulted in the formation of blowhole defects. This means, that by the application of a higher liquid metal column above the cores, not only the metallostatic head pressure had been risen, but the solidification time was also significantly changed.

References

[1] Ambidge P F and Biggins D E 1984 Environmental Problems Arising from the Use of Chemicals in Molding *Mat. Dev. for Fut. Foundry Prosp.* p. 219.
[2] Winardi L, Littleton H E and Bates C E 2007 Pressures in Sand Cores *AFS Trans.* **115** Paper 07-062(04).pdf, 303-312
[3] Erbslöh, Baier J and Köpper M 1994 Incidence and avoidance of defects attributable to molding sands *Man. of Cast. Def.* P: 51-55.
[4] Winardi L and Griffin R D 2008 Effects of coating drying methods on loi, gas evolution and core permeability *AFS Trans.* **116**:381–398
[5] Naro R L and Pelfrey R L 1983 Gas evolution of synthetic core binders: relationship to casting blowhole defects *AFS Trans* **91**:365–376
[6] Dietert H W, Graham A L and Praski R M 1976 Gas evolution in foundry materials. Its source and Measurement *AFS Trans.* **55**, pp 221- 228.
[7] Scarber P and Bates C E 2006 Simulation of Core Gas production During Mold Fill *AFS Trans.* **114** paper no. 138
[8] Orlenius J, Gotthardsson U and Diószegi A 2008: Mould and Core Gas Evolution in Grey Iron Castings, *Int. J. of Cast Met. Res.*
[9] Ravi S and Thiel J 2016 New Method for Measuring Gas Evolution in Chemically Bonded Sands *AFS Trans*
[10] Winardi L, Griffin R D, Griffin J A, Onda H, Harada A and Yoshida A 2008 Gas Evolution and permeability of shell cores in contact with Aluminum *AFS Trans*, Paper 08-107(04).pdf, 445-461
[11] Archibald J J 1993 Benchmarking the Nobake Binder Systems *AFS Trans.*, 93-14, 967-978
[12] Pilato L 2010 Phenolic Resins: A Century of Progress, *Springer* 451-502
[13] Ambidge P F and Biggins D E 1984 Environmental Problems Arising from the Use of Chemicals in Molding *Mat. Dev. for Fut. Foundry Prosp.* p. 219.
[14] Maeda1Y, Nomura H, Otsuka Y, Tomishige H and Mori Y 2002 Numerical simulation of gas flow through sand core *Int. J. Cast Metals Res.* **15**, 441-444
[15] Nastac L, Jia S, Nastac M N and Wood R 2016 Numerical modeling of the gas evolution in furan binder-silica sand mold castings *Int. J. of Cast Metals Research* **29** NO. 4, 194-201
[16] Khawaja H 2018 Semi-implicit method for pressure-linked equations (SIMPLE) ⇓ solution in MATLAB *Int. Jnl. of Multiphysics* **12** Number 4, 313-325
[17] Starobin A J and Hirt C W 2014 FLOW-3D Core Gas Model: Binder Gas Generation and Transport in Sand Cores and Molds *Flow Science Rep.* 04-14
[18] Patankar S 1980 Numerical Heat Transfer and Fluid Flow. Boca Raton: *CRC Press*,
[19] Svidró J, Diószegi A, Tóth L and Svidro J T 2017 The influence of thermal expansion of unbonded foundry sands on the deformation of resin-bonded cores *Arch. Metall. Mater.* **62** 2, 795-798
[20] Tóth L 1997 Gas evolution in foundry molding and core sand mixtures *Thesis of Ph.D. Dissertation*
[21] Winard L, Littleton H E and Bates C E 2007 Gas Pressure in Sand Cores *AFS Trans* 07-062(04)
[22] Chojecki A, Mocek J 2011 Gas pressure in sand mold poured with cast iron *Arch. of Foundry Engineering* **11**, Issue, 1/2011, 9-14
[23] Ravi S and Thiel J 2017 Prediction of Core Gas Pressure from Chemically Bonded Sand Molds Using Process Simulation Software *AFS Trans* Paper 17-097 v**125** - Page 1 of 8
[24] Bargaoui H, Azzouz F, Thibault D and Cailletaud G 2017 Thermomechanical behavior of resin bonded foundry sand cores during casting *J. of Materials Processing Technology* **246**, 30-41
[25] Pattabhi R, Lane A M and Piwonka T S 1975 Cast Iron Penetration on Sand Molds, Part III: Measurement of Mold-Metal Interfacial Gas Composition *AFS Trans.* 96-208, 1259- 1264
[26] Svidró J 2011 Transport process at the metal/molding interface - *Doctoral dissertation*
[27] Winardi L, Littleton H and Bates C E 2005 New Technique for Measuring Permeability of Cores Make from Various Sand, Binders, Additives and Coating *AFS Trans*, Paper 395-396
[28] Leitner M, Leitner T, Schmon A, Aziz K and Pottlacher G 2017 Thermophysical Properties of Liquid Aluminum- *Metallurgical and Materials Trans.* DOI: 10.1007/s11661-017-4053-6

Wetting of Non-Reactive and Reactive Metallic Substrates by Brazing Liquids

Z Weltsch

John von Neumann University, GAMF Faculty of Engineering and Computer Science, Department of Materials Technology, Kecskemét, Hungary

E-mail: weltsch.zoltan@gamf.uni-neumann.hu

Abstract. A survey of wetting as central phenomenon in soldering and brazing technologies are presented in this paper. The wettability is expressed by the magnitude of wetting angle. The equilibrium condition is widely investigated in the case of non-reactive partners. The phenomenon is less documented in the case of reactive partners, when chemical reaction (hence, local compositional change) also occur between the contacting phases. The brazing process belongs to this category. The mutual wettability is basic requirement for the formation of strong and durable bonding between the machine parts. Therefore, the correlation between the wetting angle (Θ) and the strength of the resulting bonding will be investigated. Not only the chemical concentration change, but phase transformations may also occure in such brazing procedure, when the contacting partners are reactive. As the chemical potencial of components are also change in this cases, the phase transformations has also impact on the evaluation of wetting angle between the brazing pool and alloy pieces to be joined. This phenomena will also be demonstrated in the case of non-reactive and reactive partners as well.

1. Introduction

The huge importance of wetting phenomena between liquids and solid phases is well known in the soldering and brazing technologies [1-3]. This phenomenon has gained recently increasing interest also in the development of steel production and in non-ferrous metallurgy. Accordingly, the simultaneous study of influence of metal-gas interactions on the surface tension (hence, on the mutual wettability between metallic liquids and solid phases at high temperatures) become of high technical importance [4].

The knowledge of surfaces and interfaces have also of outstanding significance due to the intensive progress in material science (the nature of interfaces between the constituent phases in various composites, or, understanding of alloying effects in ultrafine (nano-crystalline) grain assemble [5]. The complex application of high-resolution structural investigations is inevitable in this field. Similar, complex investigation is required in studying the bonding mechanism during the brazing process structural elements of electronics and also in modern body elements (dual phase steels).

The basic description of wetting phenomena between condensed phases (liquid-liquid, liquid-solid) together with the theoretical background have been elaborated in early nineteen century [6-7]. The wetting is often characterized by the sessile drop method by the measuring of equilibrium contact angle (Θ) according to Fig. 1.

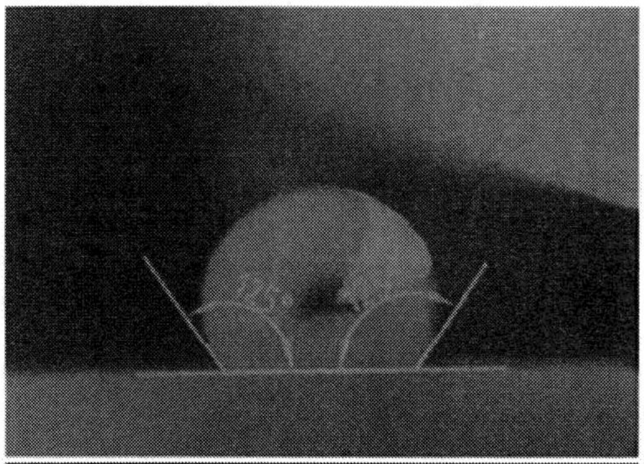

Figure 1. The real illustration of contact angle for a liquid drop on solids, with contact

In earlier study, method and equipment for the observation of the wetting angle (Θ) between non-reactive alloy melts and solid ceramics have been developed, using the sessile drop method. An instrument is also designed and built, making possible the observation Θ(T) dependence at high temperature range up to -1200°C. The essential part of equipment is the double wall thermal isolation system, for the elimination of oxygen traces in the surrounding atmosphere.

In the pesent article, we focus mainly on the study of wetting phenomena between reactive, metallic partners [7], including dual phase steels [9] and some appropriate brazing alloy liquids. Such systems have outstanding technical significance in the modern car-industry (brazing techniques for body elements). Subsequently the melting of protecting flux, the chemical reaction has started between the melt pool of filler alloy and the contacting solid partner alloy. Hence, the solid –liquid interface turns to lens-shaped, so the melt pool can be chatacterized by two (Θ_1 and Θ_2) contact angles [11-13].

As it is mentioned, one of the essential requirement for the successful brazing of machine parts is the mutual, perfect wettability between the liquid pool of the brazing alloy, and the solid metallic parts to be joined, which can also be expressed by the appropriate wetting angle. In addition, the actual wettability is strongly influenced by the atmospheric circumstances, therefore the wettability is enhanced by flux addition [5, 10, 16-17].

2. Experimental conditions

The wetting experiments were performed in home-made equipment developed for sessile drop measurements. The substrate and the solder pieces were positioned into the middle of furnace at ambient conditions. The pressure was then reduced to 0.1 Pa at room temperature in the chamber. The vacuum was replaced by a 10^5 Pa 99.999 % Ar gas. This procedure was repeated 3 times. Subsequently, the temperature was raised to melting point using a heating rate of 4 K/s. Since only a small part of the gas chamber is heated, no pressure change could be detected in the chamber (10^5 Pa) during the measuring run. The heating power has stopped around the measuring temperature. Two minutes holding time was applied before the measurement. The contact angle was determined by measuring directly the profile of the drop. Self-made automatic software in the Matlab environment is used for fixing and processing the data. Though the uncertainty of this software is below 1 degree, the total uncertainty of the measured values is higher: ±3°. At the end of measuring process, the furnace was switched off and cooled slowly to the ambient temperature (the whole cooling time is around 40 minutes). Subsequently the furnace was opened and the solidified sample was removed.

For the experiments, high-purity tin and copper were used to make the CuSn alloys, and DC01 and DP600 type steel plates were used, the chemical composition of which is shown in Table 1. and Table 2.

Table 1. Chemical composition of the DC01 sheet material

Fe (wt%)	**C** (wt%)	**Si** (wt%)	**Mn** (wt%)	**P** (wt%)	**S** (wt%)	**Cr** (wt%)	**Ni** (wt%)	**Al** (wt%)	**Cu** (wt%)
99.5	0.050	0.023	0.233	0.012	0.009	0.014	0.027	0.041	0.024

Table 2. Chemical composition of the DP600 sheet material

	Fe (wt%)	**C** (wt%)	**Si** (wt%)	**Mn** (wt%)	**P** (wt%)	**S** (wt%)
DP600	98.6	0.085	0.171	0.87	0.013	0.005

The brazing experiments were performed with a 1.2 mm diameter CuSi3 wire, the composition of which is shown in Table 3.

Table 3. Chemical composition of the CuSi3 brazing wire

	Cu (wt%)	**Si** (wt%)	**Mn** (wt%)	**Zn** (wt%)	**Sn** (wt%)	**Fe** (wt%)
CuSi3	>94	3	1	0.1	0.1	0.07

3. Results and discussions

3.1. Experiments on non-reactive substrates.

The influence of compositional change and the phase relations on the wettability between the brazing pool and non-reactive substrates (wetting study into multiphase substrate alloys) was investigated. The results are summarized in Fig. 2.

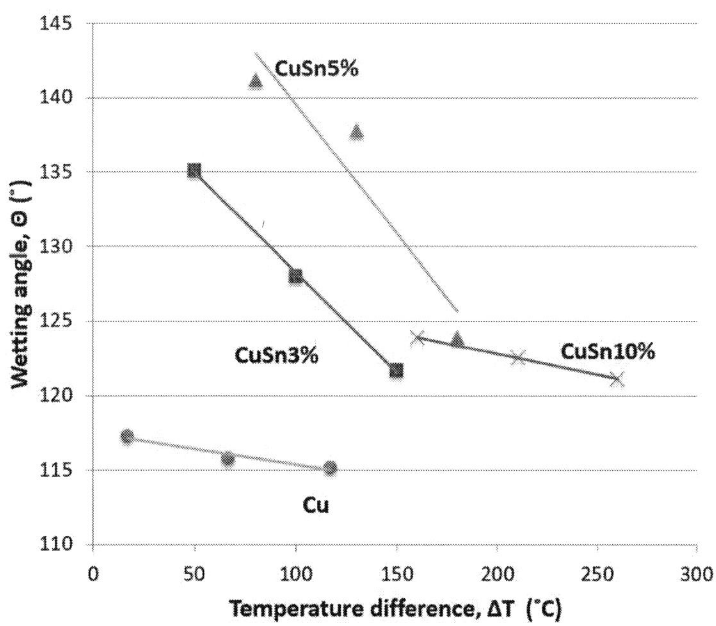

Figure 2. Θ(T) for various Cu(Sn) melts on graphite substrate

In Fig. 2. the T-dependence of Θ for various composition of Cu(Sn) melts are illustrated (measured on graphite substrate). The wetting angles are plotted versus the relative overheating of the melt drop in the figures because the higher physical relevance when the possible influence of Eötvös rule is considered [14] (defined as $\Delta T=T_{obs}-T_{liq}$, where T_{obs} is the experimentally measured temperature, T_{liq} is the liquidus temperature at the alloy composition. The slope of Θ(T) curves changes rapidly with Sn-content of the liquid, showing the sensitivity of wetting with the concentration molten phase composition. It is remarkable, that not only the Θ changes at a given superheating, but the slope of the curves exhibits significant difference. Both curves refer to the concentration range of solid solution in solid state. As the slope of curves differ significantly, one can suspect, that significant structural change (concentration gradient) can be developed as a function of temperature change.

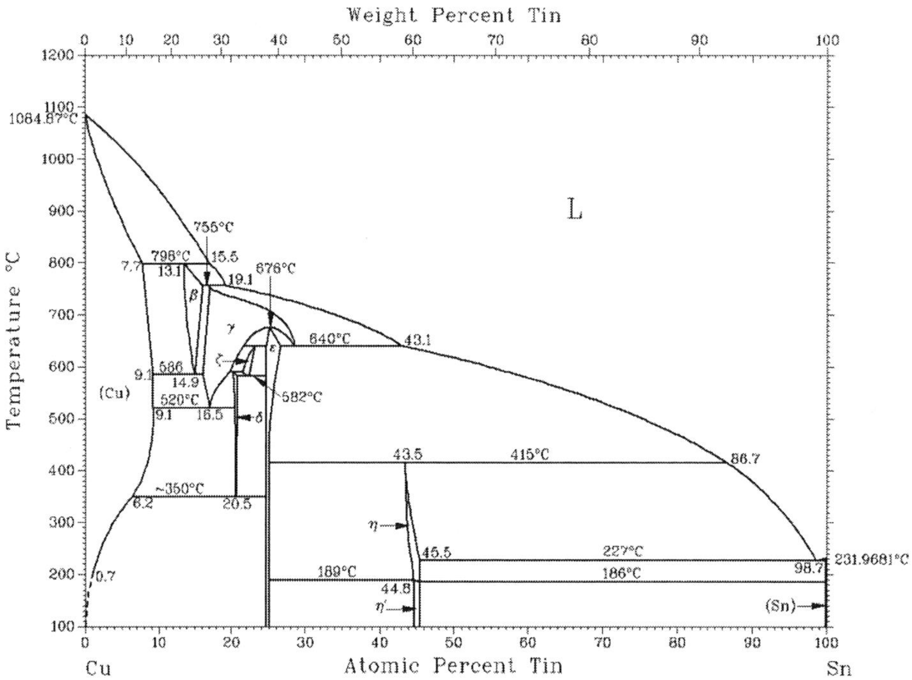

Figure 3. Cu-Sn phase diagram [15]

Another remarkable effect is the abrupt change of Θ(T) curve, when the concentration of melt exceeds the solubility limit of Sn in the terminal α (Sn) solid solution (Fig.3.). As the reaction between the liquid and the substrate is absent, this sudden change (both in the value and also in the slope of Θ(T) can be the consequence of structural rearrangements in liquid state prior to the solidification. In the case of 3 and 5 at% Sn-content liquid, solidify via primary crystallization of Cu(Sn) solid solutions, in the second case, the crystallization is isotherm, peritectic reaction. One can suspect, therefore, that the slope of the curves can be related with the type of subsequent solidification reaction.

In contrast to the previous cases, the slope of Θ(T) is negligible in CuSn10% liquid, where the solidification mechanism is peritectic, i.e. the liquid-solus transformation is isothermal process.

3.2. *Phase relations and wettability on carbon steels*

In Fig.4. the temperature dependence of wetting angle on two carbon steel are depicted.

DC01 is single phase ferritic steel, the DP600, originally „dual phase" (martensitic-ferritic) steel, for which the wettability between CuSi3 brazing melt is better in the investigated temperature region presumably due to the different C-content.

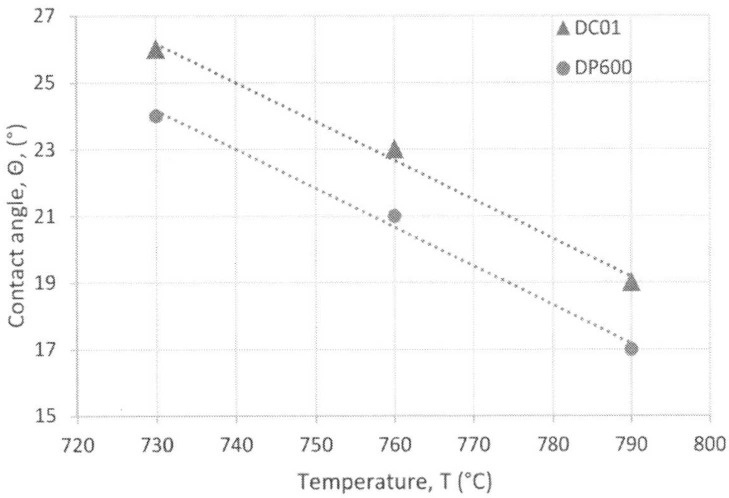

Figure 4. Temperature dependence of contact angle of CuSi3 brazing liquid on ferritic and dual phase (ferrite-martensitic) steels

3.3. Mechanical properties

In Fig. 5. the tensile strength of DC and DP in various combinations of joint formation are compared. The difference in tensile strength is obvious, so the double phase character has primary importance in the tensile strength improvement. Especially remarkable, that no sign of deterioration of strength is observed in spite of the presence of heat affected zone caused by the brazing.

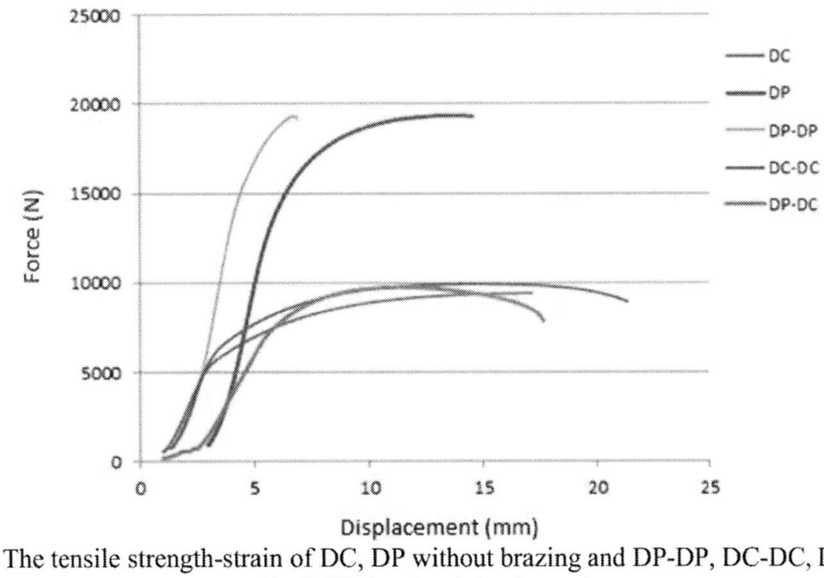

Figure 5. The tensile strength-strain of DC, DP without brazing and DP-DP, DC-DC, DC-DP steels with CuSi3 brazing joint formation.

In Fig. 6. The bending test of various sheets are compared, being previously subjected to brazing-joint formation. Neither the tensile strength, nor the bending strength suffered breakdown due to the joint formation in the case of DP sheets.

It was concluded, that mechanical properties are determined dominantly by the original structure (indirectly by the C-content, making possible the development of dual-phase structure)

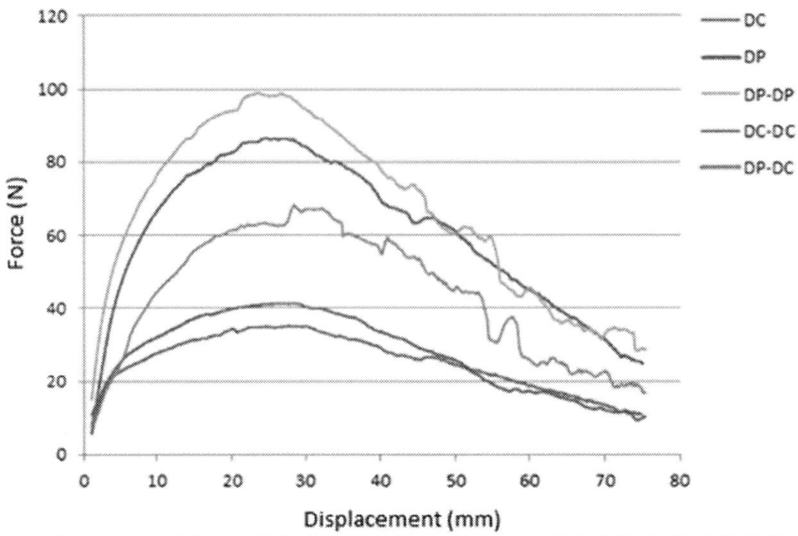

Figure 6. The bending test of DC and DP sheets without brazing, DP-DP, DC-DC, DC-DP sheets with CuSi3 brazing.

4. Conclusions

In spite of the promizing preliminary results (especially concerning the absent of deterioration of mechanical properties due to the lateral extension of heat affected zone) an extended kinetic study is required at various constant temperatures and atmospheric circumstances. In such experiments continuous monitoring of Θ(t). is required, comparing the observed phenomena with appropriate non-reactive substrate partners. (i.e. where chemical reactions between the partner substrate, is absent) so the change of Θ is not expected.

The study of local structural change in the vicinity of solidified filler alloy including also the structural consequence of heat affected zone formation, local phase transformations etc.

It is supposed, that drop- profile monitoring can supply indirect information about the local, mutual compositional changes, within and beyond the melt pool, making also possible the study the mutual relations between the extension of heat affected zone and the diffusion controlled local, compositional changes.

Further widening of knowledge is necessary about the brazing parameters and its relation with real structure of bond, allowing prediction towards the strength and duration of bonds.

Better understanding of the mechanism of brazing joint formation between the body-elements applied in modern car-industry.

Further widening of basic knowledge concerning the relation between shape of phase diagrams and the slope of Θ(T). (The type of solidification reactions: correlation between the solidification reactions and the necessary superheating of brazing environments)

Acknowledgements

The project has been supported by the European Union, co-financed by the European Social Fund. EFOP-3.6.1-16-2016-00014.

References

[1] Baumli P., Solder Materials With Micro And Nanoparticles: A Review, *Materials Science and Engineering*, Volume **40**, No. 1, 2015, pp. 42–49

[2] Baumli P. Alacsony olvadáspontú ólommentes forraszanyagok, *Anyagok Világa (Materials Word)* **1**, 2015, pp.24-34 (in Hungarian)

[3] G. Kaptay. Classification and general derivation of interfacial forces, acting on phases, situated

in the bulk, or at the interface of other phases, *J.Mater.Sci.* **40**. 2005.
[4] P. Baumli, G. Kaptay. Wettability of carbon surfaces by pure molten alkali chlorides and their penetration into a porous graphite substrate, *Mater Sci Eng A*. **495**. 2008.
[5] Siegel R.W.Ann. *Rev. Mater. Sci.* 21. 559.(1991)
[6] T. Young, Philosophical Transactions of the Royal Society of London (vol. **94**.p.65) 1805
[7] L. Eötvös: Über den Zusammenhang der Oberflehenspannung der Flüssigkeiten mit irem Molekularvolumen, *Ann. Phys*. **27**, 448 (1886)
[8] F.D. Richardson, Physical Chemistry of Melts in Metallurgy , Academic Press London, New York, 1974
[9] N. Eusthatopoulos, M.G. Nicholas,B. Drevet, Wettability at High Temperatures, Pergamon
[10] R.W Cahn, (Amsterdam-Lausanne-New York- Oxford-Shannon- Singapore-Tokyo
[11] J. Dobránszky, MTA Acad.Thesis, 2015M. Ashby, Budapest University of Technology and Economics, 2005.
[12] A. Szabo, A. Lovas, Some basic observation and considerations for the thermopower measurements used as non-destructive material testing, *Journal of Machine Manufacturing*, Volume **XLIX**. Issue E3-E5, Special Triple Edition, 2009
[13] B. Vehovszky, T. Jakubík, M.Treszkai, Thermal Examination of a Simplified Exhaust Tube-Heatshield Model, *Periodica Polytechnica Transportation Engineering,* **47(3),** (2019) 190-195.
[14] B.C. Allen: Surface Tension in Liquid Metals, in Liquid Metals (Chemistry and Physics) ed. Silvan Z. Beer Marcel Deccer, INC., New York, (1972) p.165.
[15] Alloy Phase Diagrams Centre, ASM International, Materials Park, OH, (2006)
[16] Janczak-Rusch, J., Kaptay, G. & Jeurgens, L.P.H. Interfacial Design for Joining Technologies: An Historical Perspective. *J. of Materi Eng and Perform* **23**, 1608–1613 (2014). https://doi.org/10.1007/s11665-014-0928-5
[17] Kaptay, G., Janczak-Rusch, J. & Jeurgens, L.P.H. Melting Point Depression and Fast Diffusion in Nanostructured Brazing Fillers Confined Between Barrier Nanolayers. *J. of Materi Eng and Perform* **25**, 3275–3284 (2016). https://doi.org/10.1007/s11665-016-2123-3

IOP Conf. Series: Materials Science and Engineering **903** (2020) 012036 doi:10.1088/1757-899X/903/1/012036

Lightweight Manufacturing of Automotive Parts

T Wadas[1]and M Tisza[2]

[1]University of Miskolc, Sályi István Doctoral School, Miskolc, HU
[2]Department of Materials Science and Engineering, University of Miskolc, Miskolc, HU E-mail: tatiane.wadas@uni-miskolc.hu

Abstract. In this paper, some material developments and its demands in sheet metal forming for automotive industry will be overviewed. It is not easy to match and meet good performance, low consumption, low emission and high safety with comfort simultaneously, only by applying conventional materials and processes. Due to these factors and needs, continuous material and technological developments in sheet metal forming are required.

1. Introduction

The automotive industry is the one that most drives the constant development in sheet metal forming processes. Due to the constant demand of the automotive market to reduce cost of production, customers' wishes, institutions and groups aiming at reducing pollutant gas emissions and protecting the environment, reducing the weight of automobiles is crucial, this is the reason why the application of light weight manufacturing principles has a huge importance.

The automotive sector is in a constant impasse trying to balance excellent performance, economy, comfort and reduction of polluting gases. These contradictory requirements are the fundamental impetus for the challenging advances that today's engineering needs to address.

It is well known that by increasing the strength of materials, the formability decreases considerably, however the formability of cars components or parts is regarded as one of the most important issues in car manufacturing. During the last years, significant advances and changes in high strength steels and various light weight alloys were achieved. The best examples of materials which achieved the requirements previously mentioned are, among high strength steels, the application of various dual-phase steels, while aluminum alloys are among light metals. However, the application of both aluminum alloys and light metals often leads to problems of manufacturing and formability. In this paper, it will be seen some innovations in metal forming processes and recent developments in materials related to sheet metal industry will be described [1].

2. The energy requirements for combined city/highway driving

Reduced consumption and less harmful emissions are also contradictory requirements. In order to ensure weight reduction, smaller sheet thickness should be applied, however, the requirement to increase safety is also in contradiction. The solution would be decreasing size, using high strength materials, what has unfavorable consequence, i.e. it would reduce formability [1].

In the figure 1, it can be seen the energy requirements for combined city/highway driving. In the figure the energy requirements are accounted for as part of the engine and parasitic losses [2].

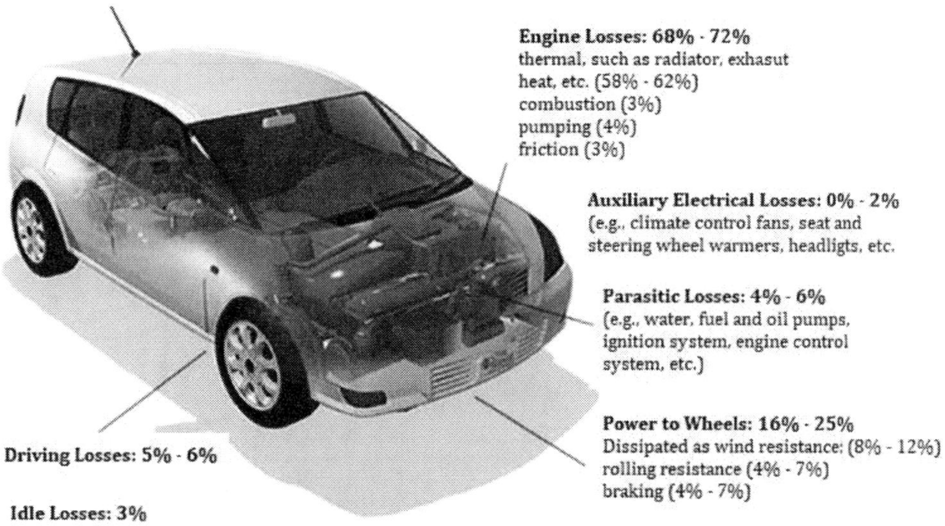

Figure 1. The energy requirements for combined city/highway driving [2]

3. Vehicle Advanced Structural Materials

A US D.O.E. comparison for the use of engineering materials shows that steel is still the most widely used material, with 725 million tons used each year. Aluminum and stainless steel are used at a rate of 20 and 14.5 million tons, respectively. Titanium is much less widely used at 0.04 million tons per year [1].

Although advanced light weighting materials are sometimes considered "high strength alloys", strength is only one factor in evaluating a particular material substitution. A better measure is the "specific strength" – the material's ultimate tensile strength normalized by its density. Additionally, many automotive components have structural stiffness specifications as the primary design requirement, so the proper parameter to consider is a material's "specific stiffness" – the material's elastic modulus normalized by its density.

Most of the engineering alloys in widespread use (steels, aluminum, magnesium, titanium) have nearly the same specific stiffness [1].

4. Effect of the material quality and the material properties

One of the most important trends in automotive industry is the application of light-weight design principles and methods. New materials are needed when new concepts in design are applied. In the last years, the application of high strength steel has been increased and new grades have been developed. However, it is known that by increasing the strength the formability decreases.

Several micro-alloyed and phosphorous-alloyed steels are currently used, with and without bake-hardening. In the figure 2 it can be clearly observed an increasing use of dual-phase, TRIP-steels, interstitial-free (IF) and ultra-low and super ultra-low carbon steels.

The figure 2 demonstrates the total elongation (A) versus ultimate tensile strength (MPa).

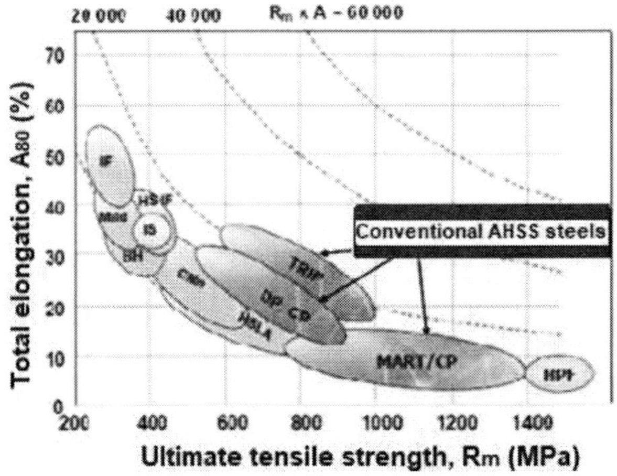

Figure 2. Tensile strength vs total elongation for conventional high strength steels [2]

From figure 2 it can be seen that the product of R_m X A80 is a constant and thus follows a hyperbolic function. The constant C (R_m X A80) for these steels changes between 10000 and 20000 and this result shows that the mass reduction was achieved during the second half of the last century. It is also important to mention that the new high strength steels increased much more significantly their strength parameters compared with the decrease of their ductility parameters. This is valid for the conventional AHSS steels present in the figure (TRIP, Martensitic Complex Phase, Complex Phase and Dual-Phase steels (DP) [2].

5. Dual Phase (DP) Steels

Though HSLA steels present incredibly good properties, their good formability is not always sufficient to produce parts that need to have a large plastic deformation. This is what motivated the creation of a new grade of high strength steel.

One example is the Dual-phase steels, which consists of ferritic matrix containing hard martensitic second phases. These are found in small martensite islands form, as it can be seen in figure 3. By increasing the volume fraction of hard martensite phases the strength is also increased. Controlling the cooling from the two phases austenite and ferrite zone, the dual-phase structure can be produced. [2]

When the ratio of martensite / ferrite is modified, the mechanical properties of DP-steels can be changed in a significant interval, where the normal percentage of martensite isles is between 10 and 20%. While the process of forming occurs, in the low-strength ferrite phases deformation is concentrated and it is surrounding the martensite isles and then extra high strain hardening of DP-steels is created. Compared to conventional HSLA steels, it is achieved a result of a much higher ultimate tensile strength allying the excellent ductility with extra high strain hardening. Another important advantage of DP-steels is the high bake hardening effect. After painting in the bake oven, their yield strength increases considerably due to the temperature [2].

In the figure 3 it can be seen both hard martensitic isles and mild ferrite matrix. Where the martensitic isles are small dark grains and the ferrite matrix are the light grains. Excellent ductility is assured by the formation of a continuous micro-structure formed by mild ferrite, while high strength of dual-phase results from hard martensite [2].

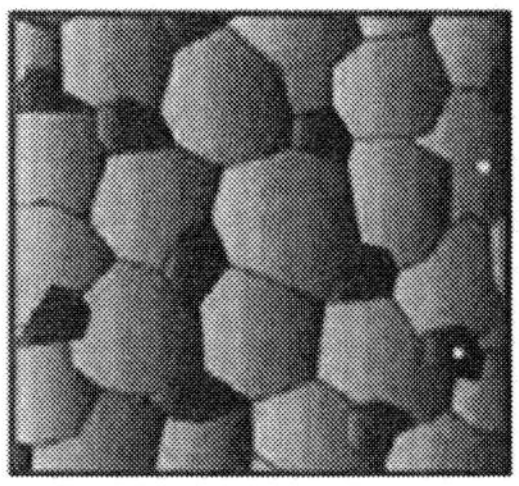

Figure 3. Schematic microstructure of DP steels [2]

6. New Generation Advanced High Strength Steels

The two most outstanding example for advanced high strength steels are the extra-advanced high strength steels (X-AHSS) and ultra-advanced high strength steels (U-AHSS), and there are many subgroups within them. There is an increasing demand on the car industry, that stimulate the development and application of X-AHSS and U-AHSS.

In the figure 4 it is demonstrated another order of magnitude represented by X-AHSS and U-AHSS. For conventional high strength steels, the constant of the product of tensile strength times total elongation was increased from 10000 to 20000. The various subgroups of both X-AHSS and U-AHSS steels cover a broader range as shown in figure 4. Taking into consideration the values C = 40000 and C = 60000 it can be noted a big achievement in AHSS steels developments. This means that, with the same value of total elongation, the strength can be twice or three times bigger, what is really important in order to meet the required strength in reducing weight in cars. The X-AHSS may be regarded as the further development of TRIP steels. The three subgroups in these steels are FB-steels, SB-TRIP and M-TRIP. The various subgroups of both X-AHSS and U-AHSS steels cover a broader range as shown in the figure 4.

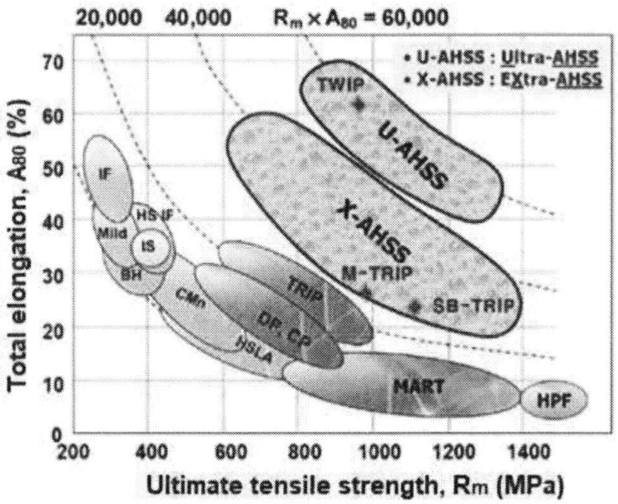

Figure 4. Total elongation (A80) vs Ultimate tensile strength (Rm) for various generations of high strength steels [2]

In the figure 5 and 6 two models made of aluminium alloys are presented (Mercedes SL and Audi A2) [4].

Figure 5. Mercedes SL aluminum body [4]

Figure 6. Audi A2 aluminum body [4]

Figure 7 shows the proportion of aluminum alloys application used within the multi-material concept.

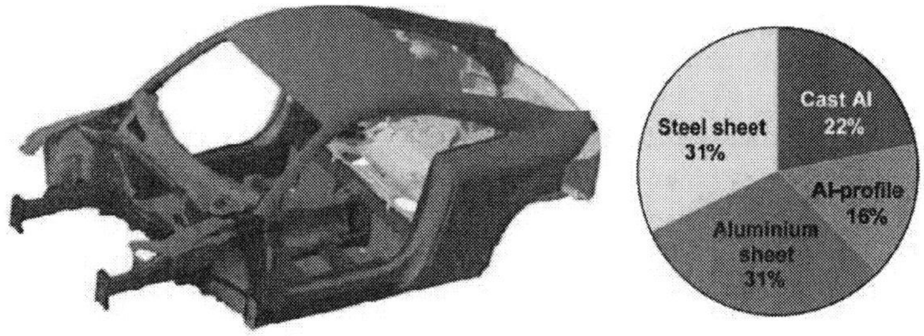

Figure 7. Multi-material body concept in car industry [4]

Many analyses also have shown that further significant weight reduction can be achieved in automobiles using fibre-reinforced composite materials. Carbon-fibre reinforced polyamide seems to be particularly suitable because it satisfies the requirements of production in large series with good mechanical strength and shape stability [4].

7. Vehicle Mass Distribution

Several studies have examined the mass distribution within typical vehicles. About 43% of the total vehicle mass, is the single heaviest group; followed by the powertrain and chassis, in almost equal proportions, at 27% and 26%, respectively. More recently, it was also found that within the body group, the unit-body or body-in-white (BIW), is the single largest component, with about 28% of the total vehicle mass, while glass is 3%, the interior 10% and 4% others [1].

8. BIW (body in white) concept

Nowadays, among all important requirements mentioned before, the formability for new materials should be strongly considered in BIW (body in white) concept.

It is of great importance to find a good balance between strength and formability properties, as using higher strength materials the formability decreases. The ultimate tensile strength and the total elongation has a hyperbolic relationship, i.e. lower strength with better formability and higher strength with lower formability. One of the main tendencies in the automotive industry is the application of light-weight design principles, what from the side of materials science, can be met by applying materials with high specific strength (UTS/ρ) and high specific stiffness (E/ρ). Considering these properties, the application of high strength steels, light metals and alloys (particularly aluminum and magnesium), as well as an increasing amount of various non-metallic materials is fundamental [1].

The figure 8 shows a scope of stamped components for BIW and chassis components [6].

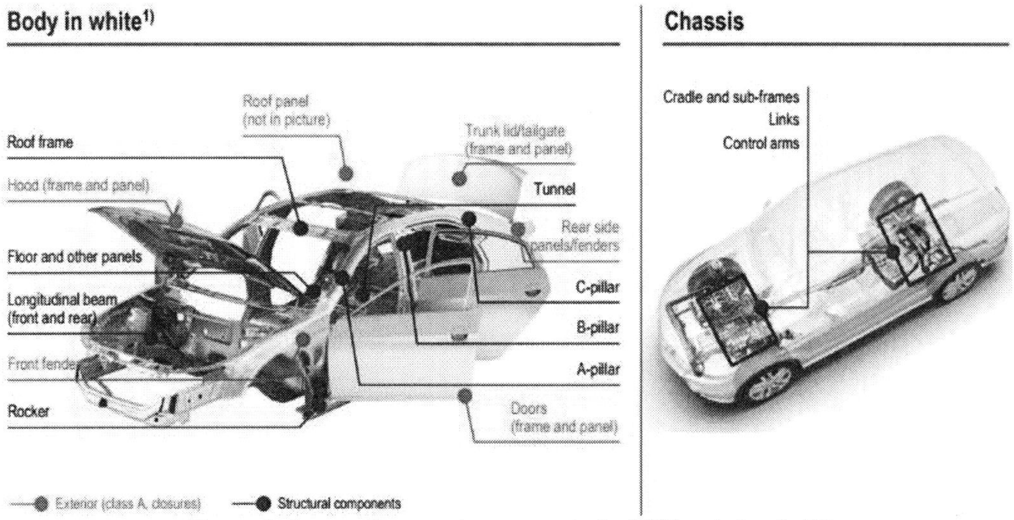

Figure 8. Scope of stamped components for BIW and chassis [6]

9. Hot forming for BIW components

Hot forming is increasingly popular for BIW components – driven by increased strength and mass reduction requirements. The figure 9 shows a comparison of share of hot-stamped steel in BIW – Sample vehicles (%) among a few models [6].

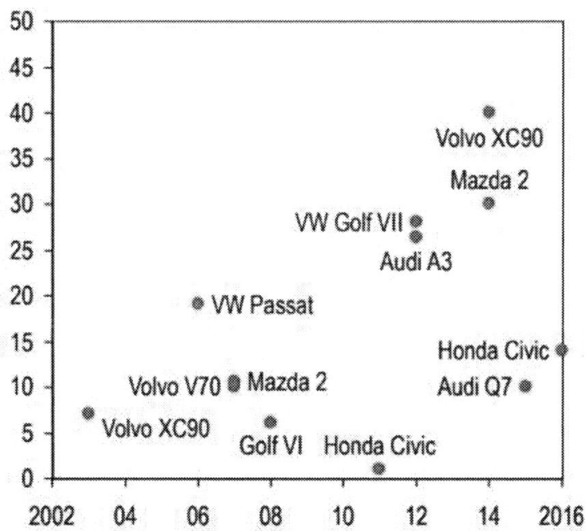

Figure 9. Share of hot-stamped steel in BIW – Sample vehicles (%) [6]

Increasing share of hot-stamped steel in BIW is driven by certain beneficial attributes, including:

- High strength and crash resistance at relatively low cost
- Low weight due to reduced material thickness
- Reduced spring-back during manufacturing process (common in cold-forming process)
- Best weight savings per additional cost compared to Aluminium and plastic composites

In addition, within specific models (figure 10): e.g. In the Mazda 2, the current model has an approximate 30% share of hot-stamped steel, beating its predecessor model (approximately 10%) – parts made of hot steel include A-pillar, roof frame and rocker panels [6].

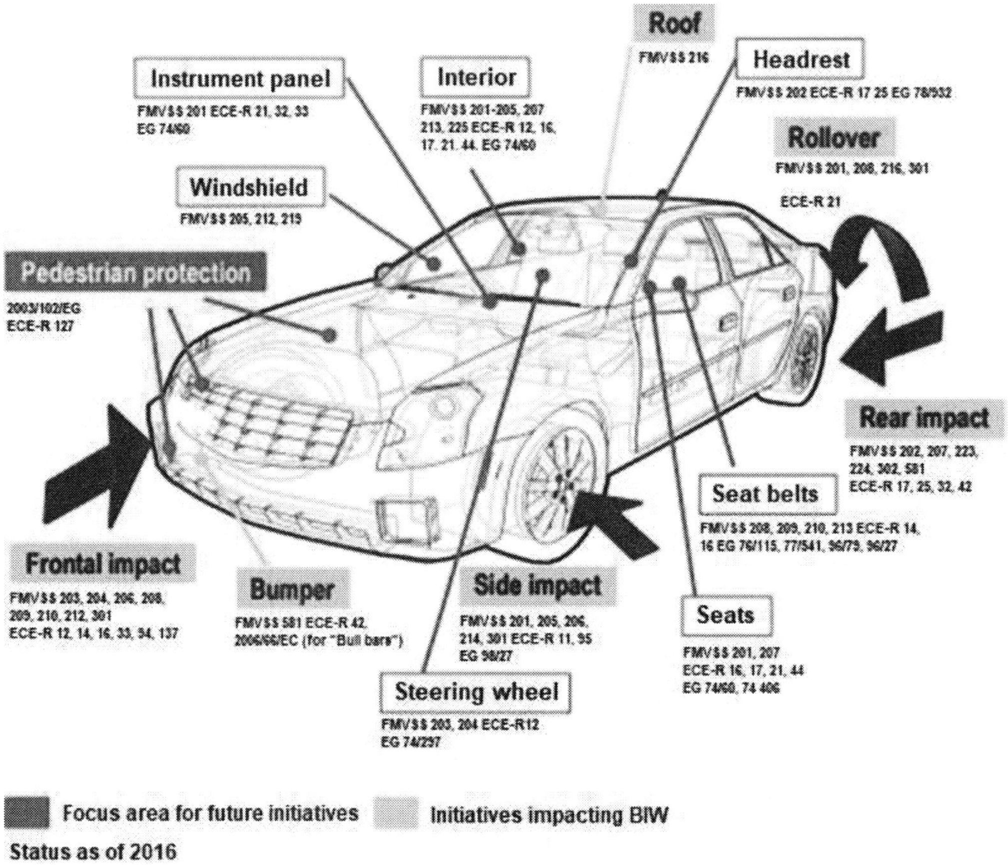

Figure 10. Crash regulations in Europe and US [6]

Past initiatives to improve vehicle safety have already led to major changes in material use and BIW component design.

In Europe and the US, the market is driven by safety performance assessment programs such as Euro and US NCAP – European car manufacturers OEMs aim for top 5-star ratings due to high customer awareness:

- In the past, Euro and US NCAP test requirements have focused on driver and passenger safety, with a high impact on BIW components – both material usage and component design were affected.
- Future tests are expected to focus more on pedestrian safety and driver assistance programs.
- This will not be majorly disruptive for BIW/chassis components, but
    - New required functionalities will have to be integrated (e.g. "intelligent hoods")
    - OEMs will be looking for materials that ensure a high level of design freedom, as design and specifications are impacted by the regulations.

10. Targets for the EU related to CO_2 emissions

In July 2012, the European Commission came forward with a regulatory proposal to set a 2020 target of 95 g/km for newly registered vehicles. The regulation was formally adopted in March 2014. It sets individual targets for manufacturers, depending on the average vehicle weight of a manufacturer's fleet, and requires all manufacturers to reduce CO_2 emissions by 27% compared to their individual 2015 targets and the regulation will be phased in later, so that only in 2021 all vehicles will be taken into account [5].

EU and NAFTA outsource BIW components more often than Asia - Overall, surface components outsourced less than other components as seen in the figure 11, *Current outsourcing rate, 2015 and future development (%)* [6]

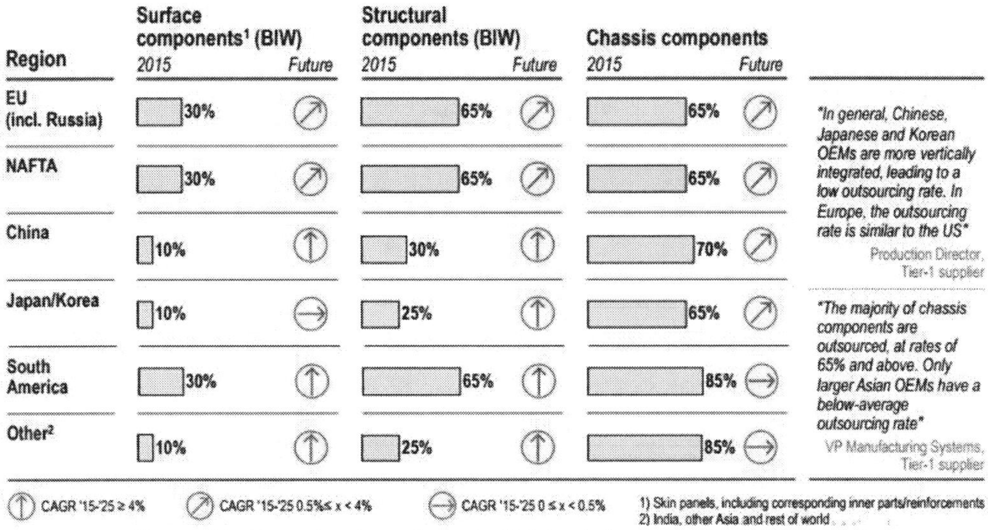

Figure 11. Current outsourcing rate, 2015 and future development (%) [6]

10.1. Future development of the market and implications [6]

As it can be seen in the figure 12, the total global market for BIW & chassis stamped components is EUR 103 bn – Approximately 15% of the total component market.

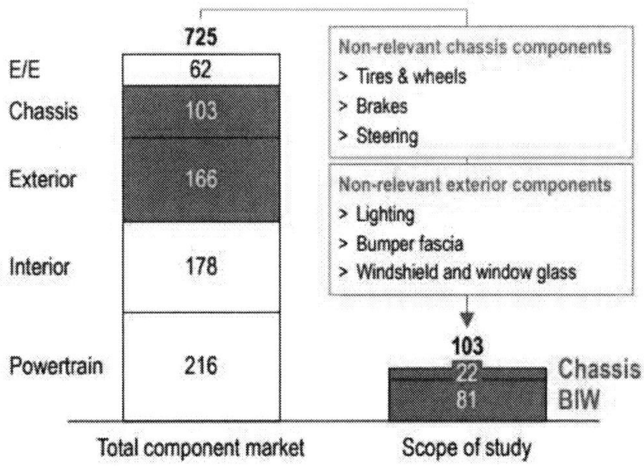

Figure 12. Global component market and total market for stamped components, 2015 EUR bn [6]

IOP Conf. Series: Materials Science and Engineering **903** (2020) 012036 doi:10.1088/1757-899X/903/1/012036

The figure 13 gives an outsourcing rate, which is expected to grow over the next decade – Global market will be worth around EUR 63 bn in 2025 [6].

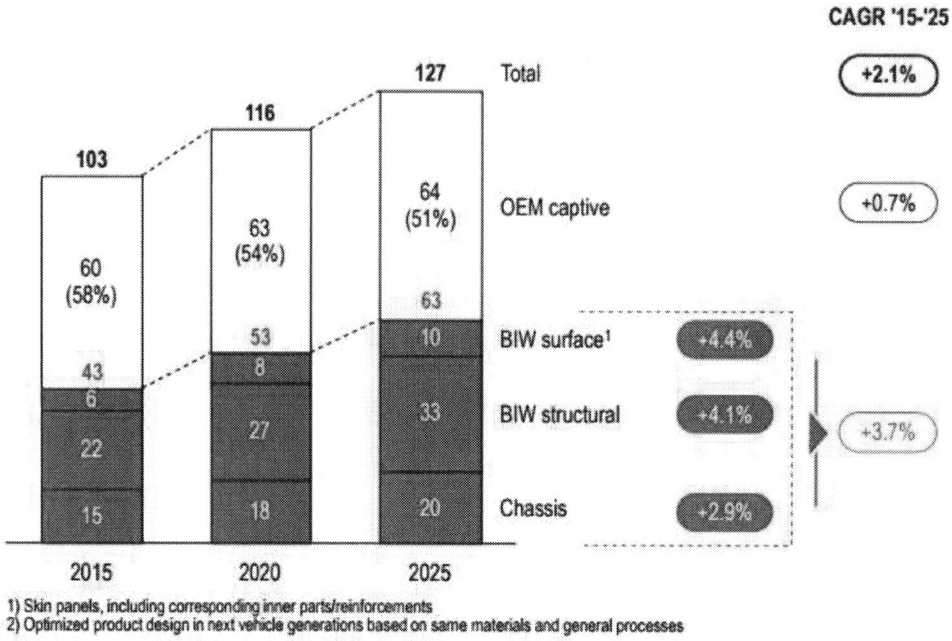

Figure 13. Total market, OEM captive vs. outsourced, 2015-25 EUR bn, (%) [6]

The total market will grow in line with global vehicle production (CAGR '15-'25 2.4%).
For suppliers, the addressable market will grow faster than the total market due to increasing outsourcing rates (42% in 2015, estimated 49% in 2020).

Component prices assumed constant over time, as material costs are typically indexed, and minor year-on-year price reductions are balanced out by evolutionary product innovations).
EU and NAFTA account for around 70% of the BIW and 50% of the chassis market – Both shares are shrinking as China expands, figure 14 - *Addressable market by region, 2015-25 [EUR bn (%)]* [6].

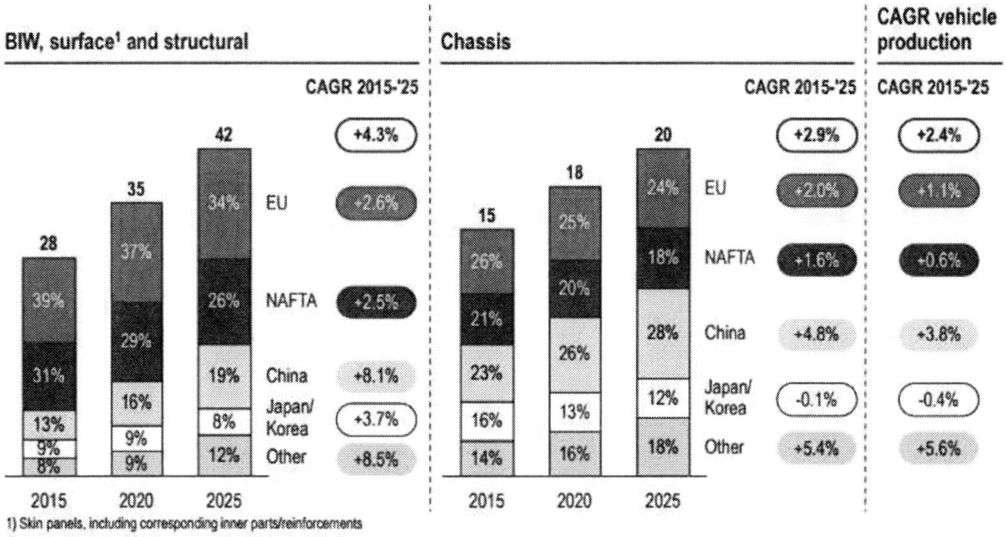

Figure 14. Comparison between BIW and chassis (prediction between 2015 and 2025) [6]

11. Share of hot stamping in total BIW market is expected to grow

The figure 15 shows the share of hot stamping in total BIW market prediction for the future. Share of hot stamped parts in total BIW (%) of market value [6].

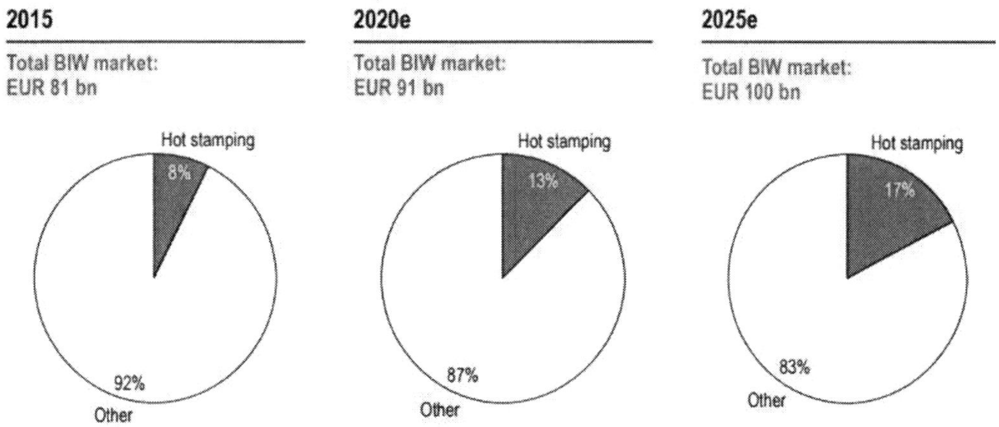

Figure 15. Share of hot stamping in total BIW market [6]

Acknowledgements

This research work started in the framework of TÁMOP-4.2.1.B-10/2/KONV-2010-0001 project and is continued in the project entitled Material Developments for the Automotive Industry within the project TÁMOP-4.2.2.A-11/1/KONV-2012-0029. Both projects are supported by the New Hungarian Development Plan and jointly financed by the European Union and European Social Fund.

References

[1] Bandivadekar K, Bodek L, Cheah C, Evans T, Groode J, Heywood E, Kasseris M, Kromer M, 2008 *On the Road in 2035 Reducing Transportation's Petroleum Consumption and GHG Emissions Laboratory for Energy and the Environment.* Report No. LFEE 2008-05 RP, Massachusetts Institute of Technology.

[2] The official U.S. government source for fuel economy information 2019. *Where the Energy Goes: Gasoline Vehicles*. Available at https://www.fueleconomy.gov/feg/atv.shtml
[3] Tisza, M., 2014 Recent development trends in sheet metal forming and Advanced Materials in Sheet Metal Forming Key Engineering Materials Vol. **581** pp. 137-142
[4] Tisza M, Kovács Z P, 2012 *New Methods for Predicting the Formability of Sheet Metals*. (Production Processes and Systems, Vol **5**). N o. 1. pp. 45-54.
[5] European vehicle market statistics: Pocketbook 2016/2017 *A statistical portrait of passenger car, light commercial and heavy-duty vehicle fleets in the European Union (EU) from 2001 to 2015*. Available at https://theicct.org/sites/default/files/publications/ICCT_Pocketbook_2016.pdf
[6] Roland Berger 2019. Automotive metal components for car bodies and chassis. Available at www.rolandberger.com

Conservation purpose material testing of corrosion products on outdoor bronze statues

Szilvia Szilvia[1], Melinda Nagy[2], Laura Juhász[1], Tamás Bubonyi[3], Péter Barkóczy[4], György Forgács[5], Eszter Szatmáriné Bakonyi[6]

[1] University of Debrecen, [2] Hungarian National Museum, [3] University of Miskolc, [4] FUX Co. Miskolc [5] Forgax Alkotóműhely kft, [6] University of Fine Arts Budapest

Email: peter.barkoczy@fux.hu

Abstract. A detailed material testing was carried out before the conservation project of the János Arany statues of Garden of Hungarian National Museum. The scope of the testing to analyse the corrosion products on the surface and help to develop the best cleaning and protection practice. Samples were taken from more areas and the samples were analysed by SEM-EDS, XRD, FT-IR methods. This article introduces the results of the testing and the relevant effect to the conservation process.

1. Introduction

Outdoor sculptures, especially metal ones, are constantly exposed to atmospheric corrosion, that is the reason why their protection is extremely important. [1,2] Corrosion of copper and its alloys is slow process; however, it could be really harmful in large time scale. [3] On the other hand corrosion product can also destroy the aesthetic value of statues, furthermore, influence its further corrosion. [4] Bronze sculptures are widespread in the world. Investigation of its corrosion and corrosion products are particularly important in case of its conservation procedure.

Method of cleaning procedure is one of the biggest questions during the conservation. [5] Composition and structure of corrosion layers are also really valuable information in this case. [6]

Knowing of the exact composition and the stability of compounds are essential for the expected chemical reactions. It is useful to know that the corroded layer's chemical composition, looking for compounds which can damage the surface of the statue during the conservation process. The corroded layer's structure influences its stability: could removed from the surface easily or not, could cause any changes in adhesion as well as which parts should be removed or can be preserved. This information can help to figure out the appropriate cleaning method. [7]

Composition of corrosion products is also a crucial question of conservation: Could any other reaction appear which can influence not only the state of the sculpture but also its conservation? [8] According to these information plan of conservation can be prepared and perform the conservation progress. Using detailed material testing methods can answer these questions only.[9]

Statue of János Arany, who was one of the greatest Hungarian poets, is situated in the Garden of the Hungarian National Museum. Janos Arany's statue is surrounded by two other sculptures, Toldi and Piroska who are the main characters of his poetry. Conservation of outdoor sculptures was also completed during the renovation of Museum Garden. Restoration of Arany's statue group was a highlighted and greatest task of this project due to its extent and its national importance.

Figure 1. The statue of János Arany and Toldi at the front of the Hungarian National Museum.

Detailed investigation of corrosion products was performed before conservation because of the previously mentioned reasons. Chemical composition of corrosion product was compared to the composition of the sculpture's material. Its aim was to determine the chlorine compounds. Presence of chlorine on the surfaces was not so probable in Budapest but we wanted also to exclude this phenomenon. Renovation in the surrounding buildings as well as in the museum can cause the presence of gypsum in the corrosion layer, that is the reason why we investigated the presence of gypsum, furthermore organic compound. In this work results of material investigation as well as results and consequences for conservation are reported. Restoration of sculptures were carried out and Janos Arany's statue group is situated in the Garden of Hungarian National Museum again.

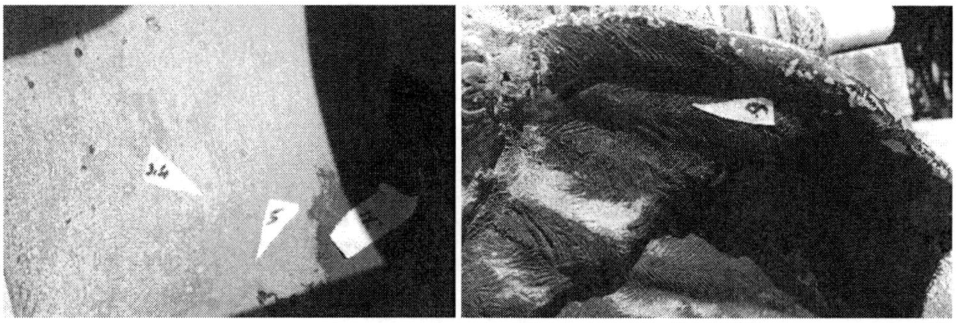

Figure 2. The different type of corrosion products on the surfaces of status. Right image shows a black and porous green layer, left image shows the continuous green and yellow-brown layer with a black covered area.

2. Materials and methods

Field portable X-ray fluorescence (pXRF) analysis was performed on the statues in several regions. The aim of this examination was to determine the elements and assumed chemical composition. Basically, four different corrosion product were identified on the surface: i, surface areas covered by black layer; ii, yellow-brown layer; iii, porous green layer on the surface; iv, continuous ductile green layer. The documentation wrote a bronze bas material. Measurements in different areas has been confirmed the documentation's copper and tin content. Silicon, aluminium and iron content also was detected in these areas. However, silicon and aluminium content were significantly higher in the black layer. Phosphorus was detected in the yellow-brown areas, which could be dangerous in terms of cleaning. Higher sulphur content was detected at the green layers, which originated to the patina first. Oxford XMET 8000 type pXRF analyser was used for the measurements. Powder samples were taken from the mentioned areas from more positions and statues. First energy dispersive X-ray spectroscopy (SEM_EDS) were made with a Hitachi –S4300 CFE to study the structure and the chemical composition of the different materials. XRD measurement were made with a Bruker D8 Advance to identify the different compounds. Only a small amount of samples could be collected, so the identification was hard. FTIR analysis was performed to find possible remains of a prior treatment with organic or polymeric materials. Samples were taken from the continuous green sample; which brittleness was tested by microhardness tests.

3. Results and discussion

Just small amount of powder, and small parts were possible to collect from the different layers. So, a detailed SEM-EDS study was made to see the microstructure of the layer, and the local chemical compositions. SEM-EDS analysis show the structure of the different layers. The black and yellow-brown layer built-up from particle agglomerates, where different particles, dust sticked together (Figure 3. right) The green layer is really a continuous one, where particles are integrated to a matrix. It was observed during the sampling, the black and yellow-brown layers are brittle, and these are easily detached from the surface. It was suggested to remove it during the conservation process due to this property. Additionally, in the case of yellow-brown layer EDS analysis showed high iron content, the colour of thin layer originates to the iron-oxide content.

Figure 3. The structure of the continuous green layer (left micrograph) and the yellow-brown layer (rignt micrograph).

IOP Conf. Series: Materials Science and Engineering **903** (2020) 012037 doi:10.1088/1757-899X/903/1/012037

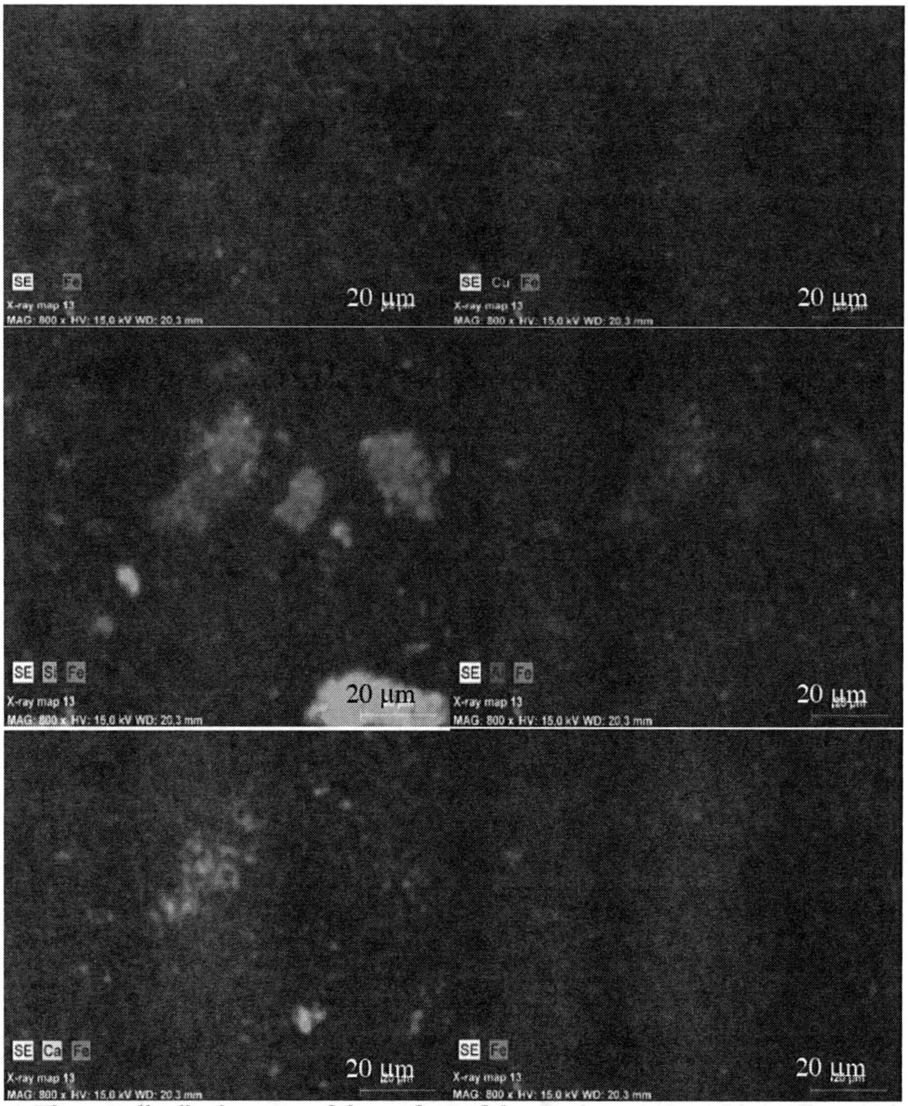

Figure 4. Element distribution map of the surface of the continuous green layer. The matrix is the patina the particles are from dust.

The green layer seems compact and it was hard to remove from the surface during the sampling. It was also analyzed, and high copper and sulfur content were measured, so this assumed as the patina layer. Figure 4 shows the distribution of different elements in a typical cross section of this layer. It reveals that the particles have a high aluminum and silicon content. These elements clearly distinguish the particles and the matrix. Calcium content detected in the particles only, which did not contain sulfur. This shows that the layer either contains calcium, but not in gypsum form, it come from the dust. An interesting observation is that the iron distribution is nearly uniform, and small particles can be seen mainly in the matrix material. Chlorine did not detect, so chlorides or other substances which contains chlorine did not exists on the surface of the statutes.

These results show that the particles are also from dust, which deposited to the surface. The matrix is the real corrosion product, the patina which growth continuously during the corrosion process. The patina layer involved the dust particles with its growth process. The results of the EDS analysis and the suggestions will compare to an XRD phase analysis later.

FTIR analysis of the samples performed to look for organic material or polymers from a possible previous treatment. Without any treatment the samples examined and neither organic nor polymeric materials were found (Figure 6.). The curves on Figure 6 cannot show any radicals or organic substances. Three samples were tested: a2, a3 and a6. Samples a2 and a3 are taken from green layer. Sample a2 was taken from Toldi's sword, sample a3 take from Toldi's helmet. sample a6 is a sample from the black layer and it was taken from Toldi's helmet. There is no significant difference in the FTIR spectra. Just a small amount of carbonates could be identify.

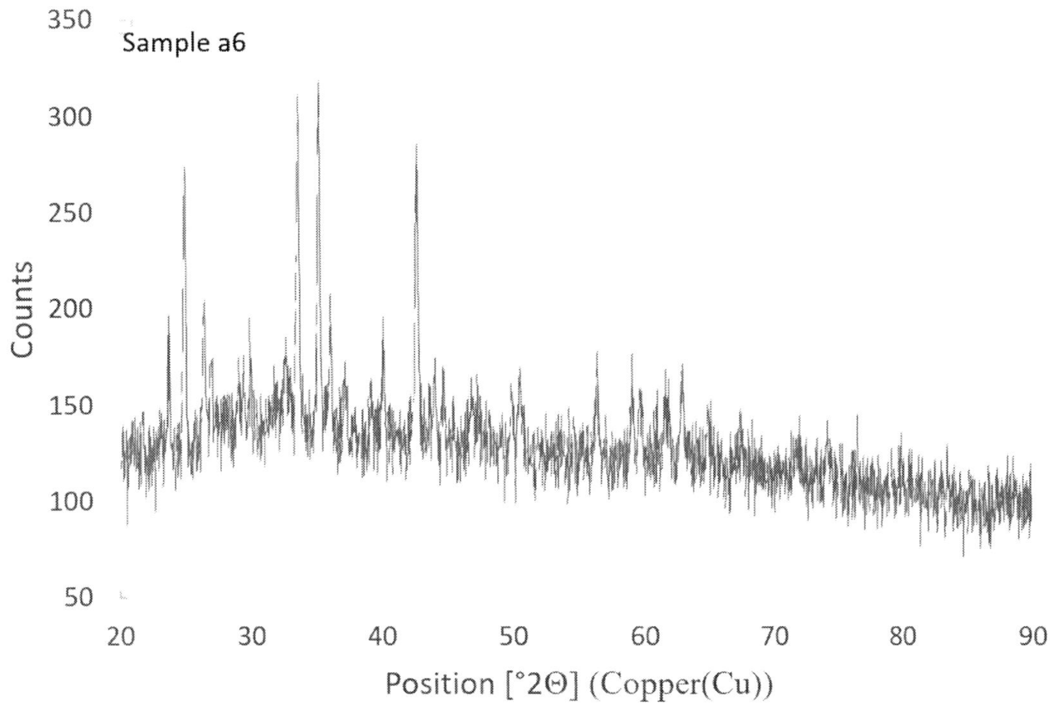

Figure 5. The XRD spectrum of the green layer.

The XRD analysis of the green layers were performed (Figure 5. sample a3). The intensity in the XRD spectra was low due to the small amount of samples but $CuSO_4{\cdot}2Cu(OH)_2$ and $CuSO_4{\cdot}Cu(OH)_2$ were clearly identified as the components of the patina layer. The SEM-EDS observations helped to identify the different compounds. $Al_3Fe_5O_{12}$+SiO_2 identified as dust particles. Next to these SnO, CuO_2 and Fe_2O_3 oxides were found in the sample. Tin and copper oxides are also a corrosion product of the raw material. The SEM micrograph revealed the distribution of the iron-oxide particles. The XRD analysis told nothing about the status of calcium due it's small concentration.

IOP Conf. Series: Materials Science and Engineering **903** (2020) 012037 doi:10.1088/1757-899X/903/1/012037

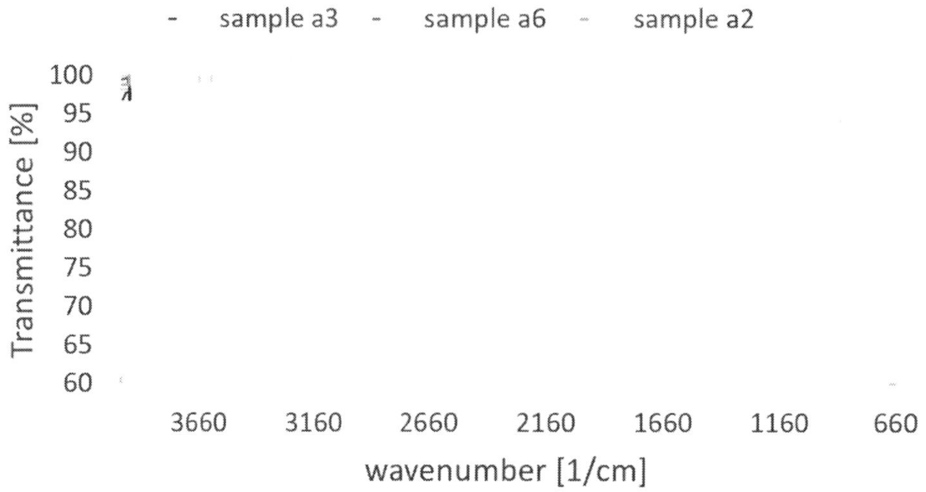

Figure 6. The FTIR spectra of three different samples of green layer.

The FTIR spectra identified only carbonates which either could be a copper compound in the patina layer or calcium-carbonate content of the dust particles. These results show another evidence the absence of gypsum in the layer.

The results suggest that it was necessary to remove the black and the yellow-brown layer from the surface of the statue. The green layers could be left over the surface and could be used a patinating process to equalize the outlook of the surface. The black and yellow-brown layers are brittle layers and easily detached from the statue, so a mechanical cleaning process (blasting) can be used. Neither polymeric nor other organic substances were found which makes necessary a special chemical treating during the cleaning. Chlorine or other substances also did not find which could cause problem during the patinating process.

Figure 7. The Toldi's shield after the claning experiment.

Based on this information cleaning probes were made to choose the best cleaning, blasting agent. The results of a $NaHCO_3$ particles blasting is showed by figure 7.

The chosen cleaning process removes all undesired material from the surface, therefore other chemical treatment was not necessary just the patinating process (Figure 8.). After the patinating process a protective thin resin layer also applied which did not change the outlook of the statues just gives an additional protection.

Figure 8. The cleaned and patinated shield.

Currently the statues can be seen at the original place and the treatment gives the original beauty of the statues. These again a great sight of the Museum Garden and of the Hungarian National Museum.

4. Conclusions

The statues of János Arany (The Poet and two main character of His poetry: Toldi and Piroska) from the Garden of the Hungarian National Museum were conserved. Before the conservation process a detailed material testing of the corrosion products were made. Different layers were identified: green layer, yellow-brown and black. Green layer was a patina, which contains silicate particles. It is a continuous intact layer, and it is hard to remove it from the surface. The other layers mainly formed by dust. These layers mainly particle agglomerates. Therefore, these are brittle and easily detach from the surface. The colour of the yellow-brown layer originates to its iron-oxide content. Detailed analysis was made to define the chemical composition, the microstructure of the layer and identify the different compounds. The main question was the existence of dangerous materials as chlorine, gypsum or organic compounds which could have a harmful effect to the conservation process or the lifetime of the conservation. Chlorine or other organic or inorganic compound did not find in the corrosion layers which can damaged the bronze material or can make reaction with the materials used by cleaning and patinating. Calcium content was detected, but in small amount. The tests show that the calcium presence as carbonate. The patina layer contains copper-sulphate and copper-hydroxide. This contains small silicate particles. The green layer is a ductile layer and sticks to the bronze material well, but the other layers are brittle. During the conservation the process the patinating chosen with the removal of the mentioned black and yellow-brown layers. $NaCO_3$ particle blasting was chosen for cleaning.

References

[1] P. R. Roberge 2008 Corrosion engineering, principle and practice, McGraw-Hill
[2] B. F. Brown, *et. al* 1977, Corrosion and metal artifacts, *U. S. Department of Commerce*
[3] S. D. Cramer, B. S. Covino Ed. 2005 ASM Handbook, Corrosion: Materials
[4] A. Doménech-Carbó, M. T. Doménech-Carbó Ed, 2009 Electrochemical Methods in Archaeometry, Conservation and Restoration, *Springer*
[5] V. N. Naudé; G. Wharton 1993 Guide to the maintenance of outdoor sculpture, *American Institute for Conservation of Historic and Artistic Works*
[6] D. Knotkova, K. Kreislova 2007 Atmospheric corrosion and conservation of copper and bronze, *WIT Transactions on State of the Art in Science and Engineering* **28** 107-42
[7] D. A. Scott 2002 Copper and Bronze in Art: Corrosion, Colorants, Conservation, Getty Conservation Institute

[8] L. Selwyn 2004 Metals and Corrosion. A Handbook for the Conservation Professional.
[9] M. B. Mcneil, B. J. Little 1992 *Journal of the American Institute for Conservation*, **31**, 355–66.

IOP Conf. Series: Materials Science and Engineering **903** (2020) 012038 doi:10.1088/1757-899X/903/1/012038

Comparison of CT and metallographic method for evaluation of microporosities of dye cast aluminum parts

Tamás Bubonyi[1], Péter Barkóczy[1], Zoltán Gácsi[1]

[1] University of Miskolc, 3515 Miskolc, Miskolc Egyetemváros, Hungary

E-mail: femhubo@uni-miskolc.hu

Abstract. Nowdays, the different material characterisation and testing methods are constantly evolving. With modern techniques, we have the ability to take a look inside the product without sample cutting, with modern CT systems. The modern techniques with their development become closer and closer to industrial applications. However, the testing protocols are sometimes 1 or 2 steps behind. This is mostly due to the cost or the complexness of the new testing methods. In this study we would like to take a quick look on the current standard industrial testing of porosities and microcracks and compare them with computer tomograph assisted testing.

1. Introduction

The die casting of aluminium products is a productive and cheap production process of complex shaped parts [1]. The enhanced cooling process is a common nature of this technology. Therefore, the crystallization process is also faster than in a gravitational sand-mould casting. During the fast crystallization gas inclusions and micropores forms in the solid metal [2]. These micropores and gas inclusions acts as microcracks during the mechanical load of the parts. Therefore, an important qualification step is the total amount and the measure of the length of the microcracks [3]. According to a classical method the microcracks are evaluated on a metallographic specimen. The micropores are dark in an optical micrograph, so computational image analysis can be used during the evaluation of these [4]. There are two basic question related to this method: the position of the sample, and the interpretation of the measured values.

The position of the sample cannot be uniform, because the microporosity strongly depends on the cross section of the parts next to the process parameters [5]. So generally, it is described by the customer, and based on the description they try to define the worst case in the part.

During the interpolation of the result it must be taken into account that the metallographic sample shows only one section of the specimen instead of spatial information. The stereology helps to transform the measured values from section to spatial values, but this conversion is statistical, and it needs more sections or more micropores to give a stable and robust result [6]. In most cases just few pores can be found in the examined section, so this statistical evaluation is not possible. But the position of the sample can be determined based on the planed stress state, and simple acceptance rules can be deformed as the maximal pore length must be less than a limit [7].

But new samples are necessary from the same part for a detailed study of the pore structure. Additionally, the metallographic examination is a destructive testing method [8], so creating statistics about the processing technology is an expensive process.

The material characterisation went through some great improvements in the last decade. The keywords in today's research field are the 3D, or non-destructive characterisation of the materials with as many details as possible, on the best resolution possible [9]. However, the industry still uses the old sample cutting method. In this case the biggest investment is some good optical microscope, and the sample preparing tools. This method could work, if the product is well known, and there are some specific failing points. But if the technology or the whole product changes, the porosities or cracks could be elsewhere. The only way to characterise the product with this method is to take a guess and try to find the worst case.

Today the 3D imaging devices like computer tomography [10] are more common and can help greatly during the investigation of the worst case in the product. Therefore, the result of a metallographic testing and a tomography is compared to show difference and parallelism of the two method through the evaluation of the microporosity of die cast aluminum part.

2. Materials and methods

An aluminum housing was chosen which was produced by die casting. The raw material of the casting is EN-AC47100. The nominal composition of the alloy is: 10.5-13.5w/w% Si, 0.7-1.2w/w% Cu and max. 1.3w/w% Fe, beside the aluminum. The housing and the typical microstructure are shown by Figure 1.

The pore structure is evaluated on the micrographs (Figure 2). Based on the tomographic image, the worst place and position of the metallographic samples are determined. The metallographic samples were cut by sawing. The examined surface of the sample was grinded mechanically by SiC particles, then polished by 3 and 1 μm diamond particles. The polished surface was etched by immersion to HF solution. Micrographs were taken by a Zeiss AxioImager M1m optical microscope. The micrographs were analyzed by computational image analysis with a self-developed Cprob software. The analysis algorithm is simple, because there is a big contrast difference between the pores and the aluminum material. The aluminum has a high contrast while the pores are dark. So, the segmentation was made based on the gray contrast on the images. Just the small digital noise was filtered by a binary open operator on the segmented image. Then the pores are indexed, and the distance of the outermost points is measured in the pores as the length. Additionally, the distance between the nearest pores also measured. The nearest pores are determined by a SKIZ operation [4], and the closest point of the nearest pores are measured as a distance. Some evaluations a distance limit is determined, and closest pores have to be handled as one big pore. In this case the length of the big pore is the sum of the pores and distances affected by this evaluation. The length of the pores and the images are compared.

3. Results and discussion

Figure 1. shows the examined housing with the position of the samples. Two samples were cut from the part for metallographic testing. The positions of the samples are selected by the tomographic image. Sections with large pores are chosen. The samples were prepared for optical microscopy, and micrographs were taken. Figure 1. also shows some micrographs where the typical microstructure also can be studied. The nominal composition is close to the eutectic composition of the Al-Si alloy system. According to it a fully eutectic microstructure can be seen on the micrograph. The eutectic structure is fine, and the silicon is not a plate like in the eutectic. The eutectic structure is modified to achieve a better mechanical property. The modification turns the silicon into small plates and globular particles. All micrograph contains micropores, which are the dark objects with different shapes. The globular objects like spherical shapes are probably gas inclusions. The others are micropores formed during the shrinkage in the solidification process.

Figure 2. shows typical sections of the tomographic reconstructed 3D images, measured with an YXLON FF35 MicroCT. The parameters of the test are summarized in Table 1. As it can be seen, the background of the images is dark while the aluminum material is bright, so the micropores – microcracks are visible also in dark color. The image shows a top-view section, and a section according to Figure 1.

IOP Conf. Series: Materials Science and Engineering **903** (2020) 012038 doi:10.1088/1757-899X/903/1/012038

On both sections the micropores are visible and can be measured by the software package of the tomography equipment.

The micropores on the sequence of optical micrographs are measured by the mentioned image analysis software. As it shown the pores are dark in the bright field illuminated images. The segmentation of the pores can be made by the intensity level of a grayscale image (Figure 1 and Figure 3). The segmented objects, the pores can be evaluated. The main measure of the pores is the length. The area, especially the area fraction of the pores on a section is also important as an impact to the static mechanical properties, but the fracture mechanics determines a maximal length based on the planned stress state of the part. This length basically the diameter of the plane section of the pores in case of a general single pore. But in a case of clustered pores the length is different. There are definition practices for the evaluation of this case. The common aspect of all method, that enumerate the clustered pores as a single large one. That pores belongs to the cluster which are closer that a critical distance. This critical distance can be derived from fracture mechanics and the stereology of the spatial pore and its plane sections. In current study we define this critical distance as 7μm, and the length of the clustered pore are measured as the distance of the farthest contour point of all pores in the cluster. There is possible descriptor based on the measured lengths: the average and the maximal length of the examined section. Therefore, the average, and the maximal length of the micropores are compared together. The values are reported in Table 2.

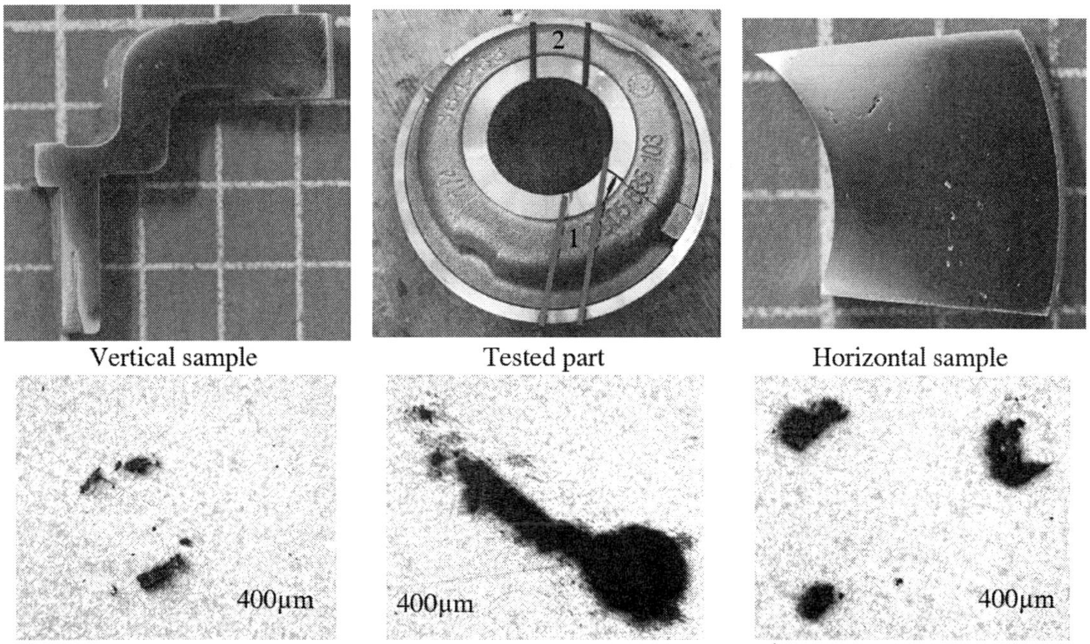

Vertical sample | Tested part | Horizontal sample

Worst case at the vertical sample | Worst case at the horizontal sample | 2nd worst case at the horizontal sample

Figure 1. The metallographic results. The tested part is in the middle, while the two sample is on its sides. Below the macro photos of the part, is the optical micrographs showing the worst cases in the vertical and horizontal sample, and the second worst case on the horizontal sample.

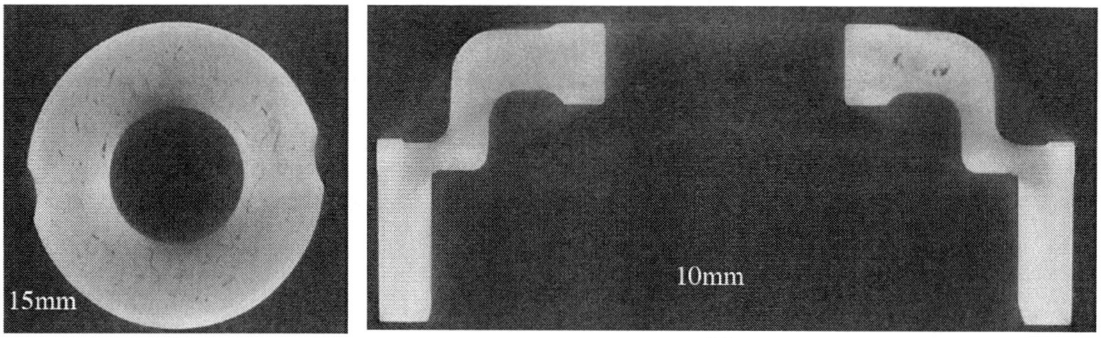

a) b)

Figure 2. CT sections of the sample. a) is the top view, b) is the front view.

Table 1. Test parameters of the CT imaging.

Test parameters			
number of projections	1080	Voxel (mm)	49.6
focal- object distance (mm)	410	accelerating voltage (kV)	150
focal- detector distance (mm)	1150	sample current (mA)	80
magnification	2.80		

Table 2. Length of the micropores measured by the different techniques.

	Metallography	CT
mean (mm)	0.175	3.9
std. dev.	0.427	1.65
maximum	2.56	6.94

The length of the pores can be measured directly on the tomographic image. Here also important the definition of the length. In current study also the so-called length, the diameter is used for the evaluation. The determination of this measure is a simple and easy process in the evaluation software of the equipment. The result of the measurement also a single value, but in the evaluation is necessary to take into account its spatial character. So the length of the micropores is larger on the CT image. A large deviation can be seen in the mean value the measured lengths. But there is a large difference in the resolution of the different images too. While the smallest object in this case in the CT images is ~50μm, the same in the optical images is around 1μm. The optical microscopic survey reveals much smaller objects than CT which causes this large deviation in the mean value. The standard deviation is in both cases so large, because there are large micropores next to small ones. The maximum value measured by CT is nearly 2.7 times larger than in optical microscopic tests. It is important when the evaluation of the part is made by the maximal length.

Plane sections of the CT image and the metallographic results are compared to evaluate the abovementioned difference in a so-called critical section.The position and the place of the examined metallographic section are chosen based on the analysis of CT images, but the metallographic is a section without depth in spatial coordinates. This can distort the section related to the CT image, and gives a smaller maximum. But the largest micropore was found and measured in the chosen section.

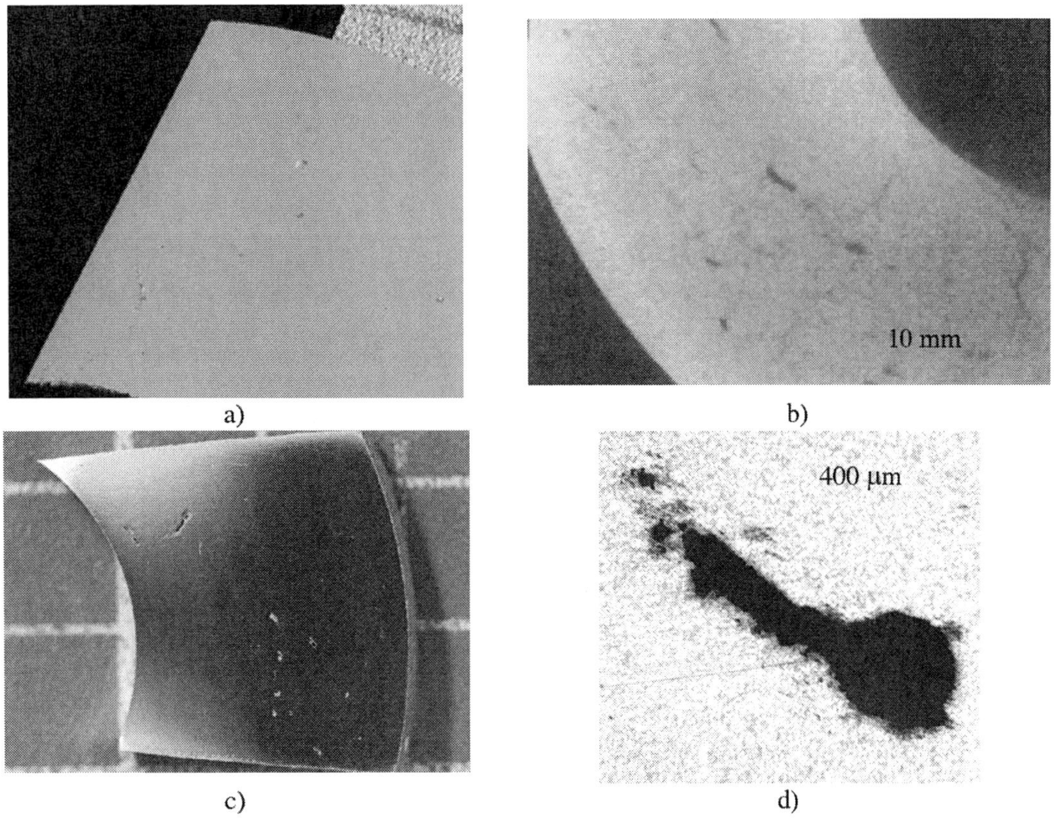

Figure 3. a) pore in 3D CT image, b) pore in 2D CT image, c) pore after sample cut and preparation, d) pore analysis on optical micrograph

To validate the results above, one object is identified on the tomographic and the same in the microscopic image (Figure 3.). The unique shape and the position of the pore was used to identified in both images. Additionally, it is found in the prepared section. The length was measured by both techniques. The CT measurement gives 1.87mm and the image analysis gives 1.64 mm. The two values are close together, which validates the two method, and the comparison. But in an automated testing the image analysis gives 2,08mm as the length of this object, because there are small pores next it, which are not present in the CT image, which is another difference between the two tested method.

4. Summary

CT and optical microscopy and image analysis are compared in the evaluation of micropores in a die cast aluminum housing. Two basic differences between the two different methods are that the CT is a nondestructive method while for optical microscopic testing a sampling is necessary. Therefore, first the CT imaging was made. Based on the CT examination, the worst position and the place for the sampling were chosen. Samples were cut from the housing and prepared to metallographic examination. Computational image analysis measured the length of the pores on the optical micrographs. Large difference was found between the means of the measured lengths. The resolution of the two type of images differs, smaller objects are identified and measured on the microscopic images. Therefore, a much larger mean of length is evaluated. Between the maximum lengths there is lower difference. The maximal length is 2,7 times larger in CT images. The metallography shows just a section of the housing; therefore, this difference can be found there. But the largest micropore is measured in the same position with both techniques. This large difference makes the validation necessary. A micropore is identified by its unique shape and position, and the measured lengths are compared. The CT and the metallography give nearly the same result, so this measurement shows that the differences really originate from the sampling and the mentioned specialty of the tests. The advantage of the metallographic tests that these can reveal the microstructure, and the solidified structure can be evaluated next to the porosity. But this comparison shows the importance of the proper determination of the examined section in case of metallographic tests.

Acknowledgements
Supported by the únkp-19-3 new national excellence program of the ministry for innovation and technology.

References

[1] K. Anderson *et al.* 2018 *Aluminum Science and Technology* ASM Handbook Vol. **2A**
[2] S. H. Davies 2001 Theory of Solidification, Cambridge University Press
[3] D. Sui *et al.* 2016 International Journal of Metalcasting, **10**, pp 32–42
[4] J.F. Friel 2000 *Practical Guide to Image Analysis,* ASM International
[5] R. Elliott 1983 Eutectic solidification processing : crystalline and glassy alloys
[6] M. Sumanasooriya, N. Neithalath 2009 *Aci Materials Journal*, pp. 428-38
[7] B. Zhang *et al.* 2005 Casting defects in low-pressure die-cast aluminum alloy wheels, *JOM*, **57** pp. 36–43
[8] G.F. Vander Voort *et al.* 2014 Metallographic Assessment of Al-12Si High-Pressure Die Casting Escalator Steps, *Microscopy and Microanalysis*, **20,** pp. 1486-93
[9] N. Pears, *et al.* 2012 *3D Imaging, Analysis and Applications*, Springer
[10] J. Baruchel *et al.* 2000 *X-Ray Tomography in Material Science,* Hermes

IOP Conf. Series: Materials Science and Engineering **903** (2020) 012039 doi:10.1088/1757-899X/903/1/012039

Effect of the homogenization and cold deformation on the mechanical performance of Al8006 aluminium alloy

J Pázmán[1], K Fehér[1], V Gonda[2], B Verő[1]

[1]University of Dunaújváros, H-2400 Dunaújváros, Táncsics Mihály str. 1/A.
[2]Óbuda University, Donát Bánki Faculty of Mechanical and Safety Engineering, H-1081 Budapest, Népszínház str. 8.

Email: pazman@uniduna.hu

Abstract. In our study, the base material was the aluminium alloy of Al8006. Its application area covers amongst others the material for multilayer tubes. This alloy has been subjected to different technological paths consisting heat treatments (homogenization) and mechanical strengthening technologies: multiple forging, equal channel angular pressing. The aim of the technological paths was to refine the primary phase structure, and to look for the technology sequence that assures the highest strength and the highest toughness at the same time. To determine this, a complex performance index has been created that integrates these property characteristics.

1. Introduction

There is a legitimate demand from members of the consumer society that the market fulfils their expectations for products and services at high quality and value for their money. Members of society can only formulate their needs and expectations on basis of their own knowledge. As an example, one may define a requirement for a passenger car for a safe transportation, without defining what the chemical composition of the passenger car's B-pillar was and what type of heat-processed steel would be best to produce it. Therefore the performance or the capacity -- in lingo we use the word "performance" -- is the concept that defines the buyer/seller relationship regarding to a product.

Obviously, the performance is embodied in the properties of the product in question, such as: material or structure. The primary task of a materials or mechanical engineer is defining the characteristics of a product to meet the needs. By defining set of properties, it is possible to select a particular material quality that carries these properties, and to determine the technology for its production. This three-element process can be described by a chain model consisting of interconnected chain links representing performance, property set and substance.

We do not have a generally accepted method for measuring performances yet, therefore in this study we propose a method for a specific material development.

In this study, the base material was the aluminium alloy of Al8006. Its application area covers amongst others the material for multilayer tubes. This alloy has been subjected to different technological paths consisting heat treatments (homogenization) and mechanical strengthening technologies: multiple forging, equal channel angular pressing. The chemical composition of this alloy contains iron as the main alloying element at 1-2wt%, manganese and silicon besides aluminium. Iron, which is the main alloying element cannot be dissolved by aluminium, therefore it will form - as well as manganese and

IOP Conf. Series: Materials Science and Engineering **903** (2020) 012039 doi:10.1088/1757-899X/903/1/012039

silicon -, primary compound phases during crystallization. These phases, which are inconsistent with the matrix, make it considerably more difficult application of the alloy is some technologies and to produce product with required performance, in this case a multilayer heating floor tube intermediate film layer [1].

To resolve it, the cast ingot is homogenized to refine the primary phases delimited by the sharp corners and to dissolve some of the alloying elements back into the aluminium [2-3]. Since the aluminium is not alloyed at very high level in this case, the mechanical behaviour is ductile, and the industry actually considers Al8006 alloy as an unalloyed material. After the homogenization, through hot and cold rolling processes, approx. 200 μm thick film/foil is produced. However, a problem remains, that in many cases the size of the primary phases is comparable to the film thickness, resulting in discontinuity or cracks during the utilization, and the lifespan of the product is strongly dependent on the alloy phase structure [4].

In the present study, we aimed to process Al8006 alloy to various technological paths and to refine further the primary phase structure, and optimize look for the technology sequence that assures the highest strength and the highest toughness at the same time. To determine this, we have been created a complex performance index that integrates these property characteristics.

2. Experiments

The chemical composition of Al8006 is shown in Table 1. From the base material, cylindrical specimens of diameter of 10 mm with a length of 60 mm were machined. By designing the experiments, we sought to study the effects of each industrial technological steps.

Table 1. Chemical composition of Al8006 alloy

Chemical composition (wt%)						
Fe	**Si**	**Cu**	**Mn**	**Zn**	**Remainder (total)**	**Al**
1,2 - 2	max. 0,4	max. 0,3	0,3-1	max. 0,1	max. 0,15	95,9 – 98,5

The initial state was the as-cast state, homogenization heat treatment as the first step of the technological route was only applied to one half of the samples. Thus, the subsequent deformation processes were performed on both as-cast and homogenized samples. Two variations of severe plastic deformation processes were chosen: (1) the multi axes forging (MF) executed on the Gleeble 3800 physical simulator, and (2) equal channel angular pressing (ECAP), wherein the cylindrical specimens were extruded using an ECAP die with a channel angle of 110° [5-6].

Table. 2. Technological paths for Al8006 alloy samples

Homogenization Heat Treatment		Severe Plastic Deformation	
Temperature (°C)	**Time (h)**	**Type**	**Strain**
-	-	- (as-cast)	-
-	-	ECAP route A 1x	0,81
-	-	ECAP route A 2x	1,62
-	-	ECAP route C 1x	1,62
605	2h	ECAP route A 1x	0,81
605	2h	ECAP route A 2x	1,62
605	2h	ECAP route C 1x	1,62
605	2h	MF	1
605	2h	MF	5
605	2h	MF	10

The multi pass ECAP processes were divided into two types. Either the route 'A' processing was performed, when the cylindrical sample after the first pass is reloaded into the extrusion die at the same position; or route 'C', when the cylindrical sample was rotated with degree of 180° along to the longitudinal axis for each subsequent passes [7]. Based on these concepts, the thermal and deformation processing of Al8006 alloy is summarized in Table 2.

To evaluate the mechanical properties of the processed specimens, samples were taken out for micro hardness testing with load of 0,3 kg (Wolper 401 MVD micro hardness tester), and tensile testing (Instron 5969) were accomplished. Out of the measured properties, the complex performance index (CPI) was defined as [8]:

$$CPI = \left(\frac{HV0{,}3}{HV0{,}3_{max.}}\right) * \left(\frac{R_{p0,2}}{R_{p0,2}^{max.}}\right) * \left(\frac{R_m}{R_m^{max.}}\right) * \left(\frac{A}{A_{max.}}\right) * 100\ (\%) \quad (1)$$

where
CPI – complex performance index (%)
HV0,3 – microhardness, load:0,3 kg
$HV0{,}3_{max.}$ – maximum hardness of all technological routes
$Rp_{0,2}$ – yield strength (MPa)
$Rp_{0,2}^{max}$. – maximum yield strength of all technological routes
R_m – tensile strength (MPa)
$R_m^{max.}$ – maximum tensile strength of all technological routes

3. Results

3.1. Mechanical properties of Al8006 alloy after different technological paths

Tensile testing was accomplished on the samples processed by different technological routes, furthermore, comparing the characteristics of cast state sample and sample deformed by traditional industrial technique (hot and cold rolling). According to the results, the hardness value of the cast state sample doubles after the first ECAP pass, furthermore the values of yield strength and tensile strength almost doubled. However, a drastic decrease in elongation values can be observed after the first pass.

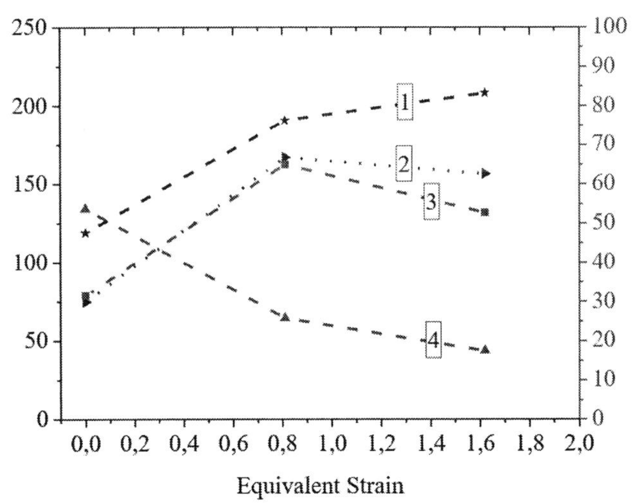

Figure 1. Mechanical properties vs. the equivalent strain for the (non-homogenized) ECAP processed samples. 1-Tensile Strength (MPa), 2-Yield Strength (MPa), 3-Hardness (HV0,3), 4-Total Elongation (%)

The second pass no longer represents a double change/increment from the first pass, but the strength values are still increasing while the toughness values are not significantly reduced (Figure 1). The ECAP routes also influence the mechanical properties, and the C route (180 ° rotation) results in better elongation

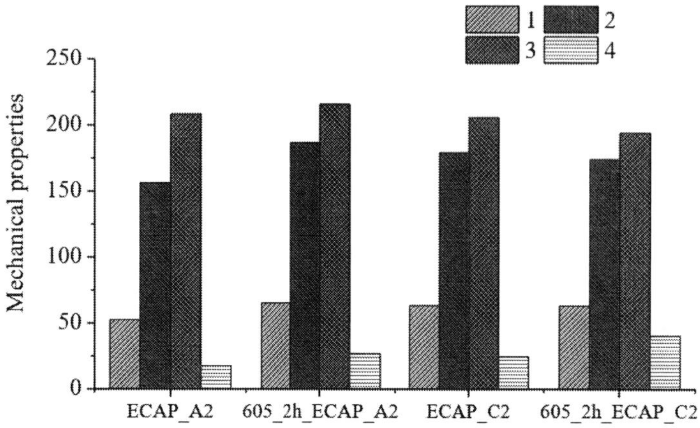

Figure 2. Comparison of mechanical properties via ECAP routes and application of homogenization

1- Hardness (HV0,3), 2-Yield Strength (MPa), 3- Tensile Strength (MPa), 4- Total Elongation (%)

values than the A route. The effect of homogenization prior to forming can also be observed in the test results. Samples were homogenized at 605°C for 2h, as a result of which the hardness, yield strength and tensile strength values are lower after the first ECAP pressing compared with the cast state, also the uniformly deformed specimens, but after the second ECAP process, they are significantly larger than their cast samples/counterparts. Elongation data were improved with 1-2%. The beneficial effect of homogenization is most pronounced in the C route formation, where both the strength values and the elongation values are far beyond the same values in the cast sample (Figure 2).

In the case of samples in homogenized state, an increase in mechanical properties with the deformation were observed. Properties of sample deformed by a strain of 5 is near the same as the value of deformation of 1,6, but increasing the strain to 10, a drastic decrease in the elongation value was found. In these cases, material might be close to its deformation limit but it has not yet been reached the strength limit, as the yield strength and tensile strength continued to increase, with only the elongation halved (Figure 3).

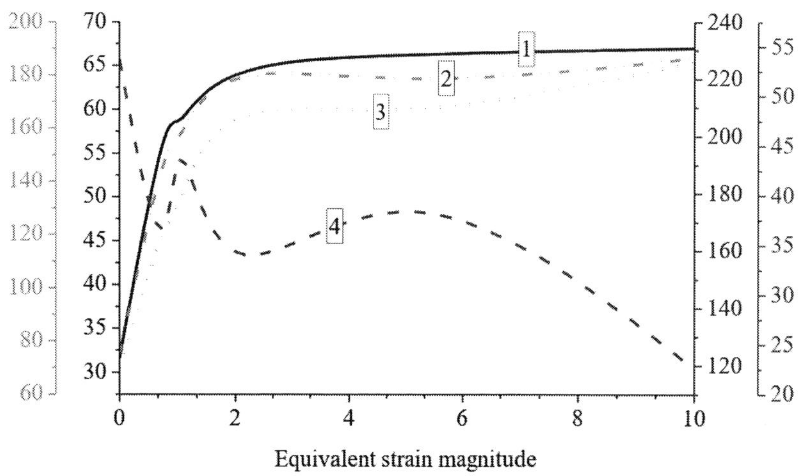

Figure 3. Mechanical properties vs. equivalent strain for homogenized samples

1-Hardness (HV0,3), 2-Yield Strength (MPa), 3-Tensile Strength (MPa), 4-Total Elongation (%)

Thus, the alloy has approached its forming limit state in selected technological paths. To classify the technological router, and to find out, which technological route leads to the performance limit of the material, we defined and calculated the complex performance index (CPI).

3.2. Complex performance index for different technological paths

Using the test results for the material properties, the complex performance index was determined by comparing the given mechanical characteristics of the samples produced by each technological path to the maximum of each parameter. On this basis, we created four ratios, then calculated the their product, where each attribute was taken with equal weight into the equation, and the performance index was generated. The CPI in percentages are plotted in the bar chart in Figure 4.

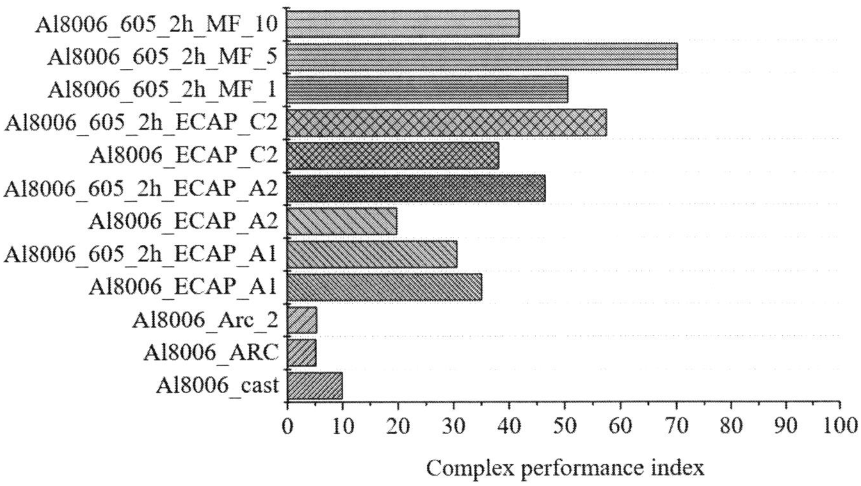

Figure 4. Complex performance index based on the material testing results

According to the CPI, the effect of homogenization on the ECAP specimens only appears intensively after the second pass, forming dominates in the first pass. By increasing the amount of strain, the complex performance index increases, and a third-order function can be fitted on the results. At the strain of 0.81 CPI is 30%, while at the strain of 1.6 it is already 46%, but increasing the strain to 5, it is already 70%, and finally at a strain of 10, a drastic decrease occurs, which is attributed to the extreme decrease in elongation values (Figure 5).

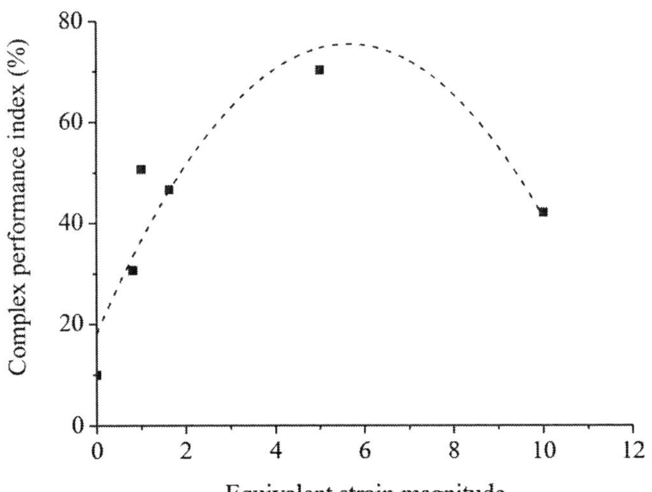

Figure 5. CPI values via equivalent strain magnitude for homogenized samples

Based on the calculated CPI results, it can be concluded, that the performance limit state of optimal strength and elongation can be found in the homogenized state sample after MF process with the strain of 5 (Fig. 4), therefore this is the technological path that leads to the performance limits of the Al8006 alloy. Of course, further increasing the diversity of forming paths would give a more comprehensive picture of the limits of this alloy. Expanding the technology paths is part of further research work.

3.3. Microstructure of Al8006 alloy

The effect of homogenization is clearly evident in the "spheroidizing" of the primary intermetallic phase and in the deformation of samples in the homogenized state (Figure 6). Partially dissolved or finer needle shaped phases during homogenization crumble during forming sequentially, hence the life of the end-product is reduced. In the non-homogenized samples, the size of the needle shape phases is comparable to the thickness of the end-product.

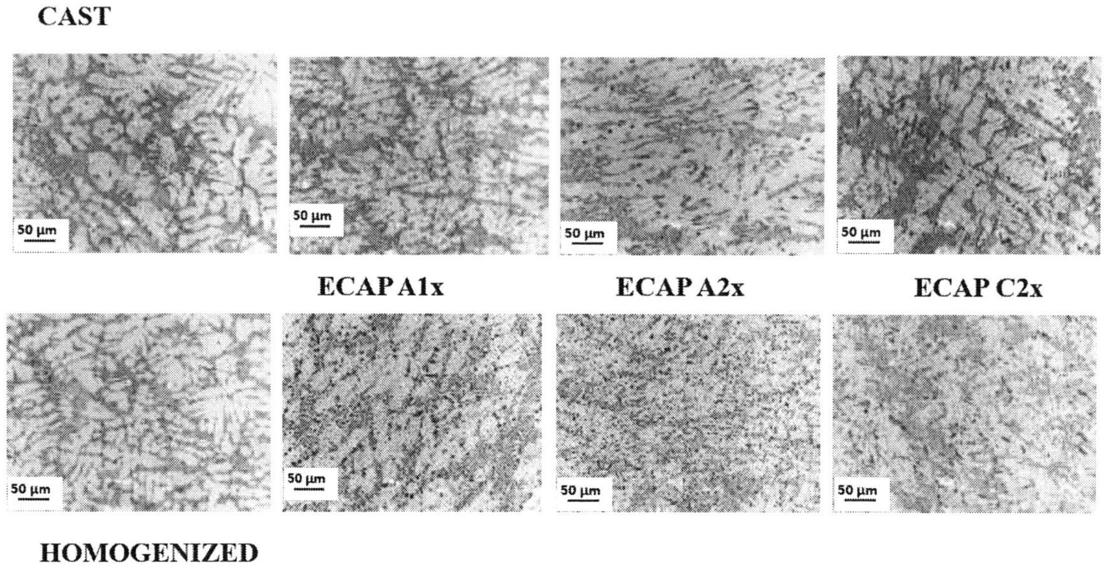

Figure 6. Phase structure of cast, homogenized and ECAPed samples

If the primary phase structure has already been refined by homogenization, and the needles characteristic of the cast structure are thinner than without homogenization. However, the actual change in the phase structure can be achieved by forming processes.

During ECAP, the cell structure of the 1x pass sample is nearly the same as that of the as-cast or homogenized ones. Deformation occurs only in the local cells and subcells with different orientations - appearing in different colours on the recordings, also the band structure. However, the primary phase structure is broken after the first pass. Homogeneous and single ECAP processed sample shows much finer phase structure than

Figure 7. Grain structure of cast, homogenized and ECAPed samples

cast and ECAP 1x sample. By increasing the number of passes, the initial cell structure disappears after the second ECAP pass samples. Elongated subcells/subgrains arranged in the direction of deformation are observed. The effect of the forming paths can also be traced on the microstructure. In fact, the phase structure does not suffer the same degree of refinement in the route C processed sample as in the route A processed sample (Figure 7).

3.4. Findings comparing with industrial data
In addition, the results were compared with the CES EDUPACK software material database. As no data was found for the test alloy, comparisons were made with Al8xxx alloys based on maximum strength (tensile strength) and maximum elongation (total elongation). Figure 8. shows the result of the comparison.

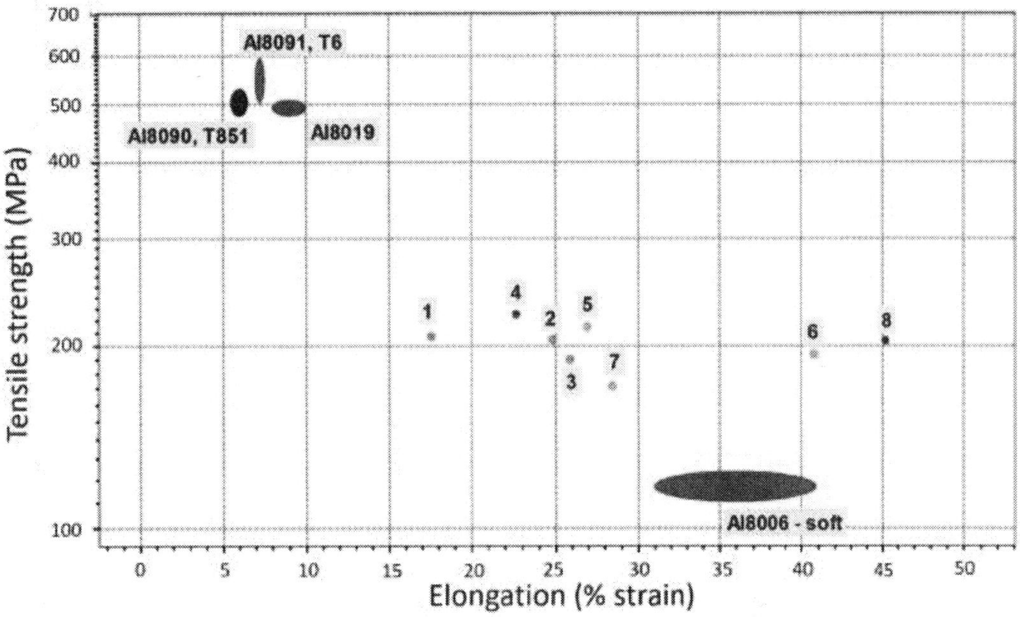

Figure 8. Comparison our results with the database of CES EDUPACK software

1 – Al8006, cast ECAP A2, 2 – Al8006, cast ECAP C2,
3 – Al8006, cast ECAP A1, 4 – Al8006, 605°C 2h, MF φ=10,
5 – Al8006, 605°C 2h, ECAP A2, 6 – Al8006, 605°C 2h ECAP C2,
7 – Al8006, 605°C 2h ECAP A1, 8 – Al8006, 605°C 2h MF φ=5

The data in the database did not apply to Al8006, but it is also clear from the test results that the homogenized samples multiple forged with the strain rate of 5 has achieved the highest tensile strength-elongation combination.

4. Conclusions

In the present study, we aimed to subject Al8006 to various technological processes and refine the primary phase structure to find out the technology sequence that provides the highest strength and toughness at the same time. To determine this, we have created a complex performance index that combines these property characteristics.

Based on the results, the following conclusions can be drawn:

1. The same amount of strength and approx. three times higher elongation values were obtained for Al8006 formed by severe plastic deformation processes than those obtained by the conventional forming technique in the H19 (full hard) state.
2. Based on the experiments and in the light of the defined complex performance indices, the Al8006 alloy has the best combination of properties with 605 ° C 2h homogenization and multiple forging with the strain rate of 5.
3. As the effect of the homogenization, the sharp/thick needles of the primary intermetallic phase structure become thinner, they can get broken by the forming processes and consequently, the ductility and usability of alloy improves.
4. As a result of the forming, the cast cell structure disappears partially (after the 1st pass) and then completely (after 2nd pass) and the sub-cell structure parallel to the forming direction is formed.
5. The complex performance index for a given alloy qualifies the applied technological path according to customer needs.
6. In the improvement of the performance index value, we take into account the elongation on necking values for the future because that is where the properties of the individual states are revealed.

Acknowledgement
The authors say thanks Mr. Péter Bereczki for the tensile testing results. The work/publication is supported by the EFOP-3.6.1-16-2016-00003 project. The project is co-financed by the European Union and the European Social Fund.

References

[1] CNBM International Corporation *Applications of Aluminium Foil*, www.icnbm.com, downloading date: 29th September 2016.
[2] Chen Z *et al* 2012 Trans. Nonferrous Met. Soc. China **22** pp 1280-1285.
[3] Engler O *et al* 2013 Mater. Charact **79** pp 60-75.
[4] Moldovan P *et al*, 2004 J. Mater Process Tech **153–154** pp 408–415.
[5] Pázmán J *et al Effect of the Multiple Forging and Aging on the mechanical properties of AlMgSi1 Alloy*, NanoSPD6, Metz, France, June 30 – July 4, 2014. poster presentation.
[6] Pázmán J *et al* 2013 *The effect of the severe plastic deformation on the aging of AlMgSi1 alloy*, OATK 2013 oral presentation
[7] Pázmán J 2010 *Equal channel angular pressing for powder metallurgical and cast AlCuSiMg alloy, diplomwork*, University of Miskolc
[8] Ashby M 2019. CES EDUPACK 2019 software, , downloading date: 20th January 2020. http://support.grantadesign.com/resources/cesedupack/2019/help/topic1.htm?rhcsh=1#t=html%2Fchart%2Fpif_about.htm, date: 13th May 2020.

IOP Conf. Series: Materials Science and Engineering **903** (2020) 012040 doi:10.1088/1757-899X/903/1/012040

Magneto-acoustic investigation on steel samples

O. Szabados, R. Baki, Zs, Csincsi, J. Molnar, Á. Pámer, P. Szabó, G. Por

University of Dunaujvaros, H-2400, Dunaujvaros, Tancsics M.str.1/A, Hungary

E-mail: porg@uniduna.hu

Abstract. Experiments have been carried out using yoke for energizing alternating magnetic field in the steel samples. We used acoustic emission sensors and ultrasound microphone for recording acoustic signal born in the steel material. Results of experiments clearly show the presence of Acoustic Barkhausen noise. We are introducing graphs presenting dependence of acoustic time response signal on magnetic signal (inductance) in form of Lissajous curve. It is claimed that this new method gives more in-sigh in the material than the widely used simple RMS measurement of acoustic response.

1. Detecting Acoustic Barkhausen Effect in steel samples

Barkhausen effect is well known in the magnetic theory and practice. The Magnetic Barkhausen Noise (MBN) was first detected by Barkhausen [1] in 1919, observing cyclical magnetization in a ferromagnetic material. Changing the polarity of magnetic field in the steel, the induced magnetic field in the material describes a hysteresis. Magnetic domains in steels turn towards the driving magnetic field abruptly causing stochastic noise, which can be transferred from measured inductivity signal into an acoustic signal. However, this is just a detection technique to demonstrate the stochastic (noise) nature of the magnetic inductivity changes. The theoretical models explain this magnetic domain changes either by real turns of domains or by dislocation of domain walls. In both cases pressure wave born on the changing domain walls should also produce acoustic emission in the material, and this is called acoustic Barkhausen noise (ABN) or some literatures magnetoelastic effect (MAE). To detect MAE one need to directly detect acoustic signal born in the material. Typical sensors, which can be used for detection of MAE are the so-called acoustic emission (AE) detectors. AE sensors are typically acceleration sensors having resonance in ultrasound frequency region. We shall demonstrate in the next sections, that similar results in acoustic measurement can be achieved using ultrasound microphone.

Earlier we have noticed (see Figure 1) that ABN can be observed during low cycle fatigue test [3]. It was not our intention to investigate that, it was just a side effect of our fatigue test, since fatigue test included not only stress but also heating and cooling parallelly with tension and pression. Heat treatment was done by strong, periodic current, which produced changing magnetic field up to 4 tesla on the skin of the sample.

This led us to investigate ABN (or MAE) in test samples using yoke. On Figure 2 we present the acoustic noise detected by AE sensors on test samples from iron and stainless steel. One can clearly see the repeating ABN burst in every half period, but with larger burst in every second half period.

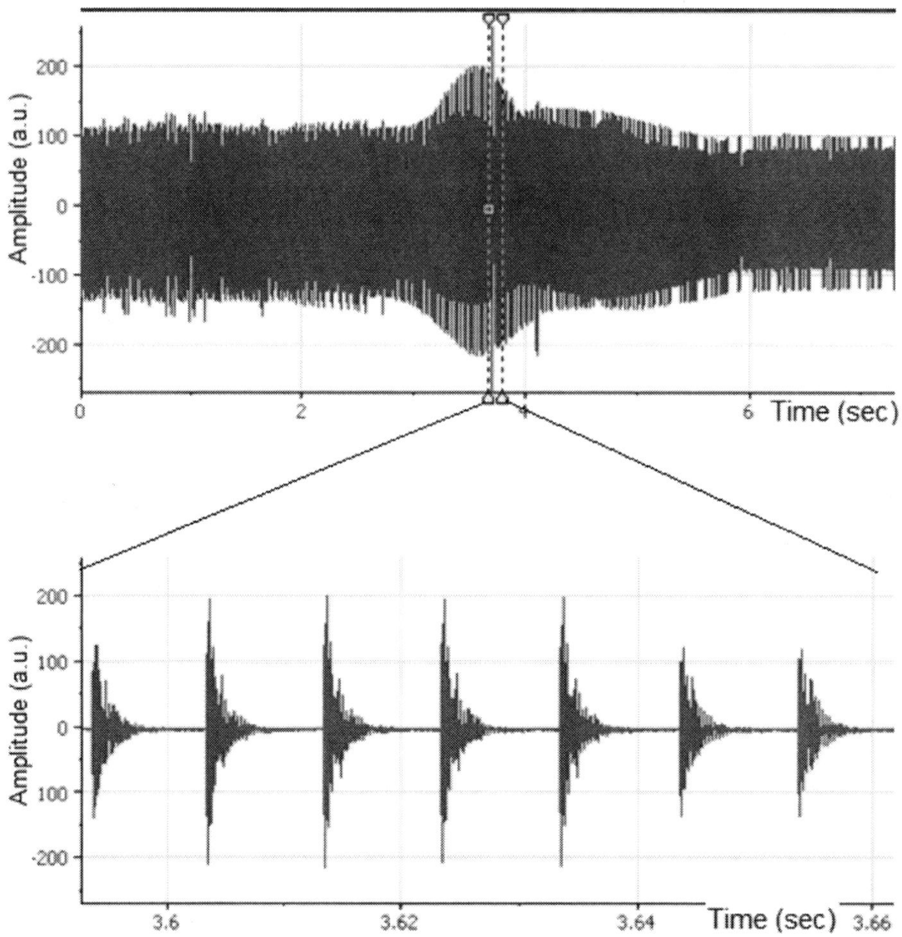

Figure 1. Acoustic signal and its zooming when heating with large AC current causing magnetic field in material up to 4 Tesla in Gleeble simulator for fatigue test. Details of experiment see in [3]

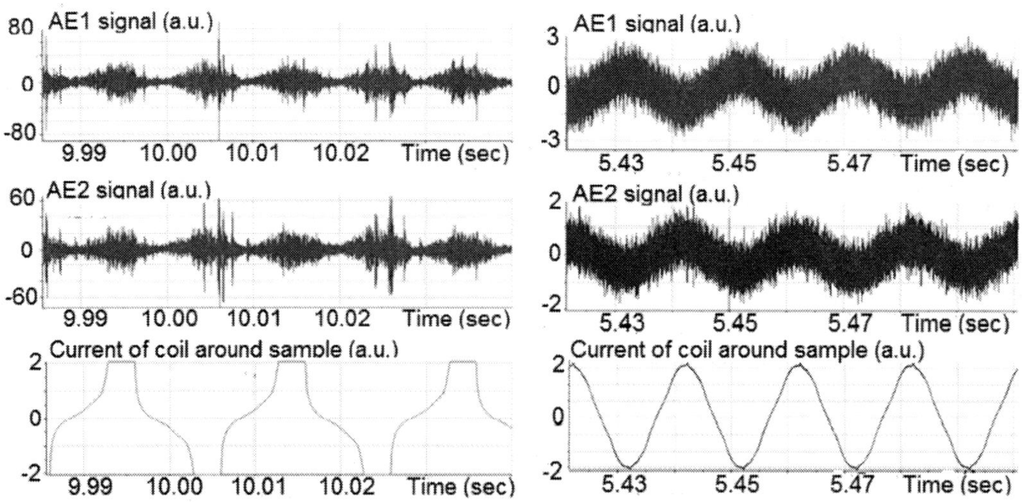

Figure 2. Left hand side: ABN bursts in sample made from reactor material 15H2MFA. Rigth hand side: in stainless steel 08KH18N10T the same yoke does not generate ABN

2. New Laboratory Experiments

2.1. Measurement arrangement

In our new laboratory experiment we used standard size steel samples for tensile test. The yoke was powered from the 50 Hz network, but the voltage was regulated by an autotransformer to produce different magnetic flux level in the sample. We were interested to find when at which saturation level Barkhausen bursts appear in the material. To measure the magnetic field, a calibrated Hall sensor was used, but it can give information only about the field around the steel plate. To get the real magnetic flux in the material we applied the old-fashioned coil technique. We installed a coil of 10 turns of wire around the tensile test sample (see No.3 on Figure 3) and a similar one around the pole of the yoke (number 4 on Figure 3). We shall see, their signals might exhibit different behavior.

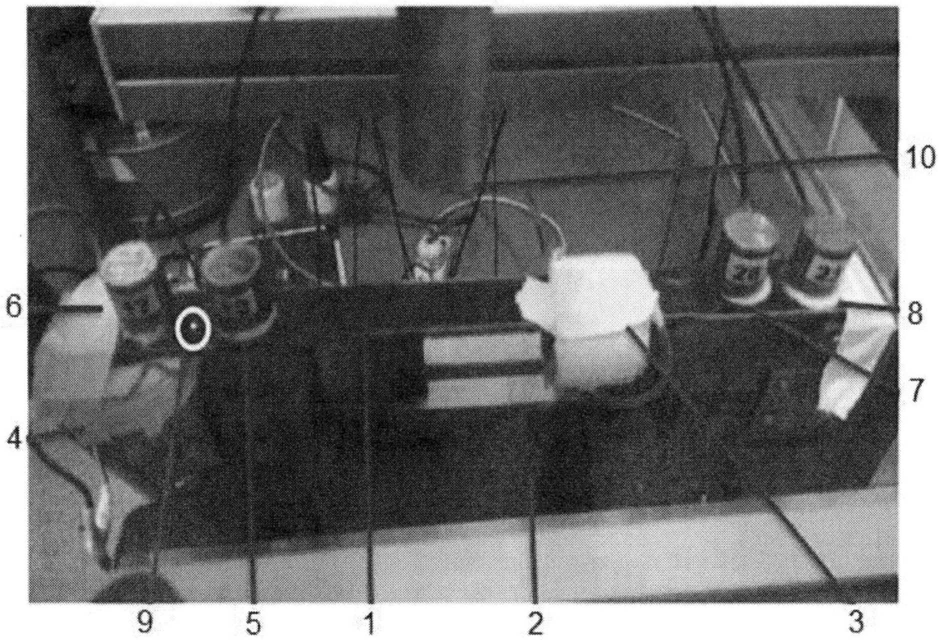

Figure 3. Laboratory arrangement for experiments. 1- sample, 2- yoke, 3-coil around sample, 4-coil on the pole of yoke, 5,6,7,8 - AE sensors 9-red dot from laser vibrometer, 10- end of tube of ultrasound microphone.

Since we noticed also mechanical vibration of the yoke and the sample contacting with that, we followed vibration using a non-contacting laser vibrometer (see red dot circled with white between AE sensors 17 and 12 in Figure 3.). Originally, we intended to register ultrasound acoustic emission in the sample due to changing magnetic field using AE sensors (No. 5,6,7,8 in Figure 3). But later it was decided to test also ultrasound microphone (No.10 on Figure 3), since it is also a non-contacting method, and this is a new direction for sensing the ultrasound response.

2.2. Developing suitable measuring software

The measuring set was programmed in LABVIEW. We were able to sample ultrasound signal with sampling rate of 1 MHz, while the slow change of magnetic field was sampled parallel with 10kHz only. Software can estimate full RMS, partial RMS including short time RMS with typical averaging time of 1-5sec.

Visual examination of the signals with zooming and comparison of different types of signals with different sampling rate and different filters are all very important part for the evaluation of the events during the test. While there are low frequency deterministic components like the magnetic field itself and some correlated vibration of the steel plate, we have stochastic burst in very short time range, and this is rather difficult to present on the same screen and on the same figure.

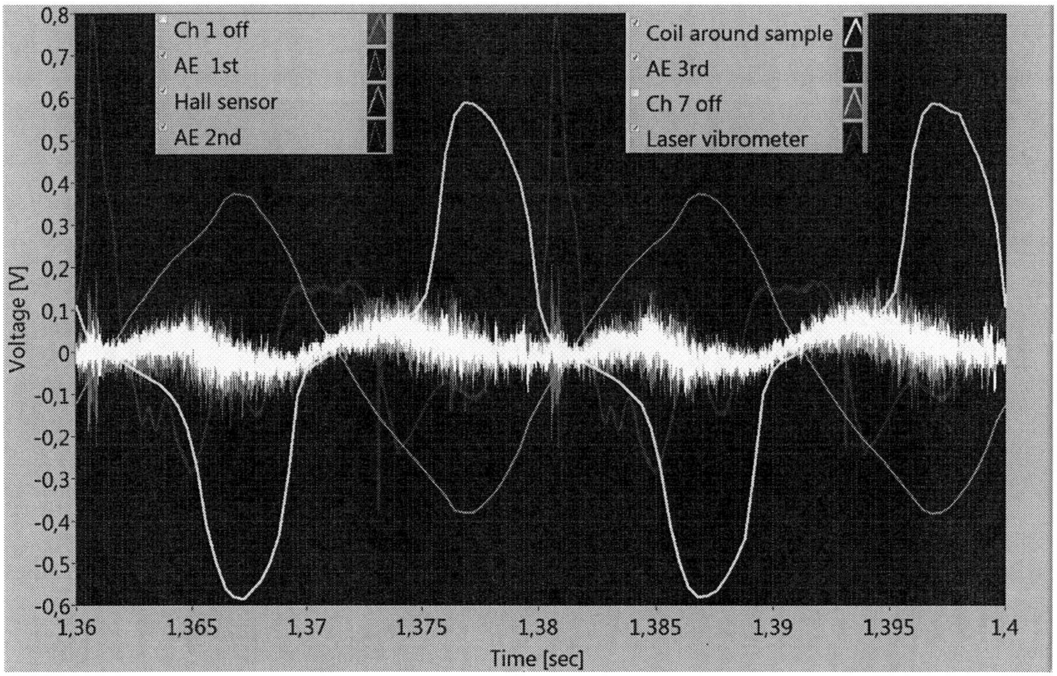

Figure 4. A typical representation of the time signals with periodic, slowly changing components (magnetic Hall and coil-green and yellow, vibration velocity-blue) , stochastic components and bursts (from AE sensors. white, purple and azure).

It is worth to pay attention to the slowly changing curves on Figure 4. The green one is the direct signal from coil around the pole of yoke. It is proportional to the inductivity, which is the derivate of the magnetic field. Due to hysteresis between the magnetic field and its induction in the steel there is not linear dependence between the field and inductivity. However, if the field is far from saturation, then the inductivity (and also it derivative, which is the measured current from the coil), will follow more or less the sinusoidal variation of the energizing 50Hz voltage applied on the yoke (see also the right side of Figure 2, where we were far from saturation due to smaller permeability of the stainless steel). In the experiment presented on Figure 3. we had coil on the pole of the yoke (green line on Figure 4) and another coil around the sample (yellow line). Obviously, the inductivity in the sample went near to saturation, since the cross section of the sample was much smaller, than the cross section of the yoke. Since the current from the coils is the derivative of the inductivity, in the next Section we calculate the real magnetic field, as well as from the velocity signal (blue line) we calculate the real displacement.

2.3. Directly measured signals and calculated ones

Many physical signals can be sampled and measured directly, presenting them as their dependence on time. For example, a simple coil around the sample or the pole of the yoke will produce current proportional to the magnetic flux derivative. If one would like to present the change of the magnetic field in time, then one must integrate that derivative dependents as it is shown on Figure 5.

One can clearly see that the time dependence of inductivity is not sinusoidal, even though the yoke was powered using fully sinusoidal 50Hz current, as we have explained that in previous Section. Comparing the acoustic emission signals with calculated inductivity (see Figure 6) we can see that strong emission occurs when there are fast changes of the inductivity, when the signal is the steepest crossing zero value.

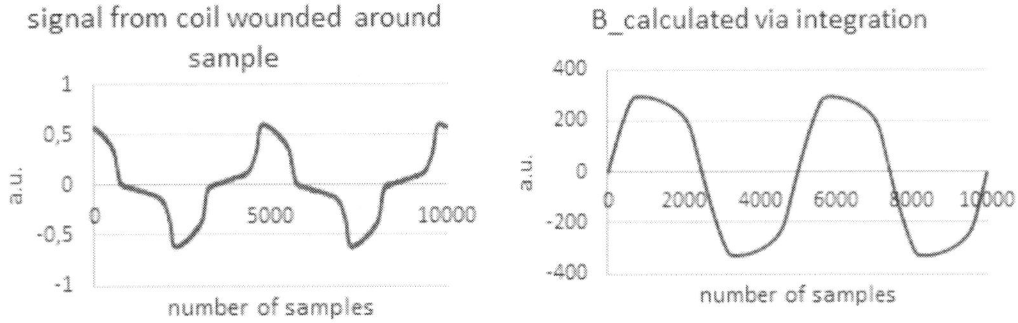

Figure 5. Left had side: The directly measured voltage drop on the input of ADC from the current of coil wounded around the sample, which is proportional to Right hand side: the time derivative of magnetic flux and B_calc- the inductivity calculated via integration from that

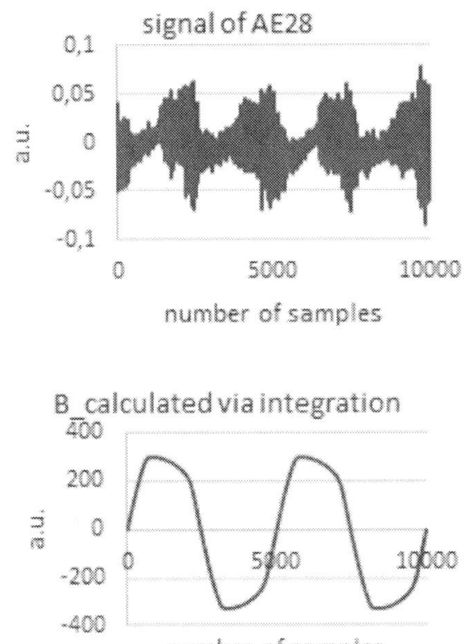

Figure 6. Correlation with the 50Hz magnetic disturbance is clearly seen on AE signals

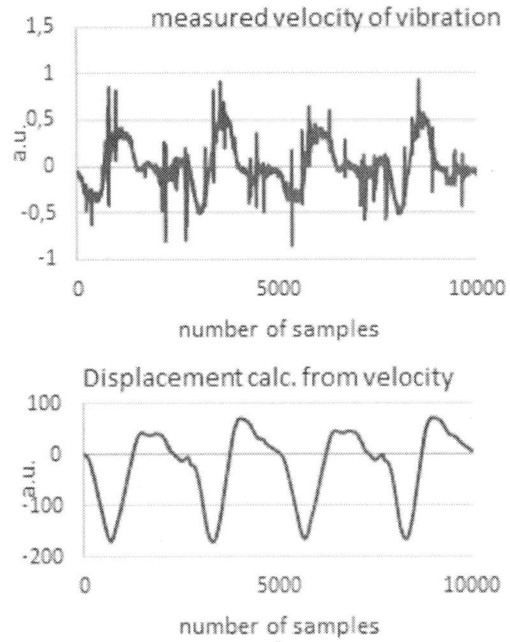

Figure 7. Vibration signal measured by contactless laser vibrometer and displacement calculated from that

This proves that this is a kind of Magnetic Acoustic Emission. It is also possible to estimate displacement signal from the measured velocity (see Figure 7.). However, if we want to see better resolution of acoustic signal of Figure 6, then we lose the behaviour of the magnetic field, since showing the burst in time of acoustic signal we cannot connect to the slow change of the magnetic signal.

2.4. New presentation of the measured signals

Due to high frequencies, relatively short records contain too many samples, therefore observation and comparison of time signals are rather difficult. We are introducing a Lissajous technique where we present direct dependence of acoustic signals on magnetic signal.

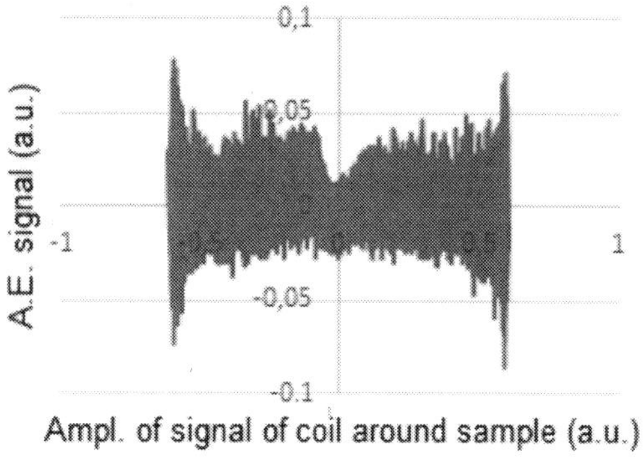

Figure 8. Direct dependence of stochastic AE signal on magnetic coil signal

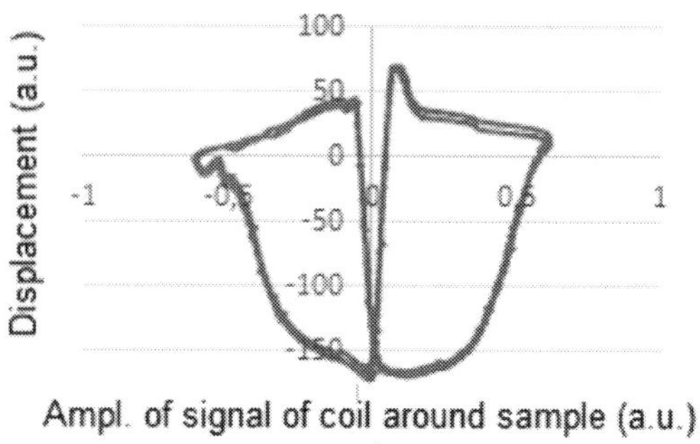

Figure 9. Direct dependence of displacement on measured current of signa of coil on Sample

One can see in Figure 8 that the maxima are in the peaks and on the halfway of the magnitude of changing magnetic field, while a smooth characteristic trajectory was formed from the displacement versus magnetic field measured by the coil wounded around the Sample (Figure 9).

2.5. Possibility to use ultrasound microphone to detect acoustic emission

On Figure 3. we marked by number 10, the end of an aluminium tube pointing toward the sample. This tube plays the role of a waveguide, leading to a Dodotronic 250 type ultrasound microphone, which samples the acoustic noise with 192kHz rate.

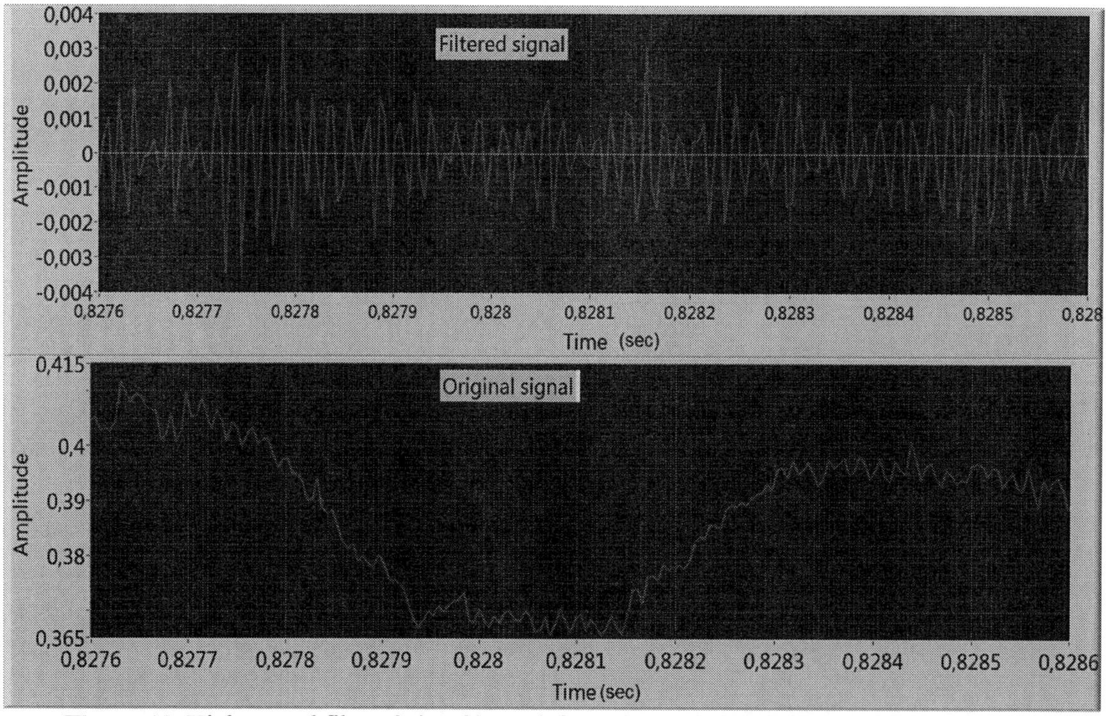

Figure 10. High passed filtered signal(upper) from the total signal(down) of the ultrasound microphone exhibits burst of acoustic emission

One can see (Figure 10), that using adequate high pass filtering we can present the burst in ultrasound frequencies. This method is a noncontact method and opens new dimensions in measuring weak AE vibration in the material reaching its surface. The long waveguide-tube ending near to the surface ensures lower disturbances from environment. Contemporary microphones with built in ADC of 16 or even 24 bits opened the possibility to follow the dynamics of high frequency part of the sound signal even if its magnitude is ten thousand time less than that of the low frequency components. Meanwhile any contacting transducer attenuate and distort the sound component, while noncontacting methods have much less influence if at all.

3. Summary and plans

We investigated the acoustic signals born in the steel plate-samples due to changing magnetic field. The main goal was to find responses in ultrasound frequency range. Two main results of our investigation are as follows: It is possible and even advisable to use ultrasound microphone as contactless sensors to detect acoustic emission instead of AE sensors. Direct dependence of the measured response signal on magnetic field can be better used to characterize the stochastic dependences than a comparison of directly measured time signals or their frequency domain presentations.

Magnetic field changes must be much slower to observe and characterize effect of magnetic domain and/or displacement effects. We plan to improve our detection method using higher frequency and sampling time for acoustic signals.

Acknowledges

This work was supported by the Hungarian Government under the grants "Thematic Excellence Programme of the National Research, Development and Innovation Office 2019".

References

[1] H. Barkhausen 1919 *Physics Z.* **29** 401
[2] Neyra Astudillo *et al* 2017 *Journal of Magnetism and Magnetic Materials* **426** 779–784
[3] G. Por *et al* 2016 World Conference on Non-Destructive Testing (WCNDT) 2016 (June Munich) https://www.ndt.net/search/docs.php3?id=19666

IOP Conf. Series: Materials Science and Engineering **903** (2020) 012041 doi:10.1088/1757-899X/903/1/012041

Deformation-softening in ultrafine-grained materials

Jenő Gubicza and Nguyen Quang Chinh

Department of Materials Physics, Eötvös Loránd University, Budapest, Hungary

Email: jeno.gubicza@ttk.elte.hu

Abstract. Plastic deformation is conventionally considered as a route for strengthening of metallic materials. During plastic straining, different crystal defects, such as dislocations, stacking faults and grain boundaries form, hardening the material. On the other hand, plastic deformation may also cause unexpected softening in ultrafine-grained and nanocrystalline materials. This paper summarizes the existing knowledge on this deformation-softening effect.

1. Introduction

Plastic deformation of metals and alloys at room temperature (RT) is usually accompanied by strain hardening, i.e., the true stress increases with increasing true strain [1]. This effect can be used for strengthening of metallic materials. The enhancement of the flow stress during plastic deformation is caused by the increase of the density of lattice defects, such as dislocations and stacking/twin faults [1]. For instance, in face centered cubic (fcc) metals deformed plastically to a strain of about 0.2 at RT, the dislocation density increased to an order of magnitude of 10^{14} m^{-2} [2]. The higher the applied strain, the higher the defect density formed during plastic deformation. Therefore, in the last decades numerous new techniques were developed for achieving large plastic strains in metals in order to obtain high strength materials. These procedures are referred to as severe plastic deformation (SPD) methods [3].

SPD techniques can achieve large plastic strains (higher than one) in metallic materials without significant change of the dimensions of the specimens. The most frequently used SPD methods are Accumulative Roll Bonding (ARB), Equal Channel Angular Pressing (ECAP) and High Pressure Torsion (HPT) [3]. Applying these techniques, the dislocation density saturates at the equivalent strain of about 1-4. The saturation dislocation density of SPD-processed materials varies between 10^{14}-10^{16} m^{-2} [4]. In addition, the grain size is refined below 1 μm due to SPD, resulting in homogeneous ultrafine-grained (UFG) or nanocrystalline microstructures with high yield strength. In the saturation state, the yield strength of SPD-processed metallic materials is one-two orders of magnitude higher than that for coarse-grained counterparts [4].

Recently, it was revealed that plastic deformation may cause softening in UFG and nanocrystalline materials (e.g., see [5,6]). In this paper, this unusual phenomenon referred to as deformation-induced softening (or simply deformation-softening) is overviewed. The research results obtained in this field can be categorized into three classes. In the first case, the deformation-softening was observed during straining of UFG materials formerly annealed at moderate temperatures. The preliminary annealing caused hardening and then subsequent plastic deformation resulted in softening. In the second class of deformation-softening phenomena, the effect was observed for fine-grained materials having low melting points. In the third case, the as-processed defect structure is decayed due to plastic deformation, thereby causing softening. Accordingly, in the following the results for deformation-softening will be overviewed in three separate sections.

2. Deformation-softening due to plastic straining of anneal-hardened UFG materials

It has been shown recently that annealing of UFG and nanocrystalline materials at moderate temperatures can further enhance the high yield strength achieved due to the formation of the very small grain size [7-13]. This effect is referred to as anneal-hardening which can be observed if the temperature of annealing is only about 0.35-0.45 × T_m where T_m is the melting point, and the duration of heat treatment is short, typically not longer than 1 hour. Anneal-hardening can yield a 10-120% increase of hardness or yield strength, depending on the chemical composition and the microstructure of the material before heat treatment. The application of a moderate annealing temperature prevents the fine-grained microstructure from recrystallization, and only a relaxation of the defect structure occurs. This means that mobile dislocations are annihilated and the remaining dislocations are arranged into low energy agglomerates [7,11]. These effects result in a harder initiation of dislocation glide during plastic deformation after annealing. In addition, the non-equilibrium grain boundaries are relaxed, thereby making the dislocation emission from the boundary more difficult [8,12,13]. The grain boundary relaxation also hinders grain boundary sliding which is an important deformation mechanism in nanocrystalline samples and also in UFG materials with low melting points [14]. In UFG and nanostructured alloys, the segregation of solute elements to grain boundaries can contribute to grain boundary relaxation, yielding a decrease of grain boundary energy [15].

Subsequent plastic deformation after moderate annealing can introduce new mobile dislocations into the relaxed UFG materials, thereby resulting in a lower flow stress (i.e., softening) as demonstrated on Al with 99.2% purity processed by ARB at RT [5,16]. After six cycles of ARB, the yield and ultimate tensile strength values were 259 and 334 MPa, respectively. The elongation to failure was obtained as 7%. The ARB-processed material had an average grain size of about 200 nm with an elongated grain shape. After annealing of the ARB-processed sample at 150 °C for 30 min, the yield strength increased from 259 to 281 MPa while the ultimate tensile strength was enhanced from 334 to 347 MPa. In addition, the elongation to failure was reduced to 1.4% due to annealing. The effect of anneal-hardening on the ultimate tensile strength and the total elongation are shown in Fig. 1. Then, the annealed sample was cold rolled to the thickness reduction of 15% which led to a decrease of the yield and ultimate tensile strength values to similar values as observed immediately after ARB. In addition, the elongation to failure increased back from 1.4% to about 6.6% (see Fig. 1). The restoration of the softer but ductile behavior due to rolling was explained by the introduction of new mobile dislocations into the anneal-hardened material. The subsequent application of annealing and rolling deformation steps can cause cyclic hardening and softening, respectively, as shown in Fig. 1, i.e., these effects are reversible. This is also valid for the elongation to failure. It is worth noting that the occurrence of deformation-softening depends on the thickness reduction applied in cold rolling [16]. Namely, if the thickness reduction is smaller than about 30% a decrease of strength was observed while above this limit deformation-softening was not observed. The latter phenomenon can be explained by the clustering of dislocations at high rolling strains into configurations causing forest hardening which suppressed the softening effect caused by the newly developed mobile dislocations.

Nanocrystalline high entropy alloys (HEA) also exhibit deformation-softening if the samples first hardened by annealing. This effect was demonstrated on CoCrFeMnNi HEA processed by HPT at RT [17]. It was revealed that annealing at 450 °C for 1 and 10 h resulted in the development of NiMn, FeCo and Cr nanosized phases which were then dissolved during subsequent deformation in an indentation experiment. This deformation-induced dissolution of the intermetallic phases was most probably caused by the increase of the energy of interfaces between the matrix and the precipitates due to the accumulation of dislocations at these interfaces formed during deformation [17]. The sample subjected to longer annealing (10 h) exhibited a more pronounced decomposition, thus the subsequent deformation caused a higher softening in this material compared to the case of the shorter annealing time (1 h).

Deformation-softening was also observed for nanosized samples, namely for single crystal nanopillars fabricated by ion milling from Au film deposited on a MgO substrate by sputtering [18,19]. The diameter of the Au nanopillars was about 300 nm while their aspect ratio was about two. First, larger nanopillars were pre-strained by compression to the strain of 35%, resulting in the formation of dislocations in the sample which then was re-milled by a focused ion beam. The dislocation density was in the order of 10^{15} m^{-2}. Annealing of these pillars at 260 °C for 7 min led to an increase of the yield strength which was caused by the development of jogs on dislocations since these jogs hindered the motion of dislocations during further compression. This was a typical anneal-hardening effect. However, vanishing of these jogs was observed during additional compression after annealing, resulting in a decrease of the yield strength to the value observed on the Au nanopillars before the heat treatment.

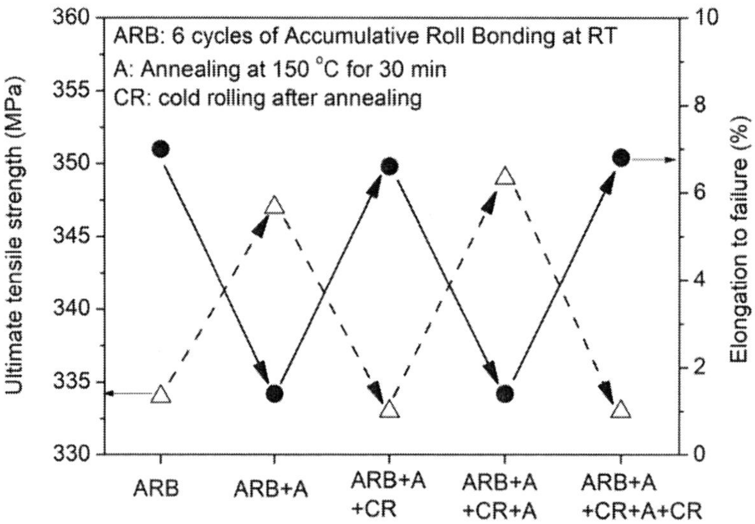

Figure 1. The variation of the ultimate tensile strength and the elongation to failure for ARB-processed Al with 99.2% purity subjected to annealing and rolling cyclically [5,16].

3. Deformation-softening in UFG metals having low melting points

Deformation-softening can also be observed for UFG materials with low melting temperatures [20-25]. In this case, preliminary anneal-hardening is not necessary for the occurrence of softening. The decrease of the yield strength or hardness at RT was observed after SPD-processing of pure metals such as Sn, Pb, In and Zn. For these metals, the melting point varies between 160 and 420 °C therefore RT corresponds to the homologous temperatures between 0.43 and 0.7. Thus, the grain refinement during SPD yielded an enhanced role of grain boundary sliding in mechanical testing due to the increased amount of grain boundaries even if the test was carried out at RT [26].

The deformation-softening effect was also observed in SPD-processed Al with a slightly higher melting point (660 °C) but only if the purity exceeded the critical level of 99.9999% (i.e., 6N) [24]. Fig. 2 shows the variation of the hardness as a function of the shear strain schematically for HPT-processed Al with three different purity levels (namely 2N, 4N and 6N purities). For 2N purity Al, the hardness increases monotonously with increasing strain until a saturation state was achieved. This state can be characterized by a dynamic equilibrium between the formation and annihilation of lattice defects such as dislocations and grain boundaries. For 4N purity Al, first the hardness increased but then decreased at high strain values [27,28]. Despite this decrease, the hardness was higher in the steady state achieved at high strains than that in the initial annealed state. For 6N purity Al, the hardness also increased at low strains but then decreased and the steady state hardness was about 12% smaller than that for the initial annealed material [24]. The hardness reduction at high strains was caused by dynamic recovery and recrystallization which was accompanied by a grain refinement. The steady state grain sizes were 400 nm, 1.3 μm and 20 μm for 2N, 4N and 6N purity Al, respectively. For 4N and 6N purity Al, the hardness was higher at low shear strains (i.e., close to the center of the HPT-processed disks) than that at high strains which can be explained by the larger dislocation density at lower strains. At high strains, the dislocation density strongly decreased due to recrystallization and this softening effect overwhelmed the hardening effect of the smaller grain size, resulting in hardness reduction. It should be emphasized that the hardness of the HPT-processed 6N purity Al sample was lower at high strains than that in the initial annealed state. In the initial material, the grain size was about 1 mm, i.e., about two orders of magnitude higher than that in the steady state for the HPT-processed sample (~20 μm). In the initial large grains, the dislocations formed during hardness testing were stored in the grain interiors (e.g., in cell walls), therefore they contributed to strengthening [24]. At the same time, in the HPT-processed fine-grained sample there is a much larger amount of grain boundaries due to the smaller grain size which acted as sinks of dislocations. In 6N purity Al, dislocations have high mobility due to the high purity level, the relatively low melting point and the high stacking fault energy. Therefore, the dislocations formed during HPT and the subsequent hardness testing can be easily annihilated at grain boundaries, therefore they do not cause hardening. Thus, the HPT-processed 6N Al sample in the steady state is softer than the initial coarse-grained material. It is noted that the contribution of grain boundary sliding to the softening of the HPT-processed 6N Al can not be ruled out as this mechanism was proved to operate during hardness testing of SPD-processed Al even at RT [26].

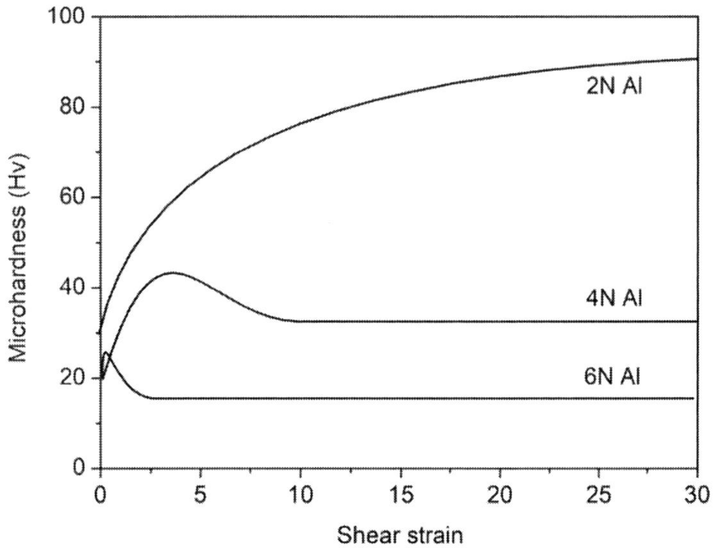

Figure 2. Schematic showing the hardness evolution as a function of shear strain for 2N, 4N and 6N purity Al processed by HPT at RT. The data were taken from [24].

The influence of Zn content on deformation-softening in HPT-processed Al was revealed in a recent study [25]. Al samples with different Zn contents up to 30 wt.% were processed by HPT until the saturation state was achieved, and the microstructure and the mechanical properties were studied as a function of the Zn content. The phase composition in the HPT-processed samples was studied by X-ray diffraction (XRD). The XRD patterns show that the samples contain two phases: the major Al(Zn) and the minor Zn(Al) solid solutions, indicating the decomposition of the initial supersaturated microstructure. Only the major Al(Zn) phase was studied in detail. Irrespectively of the nominal Zn content, the lattice constant of the Al(Zn) phase was 0.4049 ± 0.0001 nm as determined by XRD which agrees with the lattice parameter of pure Al within the experimental error. The solubility limit of Zn in

Al at RT is very low (about 2 wt. % [29]) which can explain the undetectable deviation of the lattice constant from the value of pure Al.

The Zn phase content in the HPT-processed Al-Zn samples was characterized by its intensity fraction in the XRD patterns. This quantity was determined as the ratio of the sum of the areas under the Zn peaks and the sum of the areas under all peaks after background subtraction. This intensity fraction of Zn peaks reflects the Zn precipitate volume fraction (but not equivalent to that). Fig. 3a shows this quantity as a function of the nominal Zn content. It is clear that the Zn phase fraction follows monotonously the nominal Zn content and the values suggest that for high Zn content the majority of Zn atoms are in the Zn phase. This observation is in accordance with the low solubility limit of Zn in Al (see the previous paragraph). It should be noted, however, that for Al with high Zn concentration (30 wt.%) a part of the Zn content was segregated at the grain boundaries [25]. Fig. 3b shows Zn concentration profile across a grain boundary as obtained by energy dispersive X-ray spectroscopy (EDS) in transmission electron microscopy (TEM). It is evident that the grain boundary was enriched with Zn and the full width at the half maximum (FWHM) of the concentration profile is about 3 nm.

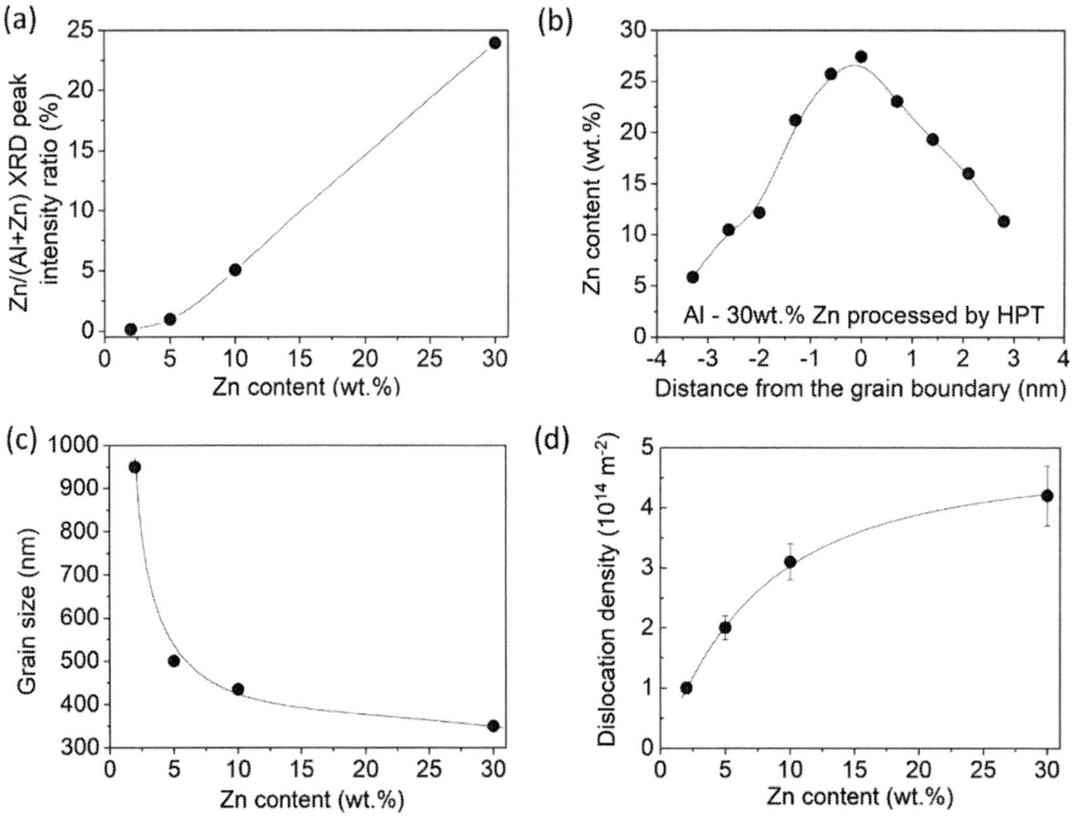

Figure 3. (a) The fraction of XRD intensity for the Zn phase versus the nominal Zn content in Al-Zn alloys processed by HPT until saturation. (b) The Zn content obtained by EDS in TEM as a function of the distance from a grain boundary in Al-30wt.%Zn alloy processed by HPT until saturation. (c) and (d) The grain size and the dislocation density, respectively, as a function of the nominal Zn content in Al-Zn alloys processed by HPT until saturation. The data were taken from [25].

The Zn content in Al processed by HPT has a significant effect on the microstructure of the Al phase. The average size of the Al grains decreased with increasing Zn content as can be seen in Fig. 3c. For the highest Zn concentration (30 wt.%) the grain size was refined to about 350 nm as determined by TEM [25]. The TEM images also revealed that the particles of the Zn phase can be found both in the triple junctions and the interiors of the Al grains. The size of the Zn particles inside the Al grains (5-15 nm) was much smaller than that in the triple junctions (100-200 nm). For low Zn concentrations, only small Zn particles in the grain interiors were observed. The dislocation density in the major Al phase was determined by X-ray diffraction line profile analysis [30]. Fig. 3d shows that the dislocation density increased monotonously with the Zn content. For the lowest (2 wt.%) and the highest (30 wt.%) nominal Zn concentrations, the dislocation density values were about 10^{14} m^{-2} and 4×10^{14} m^{-2}, respectively.

Fig. 4a shows the average hardness versus the nominal Zn content for the saturation state of the HPT-processed Al-Zn alloys [25]. For comparison, the hardness of the initial coarse-grained counterparts are also plotted. For low Zn concentrations (≤ 5 wt.%), the hardness was enhanced with a factor of about two due to HPT which can be attributed to the increase of the dislocation density. For the sample containing 10 wt.% Zn, this factor is lower, and in the case of the Al-30wt.% Zn alloy considerable softening was observed after HPT despite the fact that the highest dislocation density was formed in this material. In this case, the deformation-softening can be explained by the decomposition of the initial supersaturated microstructure and the increased role of grain boundary sliding during hardness testing since the occurrence of this mechanism requires lower stress than that for dislocation glide inside the grains. The dominance of grain boundary sliding in Al-30wt.% Zn alloy can be attributed to the increased amount of grain boundaries due to the small grain size and the formation of a Zn-rich layer in the grain boundaries. The latter effect can facilitate grain boundary sliding as the diffusivity of Zn atoms along Al/Al grain boundaries is high [25], thereby resulting in an easier plasticity (i.e., softening).

It should be noted that the dominance of grain boundary sliding during plastic deformation is often accompanied by a large ductility or even superductility. A large ductility can be achieved if the strain rate sensitivity of the material is high (0.2-0.5). Fig. 4b shows the strain rate sensitivity for the HPT-processed Al-Zn alloys determined by indentation creep technique [31] as a function of the nominal Zn content. A monotonous increase of the strain rate sensitivity with the Zn content can be observed and for UFG Al-30wt.% Zn alloy the value of the strain rate sensitivity became relatively high, suggesting the dominance of grain boundary sliding in accordance with the deformation softening shown in Fig. 4a.

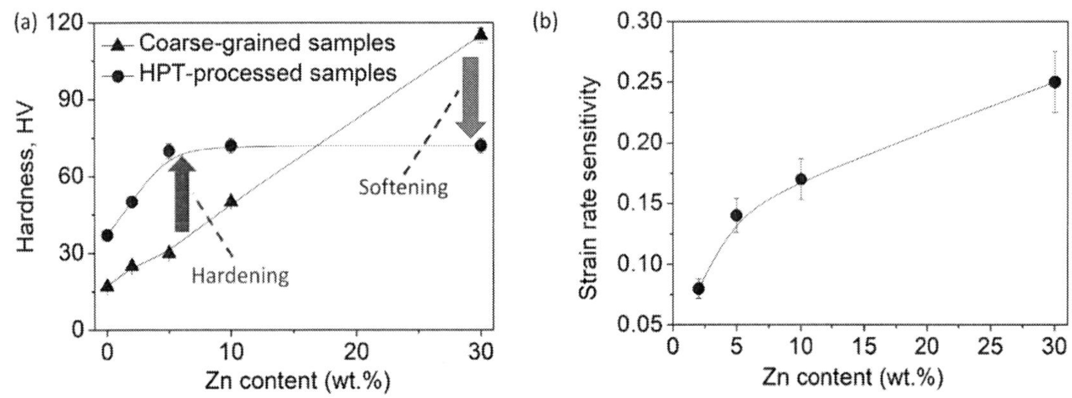

Figure 4. The hardness (a) and the strain rate sensitivity (b) versus the nominal Zn content in Al-Zn alloys processed by HPT until saturation. The data were taken from [25].

Deformation-softening was also detected in UFG Zn–22%Al and Zn–22%Al–0.3%Cu alloys with a duplex microstructure, i.e., in the material Al(Zn) and Zn(Al) solid solution phases coexisted [20]. Both phases had similar grain sizes of 300-400 nm. This microstructure formed during aging of a solution treated alloy at RT. The hardness values of the aged UFG Zn–22%Al and Zn–22%Al–0.3%Cu alloys were 400 and 900 MPa, respectively. Then, these materials were subjected to rolling at RT up to the thickness reduction of 90% which corresponds to an equivalent strain of 2.3. This deformation caused a decrease of the hardness from 400 to 200 MPa for the Zn–22%Al sample while from 900 to 450 MPa for the Zn–22%Al–0.3%Cu alloy. The grain size did not change during this rolling, however the character of grain boundaries was significantly altered since dislocations and vacancies were accumulated in the grain boundaries. This change may cause an easier grain boundary sliding, thereby resulting in the observed softening at RT.

4. Deformation-softening caused by the decay of the dense lattice defect structure

Nanocrystalline and UFG materials processed by bottom-up techniques often have microstructures with a very high density of lattice defects such as dislocations and twin faults. A part of these grown-in defects can disappear from the microstructure during plastic deformation of the as-processed samples. This effect was demonstrated on electrodeposited nanotwinned Cu and Ni alloys [32-34]. It was shown that cold rolling resulted in softening due to the decrease of the density of growth dislocations and twin faults. This reduction of defect density can be caused by recovery, detwinning or recrystallization. In the first two processes, the dislocations formed during plastic deformation interact with the growth dislocations and twin faults leading to their decay, while in the last process new defect-free grains are nucleated, leading to softening.

5. Summary and conclusions

This paper overviewed the existing knowledge on deformation-softening for UFG materials. It was found that this effect may occur during plastic deformation of UFG samples which were formerly annealed at moderate homologous temperatures for short times. The applied heat treatment can cause the annihilation of mobile dislocations, leading to anneal-hardening. Then, subsequent plastic deformation can create new mobile dislocations, thereby resulting in softening.

Deformation softening can also be observed in UFG materials having low melting points (e.g., Sn, Pb, Zn and In), processed by SPD at RT. In these samples, due to the grain size reduction the main deformation mechanism changes from dislocation glide in the grain interior to grain boundary sliding. As the latter mechanism requires lower stress, softening was detected. For Al having larger melting temperature, softening due to room temperature SPD can be observed only if the purity level is 6N or higher. In this case, the possible occurrence of grain boundary sliding and the high mobility of dislocations are the main reasons of deformation-softening. The very low impurity content leads to an elevated mobility of dislocations which then can be annihilated very easily at the grain boundaries in the SPD-refined microstructure.

It should be noted that deformation-softening was also detected in Al-Zn alloy with elevated Zn content (30 wt.%). In this case, a Zn layer was formed on the Al/Al grain boundaries which facilitates grain boundary sliding due the enhanced diffusivity of Zn along these boundaries.

The as-grown lattice defect structures in nanomaterials processed by bottom-up methods such as electrodeposition can be decayed during severe plastic straining by recovery, detwinning or recrystallization, resulting in deformation-softening.

Acknowledgements

This research was supported by the Hungarian-Russian bilateral Research program (TÉT) No. 2017-2.3.4-TÉT-RU-2017-00005. This work was financed partly by the Ministry of Human Capacities of Hungary within the ELTE University Excellence program (1783-3/2018/FEKUTSRAT).

References

[1] Kovács I and Zsoldos L 1973 *Dislocations and Plastic Deformation* (Pergamon Press, Oxford)
[2] Gubicza J, Chinh NQ, Krállics Gy, Schiller I and Ungár T 2006 *Curr. Appl. Phys.* **6** 194

[3] Valiev RZ, Zhilyaev AP, Langdon TG 2014 *Bulk Nanostructured Materials, Fundamentals and Applications* (John Wiley & Sons, Inc., Hoboken, New Jersey)
[4] Gubicza J 2017 *Defect Structure and Properties of Nanomaterials* (Woodhead Publishing, Duxford, UK)
[5] Huang X, Hansen N and Tsuji N 2006 *Science* **312** 249
[6] Chinh NQ, Szommer P, Horita Z and Langdon TG 2006 *Adv. Mater.* **18** 34
[7] Valiev RZ, Sergueeva AV and Mukherjee AK 2003 *Scripta Mater.* **49** 669
[8] Zeng W, Shen Y, Zhang N, Huang X, Wang J, Tang G and Shan A 2012 *Scripta Mater.* **66** 147
[9] Atwater MA, Bahmanpour H, Scattergood RO and Koch CC 2013 *J. Mater. Sci.* **48** 220
[10] Volkov AY and Kliukin IV 2015 *Mater. Sci. Eng. A* **56** 627
[11] Gubicza J, Pereira PHR, Kapoor G, Huang Y, Subramanya Sarma V and Langdon TG 2018 *Adv. Eng. Mater.* **20** 1800184
[12] Wang YM, Cheng S, Wei QM, Ma E, Nieh TG and Hamza A 2004 *Scripta Mater.* **51** 1023
[13] Hasnaoui A, Van Swygenhoven H and Derlet PM 2002 *Acta Mater.* **50** 3927
[14] Valiev RZ, Enikeev NA, Murashkin MY, Kazykhanov VU and Sauvage X 2010 *Scripta Mater.* **63** 949
[15] Vo NQ, Schafer J, Averback RS, Albe K, Ashkenazy Y and Bellon P 2011 *Scripta Mater.* **65** 660
[16] Huang X 2009 *Scripta Mater.* **60** 1078
[17] Lee DH, Lee JA, Zhao Y, Lu Z, Suh JY. Kim JY, Ramamurty U, Kawasaki M, Langdon TG and Jang J 2017 *Acta Mater.* **140** 443
[18] Lee SW, Han SM and Nix WD 2009 *Acta Mater.* **57** 4404
[19] Liu FX, Liu ZL, Pei XY, Hu JQ and Zhuang Z 2017 *Int. J. Plasticity* **99** 102
[20] Yang CF, Pan JH and Lee TH 2009 *J. Alloy. Compd.* **468** 230
[21] Edalati K and Horita Z 2011 *Mater. Sci. Eng. A* **528** 7514
[22] Srinivasarao B, Zhilyaev AP, Langdon TG and Perez-Prado MT 2013 *Mater. Sci. Eng. A* **562** 196
[23] Edalati K, Cubero-Sesin JM, Alhamidi A, Mohamed IF and Horita Z 2014 *Mater. Sci. Eng. A* **613** 103
[24] Ito Y, Edalati K and Horita Z 2017 *Mater. Sci. Eng. A* **679** 428
[25] Chinh NQ, Jenei P, Gubicza J, Bobruk EV, Valiev RZ and Langdon TG 2017 *Mater. Lett.* **186** 334
[26] Chinh NQ, Szommer P, Horita Z and Langdon TG 2006 *Adv. Mater.* **18** 34
[27] Kawasaki M 2014 J. Mater. Sci. **49** 18
[28] Kawasaki M, Figueiredo RB, Huang Y and Langdon TG 2014 *J. Mater. Sci.* **49** 6586
[29] Popovic S and Grzeta B 1999 *Croatica Chemica Acta* **72** 621
[30] Gubicza J 2014 *X-ray line profile analysis in Materials Science* (IGI-Global, Hershey, USA)
[31] Chinh NQ and Szommer P 2014 *Mater. Sci. Eng. A* **611** 333
[32] Ungár T, Li L, Tichy G, Pantleon W, Choo H and Liaw PK 2011 *Scripta Mater.* **64** 876
[33] Niu R and Han K 2013 *Scripta Mater.* **68** 960
[34] Ni HT, Zhu J and Zhang XY 2017 *J. Alloy. Compd* **693** 226

IOP Conf. Series: Materials Science and Engineering **903** (2020) 012042 doi:10.1088/1757-899X/903/1/012042

Preparation of soft magnetic composite from Fe-6.9wt%Si by different heat treatment strategies.

Bence Kocsis[1,2], Lajos Károly Varga[1], Ibolya Zsoldos[2]

[1] Wigner Research Center for Physics Institution for Solid State Physics and Optics, 29-33 Konkoly-Thege út, Budapest 1121, Hungary
[2] Széchenyi István University Department of Materials Science and Technology, 1. Egyetem tér, Győr 9026, Hungary

E-mail: kocsis.bence@sze.hu

Abstract. Present study investigated the effect of isothermal heat treatment strategies between 800 °C and 1150 °C on the magnetic properties of toroidal samples made from Fe-6.9wt%Si powder. The samples were prepared by classical powder metallurgy method since the classical sheet forming methods no longer work with the high silicon content. Our results presented here are part of a series of comparative experiments where we study the effectiveness of the insulating layers created during and before the compacting of soft magnetic composites (SMCs). Our goal was to create a soft magnetic composite made of ferromagnetic and inorganic insulating material with a frequency limit already in the megahertz range and a Snoek limit of few gigahertz. In the case of samples made from Fe-6.9wt%Si powder, the computed tomography results showed that significant porosity is to be expected after pressing. Its positive effect occurred during the heat treatment in the atmospheric agent, where silicon is precipitated and deposited on the surface of the particle. This coating is an electrically insulating layer at the grain boundaries. Depending on the heat treatment strategy, 1 or 2 ferromagnetic phases were observed. The frequency limit approached the target values, but due to the low value of static permeability, the Snoek limit did not reach the gigahertz range. However, there is a significant improvement in magnetic properties compared to the heat-treated samples in a protective gas.

1. Introduction

According to today's demands of high efficient, energy-saving power electronics (motors, transformers), the conversion and transfer of electromagnetic energy are possible only with the use of advanced soft-magnetic materials because they have high saturation magnetization and low magnetic loss [1]. They also have unique magnetic properties, including stable magnetic permeability at higher frequencies and magnetic anisotropy. Fe-Si is a typical type of soft magnetic alloys that has excellent soft magnetic properties but still exhibits a drastic increase in eddy current loss when excitation frequency is above 400 Hz [2]. Soft magnetic composites (SMCs) consist of a core-shell structure where the core is made up of a ferromagnetic particle and an electrically insulating layer that completely covers it. They are typically manufactured using classical powder metallurgy techniques, but recently, the 3D printed parts are beginning to appear. The main steps in classical powder metallurgy production are powder production by atomization or milling, powder pressing and final heat treatment. In order to reduce the stress anisotropy, which is caused by the high forces awaking during pressing, various heat treatment methods are used. During 3D printing, the amount of heat transmitted by the laser generates significant residual stress in the structure, which can also be eliminated by heat treatment.

In this study, the most widely used magnetic material in the industrial application, Fe-Si alloy was used, within it, Fe-6.9wt%Si with outstanding properties [3]. Differently, to the usual procedure where the coating with insulating layer precedes the compaction, our aim is to create an oxide insulating layer around the ferromagnetic particles by heat treatment after the powder compression. The advantage of this inorganic coating is that it remains stable at higher temperature ranges as opposed to organic insulators. This is needed both for special applications and for miniaturization of power electronic equipment. This can be done by increasing the frequency limit of the iron core without loss of performance. In addition, if we can minimize dissipation losses, we can improve the energy balance of the unit by either eliminating cooling or reducing performance. Expressed in terms of the energy stored in the coil (L – inductance, I_{max} – maximum current flowing in the coil) and the magnetic field (where μ_0 is vacuum permeability), we obtain the ratio of effective permeability (μ_{eff}) to volume (V), which shows that maximal permeability and volume reduction give maximum saturation induction (B_{max}).

$$\frac{\mu_{eff}}{V} = \frac{{B_{max}}^2}{\mu_0 \cdot {I_{max}}^2 \cdot L}$$

We also aimed to achieve some gigahertz values of Snoek limit, which is a number for the characterization of soft magnetic composites. In this case, the complex permeability spectrum was obtained by multiplying the static permeability and the frequency limit. This value includes the physical characteristics that are important to our development.

2. Material and Methods

2.1. Material

Iron-silicon alloys are one of the most prominent known soft magnetic materials in the fields of power electronics, telecommunications, military industry, and vehicle manufacturing. When the alloy reached 6.5-6.9wt%Si content (high silicon electrical steel) it has excellent soft magnetic properties such as high permeability, low magnetic loss and near-zero magnetostriction. Above 3wt% Si, classical rolling technologies no longer work because the plastic deformation of the material is close to zero at room temperature [4]. In this case, the Fe-6.9wt%Si bulk material was produced by a home built cold-crucible induction melting equipment. The ingot of about 20 grams was powdered by grinding with a SPEX 8000 M ball mill machine. The resulting wide-size distribution of powder was separated on a sieve shaker. We were used to the smaller powder particles than 74 μm to the final sample preparation. After the production of powder raw material, the morphology and chemical composition of the particles were investigated. Scanning electron microscopy (SEM) images of Fe-6,9wt%Si powder are illustrated in Figure 1.

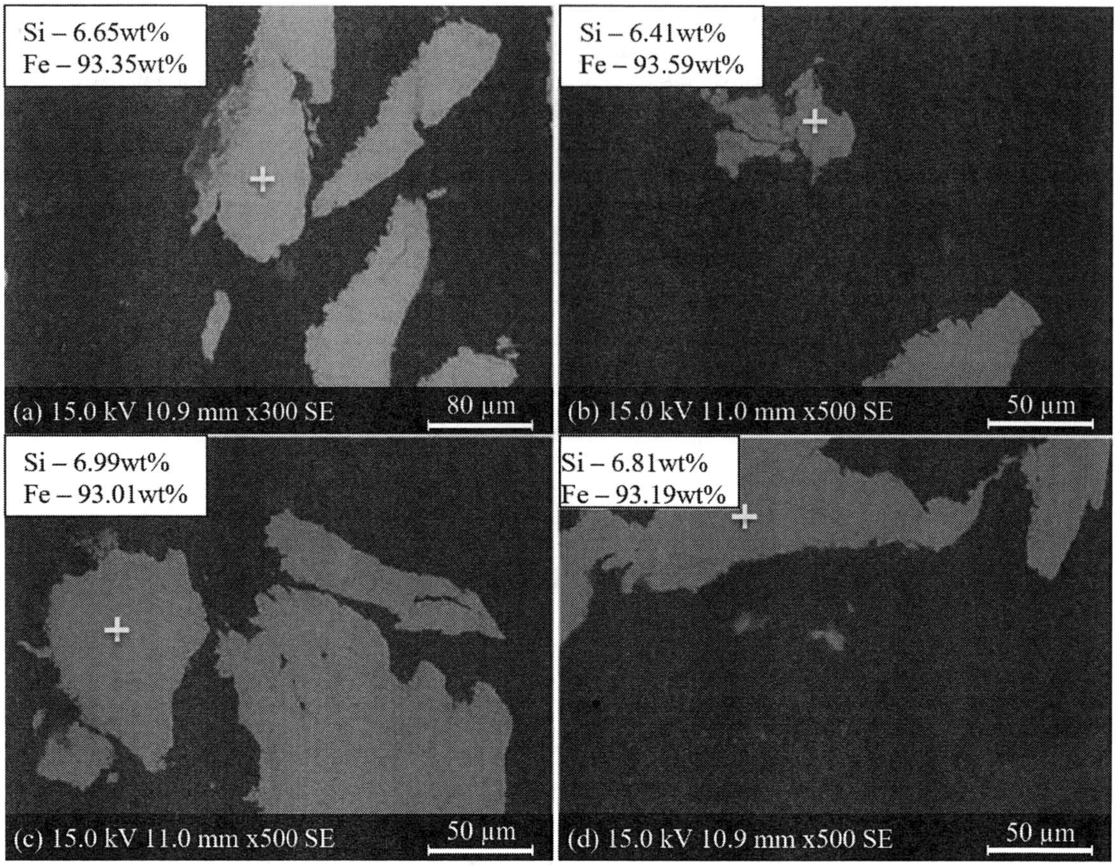

Figure 1. Fe-6,9wt%Si powder composition analysis by scanning electron microscope

2.2. Methods

Toroidal samples (D = ~ 12 mm, d = ~ 7mm; h = ~ 2.5 mm) were prepared in two steps. In the first step, the powder was compressed using a conventional hydraulic press at 1.34 GPa. In the second step, after the pressure was released, the samples were subjected to oxidation heat treatment in a chamber furnace. This procedure was always started in a heated furnace, maintaining the isotherm conditions. Heat treatment temperatures were 800 °C; 900 °C; 1000 °C; 1100 °C; 1150 °C. The annealing was repeated in the air atmosphere and in the argon shielding gas. We also examined the effect of the duration of heat treatment on the magnetic properties of the SMC. Table 1 summarizes the different heat treatment strategies.

Table 1. Heat treatment strategies

	800 °C	**900 °C**	**1000 °C**	**1100 °C**	**1150 °C**			
heat treatment time	3 h	3 h	3 h	3 h	1 h	3h	3h	5h
atmosphere	air	air	air	air	air	air	Ar	air

Heat treatment at 800 °C and 900 °C for 3 hours were performed in an air atmosphere, however, the samples did not sinter properly due to the relatively low temperature and were broken during handling.

Among the magnetic properties, we analyzed the complex permeability spectrum in the frequency range of 0.1 – 30 000 kHz, from which the product of the static permeability and the frequency limit gives the so-called Snoek limit. The Snoek limit can be considered as a figure of merit for comparing the SMCs from a magnetic point of view. Saturation magnetization was not part of our current measurement series.

3. Results and discussion

Scanning electron microscopy was used to verify the effect of heat treatment strategies. The particle morphology, coating, composition and the changes of composition were also examined.

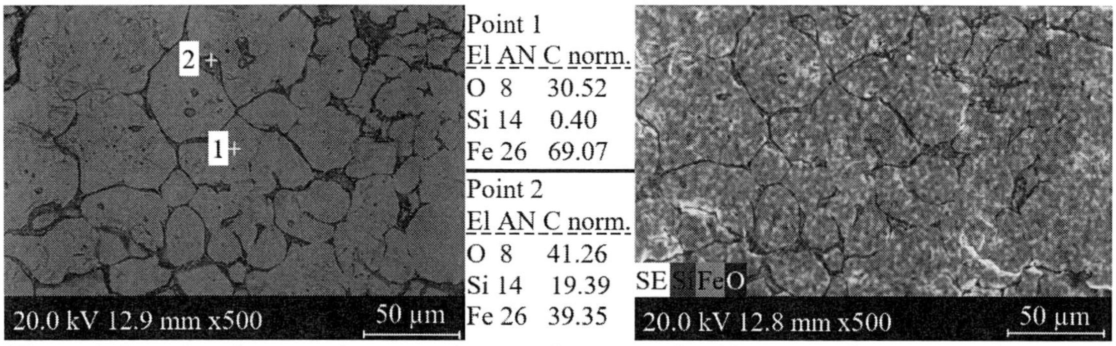

Figure 2. Scanning electron microscopy images of compacted and heat treated Fe-6,9wt%Si sample (t = 1 h, T = 1150 °C on air). Point 1 indicates the core and point 2 indicates the location of the point analysis of the shell structure. The figure on the right shows the elemental mapping results

Based on the SEM images it can be stated that a core-shell structure was obtained which core is Si-poor Fe. However, silicon and oxygen-rich coating appears on the grain boundaries and acts as an electrically insulating material. This silicon diffuses from the pre-alloyed powder particles to the surface of the grains (Fig. 2., 2. point). In addition to the silica layer on the porous sites and on the surface of the toroidal sample, a significant volume fraction of iron oxide coating is also observed. A sample (1100 °C, 3h) which was prepared by the same conditions, except the heat treatment atmosphere, it was argon, did not show any precipitation or composition change.

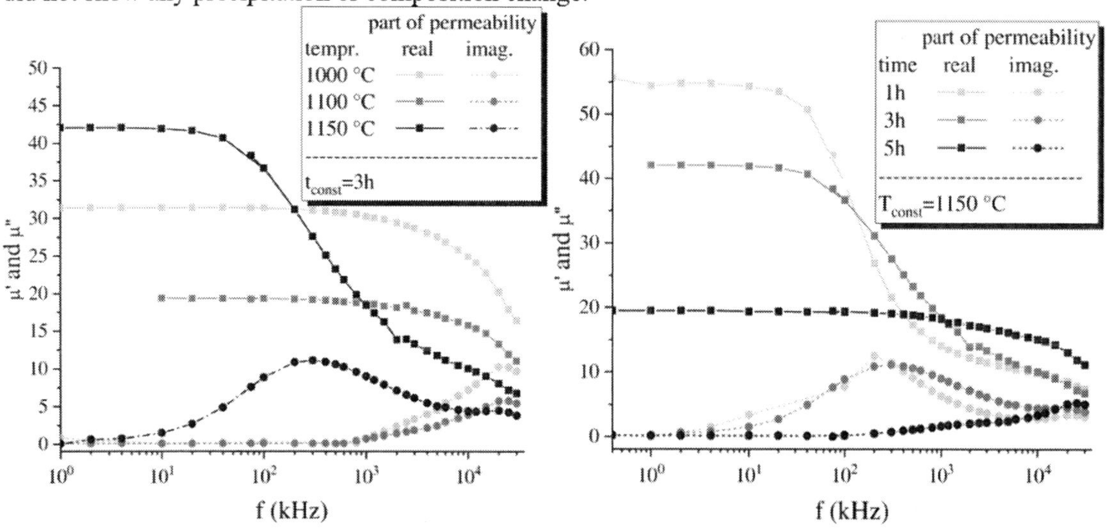

Figure 3. The heat treatment temperature (left) and time-dependence (right) of complex permeability spectra

The magnetic permeability spectra of toroidal samples were measured using the Agilent RLC meter type 4294 A measuring the series equivalent inductance (L_s) and resistance (R_s) data. The different exciting field amplitudes current in a wide frequency range between 100 Hz and 30 MHz [5]. The sample treated in argon shielding gas had a relatively high static permeability (μ_s= 954), but it's limit frequency was only f_{lim}= 1 kHz. The Snoek limit obtained by multiplying these values is only 954 kHz since SMC had only one component, which had not electrically insulating layer that separates the particles.

On the other hand, SEM images of air-annealed samples showed a composite of at least 2 components. In this case, it is clear from the permeability spectra in Figure 3, that the temperature values of the heat treatment strategies have no effect on the frequency limit. The reason for this is that in all cases a sufficient size of the insulation layer is formed. However, the static permeability values decreased from the previous 954 value to 20-55 as a result of heat treatment in air. This is presumably due to the broadening of the B-H curve. Due to the appearance of insulating layers, which resulted in a drastic reduction in iron loss, at the same time increased the internal demagnetization (N) factor ($\mu_{eff} = \frac{1}{\frac{1}{\mu}+N}$), causing a significant decrease in effective permeability.

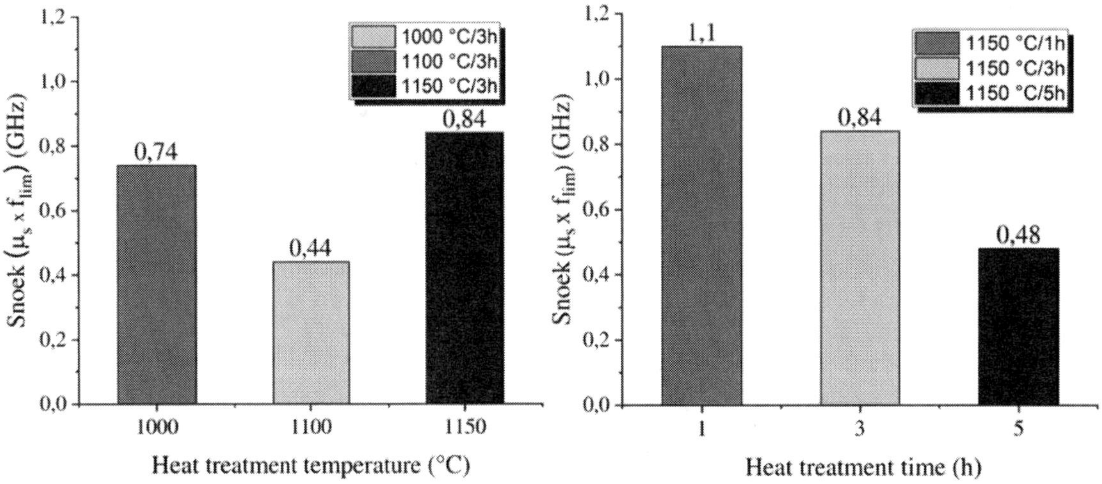

Figure 4. Snoek limit evolution as a function of heat treatment temperature and period of time in air atmosphere

Figure 4 shows the evolution of the Snoek limit in each case. In contrast to the 954 kHz sample treated with argon, the best sample treated with air was found to be 1.1 GHz. It should be mentioned that at 1150 °C, two magnetic "phases" can be observed in our sample after 1h, 3h, and 5h heat treatment. This is well illustrated by the local maxima of the imaginary part of the permeability spectrum. The 1.1 GHz Snoek limit came from the second higher limit frequency. If these values are not taken into account, the sample heat-treated at 1000 °C for up to 3h proved to be the best (Snoek limit = 0.74 GHz). It is also clear from this index that the heat treatment in the air has a positive effect on the magnetic properties of the samples.

4. Conclusions

After these experiments, it turned out that self-sustaining toroidal samples can be sintered by heat treating in the air at 1150 °C. The silicon content of the particles is reduced and a silicon oxide insulating coating is formed at the interface of the particles. It can be converted in situ to soft magnetic composites. Developing a uniform and thin coating and controlling the Si content of the particles is a rather complex task. SEM images revealed that the internal oxidation at the grain boundaries was not complete and consequently, there were regions in the sample where the powder particles were not electrically insulated from each other. This heterogeneity causes the presence of two magnetic "phases" with different insulations and different demagnetizing factors and different frequency limits. The samples treated at 1150 °C proved to be the best soft magnetic iron cores based on their Snoek limit value. With respect to the frequency limit, the temperature of the heat treatment has no significant effect, in each case approximately a value of 25 MHz was obtained. By increasing the duration of the heat treatment (see Fig. 3 right), the ratio of the first magnetic "phase" with a lower limit frequency is significantly reduced. Even with a further heat-treated sample, this magnetic phase can be completely eliminated.

In the future new ways of internal oxidation will be experimented under flowing O_2+N_2 and O_2+Ar and O_2+CO_2 gas mixtures preventing the "burning" of the samples during the long duration heat treatment necessary for homogeneous internal oxidation of each powder particles. In further experiments, we will sinter a pre-oxidized Fe-Si powder, the oxide layer of which is chemically formed under controlled conditions by the Stöber method [6]. New types of alloys and insulating materials will be introduced in the series of experiments, which will be compared with 3D printed toroidal samples.

Acknowledgements
Present study was supported by EFOP-3.6.2-16-2017-00016 "Dynamics and control of autonomous vehicles meeting the synergy demands of automated transport systems".

References

[1] Luo Z, Fan X, Hu W, Luo F, Wang J, Wu Z, Liu X, Li G and Li Y 2019 Properties of Fe2SiO4/SiO2 coated Fe-Si soft magnetic composites prepared by sintering Fe-6.5wt%Si/Fe3O4 composite particles *J. Magn. Magn. Mater.* 166278
[2] Wu Z Y, Jiang Z, Fan X A, Zhou L J, Wang W L and Xu K 2018 Facile synthesis of Fe-6.5wt%Si/SiO 2 soft magnetic composites as an efficient soft magnetic composite material at medium and high frequencies *J. Alloys Compd.* **742** 90–8
[3] Garibaldi M, Ashcroft I, Hillier N, Harmon S A C and Hague R 2018 Relationship between laser energy input, microstructures and magnetic properties of selective laser melted Fe-6.9%wt Si soft magnets *Mater. Charact.* **143** 144–51
[4] Mo Y, Zhang Z, Pan H and Xie J 2016 Improved Plasticity and Cold-rolling Workability of Fe-6.5wt%Si Alloy by Warm-rolling with Gradually Decreasing Temperature *J. Mater. Sci. Technol.*
[5] Varga L K and Kovac J 2018 Decomposing the permeability spectra of nanocrystalline finemet core *AIP Adv.* **8** 047205
[6] Stöber W, Fink A and Bohn E 1968 Controlled growth of monodisperse silica spheres in the micron size range *J. Colloid Interface Sci.* **26** 62–9

IOP Conf. Series: Materials Science and Engineering **903** (2020) 012043 doi:10.1088/1757-899X/903/1/012043

Root cause analysis of metallurgical defects in continuous cast steel slabs at ISD DUNAFERR Zrt.

K Wizner[1] and A Kővári[2]

[1] Technology department senior associate, ISD Dunaferr Zrt. Dunaújváros, Hungary
[2] University of Dunaújváros, Dunaújváros, Hungary

E-mail: wizner.krisztian@isd-dunaferr.hu

Abstract. As market demands tighten, ISD Dunaferr Zrt. is placing increasing emphasis on improving the quality of its products. Line Scanner equipment used in rolling mills can identify most surface defects, but the majority of metallurgical faults are located beneath the surface (e.g. inclusions, cavity, gas bubbles, etc.). There is no standardized device for detecting them, only ultrasonic and microscopic testing is possible. In addition, it should be noted that some of the defects detected on the surface (e.g. surface cracks) can only be determined by microscopic testing, whether they are metallurgical or due to the parameters of the rolling process.
Several pilot programs have been implemented to explore the causes. In this theme, the most significant pilot programs, as well as the evaluation of their results will be presented.

1. Introduction

As part of the continuous product development in ISD Dunaferr Zrt, several pilot programs are implemented. These programs are basically divided into three groups: testing of new raw materials, product development / test production, quality development / problem exploration. Due to the numerous technological steps, both metallurgical and rolling phases are often involved in the experiments. This study primarily presents the metallurgical aspects of experimental programs in the context of recent quality improvement and problem-solving programs.

2. Examining the causes of surface defects attributable to casting powders

The casting powder used in continuous casting performs many tasks [1]. It seals the steel surface from the environment's oxygen, provides lubrication and better heat flow between the strand and the mould surface. There is extensive literature dealing with the choice of the right powder, but due to the complex function of the powder and the different conditions in each plant, it is selected through field tests even today. Incorrect choice increases the amount and severity of surface defects originated from the mould. In the worst case, it can even cause outbreak of the strand.

Metallographic examinations of the samples of defective parts of rolled products have confirmed the metallurgical origin of defects by measurements and images, most likely due to casting powders, which were marked on the slab surface maps.

The macroscopic nature of the ruptures differs from the nature of the line defects, but in most cases penetrated oxide and decarburization around the ruptures are visible. All this suggests that the errors are originated from the slab. In some cases, mechanical damage to the slab cannot be excluded, which could have caused the surface defect.

Scanning electron microscopy in the vicinity of the defects revealed the presence of alkaline oxide elements in several cases, confirming that the origin of the defect is casting powder.

During the period under investigation Dunaferr had two casting powder for regular use (S1, PG) and two for test production (SR, AX).

Table 1 Used casting powders in Dunaferr

Casting powder code	Steel type	Usage	Investigated heat pcs	Downgraded (t)	Production (t)	Downgraded (%)
S1	Alloyed	Regular	200	1,745	3782,345	0,05%
PG	Non-alloyed	Regular	200	0	3748,22	0,00%
SR	Alloyed	Test	27	52,185	521,124	10,01%
AX	Non-alloyed	Test	42	192,336	917,323	20,97%

In the case of the casting powder used for the two non-alloyed steels (PG, AX), it is clear that the ratio of downgrading is much higher for the heats cast with the AX trial casting powder than those with the PG casting powder.

Based on the flame-scarfing inspection of the slab, the following defects were found on the surface (Table 2):

Table 2 Average surface defects on slabs

Surface	Casting powder	Inspected (pcs)	Left side			Central curve			Right side		
			Gas bubble (pcs)	Starch crack (pcs)	Longitudinal crack (pcs)	Gas bubble (pcs)	Starch crack (pcs)	Longitudinal crack (pcs)	Gas bubble (pcs)	Starch crack (pcs)	Longitudinal crack (pcs)
A	S1	8	1,25	1,38	0,00	2,88	0,00	0,00	3,25	3,13	**0,25**
	PG	16	0,88	0,44	0,06	0,63	0,19	0,00	0,81	0,75	**0,31**
	SR	2	**9,50**	**24,00**	0,00	0,00	**6,00**	0,00	0,00	**12,00**	0,00
	AX	19	0,42	2,89	0,05	0,16	**4,32**	0,11	0,37	3,37	0,00
B	S1	10	1,20	2,00	0,00	0,20	0,90	0,00	2,20	**7,00**	0,00
	PG	18	0,61	0,17	0,00	0,33	0,17	**0,17**	0,44	2,00	**0,22**
	SR	3	**7,00**	0,67	0,00	1,33	**4,00**	0,00	0,00	4,00	0,00
	AX	20	0,75	1,60	**0,25**	0,30	**5,95**	0,05	1,75	**9,75**	0,00

*__Bold__ indicates extremely high values

Both sides of the slab have not been inspected in all cases, so this is the reason of difference between the pieces tested on „A" and „B" sides.

2.1. Result of inspection

From the data, it is clear that slabs cast with AX and SR have the most metallurgical defects (typically starch cracking).

Although the surface inspection was carried out much less times with slabs cast with SR and AX casting powder, the results are consistent/in line with the statistics of downgradement.

The samples taken from 2 coils of tested heats confirmed that most of the surface defects were already present in the pushing furnace, where they opened, and were later rolled in. Other surface defects could be attributed to scale rolling-in or mechanical damage.

3. Inspection of casting powders

In the light of the results described in the previous section, the purpose of the pilot program is to develop an assay system that allows for the detection of inadequate texture / composition of casting powders. This system makes possible to eliminate errors due to the consistency / composition of the casting powder. Our study is primarily focused on the crystal structure, chemical and granulometric composition of the casting powders used.

In practice, incoming casting powders are randomly inspected by visual inspection. The casting powder delivered to the plant is sampled in every 25 tonnes. These samples are subject to sieve analysis. Casting powders are subject to chemical analysis only at random. Based on the examinations carried out so far, the majority of the supplied casting powders complied with the data on the certificate. The most typical deviation from the certificate was the granulometric composition (too high powder fraction). Based on the examinations so far, the composition can be considered as permanent and conforms to the certificate. The mineralogical composition study (X-ray diffraction) of casting powders has not been done yet at Dunaferr.

The investigation of casting powders, as already used in the literature [2, 3] and in the operating metallurgical plants, involved the observation of the parameters forming the essential properties of the casting powders.

During the experiment, the samples were taken from 158 different big-bags. Out of a total of 56 pcs of PG, 47 pcs of S1, 3 pcs of AX and 1 pc of SR casting powder, due to the low number of samples, measurement data of AX and SR casting powders were not used.

3.1. Physical and chemical homogeneity test

To qualify casting powders, it was important to investigate whether granulometric segregation or chemical composition [4, 5] changes occurred within big-bags. Samples were taken from the bottom emptied big-bags for testing at the beginning of use and approx. 90% of usage.

3.1.1. Granulometric homogeneity test

The statistical (t-test) results of the comparison of the first and second sample series of big-bags are shown in Table 3. Based on the results it can be stated that the granulometric composition of the samples can be considered to be the same, thus the examined powders are not segregated for different size fractions during filling, transport or storage.

Table 3 Two-sample t-test results (t critical two-tailed)

Casting powder			Granulometric size fraction (mm)					
			>1,0	**1,0-0,5**	**0,5-0,25**	**0,25-0,125**	**0,125-0,063**	**<0,063**
PG	t value		-0,343	-1,509	1,335	-0,714	-0,483	1,294
	variance	Equal	1,981	1,981	1,981	1,981	1,981	1,981
		Non-equal	1,986	1,993	1,983	1,981	1,982	1,982
S1	t value		0,313	0,227	-0,712	0,175	1,260	1,027
	variance	Equal	1,986	1,986	1,986	1,986	1,986	1,986
		Non-equal	1,986	1,986	1,988	1,988	1,990	1,988

3.1.2. Chemical homogeneity test

Based on the samples taken from the first 9 big-bags, similarly to the granulometric analysis, it was found that the samples did not show significant chemical differences, so from a chemical point of view the contents of the big-bags can be considered homogeneous.

3.2. Characteristic granulometric distribution of casting powders

Based on the results of the t-test, it is possible to determine the typical granulometric composition for the types of cast powder tested.

Based on the granulometric composition (Table 4) for both casting powders, the powder fraction (<0.063 mm) can be considered to be about 1%.

The average amount of the fraction below 0.125 mm for each casting powder does not exceed 10%.

Table 4 Characteristic granulometric composition of casting powder samples (w%)

Casting powder	Sample	Granulometric size fraction (mm)					
		>1,0	1,0-0,5	0,5-0,25	0,25-0,125	0,125-0,063	<0,063%
PG	1	0,00%	3,96%	62,88%	25,83%	6,44%	0,89%
	2	0,01%	3,37%	63,86%	26,35%	5,65%	0,76%
S1	1	0,01%	4,70%	65,22%	23,66%	5,59%	0,82%
	2	0,01%	9,77%	68,83%	17,87%	2,35%	1,17%

For both PG and S1 casting powders, there was a case where the granulometric fraction was significantly different from the above. For S1, the proportion of powder fraction on one occasion (<0.063 mm) was 80%. For both casting powders, it is observed in some cases that the fraction above 0.5 mm greatly increases. These phenomena can cause changes in the properties of the slag formed from the casting powder (especially higher dust content) which later can cause the appearance of surface defects on the slabs.

3.3. Chemical composition inspection

Casting powder samples were analyzed by different measurement methods according to the variety of components.

The C content of the casting powder significantly influences the rate of its melting [6], so it initially seemed appropriate to measure the organic, inorganic C content separately and the total C content.

Based on 31 big-bag samples, it was observed (~50% PG, ~50% S1 casting powder) that the C content of PG casting powder was significantly derived from organic carbon, while S1 casting powder contained higher levels of inorganic carbon. Based on the analyzes performed using the LECO RC-412 instrument, the total carbon content of both casting powders was considered to be constant.

In the remaining samples, only the total carbon content was measured with a simpler solution – using Horiba EMIA-320V instrument.

Based on the measurement results, it was observed that the **results of the two measuring instruments showed ~ 1% difference. Due to this difference, a consultation has been initiated between the supplier and Dunaferr's laboratory to harmonize the test methods (measuring instruments, methods, measuring points, etc.).**

Based on the normalized standard deviation of casting powder components (Fig. 1), it is observed that the standard deviation of the components [7] (C, Na_2O, MgO, Al_2O_3, SiO_2, K_2O, CaO, Fe_2O_3) having a significant effect on the casting powder quality is low. For MnO and F-, the value of the normalized standard deviation is slightly higher, which is due to the low values of the concentration and the inaccuracy of measurement.

Results from Dunaferr's laboratory show that the tested elements do not show a high degree of standard deviation.

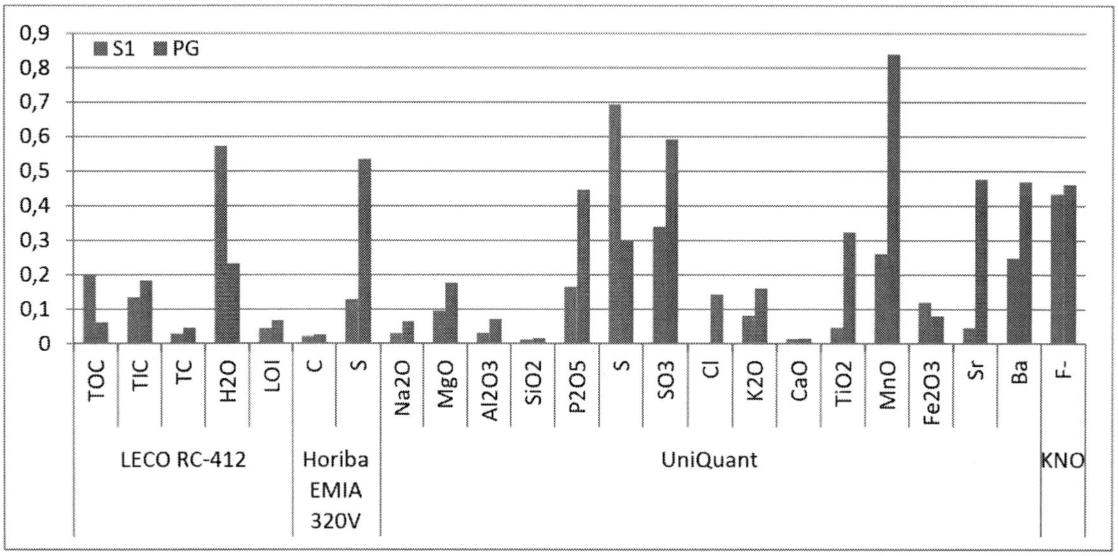

Figure 1 Normalized standard deviation values for the tested components

3.4. Measurement of casting powder slag thickness

According to the literature [8], the thickness of the slag formed from the casting powder is in relation with the conditions of the casting equipment and varies within narrow limits.

According to the data measured on the casting machine, the typical slag thickness for both casting powders (PG, S1) is ~ 9.2 ± 0.8 mm. Under operating conditions the measured values can be considered constant and no correlation can be found with the chemical or granulometric composition of the casting powder.

3.5. Changes in physical and chemical properties over time (from manufacture to use)

S1 casting powders were used after an average of 158 days, and PG casting powders were used after an average of 66 days.

Assuming homogeneous physical and chemical composition at the time of casting powder production and the fact that both casting powders were homogeneous within the big-bag at the time of sampling, it can be concluded that the elapsed time does not significantly influence the granulometric or chemical composition of the casting powder. However, it should be noted that casting powders contain compounds in a complex form (similar to hot metal and steel slags) and these compositions may change over time while the chemical composition of the casting powder remains unchanged.

Due to the homogeneity of the granulometric and chemical composition and the consistency of the molten casting powder, **no significant crystal structure differences can be expected between the various types of powder and therefore X-ray diffraction examination of the crystal structure was not justified.**

3.6. Result of the investigation

The investigation was based on 185 casting powder samples from big-bags supplied by 4 different manufacturers. Due to the low number of samples (4 pcs), casting powders from two manufacturers were excluded from the evaluation. Based on the results, it can be stated that the granulometric and chemical composition of the casting powders were homogeneous by manufacturer. The slag thickness was constant for both casting powders. In the light of the results, no X-ray diffraction crystal structure test was performed. The two casting powders are not comparable because PG is primarily used for non-alloyed steels, while S1 is used for alloyed steels. However, the results of the experiments provide a good basis for testing further new casting powders.

4. Exploration of metallurgical surface defects on continuous casted slabs

The aim of the investigation is to explore the effect of „constant" conditions, which do not change during casting, but may affect the surface texture of the strand. During the casting, the „non-constant" effects (change in casting speed, interventions in either the ladle or the tundish, mould level fluctuations, etc.) are not dealt with in detail in this investigation.

The investigation examined the sectional enrichments of 7 slabs of 5 heats, parallel to the surface of the flame cutting, in the distance of 100 mm. On the surface, 7x75-90, in total 525-630 analyzes were performed, of which 2x75-90 were as close as possible to the wide side. The sample was divided into 6 sections for ease of handling, which were further subdivided for more precise determination of the test points (Figure 2). In each „square", 3 measurements were taken in the center line parallel to the wide side. In the „squares" near the wide side, further analysis was carried out as close to the wide side as possible. Figure 3 shows a section of an examined slab divided into units.

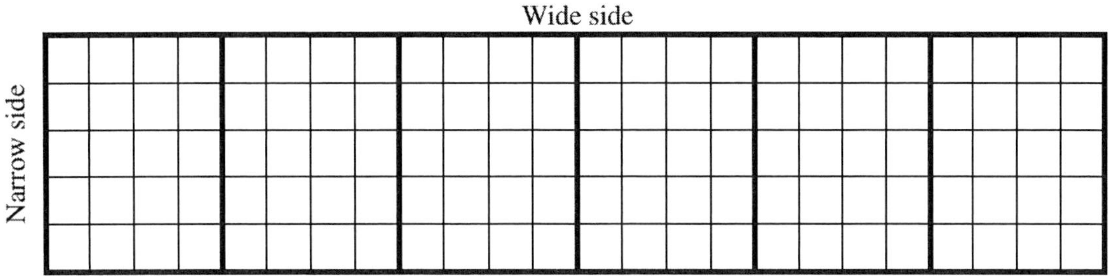

Figure 2 Division of the examined section into units of measurement

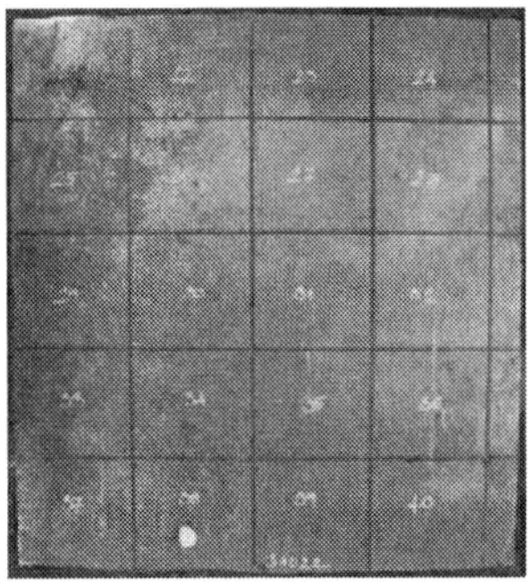

Figure 3 1/6th part of the examined surface

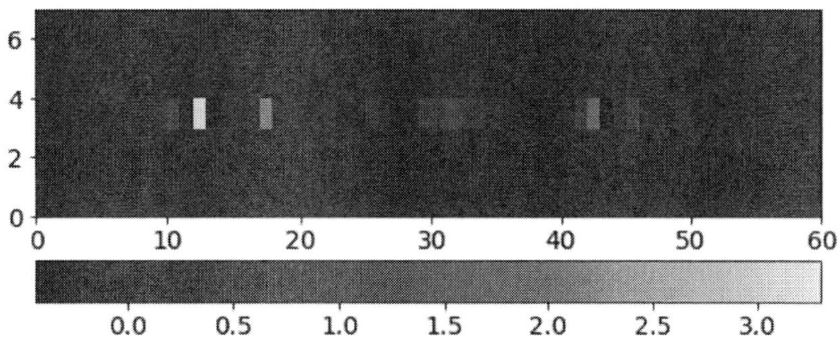

Figure 4 Deviation from average concentration

Based on the examination of the results, the cross-sectional composition of the slabs is similar in all cases (Fig. 4), regardless of the different production conditions. Both the spectroscopic chemical analyzes and the Baumann sulfur imprints made during the experiment clearly support the midline enrichment and the presence of stronger enrichment centers at the midline from the narrow edges, but no significant enrichment was observed near the surface of the slabs.

These results indicate that apart from the midline enrichment, the cross-section of the slabs can be considered homogeneous. Because of the chemical homogeneity, it is likely that downgrading is not attributable to the chemical composition / homogeneity of the cast slab.

In the reheating furnace, the surface of the slabs is oxidized and removed as scale before rolling. Copper dissolved in the iron matrix does not oxidize and can therefore be enriched on the surface of the slabs in the course of heating. The effect of the furnace goes beyond the present study.

Slight surface enrichment of trace elements in steel may also cause surface defects [9]. Based on the analysis of trace elements, it showed similar results as other alloys. Unfortunately, for several elements (Sb, Sn) the accuracy of the spectroscopic measuring equipment was not sufficient to examine the distribution of the elements. Based on the results of the investigated elements (Cu, As) it can be seen (Fig. 5, Fig. 6) that, like other elements, their distribution is almost completely homogeneous.

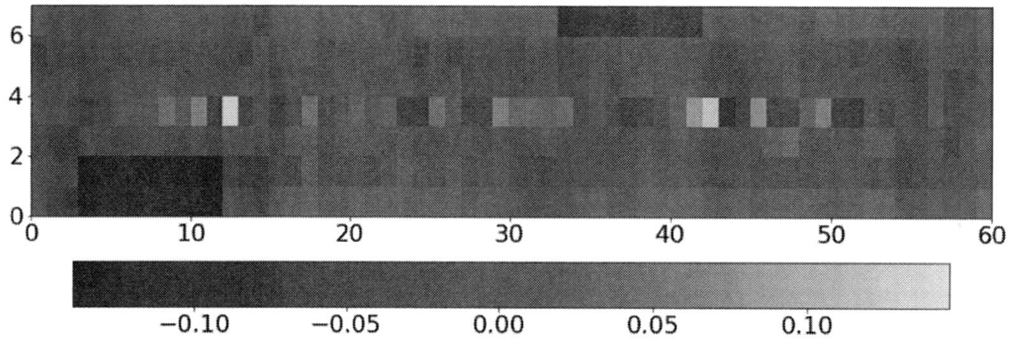

Figure 5 Deviation from average concentration (Cu)

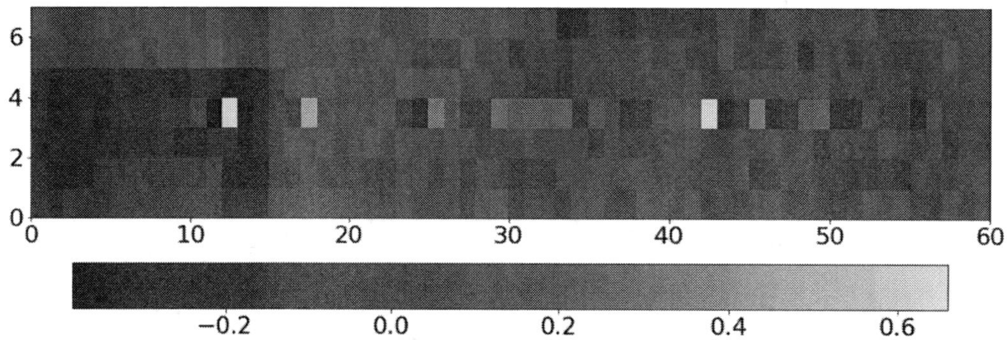

Figure 6 Deviation from average concentration (As)

According to the literature, the surface quality of steel can be greatly influenced by the concentration of some trace elements near the surface. In the case of surface defects, copper is the most important element. Copper enrichment is primarily responsible for the formation of surface scales and surface cracks, and above a certain temperature it forms a liquid phase which causes surface cracks. This liquid phase occurs when the concentration of copper in the austenite exceeds the solubility limit [10]. Copper precipitation typically occurs during reheating. Some elements affect the solubility of austenite in copper, thereby counteracting or enhancing the effect of copper on surface quality. The copper solubility of austenite is improved by its nickel content. The 2:1 Ni:Cu ratio completely eliminates the copper surface defect. Even with a 1:1 ratio, surface cracking can be prevented, but this depends on the reheating temperature.

In steel, Sn and Sb significantly reduce the copper solubility in austenite (up to a third by Sn).

4.1. Result of the investigation

A cross-sectional chemical study of the slabs shows that apart from the midline enrichment, the slabs are considered to be homogeneous. Because of the chemical homogeneity, it is likely that downgrading is not attributable to the chemical composition / homogeneity of the cast slab. However, the slab to be rolled may be subject to heat effects in the reheating furnace which can cause enrichment in the surface layers of the slab.

Primary downgrading cannot be unambiguously paralleled with the parameters examined due to the small number of samples (7 pcs slabs).

5. Summary

Implementing and evaluating pilot programs based on factory measurements is an integral part of Dunaferr's continuous quality and product development. The experiments described above provide only insight into the numerous experimental programs to reduce defects of metallurgical origin.

Studies have shown that a large proportion of the surface defects on the slab are of casting powder origin. Most of the surface defects were already present in the pushing furnace and, when opened, were later rolled in. The remainders were due to scale rolled in or mechanical damage.

For two experimental casting powders (AX, SR), an increased amount of downgraded products was clearly detectable.

Chemical and granulometric analysis of the tested casting powders can be considered as homogeneous. Due to the homogeneity and the same casting powder slag thickness, the crystal structure composition is assumed to be homogeneous in the casting powders under investigation.

Operationally used PG and S1 casting powders are not comparable because PG is primarily used for non-alloyed steels and S1 casting powders for alloy steels. However, the results of the experiments provide a good basis for testing further new casting powders.

A cross-sectional chemical study of the slabs shows that apart from the midline enrichment, the slabs are considered to be homogeneous. Because of the chemical homogeneity, it is likely that downgrading is not attributable for the chemical composition / homogeneity of the cast slab. However, the rolled slab may be subjected to heat effects in the reheating furnace which can cause enrichment in the surface layers of the slab.

Primary downgrading cannot be unambiguously paralleled with the parameters examined due to the small number of samples (7 pcs of slabs).

Acknowledgment

The study is funded by „EFOP-3.6.1-16-2016-00003 K+F+I folyamatok hosszú távú megerősítése a Dunaújvárosi Egyetemen" project.

References

[1] Mazumder B 1999 A Process for Developing Continuous Casting Mould-Powder for Steel Industries *Journal of Scientific & Industrial Research* **58** pp 773-780

[2] Reeb P and Schmitz H-U and Scher I 2015 Complexity of reception inspection for casting powders at Dillinger Steel Works *Cetas-Conference* Düsseldorf

[3] Lis T and Kania H and Nowacki K. 2012 The impact of the chemical composition of continuous casting moulds on their physical properties *Journal of achievements in materials and manufacturing engineering* **55**(2) pp 345-348

[4] Benavidez E. et al 2012 Influence of different oxides on the viscosity of fluorine-free mold fluxes *Procedia Materials Science* **1** pp 389-396

[5] Benavidez E et al 2015 Microstructural characteristics of mold fluxes associated to their termal behavior *Procedia Materials Science* **8** pp 218-227

[6] Wei E et al 2006 Effect of carbon properties on melting behavior of mold fluxes for continuous casting of steels *Journal of Iron and Steel Research International* **13**(2) pp 22-26

[7] Görnerup M et al 2004 Mould fluxes in continuous casting of steel–characterization and performance tuning. *Proc. VII International Conference on Molten Slags Fluxes and Salts, African Institute of Mining and Metallurgy* pp 745-752

[8] Pradhan N et al 1999 Prediction of slag pool thickness in continuous casting mould *ISIJ international* **39**(8) pp 804-808

[9] Herman J C and Leroy V 1996 Influence of residual elements on steel processing and mechanical properties *Iron & steelmaker* **23**(12) pp 35-43

[10] Leroy V et al 1995 Effects of tramp elements in flat and long products *European Commission technical steel research*

IOP Conf. Series: Materials Science and Engineering **903** (2020) 012044 doi:10.1088/1757-899X/903/1/012044

On the necessity of domestic research on nuclear materials science

Peter Trampus

Professor emeritus, University of Dunaújváros, Dunaújváros, Hungary

Email: trampus@uniduna.hu

Abstract. Globally, systematic research is going on to better understand the operation induced degradation effects of reactor structural materials and to enhance the effectiveness of their in-service monitoring. In Hungary, there are four reactor units in operation and two new ones are soon to be built. Some of the construction features of the new units differ from that of the operating ones. In order to be prepared in time for efficiently managing materials science related questions, it is necessary to establish a coordinated research program utilizing the capabilities of domestic research institutions. This paper describes the major elements of this potential program. Implementation of the program can assist ageing management activities of the operating units.

1. Introduction

For more than 40 years, Hungary has belonged to that relatively small, elite group of countries, in which nuclear power plants (NPPs) are in operation. The Hungarian energy strategy plans on long term use of nuclear power; consequently, the currently operating units will be replaced by nuclear based capacity after their retirement. The four "old" reactors at Paks site has run for more than 30 years, i.e. the originally licensed term, thus, structural materials of mechanical components and structures are ageing. The term ageing means the continuous change in material properties due to service life and/or operation loading and the environment. Construction of two "new" reactors at the same site has already started; the components of these units are intended to serve for 60 years.

Having accumulated some twenty thousand reactor year experience worldwide, the civil nuclear technology is unquestionably mature. Despite of matureness, the prospects of nuclear energy expansion is not considered positive in every region of the world. Global climate change and energy generation are in strong connection because generation related emissions take a significant portion of greenhouse gas emission. NPPs can mean a vital option for decarbonisation of energy systems. As a consequence of all these global issues, the value of operating reactors is increasing and life management, service life extension – nowadays called: long-term operation (LTO) – are playing a central role in the world's nuclear power generation. We are close to the truth when saying that LTO is not on the agenda only in countries where it is hampered by political will. The major characteristic of the new nuclear power projects is that more and more developing countries plan to embark in nuclear energy utilisation.

2. The role and importance of materials science and engineering

To assess structural integrity, i.e. the safety of mechanical components and structures, knowledge in materials science is indispensable. The basic elements of materials science and engineering, which create an intellectual foundation are the following: (1) composition and structure, (2) synthesis/processing, (3) properties, and (4) performance under operating circumstances. Nature laws control the individual elements and their coherence. When building new NPPs and during long-term plant operation, the key for both engineering and operational safety is the integrated application of each element of materials science.

Since 1970s, the operating license of the nuclear reactors has been issued by regulators for a predetermined period. This period was typically 40, in the case of Soviet designed units 30 years, determined by design considerations (also called design life) or – in the USA – by antitrust law. The fact that a nuclear power plant can be operated beyond its originally licensed term, even for double period, is acceptable if it is properly justified technically. The world's nuclear industry has elaborated and routinely been using the service life extension methodology [1]. In the light of the accumulated experience, one could even state that there is enough knowledge available on structural materials' behaviour under normal operating and potential off-normal conditions. However, this statement may be questioned, because in the past decade more and more reactors entered into the LTO phase. Figure 1 illustrates the age of the world's operating reactors [2].

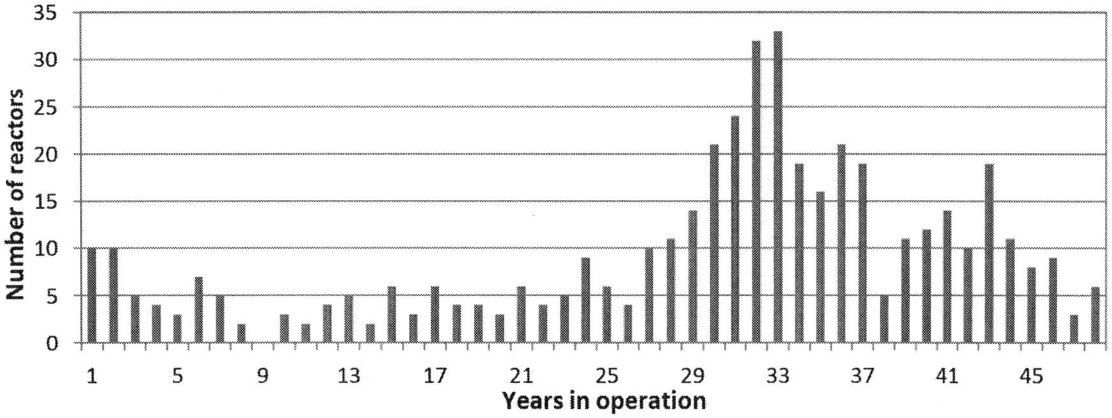

Figure 1. Distribution of operating reactors by age (as of December 12, 2016.)

The direct consequence of operation beyond design life is an increase of ageing effects on the components. Moreover, some ageing processes can accelerate and even degradations not considered by the designer may occur. Many factors influence reactor materials' ageing. Chemical composition and microstructure determine the initial mechanical properties (e.g. tensile strength, fracture toughness). Degree of loading (e.g. stresses, number of cycles) and its mode (e.g. static, dynamic, cyclic), as well as environmental effects (e.g. temperature, radiation, corrosive agent) can initiate the degradation processes, which lead to a gradual decrease in initial properties. Once the degradation exceeds a certain limit, the component fails. Cracks or crack-like flaws in the material have a significant influence on the degradation processes and thus on structural integrity. The aforementioned factors have different weight in components' degradation and usually a single factor leads to failure. It is possible, however, that more than one degradation processes are active at the same time, manifesting in a synergistic effect (e.g. environmentally assisted fatigue, irradiation accelerated stress corrosion cracking).

Mechanical components of NPPs are usually made of conventional structural materials (steels, nickel-based alloys). Materials development is evolutionary; enhancement of their operating features is achieved through the improvement of manufacturing processes or optimisation of operation and

IOP Conf. Series: Materials Science and Engineering **903** (2020) 012044 doi:10.1088/1757-899X/903/1/012044

maintenance. Structural materials of the new reactors (so called fourth generation, fission based) most probably will not have significantly more rigorous requirements.

By extending plant life management, i.e. power uprate and/or LTO, the attention of research tends to focus on reactor structural materials' behaviour. In this context, one of the most challenging tasks is the determination of the actual components condition and, based on this, the estimation of their technically achievable service life. This requires an intensification of the knowledge on degradation mechanisms, development of physically based degradation models, improvement of extrapolation methods, as well as the improvement of destructive and non-destructive testing methods and techniques to monitor materials' condition and elaboration of new maintenance and repair technologies.

3. The new reactor project in Hungary

Hungary's four old reactors (Russian design, VVER-440/V-213, current electric output: 500 MW each) are planned to shut down between 2032 and 2037. The lost capacity will be replaced by two new reactor units (also Russian design, VVER-1200 with an electric output of 1200 MW each).

3.1. Historical retrospection

Before investigating the materials science related questions of the new reactors, let us briefly look back to the situation of the old units' construction phase. In 1980, the Hungarian Academy of Sciences (HAS) established a program on „Research tasks for serving safe operation of NPPs", under the leadership and coordination of the Central Research Institute for Physics. One of its sub-programs was dedicated to „Strength analyses and in-service inspections to determine components condition and estimated service life", led by the Research Institute for Ferrous Metallurgy. The sub-program budget was approximately 5 billion HUF (converted into current value). Within the framework of this sub-program, research was conducted in the following areas between 1981 and 1985 [3]:

- Modelling and investigating thermal ageing and low-cycle fatigue in structural materials of the reactor coolant system, and determination of crack initiations.
- Investigating neutron irradiation embrittlement of reactor pressure vessel (RPV) material and developing methodology for monitoring the embrittlement process.
- Investigating crack stability in the pressurised components of the primary circuit. Creating data bank to estimate system reliability (fracture probability).
- Acoustic emission testing to identify defects in pressurised components to increase reliability of the evaluation of primary circuit components' inspections.
- Developing scientific basis for ultrasonic testing of primary circuit pressurised systems.
- Selecting numerical and experimental methods to determine stress-strain conditions of pressurised components, taking into account operating and transient loading and defects with various shape and size.
- Investigating forms and extension of cracks due to corrosion degradations, stress corrosion and corrosion fatigue for RPV pressure retaining and cladding materials, nozzles, weld materials, and welded joints.
- Comparing the conditions of stress corrosion and corrosion fatigue crack initiation and extension using fracture mechanics testing and numerical models.

The sub-program contributed to the development of a broad, local expertise in the subject, which helped to handle any material related problems during construction of the units. At the same time, the knowledge acquired could be well applied in the operation period. To the best of the author's knowledge, there is no existing or planned coordinated research program dedicated to materials' integrity issues of the new reactors. The solution to this question does not depend on the materials science community only, but rather on political decision makers.

3.2. *Justification of the research program's necessity*

Since the coordinated nuclear materials science research program described above was established, a new generation of materials scientists and engineers have grown up. Their individual knowledge is unquestionable, but it is also undoubted that an integrated knowledge base established by a properly coordinated program worth much more than the simple sum of knowledge owned by individual experts.

According to the author's opinion, the main reason for the need of establishing a coordinated, domestic, material related research program is that the construction of the new reactors differs in some – not negligible and not properly investigated – solutions from the operating ones. One difference is the RPV material: the manufacturer uses nickel as an alloying element. The effect of nickel is known from the literature [4, 5], but neither data for LTO is yet available, nor domestic practical experience and experimental database exist. Nickel as an alloying element – in certain circumstances – can reduce the resistance of steel against irradiation embrittlement. Another, more general difference is the method of the RPV structural integrity assessment. Currently, there are two engineering methods, namely, the internationally accepted Master Curve [6] and the Russian Unified Curve [7] that allow the construction of the time-dependence curve of fracture toughness for various RPV steels. The Unified Curve differs from the Master Curve in that it takes into account the transformation of the shape of the curve as a function of the embrittlement of a material.

Another fundamental difference between the new and old reactors is related partly to material selection and partly to manufacturing. The new reactor coolant pipeline is made of low-alloy steel equipped with stainless steel cladding in its internal surface. This solution is new considering both the corrosion resistance behaviour and the in-service inspection, i.e. periodic non-destructive examinations. Dissimilar metal welds (DMWs), like cladding and base metal interface, may have an increased corrosion sensitivity. The production of welded joint is more than a routine procedure because of the difficulties resulting from different chemical composition, microstructure, and thermal-physical characteristics of the three materials. These also make the DMW non-destructive examination more difficult. Figure 2. shows an ultrasonic testing simulation on a DMW illustrating some anomalies in sound beam paths. Regarding crack initiation sensitivity and in-service inspection problems of the cladding and especially the underclad area, the operator of the old reactors gained some experience on RPVs.

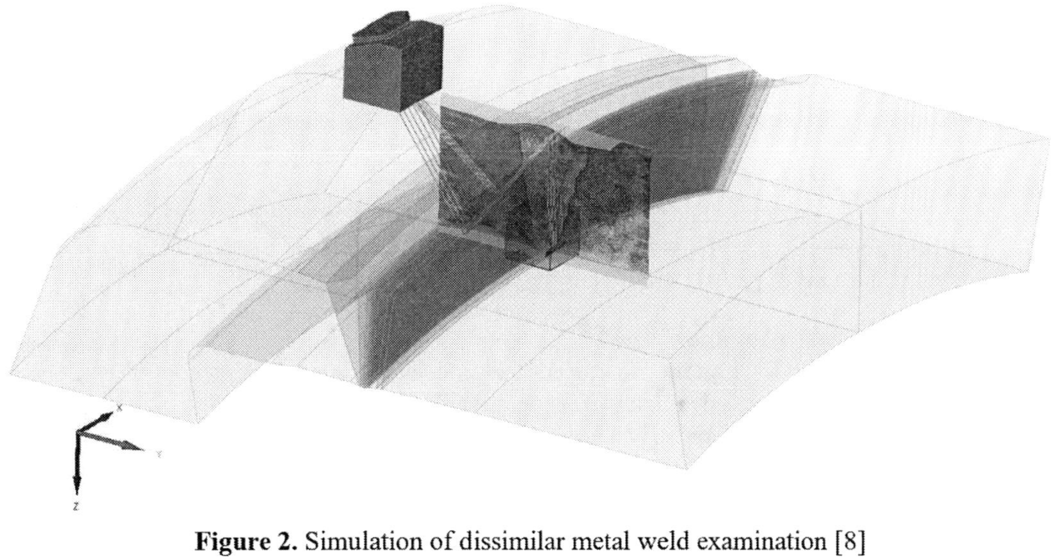

Figure 2. Simulation of dissimilar metal weld examination [8]

If the primary coolant pipelines (except for the reactor coolant) is made of the same austenitic steel as in the old units, it is advisable to apply the practical experiences gained by plant operators during the third decade of operation [9]. Experience showed that there is still not enough knowledge on special corrosion phenomena; such process is the microbiologically influenced pitting corrosion.

As mentioned above, according to the design intent, the new reactors will operate for 60 years. This gives another strong argument for the coordinated program, i.e. creation of an up-to-date database meeting Industry 4.0 requirements must be started during or preferentially before the construction period. The lack of pre-service condition experience of mechanical components in the "old" units' operation often led to serious difficulties.

3.3. *Objectives of the research program*

To be prepared for material related issues of the new reactors, the author recommends to consider the establishment of a state-sponsored, coordinated research program involving competent institutes and experts of Hungary. Ideally, duration of the research program should last for at least 5 years. The minimum objectives should be the following:

3.3.1. Establishing knowledge base. It is necessary to be familiar with the properties of the structural materials and production technologies of the new reactors, and it is necessary to understand the physical basis of ensuring and maintaining necessary properties. The best way to achieve this objective is to collect results of our own research within the coordinated program.

3.3.2. Creating foundation of the ageing management program. Today, one if not the most important criterion of safe and reliable NPP operation is to establish and maintain a systematic and integrated ageing management program. Besides knowing the fundamental characteristics of structural materials, dominant, service induced degradation processes have to be identified and understood from the point of view of substantiate the components' ageing management program. Since the ageing management program has to be available from the start-up of the new reactors, the coordinated research program provides a unique opportunity to lay down its foundations.

3.3.3. Preparing the domestic institutes for solving material issues. During the construction and in the first decade of operation of the old reactors, the law required notification of a leading institute on materials science. The licensee had to ask for the institute's expert opinion in subjects like material selection, material procedures, especially welding, material testing, failure assessment; the nuclear regulator has always considered this expert opinion as a gold standard when taking decisions. This mechanism is still present in the Russian normative technical documents but no longer legally binding in Hungary. There is no intention, of course, to re-establish this mechanism, but without any doubt, the task itself – with its complexity and diversity – will permanently come up and has to be solved.

3.3.4. Contributing to long-term, safe and reliable plant operation. All materials science related issues, without exception, contribute to the assurance of the mechanical components' structural integrity. Structural integrity as an important pillar of the "defence in depth" concept [10] is one of the most important conditions of plant operational safety as well as economic efficiency.

3.4. *Organisation of the research program*

The author of this paper is strongly convinced that the establishment of a coordinated research program with the aforementioned objectives is reasonable and necessary to be able to solve this complex and large-scale task sufficiently. Currently no single institute of Hungary accumulates such a knowledge base that could cover all relevant scientific areas presented here. Therefore, it seems advisable to delegate the elaboration and then the coordination of the coordinated research program to the Committee for Materials Science and Technology of HAS. The HAS Committee must be capable to gather and coordinate experts internationally recognized. The best form of the program's operative implementation is still to be discussed and agreed upon. One possible option could be the establishment of a so-called 'virtual' institute composed of assigned experts of relevant scientific areas.

4. Materials challenges of the operating reactors

As mentioned earlier, ageing phenomena and related issues get into the focus of LTO. NPPs one after the other reach a stage of their operating life, that no single NPP in the world has ever reached yet. It is not a coincidence that worldwide there is ongoing systematic research activity to better understand the operation induced degradation effects of reactor structural materials and to enhance the effectiveness of their monitoring.

On the basis of research results and possessing plentiful operation experience, material scientists agree that structural materials of the reactors in operation or under construction face the following significant ageing challenges [11, 12]:

- Neutron irradiation embrittlement and loss of toughness.
- Corrosion, especially its localised forms, e.g. stress corrosion cracking (SCC) and irradiation-assisted SCC.
- Environmentally assisted fatigue.

The ductile-to-brittle transition temperature shift and decrease in toughness (usually known as embrittlement) are typical means of degradation of RPV pressure retaining material. The origin of this issue stems from the impact of the increased fast neutron fluence on the ageing process due to extended service life. Figure 3 shows the scanning electron-microscopic (SEM) image of the broken surface of a half Charpy specimen; brittle fracture is caused by loss of grain boundary cohesion due to phosphorus segregation on grain boundaries (fluence: $1.07 \cdot 10^{25}$ $n \cdot m^{-2}$, E>0.5 MeV). In this regard, it is still necessary to investigate the influence of impurities of the RPV steels on microstructure changes (e.g. phosphorus), or the influence of nickel on formation of Ni-rich, so called late blooming phases and the impact of these phases on the embrittlement process. The RPV structural integrity assessment requires further research: The Master Curve concept enabling direct measure of fracture toughness still needs quantitative validation (see earlier). LTO also requires to keep the thermal embrittlement investigation on the research agenda. Increased attention, and consequently, research is necessary in the area of welded joints integrity with special regard to the behaviour of DMWs, joining alloys with different chemical compositions, microstructures and thermal-physical characteristics. In the case of DMWs, corrosion can play an important role.

Figure 3. Cleavage fracture on irradiated Russian RPV steel (SEM)

Corrosion attacks the components and pipelines made of stainless steel, e.g. RPV cladding, reactor internal structures that control hydraulic flow, support fuel assemblies, and control rods. SCC degradation mainly develops in the steam generator heat exchanger tubing made of nickel-based alloy. SCC can appear in stainless steel components, too (e.g. steam generator tubing). Serious consequences of SCC can occur in reactor internals: in the case of the baffle bolts in core basket, as a result of a highly complex degradation effects, the irradiation assisted SCC leads – in extreme case – to the fracture of the cracked bolt head. Figure 4 shows the relations of loading situation and degradation effects.

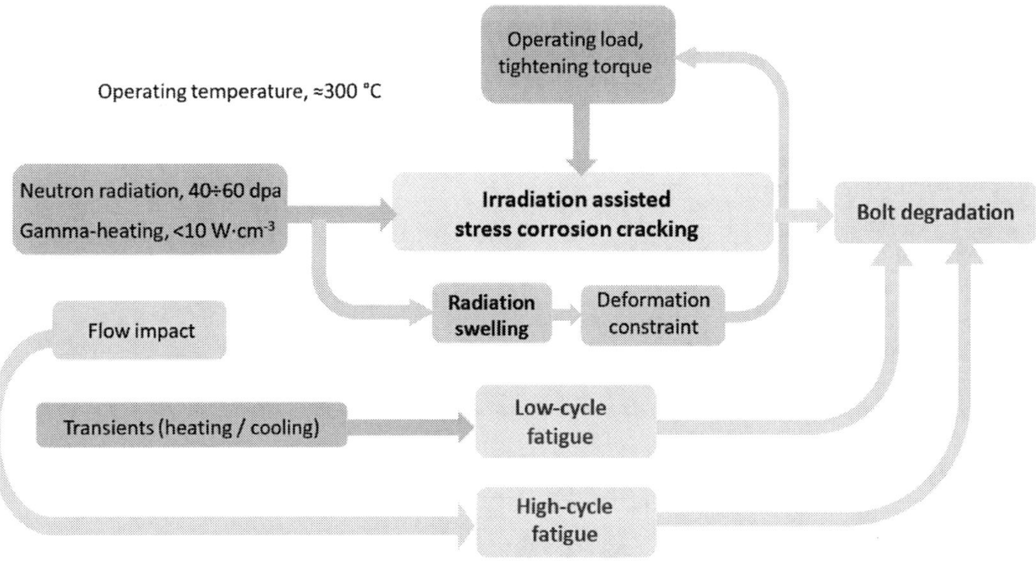

Figure 4. Loading of and their impact on baffle bolts [13]

In LTO, the austenitic stainless-steel pipes of the reactor cooling system dwell longer on the operation temperature, than they were originally expected by the designer. During the longer dwell-time, thermal activation process leads to precipitation of grain boundary carbides and thus to sensibilisation of the grain boundary area against intergranular attack (IGA). Even though the operation temperature is relatively low, depending on other factors such as time, C-content, Ni-content the austenitic steels may be sensitive to IGA. Figure 5 shows an example: dwell time = 225 000 hours, temperature = 300°C, C = 0.08%, Ni = 10%). The grain boundary carbide precipitation is clearly visible [14].

Environmentally assisted fatigue (EAF), sometimes called as corrosion fatigue, concerns the reduction in fatigue life that is observed in a reactor water environment compared to room temperature air. EAF involves two primary elements: 1) the effects of reactor water on the overall fatigue life of reactor components (as represented by either multiplying the fatigue usage factor by an 'environmental factor' to account for aqueous corrosion effects or the use of an environment-adjusted fatigue design curve), and 2) the potential

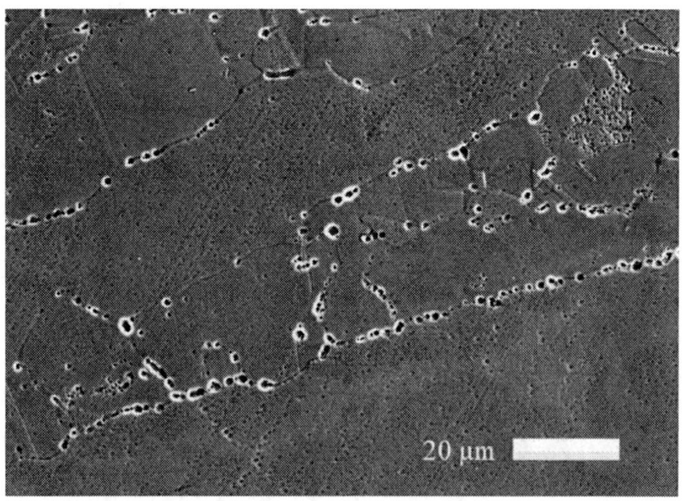

Figure 5. Grain boundary segregation in stainless steel (etching: ASTM A262 Practice A, SEM)

accelerated growth of an identified defect caused by exposure to reactor water environments. Fatigue crack initiation and growth resistance is governed by a number of materials, structural and environmental factors, such as stress range, temperature, dissolved oxygen or hydrogen concentration, mean stress, loading frequency (strain rate), surface roughness, and number of cycles [12].

5. Conclusions

It is obvious that both building of the two new reactors and long-term operation of the four old units define numerous issues for the Hungarian materials science community in the coming decades. The author of this paper is strongly convinced that establishment of a coordinated research program is reasonable and necessary to be able to solve this complex and large-scale task sufficiently. The major objectives of the research program were presented here. It seems advisable to delegate the elaboration and coordination of the program to the Committee for Materials Science and Technology of HAS. Currently no single institute of the country accumulates such an expert system that could cover all relevant scientific areas presented here. The HAS, however, is capable to gather and coordinate experts recognized internationally and provide them with appropriate financial support. The form of the program's operative implementation is still to be discussed and agreed upon. One possible scenario includes the establishment of a so-called 'virtual' institute composed of assigned experts of relevant scientific areas.

Acknowledgment

The author would like to express his gratitude to professors L. Tóth and B. Verő for their invaluable suggestions to the research program, and to F. Oszvald for performing the electron-microscopic investigations.

References

[1] Plant Life Management Models for Long Term Operation of Nuclear Power Plants *NP-T-3.18* 2015 (Vienna: International Atomic Energy Agency)
[2] Nuclear Technology Review 2017 (Vienna: International Atomic Energy Agency)
[3] Szunyogh L 2004 *GÉP* **LV** 2 28-31 (in Hungarian)
[4] Kryukov A M *et al* 1997 *Nucl. Eng. Design* **173** 333-339
[5] Buswell J T *et al* 1995 *J. Nucl. Materials* **225** 196-214
[6] Merkle J G, Wallin K and McCabe D E 1998 *NUREG/CR-5504 (ORNL/TM-13631)*
[7] Margolin B Z *et al* 2003 *Int. J. Pres. Ves. Piping* **80** 817-829
[8] Erdei R, Bézi Z and Takács Cs 2019 XI. Hungarian NDT Conf. Eger, 20-22.03.2019 (in Hungarian)
[9] Trampus P *et al* 2017 *Materials Science Forum* **885** 92-97
[10] Defense in depth in nuclear safety INSAG-10 1996 (Vienna: International Atomic Energy Agency)
[11] Zinkle S J and Was G S 2013 *Acta Materialia* **61** 735-758
[12] EPRI Materials Degradation Matrix Rev 4 2018 (Palo Alto: EPRI)
[13] Székely L Cs and Trampus P 2018 XVII. Nucl. Techn. Symp. Szekszárd, 30.11.2018 (in Hungarian)
[14] Oszvald F 2019 *Private communications*

IOP Conf. Series: Materials Science and Engineering **903** (2020) 012045 doi:10.1088/1757-899X/903/1/012045

Mechanical properties of AlSi10MnMg matrix syntactic foams filled with lightweight expanded clay particles

A Szlancsik[1,2], D Kincses[1], I N Orbulov[1,2]

[1]Department of Materials Science and Engineering, Budapest University of Technology and Economics, Budapest, Hungary
[2]MTA–BME Lendület Composite Metal Foams Research Group, Budapest, Hungary

E-mail: szlancsik@eik.bme.hu

Abstract. Compression tests measured the mechanical properties of AlSi10MnMg matrix syntactic foams filled with lightweight expanded clay particles (LECAPs). The metal matrix syntactic foams were produced by low-pressure liquid state infiltration method. Two types of LECAPs were used during production. One with ~3 mm diameter and the other type with ~9 mm diameter. The samples were machined and heat treated. The T6 heat treatment caused a significant increase in the properties in terms of plateau strength and absorbed energy. Also, the heat treatment caused a different failure mode in the case of the LECAP-9, which is why the absorbed energy increased by ~50% compared to the as-cast sample.

1. Introduction

Metal matrix syntactic foams (MMSFs) are developed from the polymer matrix syntactic foams [1]. In these foams, the porosity is created by previously produced materials such as hollow spheres or lightweight particles. Due to their low density, the syntactic foams are getting more attention because the parts created by these kinds of materials are usually more energy-efficient than the bulk ones, which is one of the main goals nowadays [2].

MMSFs are widely investigated in the literature. Mainly the compressive properties are measured [3-12], but in some cases, the tension [13], wear [14-16] or bending properties [17-18] are also determined. Another large investigation field is the thermal properties of these kinds of materials [19-22].

Regarding the filler materials, mainly the relatively expensive ceramic hollow spheres are used. For example, Su et al [23] investigated an alumina-aluminum matrix syntactic foam filled thin-walled tubes. Three different sphere sizes were investigated, namely 1.0–1.5 mm (a), 1.5–2.0 mm (b) and 2.5–3.0 mm (c). The matrix material was AlSi9Cu2Mg (ZL111) aluminum alloy which can be heat treated, and they also investigated the effect of the T6 heat treatment on the mechanical properties of the produced foams and structures. The tube's material was AlMg1SiCu (A6061) aluminum alloy, while the ceramic sphere's material was Al_2O_3. Quasi-static axial and radial compression tests were performed on the syntactic foams, on the hollow tubes and also when a syntactic foam sample was placed into the tube. Every test was repeated three times, and the results were averaged. The density of the produced foams are as follows: 2.01 ± 0.03 g/cm^3 (a), 1.85 ± 0.01 g/cm^3 (b) and 1.70 ± 0.01 g/cm^3 (c). It can be seen that with the larger sphere, a lighter structure is achieved. The authors concluded from the mechanical tests that the addition of the thin-walled tubes under axial compression the deformation and fracture mode were influenced and changed. It is caused a significant increase in energy absorption compared to the sum of the individual components. They proved that this kind of material compared to the literature's data is a good choice for energy absorbers.

In the case of cheaper syntactic foams, one has to change the filler material. A good alternative for the hollow spheres is the lightweight expanded perlite. Movahedi et al. [24] investigated the influence of particle arrangement on the compression of functionally graded metal syntactic foams. Two different filler materials were used, namely expanded perlite (EP) and activated carbon (AC). The matrix material was a Zn-27Al-2Cu-0.015Mg (ZA27) zinc-aluminum alloy due to its high strength and relatively low density (ρ = 5.00 gcm^{-3}). Counter-gravity infiltration casting technique was used in order to produce metal matrix syntactic foams. Four different types of samples were produced based on the location and mixture of the filler materials: (a) two layer longitudinally-graded, (b) six layer longitudinally-graded, (c) radially-graded and (d) hybrid samples. Quasi-static compression tests were performed on these samples until ~80% strain. They found that the volumes of the samples which contain EP particles were undergone ductile layer-by-layer collapse while the regions with AC particles exhibit brittle fracture. Radial and hybrid samples had superior initial strength due to the fact that in these samples, the stronger AC particles must deform with the EP particles while in the layered structure the EP particles can deform separately.

Based on the previously presented studies, our aim was to find a low-cost alternative for the ceramic hollow spheres. Also, determine the influence of the heat treatment on the foams mechanical properties and deformation methods.

2. Materials and methods

2.1. Materials

AlSi10MnMg aluminum alloy was applied as matrix material due to the fact that this material is widely used in the automotive industry. The filler material was a set of lightweight expanded clay particles (LECAPs) which were obtained from Liapor GmbH & Co. KG. Two different sizes were investigated, namely ~3.2 mm diameter (LECAP-3) and ~9.0 mm diameter (LECAP-9). The size distribution was measured with an Olympus SZX16 stereomicroscope on 100 particles from each type, and the results are depicted in Figures 1. and 2., respectively. The materials chemical composition is listed in Table 1.

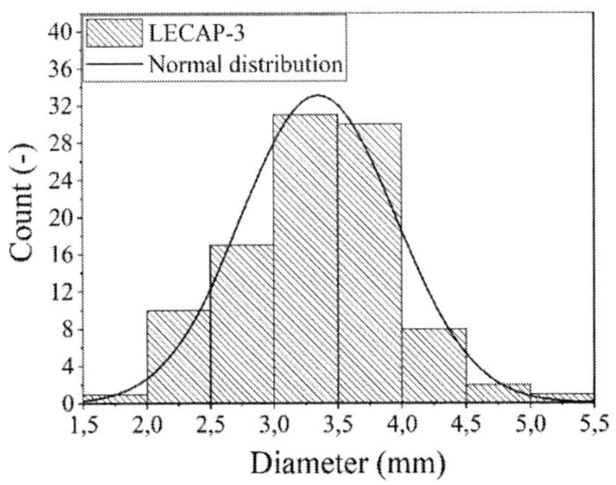

Figure 1. The size distribution of the LECAP-3 filler. The average diameter is 3.2 mm.

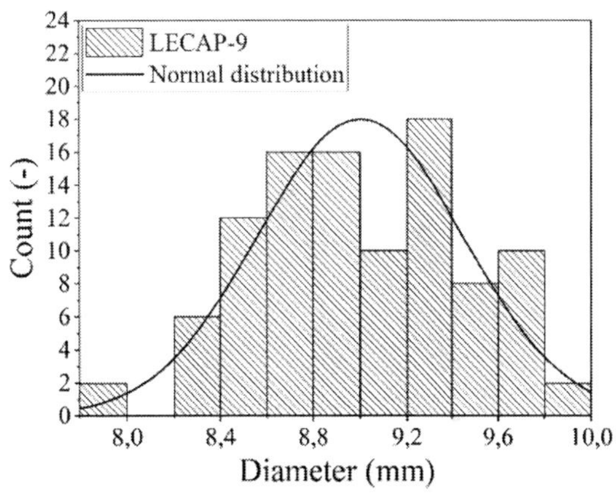

Figure 2. The size distribution of the LECAP-9 filler. The average diameter is 9.0 mm.

The production of the syntactic foam was made by a low-pressure infiltration process which is described in details in [25]. The mold was preheated at 300°C, and the aluminum alloy was heated with

IOP Conf. Series: Materials Science and Engineering **903** (2020) 012045 doi:10.1088/1757-899X/903/1/012045

an induction furnace. The samples for the compression tests were machined according to the ruling standard for porous and cellular metals [26]. The LECAP-9 samples were 70×70×70 mm cubes and the LECAP-3 had 30 mm diameter and 30 mm height. The difference between the shape of the samples is not relevant in terms of the results. Both the rectangular and cylindrical shaped specimens are satisfying the requirements of the standard [26]. The main reason behind the different shapes, is the difficulty of the machining, and these shapes could be produced with the least amount of machining.

Table 1. The chemical composition of AlSi10MnMg aluminium alloy and the LECAP in weight percent

	Chemical elements in weight percent							
Matrix material:	Si	Mn	Mg	Fe	Cu	Zn	Ti	Al
AlSi10MnMg	10.4	0.7	0.4	0.15	0.03	0.07	0.15	Balance
Filler material:	O	Al	Si	Fe	Other			
LECAP-3	55.4	32.6	7.6	2.4	Balance			
LECAP-9	49.1	12.8	22.8	7.7	Balance			

Half of the produced samples were measured as cast, and the other half in heat treated condition. The first step was a solution heat treatment on 500°C for 3 hours and cooling in water. The second step was aging at 190°C and cooling in air.

2.2. Compression test procedure

The samples were measured on a 400 kN ZWICK/Roell Z 400 electromechanical test machine. Every compression test was performed under quasi-static conditions; the deformation rate was 0.01 s^{-1}. A compression tool was used to maintain the uni-axial deformation state during the whole compression until the samples reached 50% strain. Figure 3. shows the measurement setup in the case of a LECAP-9 sample. During the measurement, the force – displacement curve was registered and later the structural stiffness (S), the yield strength (σ_Y), the plateau strength (σ_{plt}) and the absorbed energy until 50% strain ($W_{50\%}$) were evaluated according to the standard [26].

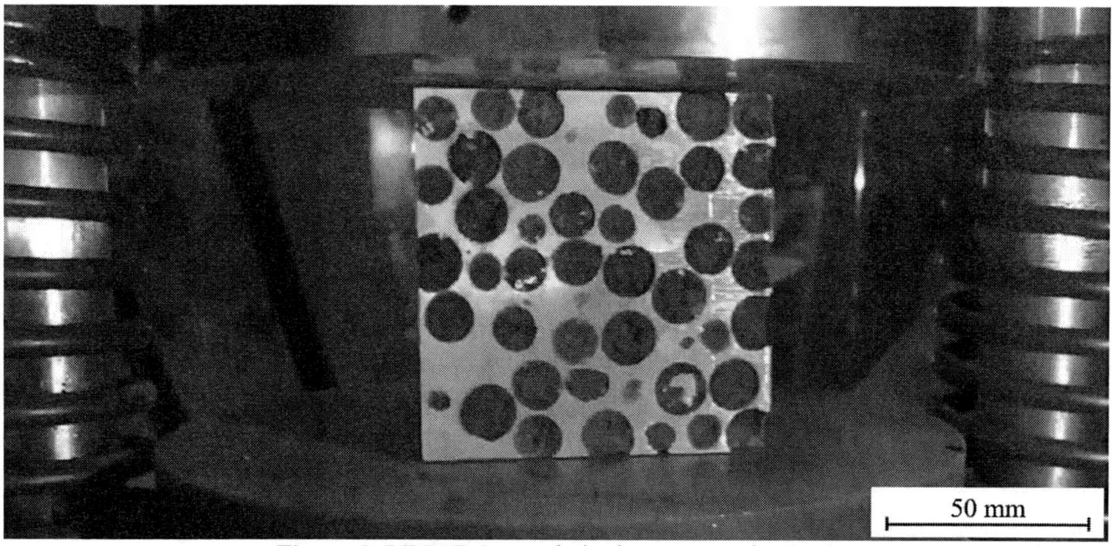

Figure 3. LECAP-9 sample in the compression tool

3. Results and discussion

Three samples were measured from every different type and heat treated foams, namely LECAP-3 (as cast), LECAP-3-T6, LECAP-9 (as cast), LECAP-9-T6. Figures 4. and 5. depict the results from the compression tests. It can be seen that the T6 heat treated samples have superior mechanical properties than the ones without heat treatment. Also, the foams with the larger filler particles have better mechanical properties.

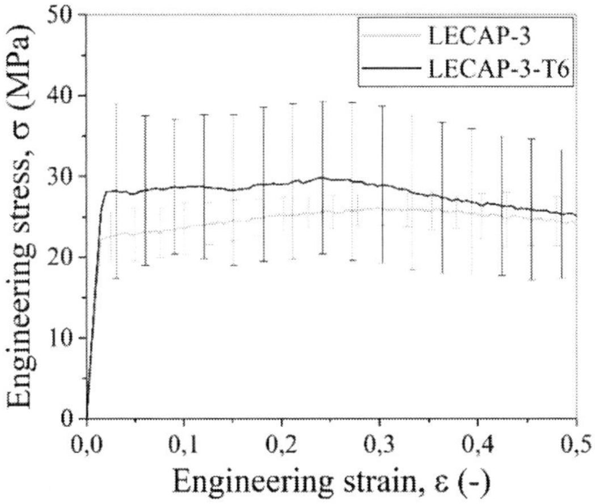

Figure 4. Engineering stress – engineering strain curve of the LECAP-3 and LECAP-3-T6 samples with error

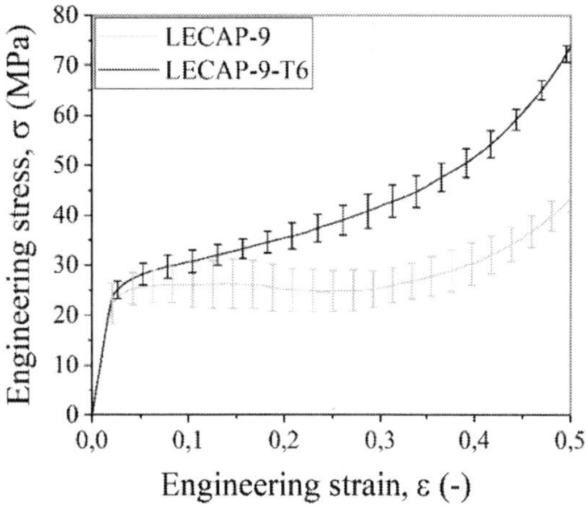

Figure 5. Engineering stress – engineering strain curve of the LECAP-9 and LECAP-9-T6 samples with error

It can be seen that in the case of the LECAP-3, the error is much larger, which can be caused by the fracture of the samples. Every type of sample fractured during the compression except the LECAP-9-T6 samples, that is why the error is much lower in that case. Also, because of the different deformation method, it exceeds higher stresses than the other samples. Table 2. contains the characteristic values and the coefficients of standard deviation (CoSD) calculated from the engineering stress – engineering strain diagrams according to the standard [26].

Table 2. The mechanical properties of the LECAP filled syntactic foams according to the standard [26]

Name	S (MPa)	CoSD (-)	σ_Y (MPa)	CoSD (-)	σ_{plt} (MPa)	CoSD (-)	$W_{50\%}$ (J/cm^3)	CoSD (-)
LECAP-3	1563 ± 147	0.09	22.7 ± 3.4	0.15	25.7 ± 2.1	0.08	12.2 ± 1.3	0.11
LECAP-3-T6	1582 ± 321	0.20	28.1 ± 10.6	0.38	28.5 ± 9.5	0.33	13.8 ± 4.5	0.33
LECAP-9	1162 ± 91	0.08	24.3 ± 3.6	0.15	26.8 ± 3.9	0.15	14.5 ± 1.8	0.12
LECAP-9-T6	1179 ± 64	0.05	27.1 ± 2.1	0.08	43.5 ± 3.0	0.07	20.3 ± 1.2	0.06

The structural stiffness was almost similar for the as cast and heat treated samples which means it does not depend on the heat treatment, but it depends on the filler size. In the case of the yield strength, the heat treated samples have higher values which were expected because the matrix material has higher yield strength with the T6 heat treatment. The plateau strength was following the same trend as the yield strength, but the LECAP-9-T6 exceed the expected values with 15 MPa, which can be caused by more plastic behaviour. This behaviour also influences the absorbed energy, which also became the highest in the case of the LECAP-9-T6. Figures 6. and 7. shows a LECAP-3-T6 and LECAP-9-T6 sample after the test, respectively.

Both samples got fractured, but from the LECAP-3-T6 samples, parts are missing due to the fractures. Every other type got the same failure type like this one. The integrity of the samples was lost during the compression, but the LECAP-9-T6 samples were not. In Figure 7., one can see that the walls between the fillers fractured, and the matrix material deformed plastically, but the integrity of the sample remained. This fracture mechanism is really useful in the perspective of energy absorption applications. Also, the high plateau strength and absorbed energy indicate that this low-cost filler material with the AlSi10MnMg aluminium alloy matrix material is a good choice for further investigations.

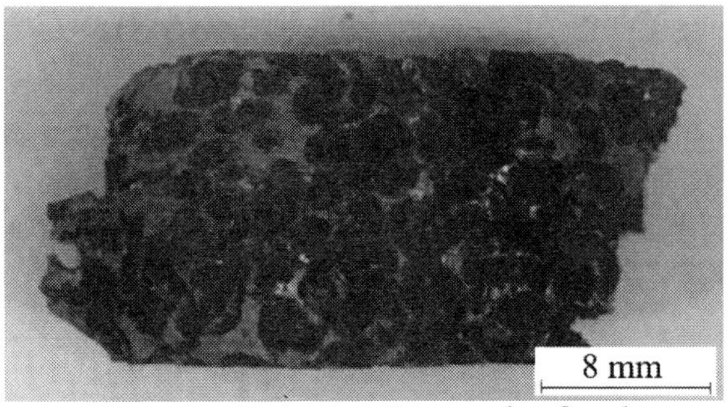

Figure 6. LECAP-3-T6 sample after the compression test

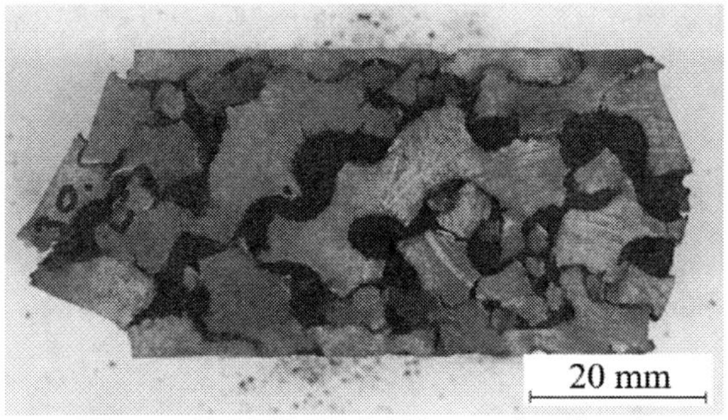

Figure 7. LECAP-9-T6 sample after the compression test

4. Conclusions

From the presented work, where two different particle ranges were investigated in as cast, and in T6 heat treated conditions, the following conclusions can be drawn:

(a) Expanded clay particles filled AlSi10MnMg aluminium alloy matrix syntactic foams were successfully produced by low-pressure infiltration technique.
(b) The T6 heat treated samples showed higher yield strength, plateau strength and absorbed energy.
(c) The LECAP-9-T6 samples had a different type of failure mode than the other types; namely, these samples got fractured at the thin walls between the filler particles, but they have preserved their integrity.

Acknowledgement

This paper was supported by the ÚNKP-19-3 New National Excellence Program of the Ministry for Innovation and Technology.

References

[1] Kallas DH and Chatten CK 1969 Buoyancy materials for deep submergence *Ocean Engineering* **4** pp 421-424
[2] Gupta N and Rohatgi PK, 2014 Metal Matrix Syntactic Foams *DEStech Publications*, Inc., Lancaster, Pennsylvania, USA
[3] Katona B and Orbulov IN 2017 Structural Damages in Syntactic Metal Foams Caused by Monotone or Cyclic Compression *Periodica Polytechnica Mechanical Engineering* **61(2)** pp 146–152
[4] Goel MD, Mondal DP, Yadav M S and Gupta S K 2014 Effect of strain rate and relative density on compressive deformation behavior of aluminum cenosphere syntactic foam *Mater. Sci. Eng.* A **590** pp 406–415
[5] Zhang LP and Zhao YY 2007 Mechanical Response of Al Matrix Syntactic Foams Produced by Pressure *Journal of Composite Materials* **41 (17)** pp 2105–2117
[6] Daoud A 2009 Effect of strain rate on compressive properties of novel Zn12Al based composite foams containing hybrid pores *Materials Science and Engineering A* **525** pp 7–17
[7] Tao XF and Zhao YY 2012 Compressive failure of Al alloy matrix syntactic foams manufactured by melt infiltration *Materials Science and Engineering A* **549** pp 228– 232
[8] Movahedi N, Conway S, Belova IV, Murch GE and Fielder T 2019 Influence of particle arrangement on the compression of functionally graded metal syntactic foams *Materials Science & Engineering A* **764** 138242
[9] Puga H, Carneiro VH, Jesus C, Pereira J and Lopes V 2018 Influence of particle diameter in mechanical performance of Al expanded clay syntactic foams *Composite Structures* **184** pp 698-703
[10] Mankovits T, Varga TA, Manó S and Kocsis I 2018 Compressive Response Determination of

Closed-Cell Aluminium Foam and Linear-Elastic Finite Element Simulation of μCT-Based Directly Reconstructed Geometrical Models *Strojniški vestnik - Journal of Mechanical Engineering* **64** pp 105-113

[11] Zhang Y and Zhao Y 2019 Hysteretic energy dissipation in aluminium matrix syntactic foam under intermittent cyclic compression *Materialia* **6** 100286

[12] Movahedi N, Murch GE, Belova IV and Fiedler T 2020 Manufacturing and compressive properties of tube-filled metal syntactic foams *Journal of Alloys and Compounds*

[13] Ming Y, Ping Z and Yingqi M 2012 Experimental study and numerical prediction of tensile strength properties and failure modes of hollow spheres filled syntactic foams *Computational Materials Science* **63** pp 232–243

[14] Májlinger K, Bozóki B, Kalácska G, Keresztes R and Zsidai L 2016 Tribological properties of hybrid aluminum matrix syntactic foams *Tribology International* **99** pp 211–223

[15] Szalóki I and Viharos Zs J 2018 Surface Structure Analysis of Syntactic Metal Foams Machined by Milling *Periodica Polytechnica Mechanical Engineering* **62 (1)** pp 26–32

[16] Májlinger K 2015 Wear properties of hybrid AlSi12 matrix syntactic foams *Int. Jour. of Mat. Res.* **106 (11)** pp 1165–1173

[17] Taglivia G, Porfiri M and Gupta N 2010 Analysis of flexural properties of hollow-particle filled composites *Composites: Part B* **41** pp 86–93

[18] Vesenjak M, Duarte I, Baumeister J, Göhler H, Krstulović-Opara L and Ren Z 2019 Bending Performance Evaluation of Aluminium Alloy Tubes Filled with Different Cellular Metal Cores *Composite Structures*

[19] Fiedler T, Belova I V and Murch G E 2015 On the thermal properties of expanded perlite – Metallic syntactic foam *International Journal of Heat and Mass Transfer* **90** pp 1009–1014

[20] Mondal D P, Jha N, Badkul A, Das S and Khedle R 2012 High temperature compressive deformation behaviour of aluminum syntactic foam *Mater. Sci. Eng.* A **534** pp 521–529

[21] Ozkutlu M, Dilek C and Bayram G 2018 Effects of hollow glass microsphere density and surface modification on the mechanical and thermal properties of poly(methyl methacrylate) syntactic foams *Composite Structures* **202** pp 545–550

[22] Linul E, Lell D, Movahedi N, Codrean C and Fiedler T 2019 Compressive properties of zinc syntactic foams at elevated temperatures *Composites Part B* **167** pp 122–134

[23] Su M, Wang H and Hao H 2019 Axial and radial compressive properties of alumina-aluminum matrix syntactic foam filled thin-walled tubes *Composite Structures* **226** 111197

[24] Movahedi N, Conway S, Belova IV, Murch GE and Fiedler T 2019 Influence of particle arrangement on the compression of functionally graded metal syntactic foams *Materials Science & Engineering A* **764** 138242

[25] Szlancsik A, Katona B, Bobor K, Májlinger K and Orbulov IN 2015 Compressive behaviour of aluminium matrix syntactic foams reinforced by iron hollow spheres *Materials & Design* **83** pp 230–237

[26] Mechanical testing of metals - Ductility testing - Compression test for porous and cellular metals ISO 13314:2011

IOP Conf. Series: Materials Science and Engineering **903** (2020) 012046 doi:10.1088/1757-899X/903/1/012046

Joining aluminium and poly(lactic acid) specimens by laser beam – a feasibility study

T Temesi[1] and T Czigany[1,2]

[1] Department of Polymer Engineering, Faculty of Mechanical Engineering, Budapest University of Technology and Economics, Műegyetem rkp. 3, H-1111 Budapest, Hungary

[2] MTA-BME Research Group for Composite Science and Technology, Műegyetem rkp. 3, H-1111 Budapest, Hungary

E-mail: czigany@eik.bme.hu

Abstract. In this article, our main aim is to demonstrate and prove that it is feasible to join aluminium and poly(lactic acid) (PLA) specimens by laser beam. We investigated the effects of structuring the surface of the aluminium specimens with corundum blasting and joining speed and used three types of PLA (with the same D-lactide content but different average molecular weights) to investigate the effect of the viscosity of the polymer melt on the load-bearing capacity of the joints. Joined aluminium-PLA specimens were successfully manufactured and examined with standard lap-shear tests. We found that both surface structuring, joining speed and the type of PLA material influenced the load-bearing capacity of the aluminium-PLA joint.

1. Introduction

An important aim of industrial research and development is to integrate numerous structural elements into one complex part. Costs can be decreased with the use of automatable technologies and low-density materials like aluminium, polymers and polymer composites. In the vehicle industry, this results in less weight, and thus less fuel consumption and less harmful gases emitted during the lifetime of the vehicle [1-3]. Another trend in the vehicle industry is to use fewer types of materials to increase the effectiveness of recycling. The 2000/53/EC directive requires that at least 85 wt% of the material in a scrap vehicle must be reused or recycled [4]. Thus, in the last decade, much research has been focused on the development of materials that are easy and cheap to manufacture and recycle. One such polymer material is poly(lactic acid) (PLA), which is biodegradable and can be synthesised from renewable resources. It is widely researched, as it is a promising material in certain applications (such as containers and food packaging). Its mechanical properties, however, do not yet permit its use in products with a long lifetime [5-7].

The R&D of direct joining techniques of polymers and metals (without the use of any additional materials or technological steps) has been a hot topic in recent years [8, 9]. There is a special focus on researching the applicability of easily automatable, quick polymer welding techniques (such as techniques based on friction, and ultrasonic and laser welding). In all of these, the polymer material is heated, melted and pushed into the surface structures of the metal, promoting the formation of strong, shape-connected joints [10, 11].

As metals and polymers have different chemical structures and melting temperatures, only adhesive joints can be created between them with welding techniques. Besides chemical connections creating the necessary adhesion, the surface structure of the metal can also significantly influence the strength of the joint. But, as numerous publications prove, it is possible to manufacture strong joints between metal and polymer materials with joining technologies such as friction stir welding, ultrasonic welding and laser welding. In these, mostly steel and aluminium, engineering thermoplastics and thermoplastic composites (e.g., unreinforced and glass fibre-reinforced polyamide (PA, PA-GF), polycarbonate (PC), and polyethylene terephthalate (PET)) were used and examined [12-14]. However, the joinability of aluminium and PLA has never been investigated before.

In this feasibility study, we are going to prove that strong joints between aluminium and PLA can be manufactured. We chose an aluminium alloy that is widely used, even in the vehicle industry, and used grit blasting to create structures on its surface. We also used three different PLA materials and three different joining speeds to examine the effect of multiple parameters on the load-bearing capacity of the aluminium-PLA joints.

2. Materials and Methods

2.1. Materials

In our experiments, we used 116 mm × 46 mm × 1.4 mm (length × width × thickness) AA1050A-type aluminium specimens. The chemical composition of the aluminium (Table 1) was confirmed by optical emission spectrometry (OES) and energy-dispersive spectroscopy (SEM-EDS). The aluminium specimens were grit blasted with corundum of two different sizes (80-100 μm and 300-400 μm). The surface roughness of the aluminium specimens (in factory-rolled "as-received", and corundum-blasted states) was measured with a surface roughness measuring device (Mitutoyo SJ-400) and by scanning electron microscopy (SEM).

Table 1. Chemical composition of the AA1050A-type aluminium alloy

	Chemical composition (wt%)							
	Al	**Mg**	**Si**	**Cu**	**Mn**	**Ti**	**Fe**	**Other**
AA1050A	99.50	0.15	0.12	0.04	0.03	0.01	0.01	0.14

Three types of PLA (Ingeo Bioplastics 2500HP, 3100HP and 3260HP) with different average molecular weights were used (Table 2). The PLA granules were dried in a drying oven at 80°C for 6 hours. After drying, flat, square-shaped specimens (80 mm × 80 mm × 2 mm in size) were injection moulded on an Arburg AllRounder Advance 270S 400-170 injection moulding machine (an ascending temperature profile with 5 °C increments was used, the temperature at the screw tip was set to 200 °C). From these, 80 mm × 25 mm × 2 mm rectangular specimens were cut with the use of a VersaLaser VLS 2.30 type CO_2 laser cutter. The average transparency of the PLA specimens was measured with a Perkin-Elmer Lambda 1050 spectrophotometer (Table 2).

Table 2. Average molecular weight and MFI values of PLA materials
(* MFI values are taken from the Technical Data Sheets of the materials)

PLA material	Weight-average molecular weight (kDa)	Melt Flow Index* (MFI, 210 °C, 2.16 kg) (g·10min^{-1}))	Average total transparency at 950 nm (%)
Ingeo 2500HP	175	8	90
Ingeo 3100HP	138	24	89
Ingeo 3260HP	105	65	90

2.2. Joining process and measurement method

To join the aluminium and PLA specimens, we used a Trumpf TruDiode 151-type laser welding machine (wavelength was 950 nm) at 150 W power output. Both the aluminium and polymer specimens were cleaned in an ultrasonic bath for 5 minutes in methanol before the joining process. The specimens were joined in an overlapped configuration: the polymer was placed on top of the aluminium, and then both materials were irradiated by the laser beam (transmission joining; Figure 1). During this process, the laser beam passed through the PLA specimens, and its energy was absorbed in the aluminium, producing heat. It is crucial that the transparency of the polymer specimens is high enough to let most of the laser beam pass through them with the least amount of deflection and reflection possible. Some of this heat was then conducted to the PLA specimens, which were melted as a result. The melted PLA then wetted the surface of the aluminium specimens, filled in the surface structures and after cooling down, shape-connected joints were formed.

The specimens were clamped together during the joining process with a weight plate (m=1.25 kg). In this plate, a window with an appropriate size was cut, so the laser beam could irradiate the materials below the plate. The length of the path that the laser beam irradiated was set to 169 mm, the laser beam started from the spot marked with an X, and its path is shown as a red, dashed line in Figure 1.

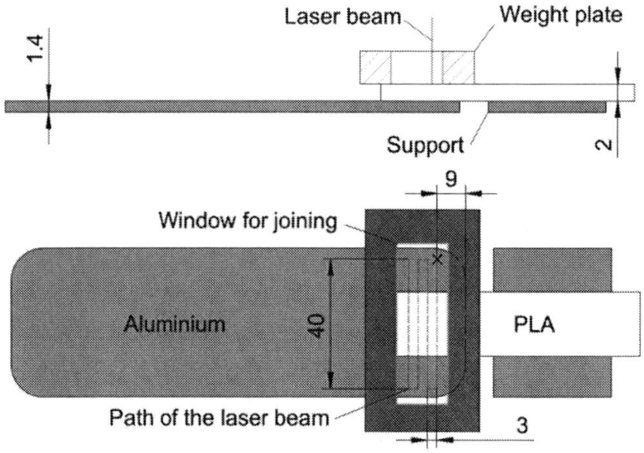

Figure 1. Overlapped configuration of specimens for joining and the path of the laser beam in the joining process (all numerical values in this figure are in mm)

The variable parameter of the joining process was joining speed, which was set to 1, 2 and 3 $mm \cdot s^{-1}$, as at these speeds, lasting joints were formed with all three PLA materials. Slower joining speeds meant that the laser beam irradiated the specimens longer, thus more heat was formed, and the PLA specimens melted more. At higher joining speeds, the strength of the joints became inconsistent: in some cases, joints did not even form, while in other instances, joints failed after the joining process finished.

We used a Zwick Z005 universal testing machine to measure the load-bearing capacity of the joined specimens under shear load, at room temperature. Measurement speed was set to 5 $mm \cdot min^{-1}$.

3. Results and discussion

3.1. The surface roughness of aluminium

Firstly, we tried to join the PLA specimens with aluminium specimens without any surface structuring (in "as-received" state). We found that no joint can be manufactured at any of the joining speeds because the PLA did not properly wet the surface of the aluminium specimens. Our literature review revealed that certain surface preparation techniques, such as grit blasting, can solve this problem by forming structures on the surface of the aluminium. This increases the size of the area where shape-connected joints can form and thus, the load-bearing capacity of the joints increases. We used corundum of two different sizes (the average diameter of the corundum was 80–100 μm and 300–400 μm) to structure the surface of the aluminium specimens before joining. Average surface roughness (Ra) and average roughness height (Rz) values were measured (Table 3).

Table 3. Average surface roughness (Ra) and average roughness height (Rz) for corundum-blasted aluminium specimens

Average corundum size	Ra (µm)	Rz (µm)
"As-received" surface	0.47	3.15
80-100 µm ("Smooth")	1.89	15.73
300-400 µm ("Rough")	6.00	41.97

By grit blasting with the two different types of corundum, we were able to increase the average roughness of the surface by a factor of 4 ("smooth") and 12 ("rough"). Figure 2 shows SEM images of the surfaces of the "as-received" and the corundum-blasted specimens. The "as-received" surface (Figure 2a) shows the imprint (parallel lines) of the rolling device used in manufacturing the aluminium sheet. These lines were parallel on the whole surface of the specimens, and thus, there was little chance for shape-connected joints to form between the aluminium and the PLA specimens. Figures 2b and 2c show the corundum-blasted surfaces of aluminium specimens. During the grit blasting process, surface structures (ridges and trenches) formed. The polymer melted during the joining process was able to flow into these structures. This made it possible for shape-connected joints to form, which also influenced the global strength of the aluminium-PLA joints.

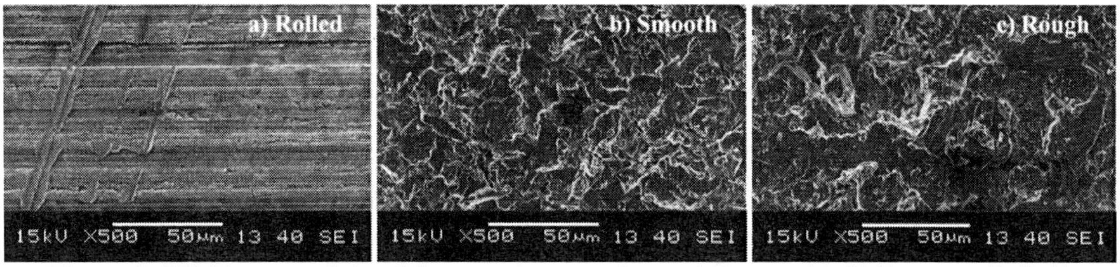

Figure 2. SEM images of aluminium specimen surfaces, a) "as-received" (factory rolled), b) "Smooth" (after surface prep. with 80-100 µm corundum), c) "Rough" (after surface prep. with 300-400 µm corundum)

3.2. Shear tests

During the shear tests, the clamping jaws were offset so that no additional torque (caused by the clamping) loaded the specimens. Results show that strong joints can be manufactured. The shear load-bearing capacity (Figure 3) of the aluminium-PLA joints reached about 66% of the average tensile load-bearing capacity of the 3260HP PLA material. This is the result of tensile tests on three 80 mm × 25 mm × 2 mm specimens. The tensile tester was the same Zwick Z005 testing machine and testing speed was 5 $mm \cdot s^{-1}$.

Both the type of the PLA material and joining speed influenced the quality and load-bearing capacity of the joints (Figure 3). The optimal parameter combination for maximum load-bearing capacity of the aluminium-PLA joint seems to be the "smooth" corundum-blasted aluminium, 2 $mm \cdot s^{-1}$ joining speed and 2500HP PLA. At a joining speed of 1 $mm \cdot s^{-1}$, the reduction in load-bearing capacity of the joints was most probably caused by the degradation of the PLA because of the high heat input. High heat promotes the thermal degradation of the material through random chain scission, cis-elimination and trans-esterification reactions. As the PLA specimens were not dried before the joining process took place, water molecules were also present inside the material, which further increased the molecular degradation of PLA through hydrolysis, as discussed in [15, 16]. At 3 $mm \cdot s^{-1}$, the heat input was too low (because joining speed was too high), which again caused the load-bearing capacity of the joints to decrease.

In almost all of the specimens manufactured with 1 and 2 $mm \cdot s^{-1}$, the PLA specimen failed cohesively, at the edge of the aluminium specimen. This was caused by the high temperature gradient in and the low thermal conductivity of the PLA, which probably promoted the degradation of the material. The failure mode of the joints made with a joining speed of 3 mm·s^{-1} varied between adhesive peeling of the PLA from the surface of the aluminium and cohesive failure of the PLA.

At 1 mm·s^{-1}, the melt viscosity of the PLA did not influence the load-bearing capacity of the joints. This was probably because the effect of shape-connected joints on load-bearing capacity was compensated by the degradation of the PLA material. Our theory was that if we used a PLA with a high melt viscosity, the polymer melt could more easily fill the surface structures of the aluminium specimen, and thus, a stronger joint could be manufactured. This theory, however, was contradicted by joints manufactured at 2 mm·s^{-1}. In these tests, the joints with the highest load-bearing capacity were manufactured with the PLA with the lowest melt viscosity. We will further examine this phenomenon and its causes in the future. At a joining speed of 3 mm·s^{-1}, it seems that the melt viscosity of the PLA did not have a significant effect on the load-bearing capacity of the joints.

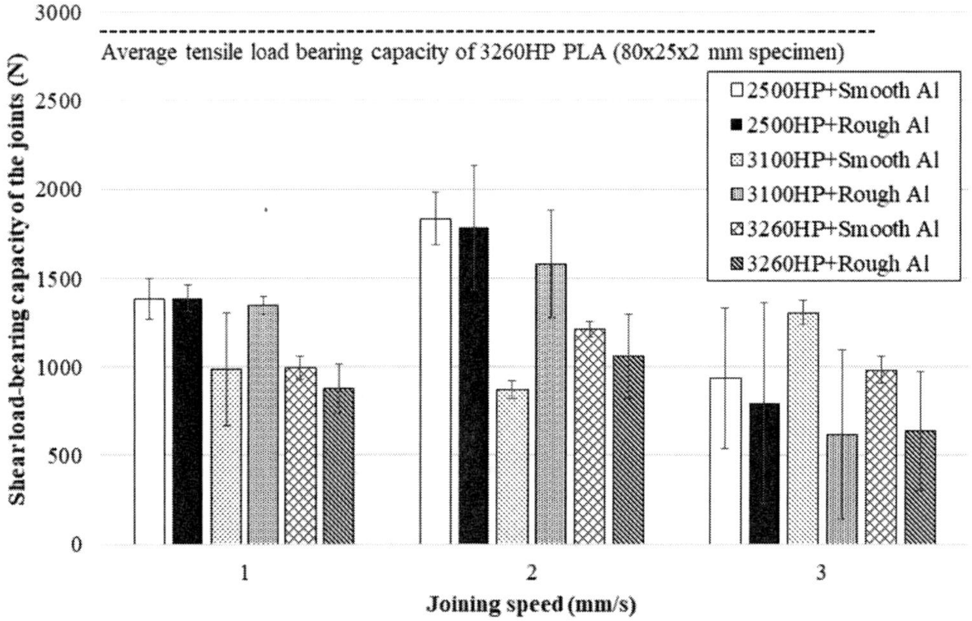

Figure 3. The shear load-bearing capacity of the aluminium-PLA joints

In almost all the cases, the joints manufactured with the "Smooth" aluminium specimens had an equal or better load-bearing capacity compared to the joints manufactured with the "Rough" aluminium surface. This was probably because in the case of "Smooth" aluminium specimens, the surface ridges and trenches were shallower, but more structures were formed on the surface compared to the "Rough" aluminium specimens. The polymer melt can easily fill shallow surface structures, forming many shape-connected joints. At the bottom of deep surface trenches, however, air bubbles can remain under the polymer, decreasing the global strength of the joint (as reported in [17]).

In some cases, near the hole in the weight plate that we used to clamp the materials together, the PLA material was partly or fully displaced (the PLA material formed an arch, marked with a red curve in Figure 4), in the region where the materials were not pressed together. This led to high standard deviations of the measurement results, as the specimens and the area where the PLA properly wetted the aluminium differed from specimen to specimen, even when the process control parameters were the same.

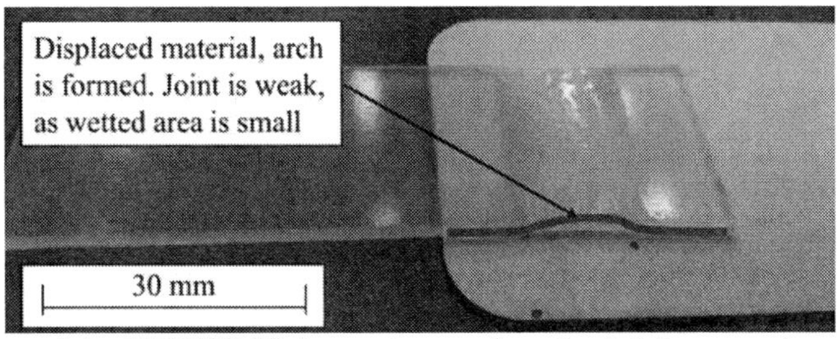

Figure 4. Arching of the 3260HP PLA specimen when the joining speed was set to 3 mm·s^{-1}

4. Conclusions and future plans

We showed that it is possible to join aluminium and PLA specimens by laser beam and that the joints can be strong if the surface of the aluminium specimens is properly structured. In our experiments, we used corundum blasting to change the surface structure, and three different types of PLA materials (these had different average molecular weight, but the same D-lactide content). Our results showed that the most durable joints were manufactured with the use of corundum-blasted aluminium and the PLA with the highest average molecular weight, at a joining speed of 2 mm·s^{-1}. In the future, the clamping system must be improved, and further tests are necessary to evaluate the load-bearing capacity of the joints.

We are currently working on a clamping system, in which the specimens can be clamped together between a highly transparent glass plate and a plate with low thermal conductivity. With this clamping system, we will be able to prevent the arching (displacement) of the polymer material. We are also going to investigate other polymer materials used in the vehicle industry (for example, polyamide and polypropylene and composites based on these materials) and other surface structuring techniques (for example laser ablation of the aluminium specimens).

Acknowledgements

This work was supported by the Higher Education Excellence Program of the Ministry of Human Capacities in the framework of Nanotechnology research area of Budapest University of Technology and Economics (BME FIKP-NAT) and by the National Research, Development and Innovation Fund (TUDFO/51757/2019-ITM, Thematic Excellence Program). The authors would like to thank Szabolcs Bella and AedusSpace Kft. for providing the laser welding machine and equipment, and Balázs Farkas for and Csongor Bukor for his help during measurements.

References

[1] Hirsch J 2014 *Trans. Nonferrous Met. Soc. China* **24** 1995–2002
[2] Taub A et al 2019 *Annu. Rev. Mater. Res.* **49** 327–59
[3] Lehmhus D et al 2019 *Adv. Eng. Mater.* 1900056
[4] Directive 2000/53/EC of the European Commission
[5] Gere D, Czigany T 2018 *Waste Manage.* **76** 190–8
[6] Taiwah B et al 2019 *Express Polym. Lett.* **13** 697–712
[7] Hajba S, Tábi T 2019 *Period. Polytech., Mech. Eng.* **63** 270–7
[8] Zhou J, Wan X, Li Y 2015 *Mater. Today: Proc.* **2** 5015–22
[9] Marsh G 2014 *Reinforced Plastics* **58** 38–42
[10] Feistauer E E et al 2016 *Mater. Lett.* **170** 1–4
[11] Chen Y J, Yue T M, Guo Z N 2016 *Mater. Des.* **110** 775–81
[12] Farazila Y et al 2011 *J. Laser Micro/Nanoeng.* **6** 69–74
[13] Borrisutthekul R, Saengsai A, Mitsomwang P 2016 *Key Eng. Mater.* **719** 142–48
[14] Rodríguez-Vidal E et al 2016 *Phys. Procedia* **83** 1110–7
[15] Speranza V, De Meo A, Pantani R 2014 *Polym. Degrad. Stab.* **100** 37–41
[16] Ozdemir E and Hacaloglu J 2016 *Fibers Polym.* **17** 66–73
[17] Chan C-W and Smith G C 2016 *Mater. Des.* **103** 278–92

IOP Conf. Series: Materials Science and Engineering **903** (2020) 012047 doi:10.1088/1757-899X/903/1/012047

Influence of pack cementation time on the microstructure of Cu nanofoams processed by dealloying

Péter Jenei[1*], Gigap Han[2], Pham Tran Hung[1], Heeman Choe[2] and Jenő Gubicza[1]

[1] Department of Materials Physics, Eötvös Loránd University, Budapest, Hungary
[2] School of Materials Science and Engineering, Kookmin University, Seoul, Republic of Korea

E-mail: jenei@metal.elte.hu

Abstract. Copper nanofoams were processed by dealloying for a prospective application as a Li-ion battery anode. The precursor material was produced by pack cementation. The effect of length of pack-cementation time on the pore structure and the microstructure of the struts was investigated in this study. It was found that for high pack-cementation times a hierarchical pore structure with nano- and micro-sized pores was formed. In addition, the longer the pack-cementation time, the lower the crystallite size and the dislocation density in the struts. A very high twin fault probability was also detected, irrespective of pack-cementation time.

1. Introduction

Metallic foams have a wide range of practical applications, such as energy absorbers in the automotive industry [1], non-flammable acoustic absorbers in vehicles and houses [2], filters [3], heat exchangers [4], implants in medicine [5], and anode materials in batteries [6]. The properties of these foams can be tailored with an appropriate design of pore structures [7]. Among metallic foams, nanofoams attract significant attention due to their extremely high specific surface area, since these materials have pore sizes <100 nm. The very high specific surface area can yield improved performances in some applications, such as for battery electrodes. Pore size and distribution can be designed by a careful selection of production method and processing conditions.

Various methods have been developed for the synthesis of nanofoams, such as the different dealloying processes (chemical, plasma, or metallic melt dealloying methods) [8-10]. During dealloying, the less noble metal in the alloy is dissolved by dealloying solutions, thereby creating pores in the final product. In many dealloying procedures, the precursor alloy is produced by powder metallurgy [8,11]. In this case, there is a high risk of crack formation during dealloying as a consequence of the incomplete sintering. Recently, a novel dealloying procedure has been developed in which the precursor material is processed without powder consolidation [6]. Namely, for manufacturing of the novel Cu nanofoams, first a precursor material was produced by pack-cementation technique, in which Al was coated onto a Cu foil that was subjected to annealing at a temperature between the melting points of Al and Cu. Then, Al diffused into Cu, forming Al and Cu intermetallic compounds. For long pack-cementation times, a pure Al phase also formed due to the high Al content of the material. Finally, Al atoms were removed from the sample by etching, and a Cu foam with fine pore structure was produced. If the precursor material contained both Al_2Cu and Al phases, a hierarchical pore structure with nano- and micro-sized pores was obtained [6]. The larger pores developed at the places of the former Al phase. The Cu foams with hierarchical pore structures exhibited an improved electrochemical performance as anode in Li-ion batteries (e.g., showed a higher capacity compared to simple nanofoams [6]).

The pore size distribution and the microstructure of the foam struts strongly influence the functional properties of metallic foams. Therefore, it is very important to elucidate the effect of the processing conditions on the microstructure of nanofoams. In this paper, the influence of pack-cementation time on the microstructure of Cu nanofoams processed by dealloying is studied. The microstructure is characterized by scanning electron microscopy (SEM) and X-ray line profile analysis (XLPA). The latter method allows determination of the density of lattice defects (e.g., dislocations and twin faults) in the strut material.

2. Experimental materials and procedures

2.1. Sample preparation

In the present study, Cu nanofoams were processed using a dealloying technique. The precursor material was produced by pack cementation. First, a disk was cut from a Cu plate with a diameter of 11 mm and a thickness of 0.25 mm. Then, a blend of Al, Al_2O_3 and NH_4Cl powders were stacked on the Cu plate in a stainless steel container. In the pack-cementation step, the Cu plate in the powder bed was heat-treated at 800 °C for different times: 3, 6, 12, and 15 h. During annealing, the Al particles were melted and Al atoms diffused into Cu, forming Al_2Cu phase [6]. For long pack-cementation times, an additional Al phase also formed. Al_2O_3 is a filler material in the powder blend, which slowed the process by decreasing the effective surface of the plate [6]. After pack cementation, the material was annealed in two stages, first at 700 °C for 9 h and then at 500 °C for 6 h, in order to homogenize the Al distribution in the Al_2Cu phase. It should be noted that the thickness of the samples increased with increasing pack-cementation time due to the incorporation of a higher amount of Al. For the 15 h time, the specimen thickness was enhanced from 0.25 mm to 1.1 mm [6]. In the dealloying process, in order to remove Al atoms from the material, the samples were chemically etched in water solution of HCl (using deionized water) at 45 °C for 12 h. The pH of the HCl solution was set as 0.25. After dealloying, the etched disks were washed in ethanol and dried at 0.8 bar for a few hours. In the following, the samples are denoted according to the pack cementation time (3, 6, 12, and 15 h), however it should be emphasized that in this study the Cu nanofoams were studied after dealloying.

2.2. Methods for the characterization of the microstructure

The microstructure was investigated combining scanning SEM equipped with energy dispersive X-ray spectroscopy (EDS) and X-ray diffraction (XRD). The pore structure of the foam was studied by SEM using a FEI Quanta 3D electron microscope. The chemical composition of the foam struts were investigated by EDS, which was carried out with an electron energy of 20 keV. The phase composition of the foams was investigated by XRD using a Rigaku Smartlab X-ray diffractometer in Bragg–Brentano geometry with CuKα radiation (wavelength: 0.15418 nm). Moreover, XLPA was performed to examine the microstructure of the copper phase after dealloying. The X-ray line profiles were measured with a high-resolution rotating anode diffractometer (manufacturer: Rigaku, type: RA MultiMax-9) using $CuK\alpha_1$ radiation (wavelength: 0.15406 nm). The Debye–Scherrer diffraction rings were detected with two-dimensional imaging plates. The line profiles were determined by integrating the intensity along the rings, and the X-ray diffraction patterns were evaluated using the extended convolutional multiple whole profile (eCMWP) fitting method [12,13]. In this procedure, the diffraction pattern is fitted by the sum of a background spline and the convolution of the instrumental pattern and the theoretical line profiles related to the crystallite size, dislocations, and twin faults. Due to the very broad diffraction peaks, instrumental correction was not necessary. This evaluation method gives the area-weighted mean crystallite size ($\langle x \rangle_{area}$), the dislocation density, and the twin fault probability, where the twin fault probability in face-centered cubic (fcc) crystals is defined as the fraction of twin faults among the {111} lattice planes.

3. Results and discussion

3.1. Phase composition of the foams

Figure 1 shows the XRD patterns obtained on the Cu foams processed with different pack-cementation times. XRD analysis revealed the presence of Cu_2O and a very low amount of Al_2O_3 phase, beside the main fcc Cu phase, in the samples. The secondary phase content was characterized by the fraction of the integrated intensities of the peaks (i.e., the area under the peaks) related to the different phases in the XRD patterns after background subtraction. The fractions of the main and secondary phases are listed in Table 1. The Cu_2O phase was formed due to oxidation of the surface of the Cu foam struts. The fraction of this phase was 7–8% for the 3, 6 and 12 h pack-cementation times. For 15 h, the Cu_2O fraction increased to about 17%. The Al_2O_3 phase was a filler material in the pack-cementation process and a very small amount of this phase ($\leq$ 1%) remained in the final dealloyed samples. In the main Cu phase, the remaining solute Al content was less than 2 at.%, as revealed by EDS, indicating that most of the aluminum was removed during the dealloying process.

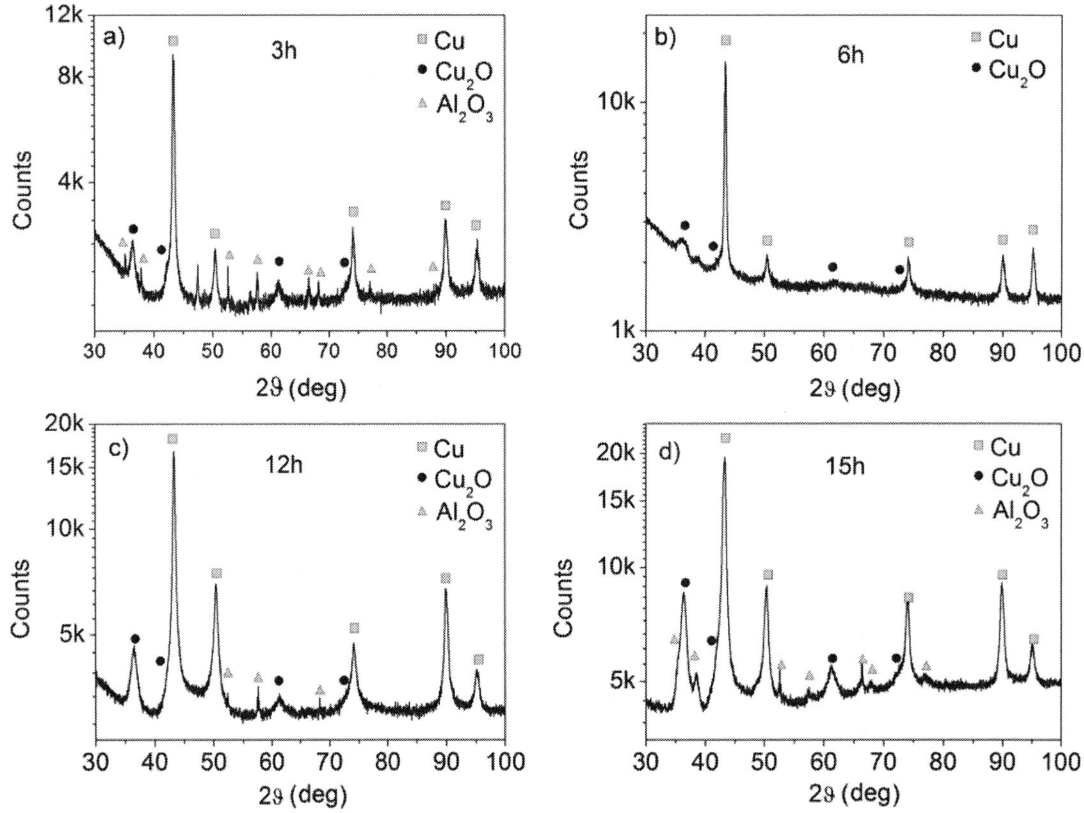

Figure 1. X-ray diffractograms for the Cu foam samples with pack cementation times of (a) 3 h, (b) 6 h, (c) 12 h, and (d) 15 h after dealloying.

Table 1. The chemical compositions of the bulk samples after pack-cementation but before dealloying, and the phase compositions of the dealloyed foams. The average ligament widths of the foams evaluated from SEM are also shown.

Pack-cementation time (h)	Composition of the samples after pack cementation (at.%)		Phase fractions in the nanofoams (%)			Ligament width (nm)
	Al (at.%)	Cu (at.%)	Cu	Cu_2O	Al_2O_3	
3	52.2 ± 0.6	47.8 ± 0.6	92 ± 3	7 ± 2	1 ± 1	200 ± 64
6	60.7 ± 3.1	39.3 ± 3.1	93 ± 2	7 ± 2	-	158 ± 26
12	75.0 ± 3.3	25.0 ± 3.3	92 ± 2	8 ± 2	<1	210 ± 101
15	79.6 ± 2.9	20.4 ± 2.9	83 ± 2	17 ± 2	<1	99 ± 47

3.2. Study of the pore structure by SEM

Figure 2 shows SEM images obtained for the lowest (3 h) and the highest (15 h) pack-cementation times. It can be seen that the morphologies of the pore structures in the two samples are very different. For low pack-cementation time (3 h), a homogenous nanoporous Cu foam was formed. Similar pore structure was observed for the sample processed by pack-cementation for 6 h. At the same time, for the long pack-cementation time (15 h), large micro-sized pores were also formed between the nanoporous struts of the foam. The same feature was found for the 12 h sample.

Figure 2. SEM images with different magnifications taken on the surfaces of the nanofoam samples processed with the pack-cementation times of (a-b) 3 h and (c-d) 15 h.

Difference in pore structures for short and long pack-cementation times is due to different amounts of Al diffused into Cu during pack-cementation. Table 1 shows the chemical composition of the bulk samples after pack cementation but before dealloying. It seems that the increase in pack-cementation time yielded a higher Al content. This can be easily understood if we consider that during pack cementation, Al atoms diffused from the molten Al into the solid Cu plate, and that, therefore, the longer pack-cementation time resulted in a higher Al concentration in the bulk samples before dealloying. For lower Al contents, only Al_2Cu intermetallic compound formed in the pack-cemented samples (specimens 3 h and 6 h). For high pack cementation times (12 and 15 h), when the Al:Cu ratio was higher than 2:1, a pure Al phase necessarily formed in the material. The dissolution of this pure Al phase caused the development of micropores after dealloying in the samples 12h and 15h, with mean micropore sizes of 5.5 ± 2.2 μm and 5.9 ± 2.4 μm, respectively. For all the four studied samples, the foam struts were built up from ligaments and the nanopores can be found between these ligaments. Therefore, the fineness of the nanopore structure in the struts can be characterized by the ligament sizes, which are listed in Table 1. The average ligament width was about 100–200 nm for the studied specimens.

3.3. Microstructure from XLPA

As an example, Figure 3 shows an eCMWP fitting of the XRD pattern for the sample 3h. The microstructural parameters obtained from XLPA are listed in Table 2. The crystallite size varied between 18 and 60 nm for the studied samples. These values are much smaller than the particle (ligament) sizes observed in the SEM images (see Table 1). This difference suggests that the ligaments consisted of smaller crystallites/grains. The Cu ligaments were formed during dealloying when Al atoms were dissolved from the Al_2Cu phase, and the remaining Cu atoms were clustered in an fcc Cu structure [14,15]. Table 2 shows that the Cu ligaments contained a significant amount of lattice defects, such as dislocations and twin faults. The order of magnitude of the dislocation density was 10^{14} m^{-2}, while the twin fault probability varied between 3.5 and 5.2%. It can be seen that the crystallite size and the dislocation density significantly decreased for high pack-cementation times (12 and 15 h). A less considerable change was observed for the twin fault probability.

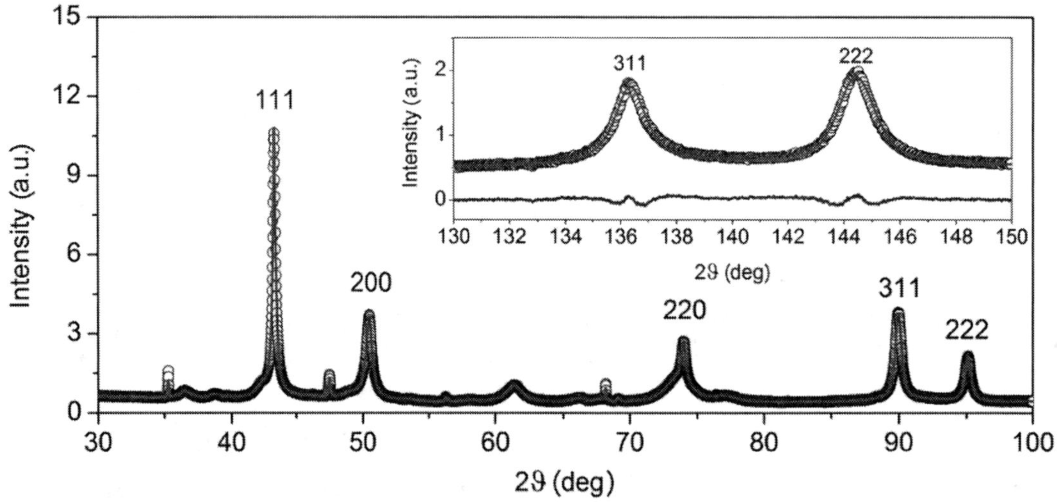

Figure 3. eCMWP fitting of the XRD pattern taken on the Cu foam processed with the 3 h pack-cementation time. The inset shows a magnified part of the diffractogram for a better visibility of the quality of fitting. The open circles and the red solid line represent the measured and the fitted XRD patterns, respectively. The blue curve in the inset shows the difference between the measured and fitted diffractograms.

IOP Conf. Series: Materials Science and Engineering **903** (2020) 012047 doi:10.1088/1757-899X/903/1/012047

Table 2. The area-weighted mean crystallite size ($\langle x \rangle_{area}$), the dislocation density (ρ), and the twin fault probability (β) for the Cu phase of the foams produced with different pack-cementation times.

Pack-cementation time (h)	$\langle x \rangle_{area}$ (nm)	ρ (10^{14} m^{-2})	β (%)
3	60 ± 6	8.6 ± 0.9	5.2 ± 0.5
6	60 ± 6	7 ± 0.7	3.5 ± 0.4
12	29 ± 4	2.6 ± 0.3	3.7 ± 0.4
15	18 ± 4	1.4 ± 0.3	3.6 ± 0.4

The dislocations and the twin faults observed in the Cu struts were grown-in defects formed during the nucleation of Cu ligaments in the dealloying step of sample processing. Dislocations can decrease the mismatch stresses between the neighboring Cu grains in the struts. For long pack-cementation times, the samples contained many Cu grains located on the surfaces of the struts at the large micro-sized pores, and these grains had lower numbers of neighbors. Therefore, these materials have lower dislocation densities.

During the dealloying process, when a new Cu grain nucleates on a pre-existing ligament non-epitaxially, the new grain is preferably separated from the older one by twin fault as this has the lowest energy among the grain boundaries in fcc crystals. Therefore, a very high twin fault probability was detected in the present Cu nanofoams, almost irrespective of the pack-cementation time. It should be noted that further research is planned to study the influence of the pore structure and the lattice defects on the mechanical behavior of the samples manufactured with different pack-cementation times.

4. Summary

The present study shows the influence of pack-cementation time on the microstructure of Cu nanofoams processed by dealloying. To the authors' best knowledge, this is the first investigation in the literature, which quantifies the lattice defect density in the struts of foams obtained by dealloying. The following conclusions were drawn:

1) The pore structure can be tuned with an appropriate selection of pack-cementation time. The higher pack-cementation time induced hierarchically structured materials due to the larger Al content. The ligament width was about 100–200 nm, while the size of the large pores was about 5–6 μm.
2) A significant amount of lattice defects (dislocations and twin faults) was detected in the main Cu phase of the foams, irrespectively of the pack-cementation time. It was found that the dislocation density and the crystallite size decreased considerably with increasing pack-cementation time. The twin fault probability in the Cu phase was very high (3.5–5.2%) for all studied samples, as these boundaries have low energy and, therefore, were preferably formed during growing of the ligaments for the accommodation of misorientations inside the struts.

Acknowledgements

This research was supported by the Hungarian-Korean bilateral Research program (TÉT) No. 2018-2.1.17-TÉT-KR-2018-00003. This work was financed partly by the Ministry of Human Capacities of Hungary within the ELTE University Excellence program (1783-3/2018/FEKUTSRAT). H. Choe also acknowledges support from the International Research & Development Program of the National Research Foundation of Korea (2018K1A3A1A39086825). P. Jenei acknowledges the financial support of János Bolyai Research Scholarship of the Hungarian Academy of Sciences and ÚNKP-19-4 New National Excellence Program of the Ministry for Innovation and Technology.

IOP Conf. Series: Materials Science and Engineering **903** (2020) 012047 doi:10.1088/1757-899X/903/1/012047

References

[1] Kretz R, Hausberger K, Götzinger B 2002 *Adv. Eng. Mater.* **4** 781–85
[2] Miyoshi T, Itoh M, Akiyama S, Kitahara A 2000 *Adv. Eng. Mater.* **2** 179–183
[3] Kim S, Lee C W 2014 *Proc. Mater. Sci.* **4** 305–309
[4] Evans A G, Hutchinson J W, Ashby M F 1998 *Prog. Mater. Sci.* **43** 171–221
[5] Choi H, Shilko S, Gubicza J, Choe H 2017 *J. Mech Behav Biomed Mater* **72** 66–73
[6] Han G, Um J H, Park H, Hong K, Yoon W S, Choe H 2019 *Scripta Materialia* **163** 9–13
[7] F García-Moreno 2016 *Materials* **9** 85–27
[8] Jo H, Cho Y H, Choi M, Cho J, Um J H, Sung Y E, Choe H 2014 *Mater. Chem. Phys.* **145** 6–11
[9] Wada T, Yubuta K, Inoue A, Kato H 2011 *Mater. Lett.* **65** 1076–78
[10] Lee G H, An S, Jang S W, Hwang S, Lim S H, Han S 2017 *Thin Solid Films* **631** 147–51
[11] Chen F, Chen X, Zou L, Yao Y, Lin Y, Shen Q, Lavernia E J, Zhang L 2016 *Mater. Sci. Eng. A* **660** 241–250
[12] Ribárik G, Gubicza J, Ungár T 2004 *Mater. Sci. Eng. A* **387–389** 343–347
[13] Balogh L, Ribárik G, Ungár T 2006 *J. Appl. Phys.* **100** 023512
[14] Liu W, Zhang S, Li N, Zheng J, An S, Li G 2012 *Int. J. Electrochem. Sci.* **7** 7993–8006
[15] Erlebacher J, Aziz M J, Karma A, Dimitrov N, Sieradzki K 2001 *Nature* **410** 450–453

Flame retardancy of recycled PET foam

V A Szabó[1] and G Dogossy[1]

[1]Department of Materials Science and Engineering, Széchenyi István University, Győr, Hungary

E-mail: szabo.veronika.anna@ga.sze.hu

Abstract. Although significant marketing efforts have been made in recent years to reduce polymer use, the number of plastic bottles being discarded is increasing worldwide. The global environmental-socio-economic problem posed by polyethylene terephthalate (PET) bottles can only be solved by expanding large-scale recycling opportunities, while reducing the use of pure raw materials. In this article on the large quantities product of PET was upcycled with chemical foaming. During the experiments 2 m% chain extender and 4 m% chemical blowing agent and different amount of brominated flame retardant was used. The invested materials were examined with standard mechanical tests, scanning electron microscopy and UL-94 standard flammability test. After the investigation it was found that the crystallized blue PET bottle re-granulate can be used forming a flame retardant closed cell foam structure.

1. Introduction

The use of PET, due to its properties, is most noticeable in the packaging industry. Its low density, transparency, impact resistance and carbon dioxide barrier make it the most important raw material for the packaging industry. However, one-way packaging materials become very quickly waste and have a serious impact on the environment. The natural degradation of PET is over half a thousand years. Therefore, the need for recycling PET waste is permanently increasing [1].

Due to the unmanageable nature of solid plastic waste, its volume is constantly increasing and burdens the nature. In Europe, the volume of collected PET bottles participating in the greenhouse effect increased from 0.2 Mt in 1998 to 1.26 Mt in 2008, with an annual growth rate of 19% [2]. In 2016, the number of PET waste increased to 9.6 Mt. According to current forecast, by 2021 only PET bottles will account for 11.6 Mt worldwide [3].

The primary solution to the environmental burden of PET bottles is the production from recycled materials. PET recycling involves three different chain breakage, which can occur as a result of mechanical- and thermal loading and hydrolysis [4]. This behaviour is a serious problem when using recycled polyethylene terephthalate (rPET) as the quality of both the raw material and the finished product is deteriorated. In our experiments, chemical foaming is used to form an integral foam structure from waste bottles using a conventional injection moulding machine. In the case of PET, the foaming depends mainly on the IV value of the base material, since the viscosity of the molten material to a large extent influences the formation of the foam structure. Limits for foam structure formation are found in the literature to be above of 0.8 dl/g, whereas for low density foams require an IV value above 0.95 dl/g. Thus, it is essential to formulate a chain extender to foam rPET [5].

Coccorullo et al. have used a chain-extender additive in their PET recycling research for better applicability. Due to the low IV value of the starting material (0.48 dl/g) it was not suitable in its original state. As a result of their experiments, it has been found that the increasing use of a chain extender increases the viscosity of the molten material, while at higher shear rates, small differences in the ratio of the chain extender additive are found. In their experiments they were able to increase the IV value of the starting material by 37.7% [6].

The amount and rate of chain-extender additives were also investigated in case of rPET by Turfa et al. The use of 1% additive is sufficient to increase the length of the polymer chains, but at 2% the frequency of branching within the molecular chain also increases. At higher rates of application, unwanted crosslinking is caused by the chain extender, which can lead to crystallising leading to degradation of processing properties [7].

Compared to extrusion foaming [6-9], the production of injection moulding foamed products is a less researched area. During the production of the foam structure on the injection moulding machine, the porosity is greatly influenced by the injection moulding parameters.

Gómez et al. investigated the effect of injection moulding parameters on glycol modified polyethylene terephthalates (PETG) microcellular foam structure. Their results show that the tool temperature and the batch volume have the greatest influence on the foam structure formed. It has been shown that increasing the tool temperature increases the porosity of the samples [10].

If the foam structure of rPET is to be widely used, care must be taken to reduce the rate of inflammation and flame propagation. The main raw material for polymer production is petroleum, so the calorific value of the products is almost the same (40-45 MJ/kg). The primary purpose of incorporating flame retardants is to reduce flammability [11].

In their research, Bocz et al. investigated the development of flame retardant, fabric reinforced composites from mixed plastic waste. During their work, the research team has developed a multilayer, value-added composite based on high polyolefin-containing automotive shredder light fraction. The mechanical properties of the recycled composites are provided by the fabrics made of polypropylene. The reduced flammability is provided by a matrix layer of secondary raw material, a phosphorus-containing flame-retardant additive. Based on the results, it was found that high technical value plastics, which can be recovered from secondary raw materials, were eventually obtained by self-reinforcing composite with added flame retardancy [12].

Xue et al., in their research, produced magnesium hydroxide and poly (ethylene terephthalate) microencapsulated CMSs by polymerization and liquid phase deposition. The PET used was produced by blending with the flame retardant using a direct molten blend. Based on their results, the developed material showed a LOI of 27.2%. Developed CMSs suppress smoke production [13].

The design of the foam structure affects the flammability of the plastics. An important task in developing the foam structure is to increase the resistance to combustion. In their research, Wu et al. developed a novel and simple strategy for the composite material of high flame-retardant polymer foams using a silicone resin coating and a silica gel self-extinguishing layer. An organic / inorganic silicone polymer coating was formed on the surface of the polymer foam based on condensation polymerization and they were able to minimize the oxygen concentration. During the UL94 test, the developed foams completed a complete self-extinction within 30 seconds.

In their research, Carosio et al. investigated the layered structure of flame retardants in a closed cell foamed PET structure. The flame-retardant properties of ammonium polyphosphate (APP) have been compared with that of deoxyribonucleic acid (DNA). Inflammation and cone calorimetry studies clearly demonstrated the superior performance of APP-coated coatings over DNA-based coated foams. As a result of their experiment, they have found that APP-based coatings can suppress the melt-drip behaviour characteristic of PET and that they develop a 25% reduction in heat release peak [14].

At the beginning of our research the aim was to produce recycled PET integral foam structure by chemical foaming. As a result of the first experiments, we managed to produce closed cell foam on a conventional injection moulding machine using chemical blowing agents [15]. In order to improve the foam structure, the ability of various foaming agents to form the foam structure of rPET was tested. Degradation due to chain breakage greatly reduced foam formation; we increased the length of the molecular chains using CESA Extend. To improve the mechanical strength of the foam structure, 10%

impact modifier additive was added to the blend, which improved the Charpy impact strength by 18 %. As a further development of the material, various flame retardants have been tested to improve the flame resistance of the foam [16]. The results of the new blend showed that even a small amount of flame retardant improved the Charpy impact strength of the samples by a considerable amount. The flame retardant increased the porosity of the samples and reduced the cell diameters, so the results of all mechanical tests were modified with the addition of the additive. Thus, the additive used for flame retardancy showed a greater improvement in the mechanical properties of the blend than the master blend for impact resistance [17]. Based on the results, the question raised; is there a need to use impact modifier during manufacture? In this article, we investigate the properties of rPET foam developed with the use of a flame retardant without the addition of impact modifier.

2. Applied materials and methods

The used matrix material was the blue crystallized PET regranulate (rPET) which supplied by Fe-Group Invest Zrt (Hungary) (intrinsic viscosity (IV) 0.8 dL/g). The used chain extender additive was CESA Extend NCA0025531-ZA that provided by Clariant (Switzerland), which contains Joncryl ADR 4368 type epoxy-based styrene-acrylic multifunctional oligomer reagent. The used flame retardant additive was ICH Fl. Ret. 01904 which donated by ICC-Chemol (Hungary) with 85% dispersion of brominated organic compound in ethylene co-polymer. The used chemical blowing agent was Tracell IM 7200 that granted by Tramaco (Germany) endothermic compound with 120 ml/g gas expansion and 70% blowing agent content.

The matrix was dehydrated for 6 h at 120 °C, and then blended with the additives. The twin-screw extruder type LT 20-440 (Thailand) was used for the melt mixing with 265 °C melt temperature. The produced re-granulate was dehumified for another 6 h at 120 °C, then 4x10 mm^2 cross-section dumbbell-shaped specimens were prepared by an Arburg Allrounder Advance 420C Golden Edition (Germany) injection moulding machine with breath moulding technique, with 265 °C melt temperature, 45 cm^3/min injection speed and 35 °C mould temperature. The injected dumbbell-shaped specimen was used for tensile and flexural test. For impact test both ends were cut off to get standard 4x10x80 mm specimen. The SEM images were took from the fracture surface of impact test specimen. For UL-94 test the dumbbell-shaped specimen's one end was cut off. Table 1 shows the composition of the test samples. The notation B used indicates that this is a continuation of an earlier series of experiments, the results of series A are in our earlier publication [17].

Table 1. The compositions of the test samples

	ref	B-0	B-5	B-10	B-15	B-20
	parts per hundred (phr)					
rPET	100	100	100	100	100	100
Chain Extender	0	2	2	2	2	2
Chemical Blowing Agent	4	4	4	4	4	4
Flame Retardant	0	0	5	10	15	20

The melt volume rate (MVR) was determined using a modular MVR tester type Ceast MF-30 (Italy) with 2.16 kg load at 210 °C. The density was determined based on the Archimedes principle using analytical balance type Ohaus Explorer (USA). The inner structure of parts was investigated with a Jeol JSM 6380LA (Japan) type scanning electron microscope (SEM).

The EN ISO 527 standard tensile tests were executed using universal testing machine type INSTRON 5582 (USA) with 100 mm grip length and 1 mm/min test speed when determining the modulus and 5 mm/min when determining the strength. Three-point-bending tests were carried out using universal testing machine type INSTRON 5582 (USA) as stated by EN ISO 178 standard with 10 mm/min test speed and 64 mm support span. Charpy impact tests, as claimed by EN ISO 179 standard, were conducted by impact equipment type CEAST 65-45,000 (Italy), using unnotched specimens, with a pendulum of 15 J and 62 mm support span.

For the determination of mean value and standard deviation five specimen was used in case of every test. The tests were performed at room temperature with relative humidity of 50%.

Standard UL-94 tests were implemented in a chemical fume hood with methane gas on 4 mm thickness specimen. UL-94 classification is used to determine dripping and flame spreading rates. First, vertical burning tests were carried out. If the samples did not reach any class (V-0, V-1 or V-2), then the horizontal burning test was conducted as well.

3. Results and discussions

The following section presents the test results for the samples. The aim of our study was to analyse the effect of the flame retardant on the formation of foam structure and its mechanical properties.

3.1 Morphology

The foam structure is highly dependent on the fluidity of the material. In the 1. figure, sample densities are compared with MVR values. The results show that the density of the samples increases with increasing flow rate.

Increasing the proportion of flame retardant, the density increased steadily. This contrasts with the results of samples produced with impact resistance enhancers. Increasing the proportion of flame retardant, fewer cells were formed in the material. This suggests that, although the impact modifier itself did not promote foaming, it had a significant effect on the flowability of the material when mixed with the flame retardant. The MVR value of the samples with impact modifier increased continuously with increasing the flame retardant ratio, thus increasing the viscosity of the blends in inverse proportion [17]. In contrast, the viscosity of the presently tested blends shows a steady decline as the rate of flame retardant increases. Samples with a lower MVR value of 5 m% (B-5) and 10 m% (B-10) flame retardant, respectively, showed greater loss of density than previous samples with impact modifier.

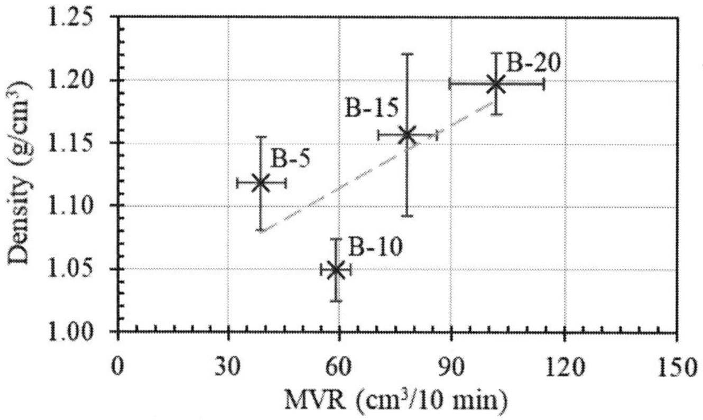

Figure 1. The melt volume rate effect on density of the different flame retardant content samples

Comparing the density and flow rate results with the fractures microscope images (Fig. 2.) shows that not only did the cell number decrease with the increase of the flame retardant ratio, but also the cell diameters increased. The large reduction of the density in the samples containing 5 m% (B-5) and 10 m% (B-10) flame-retardant was greatly influenced by the thinner shell of the samples. The melt in contact with the temperature of the tempered tool wall is continuously solidified to form an external pore-free shell. The skin layer on the sample containing 10 m% (B-10) flame retardant is less than half that of the samples containing 20 m% (B-20) flame retardant. For mixtures with lower MVR values, a more ordered cellular structure is obtained, and indeed an integral foam is formed in which the expansion ratio increases towards the centre of the section.

Fragmentation of the samples can be compared to the measured density values. Lower flow rates resulted in higher porosity and more ordered cellular structure. Even with higher flow rates, less cells with large diameters were formed. There is a risk of damage to the closed-cell foam structure as the foam formation of mixtures with lower viscosity may cause the cells to join.

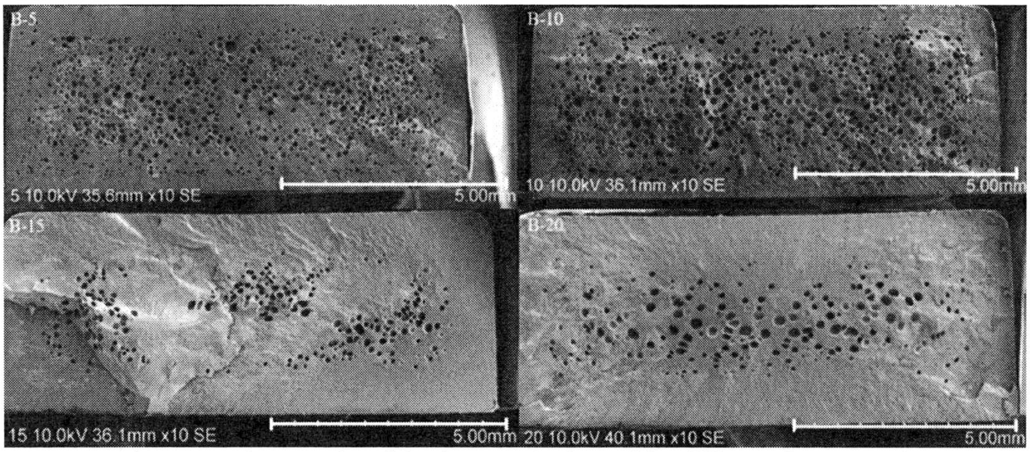

Figure 2. The SEM micrographs of the fracture surface of the different flame retardant content samples

3.2 Mechanical properties

The foam structure forming turns the tough material brittle. The results of the tensile test are shown in Figure 3. The results indicate that the chain extender had a positive effect on both tensile strength and Young modulus. The primary reason for this is that the use of a chain extender by coupling the molecular chains facilitated the formation of a closed cell foam structure. A more ordered cell structure was obtained than with no additive, which is supported by the lower standard deviation. The results of the study show that although the value of the modulus of elasticity increased due to the high value of the standard deviation, there is no significant difference between the results of the flame retardant samples. The higher modulus of elasticity can be related to the lower cell number seen on SEM images. Samples with 15 m% (B-15) flame retardant with the highest modulus of elasticity had a small number of cells with uneven distribution. The distance between the cells was the largest on these samples. As a result, while samples with high porosity were rapidly fractured, samples with low cell number were capable of elastic deformation.

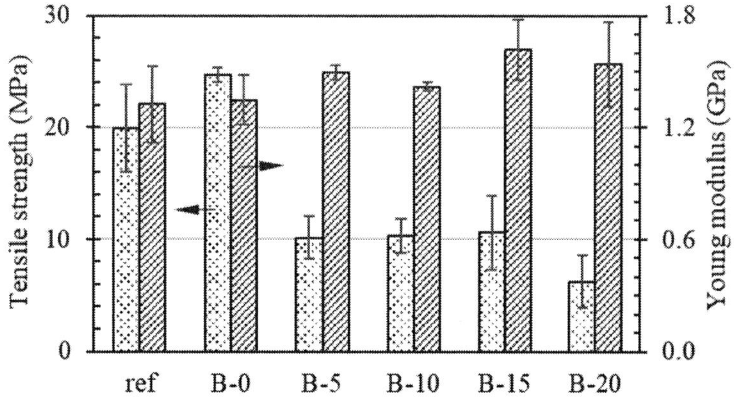

Figure 3. The effect of flame retardant to tensile properties

The results show that the tensile strength of the samples containing the flame-retardant is continuously decreasing with increasing the proportion of flame retardant. The standard deviation of the samples containing 15 m% (B-15) and 20 m% (B-20) flame retardant was high even with the use of chain extender. During the test, several samples had joined cells. This can be related to the higher

MVR value. In the test, joined cells facilitated rupture as stress collection areas, even though these samples had a much larger shell thickness. As a result of the thinning or rupture of the intercellular material, the whole sample soon broke. The high standard deviation values are also due to these incomplete cells.

The more arranged foam structure observed in the samples containing 5 m% (B-5) and 10 m% (B-10) flame retardant, respectively, resulted in a reduction in the inter-sample variation in both tensile strength and elastic modulus results. The results indicate that the use of the flame retardant had a negative effect on the tensile stress resistance of the material.

The results of the three-point bending test are shown in Figure 4. The foamed samples of rPET soon underwent brittle fracture after minimal bending. The use of flame retardant reduces the results of the bending elastic modulus. The results of the bending test can also be compared with the shell thickness seen on the fracture images. Both the flexural modulus and the flexural strength were greatly reduced using a chain extender. In contrast, a more ordered cell structure resulted in lower standard deviation values during sample analysis.

At B-5 and B-10 samples high cell density foams showed low flexural modulus. During the three-point bending test, the thick shell showed greater elastic deformation. Another reason is that where the samples containing 15 m% (B-15) and 20 m% (B-20) flame retardant still form a thick solid shell, for the samples containing 5 m% (B-5) and 10 m% (B-10) flame retardant, only the intercellular matrix material can withstand the stress. These thin solid units crack between the cells so that the elastic deformation of the specimen itself is lower. In contrast, their flexural strength is much higher for samples with larger shells containing 15 m% and 20 m% flame retardant, respectively. This is due to the larger cell diameter experienced with the higher flame retardant content and to the joined cells. As a result of the stress, samples with larger cavities suffered fractures with low flexural strength in which the highest standard deviation was observed.

High cell density B-5 and B-10 specimens were broken only after cracking and breaking of matrix between cell units, resulting in higher flexural strength values.

If samples without a flame retardant are considered, the use of a chain extender, while providing a more ordered cellular structure and higher porosity, results in approximately one-third of the flexural strength of the samples. At this value, the use of a flame retardant at 5 m% (B-5) also showed a more than fourfold improvement. This can be paralleled with MVR values at lower flame retardant contents. The absence of impact modifier had a positive effect on the flexural strength of the material because the additive increased the stiffness of the samples.

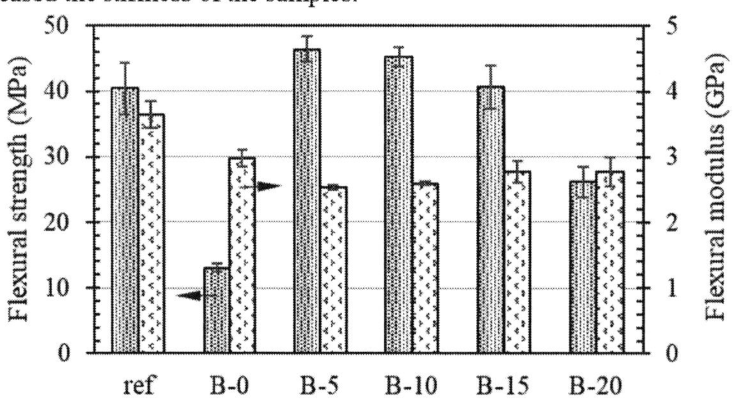

Figure 4. The effect of flame retardant to flexural properties

In our research results, the application of impact modifier has reduced the resistance of the samples to tensile and flexural strength. In contrast, improved impact strength in Charpy impact test [17]. The results of samples prepared without impact modifier are illustrated in Figure 5. The figure shows that the use of a chain extender, which has resulted in a higher cell density, nearly halves the impact strength of the specimens. Contrary to increasing the flame retardant ratio, the Charpy impact strength of the samples is reduced. Based on the fracture recordings, the chain extended samples result in fast fracture

IOP Conf. Series: Materials Science and Engineering **903** (2020) 012048 doi:10.1088/1757-899X/903/1/012048

due to the high porosity. Even samples with a 15 m% (B-15) and 20 m% (B-20) flame retardant with a thick skin layer have low toughness. This can be compared to thinned cell material, which decreased in proportion to the increase in MVR.

The high standard deviation observed during the measurements is primarily due to the disordered cell structure. In the test, the foamed PET specimens produce high dispersion in all cases. This is due to the cell density in the middle section and the size of the cell diameters. If we consider the cell distribution seen on the SEM images, we found that the 15 m% flame retardant (B-15) had the lowest cell density in the middle section.

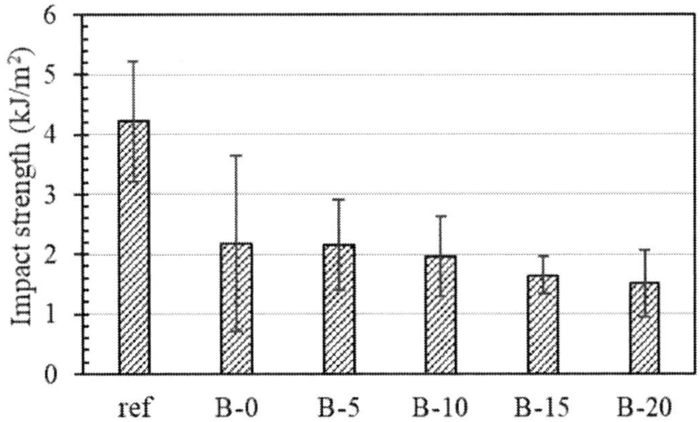

Figure 5. The effect of flame retardant to impact properties

3.3 Flammability properties

The samples were subjected to the UL-94 flammability test (Table 2.). We started the study with the more rigorous vertical burning. All samples were degraded by drip burning during the study. Thus, samples containing 15 m% (B-15) and 20 m% (B-20) flame retardant, respectively, only met the criteria for classification V-2. This can be compared to the high MVR value measured during the flow test. All in all, it can be concluded that although the impact modifier does not have a direct effect on the material's resistance to burning. The application increases the viscosity of the samples, resulting in thickening of the intercellular areas, which play a major role in the quenching process. A large volume of burning was observed during the heat transfer, which also dripped from the sample. In contrast, samples with impact modifier agent exhibited much lower levels of degradation due to the release of bromide free radicals from cell walls. Furthermore, the lower MVR value reduced the possibility of dripping. Thus, these samples met the requirements of flammability class V-0 [17].

Table 2. Flammability characteristics of samples

Samples	UL94 rating	Burning rate (mm/min)
ref	No rating	-
B-0	No rating	-
B-5	HB	19.5
B-10	HB	16.2
B-15	V-2	-
B-20	V-2	-

Samples containing 5 m% (B-5) and 10 m% (B-10) flame retardant could not be classified in the vertical test. Therefore, they were subjected to the UL-94 test in a horizontal configuration. The B-5 sample burned at 19.5 mm/min while the B-10 sample burned at 16.2 mm/min. None of the specimens reached the maximum burning rate of 40 mm/min and therefore the samples can be classified in the HBF flammability class. Overall, it can be stated that B-5 or B-10 material may be suitable for industrial use, but its use is limited. The non-flame retardant rPET foams could not meet the criteria for any classification.

4. Conclusion

After analysing the specimens, it was found that the available crystallized blue bottle re-granulate can form a closed cell foam structure with flame-retardant, without the use of impact modifier. In contrast, the impact modifier plays an important role in the foam texture because it influences the flowability of the blends. The impact modifier reacts with the flame retardant during extrusion and doubles the viscosity of the blends. In the higher viscosity material, the solid matrix between the gas cells formed by the carbon dioxide is not damaged, thus providing a closed cell foam structure. By omitting the impact modifier, the cells combine in the high flow samples. Thus, in the samples containing 15 m% (B-15) and 20 m% (B-20) flame retardant, the merged large cells formed cavities that function as a stress collection point. However, around such a large hole, the matrix formed a thick solid unit. As a result, the overall density of the samples hardly decreased. On the other hand, mechanical tests showed a steady decline with the increase of the flame-retardant ratio. The disordered cellular structure formed also reduced the resistance of the samples to burning. In addition, the high MVR in the UL-94 tests resulted in drip burn, so that even a 20 m% flame retardant sample (B-20) could only meet the criteria of theV-2 flammability class. The results achieved are moving in the right direction for the E&E industry, for example injected TV parts [18], but further development is still needed.

Acknowledgement
The research was carried out as part of the EFOP-3.6.2-16-2017-00016 project in the framework of the New Széchenyi Plan.

References

[1] Torockay K 2009 *Műanyag és Gumi* **46** 442-4
[2] Shen L, Worrell E and Patel MK 2010 *Resour., Conserv. Recyc.* **55** 34-52
[3] Sebők O 2017 *Piac & Profit* **21** pp 17-8
[4] Paci M and La Mantia FP 1998 *Polym. Degrad. Stabil.* **61** 417-20
[5] Matuana L, Faruk O and Diaz C 2009 *Bioresource Technol.* **100** 5947-54
[6] Coccorullo I, Di Maio L, Montesano S and Incarnato L 2009 *Express Polym. Lett.* **3** 84-96
[7] Turfa E, Dogossy G and Ronkay F 2013 *Anyagok világa* **2** 50-8
[8] Kmetty Á, Litauszki K and Réti D 2018 *Applied Sciences-Basel* **8** 1960
[9] Bocz K, Tábi T, Vadas D, Sauceau M, Fages J and Marosi Gy 2016 *Express Polym. Lett.* **10** 771-9
[10] Gómez J, David A, Sanchez-Soto M and Martinez A 2013 *J. Cell. Plast.* **49** 47-63
[11] Lyon RE and Janssens ML 2015 Polymer Flammability (*Encyclopedia of Polymer Science and Technology* 15 Volume Set) ed Herman FM (New York: Wiley) chapter 1 pp 1-70
[12] Bocz K, Toldy A, Bárány T and Marosi Gy 2011 *Műanyag és Gumi* **48** 84-7
[13] Xue B, Niu M, Yang Y, Bai J, Song Y, Peng Y and Liu X 2018 *Appl. Surf. Sci.* **435** 656-65
[14] Carosio F, Cuttica F, Di Blasio A, Alongi J and Malucelli G 2015 *Polym. Degrad. Stabil.* **113** 189-96
[15] Szabó VA and Dogossy G 2018 *Acta Technica Jaurinensis* **10** 157-67
[16] Szabó VA and Dogossy G 2018 *IOP Conf. Ser.: Mater. Sci. Eng.* **426** 012043
[17] Szabó VA and Dogossy G 2020 *Period. Polytech., Mech. Eng.* **64** 81-7
[18] Ronkay F, Molnár B, Szalay F, Nagy D, Bodzay B, Sajó IE and Bocz K 2019 *Polymers* **11** 233

IOP Conf. Series: Materials Science and Engineering **903** (2020) 012049 doi:10.1088/1757-899X/903/1/012049

Investigation of the interfacial adhesion of glass bead-filled multicomponent injection moulded composites

A Suplicz[1], R Boros[1] and O V Semperger[1]

[1] Department of Polymer Engineering, Faculty of Mechanical Engineering, Budapest University of Technology and Economics, Műegyetem rkp. 3., H-1111 Budapest, Hungary

E-mail: suplicz@pt.bme.hu

Abstract. Polymeric materials are often combined with fillers and reinforcements, which modify their properties. The goal can be for example the improvement of thermal and mechanical properties or a reduction in costs. On the other hand, these materials can have negative effects too. In multicomponent injection moulding or in the case of products with weld lines, bonding problems can occur. In this research project, we analysed the effect of glass fibre and glass beads, and the most important technological parameters, melt temperature and holding pressure on the bonding strength between multicomponent injection moulded parts. The test samples were produced with a special injection mould, and the tear-off tests were performed on a tensile testing machine with a grip we developed.

1. Introduction

In the last decades, the plastic industry has been one of the most dynamically developing sectors. In 2018, more than 350 million tons of polymer material was produced. The most important and versatile polymer processing method is injection moulding, which is supported by various technological solutions, processing techniques and a great variety of raw materials [1, 2].

Polymeric materials are often compounded with additives, fillers and reinforcements as many industries, such as the automotive or medical industry have strict requirements for plastics. In most cases, the goal of using additives is to improve mechanical properties, colour the polymer or simply decrease costs. The additives can have negative effects, too. During injection moulding, they can cause welding or bonding failures in multicomponent, overmoulded parts or in parts containing weld lines. This problems can decrease the lifetime of the parts [1, 3-8].

Interfacial adhesion is influenced by many factors. The theories are grouped in the literature into seven types: mechanical coupling (interlocking), diffusion, thermodynamic, electrostatic interaction, weak boundary layer, polarization and chemical bonding theory. The mechanical coupling theory means that the overmoulded material interlocks into the surface irregularities of the other part. According to the diffusion theory, adhesion forms when the two surfaces come into contact to each other at appropriate temperature and clamping pressure, and the macromolecules diffuse through the interface. In the thermodynamic theory, adhesion depends on the wettability of the materials. The strength of adhesion is affected by the surface energy, the chemical structure and the interactions of the components. Finally, the chemical bonds theory means that chemical bonds, like covalent bonds can be formed between the surfaces [9-10].

Adhesion between the components in overmoulding is a little researched topic, only a few articles can be found in this area. In most cases, the publications focus the development of a test mould or an adhesion test, or the analysis of the adhesion between different overmoulded materials [11-14]. Therefore, the aim of our work was to analyse the bonding strength of the overmoulding process. First, we analysed the effect of the main injection moulding parameters (melt temperature and holding pressure) on unfilled samples. After that, unfilled polypropylene was overmoulded on samples filled with glass fibre and glass beads, and the effect of these fillers was analysed. We used a special grip in the evaluation of bonding strength and explained the results using microscopic images.

2. Materials and methods

2.1. Materials

For the matrix, we used Tipplen H145 F homopolymer polypropylene (PP) (MOL Petrochemicals Co. Ltd., Hungary). The reinforcement was Camelyaf BMC-1 (Şişecam Chemicals Group, Turkey) 6 mm long chopped glass fibre (GF). We also used glass beads (GB) (Cerablast GmbH & Co. KG, Germany) in two different sizes to make PP/GB compounds. According to the producer, the chemical composition of the GB is the following: SiO_2: 68–75%, Na_2O: 12–18%, CaO: 7–12%, MgO: max. 5%, Al_2O_3: max. 2.5%, K_2O: max. 1.5% and Fe_2O_3: max. 0.5%.

2.2. Sample preparation

First, we prepared the compounds of the preforms for the overmoulding tests. Polypropylene with 30 m% glass fibre was compounded on a Labtech LTE 25-30/C (Labtech Engineering Co., Ltd., Thailand) single-screw extruder. We used a single-screw extruder to minimize the breakage of the reinforcement and fillers. The temperature of the zones were 210 °C; 205 °C; 200 °C and 195 °C, the temperature of the die was 210 °C and screw rotation speed was 25 1/min. The PP/GB compounds were prepared in the same way, but before compounding we separated the beads into two fractions (75-125 µm and 125-250 µm) with a Cisa BA 200N sieve shaker (CISA Cedaceria Industrial, Spain). PP/GB compounds were produced with 10 m%, 25 m% and 40 m% GB concentration.

After that the 80 mm x 80 mm x 2 mm flat preforms were injection moulded from the compounds with an Arburg Allrounder 370 S 700-290 injection moulding machine. The main injection moulding parameters of the preforms can be found in Table 1.

Table 1 Injection moulding parameters

Melt temperature	Value	Unit
Temperature of the heating zones	210, 210, 205, 200, 195	(°C)
Mould temperature	40	(°C)
Volume	48	(cm^3)
Switchover point	6.6	(cm^3)
Holding pressure	800	(bar)
Holding time	20	(s)
Injection rate	20	(cm^3/s)
Plasticizing speed	20	(m/min)

The overmoulding tests were performed with a special mould (Figure 1), presented in our earlier paper [14]. With this special cold runner tool, we can overmould a 70 x 50 mm rib onto the 80x80 mm preforms. We placed two pressure sensors and three temperature sensors near the bonding area. The temperature of the mould can be controlled with a temperature controller up to 100 °C. During the overmoulding test, mould temperature was 40 °C, the injection rate was 20 cm^3/s and holding time was 5 s. Melt temperature and holding time were varied according to the DoE (Design of experiments) presented in the next chapter.

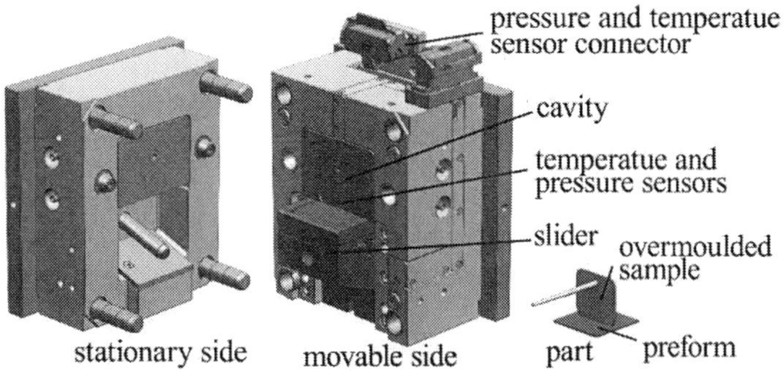

Figure 1 Special mould for overmoulding test

2.3. Design of Experiments

We performed three different designs of experiments to analyse interfacial adhesion between the preforms and the overmoulded material. In the first experiment, melt temperature and holding pressure were varied, as these can affect adhesion. During the tests, the other parameters were kept constant.

Table 2 DoE for analysing the effect of the main injection moulding parameters

Settings	Melt temperature (°C)	Holding pressure (bar)
Melt temperature test	190; 215 or 240	450
Holding pressure test	215	250; 450 or 650

In the second experiment, the effect of the reinforcement was analysed. First the unfilled polypropylene was overmoulded on unfilled flat preforms, then the unfilled PP was joined to the polypropylene preforms filled with 30 m% glass fibre. The tests were performed at three different melt temperatures: 190 °C, 215 °C and 240 °C.

During the third DoE, the unfilled PP was overmoulded on glass bead-filled polypropylene samples. In this experiment, glass bead sizes and filler concentration were varied. The first bead size was between 75 and 125 μm, and the second was between 125 and 250 μm. Melt temperature was 215°C and holding pressure was 450 bar.

2.4. Characterization of the samples

Bonding strength was analysed with the use of a special grip (Figure 2.) in tensile test mode, designed for the T-shaped injection moulded samples. The tensile tests were performed on a Zwick Z020 (Zwick GmbH & Co. KG, Germany) universal testing machine. During the tests, the injection moulded samples were laid upside down on the upper plate and the overmoulded rib was put into the gap of the plate. After that the rib was clamped with a grip. Then the upper half of the grip was displaced vertically. We determined bonding strength (tear-off strength) by dividing the maximum tear-off force with the connecting surface area (120 mm^2).

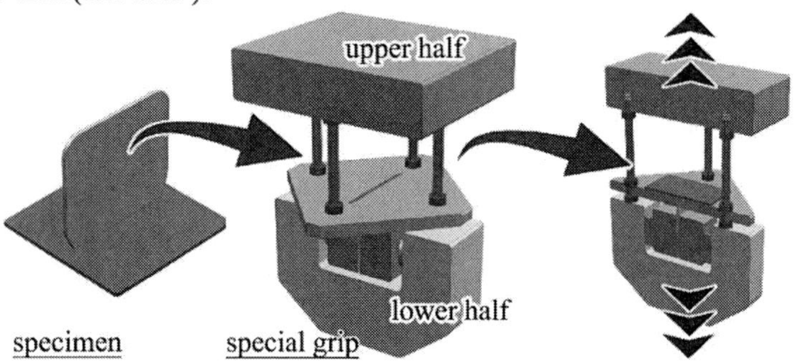

Figure 2 Special tool to measure the tear-off strength

The fracture surfaces of the samples were analysed with a JEOL JSM 6380LA scanning electron microscope (SEM) (Jeol Ltd., Japan). Before the analysis, the surface of the samples was gold spur coated with a JEOL FC 1200 device. Glass bead distribution was analysed with a Keyence VHX-5000 (Keyence Corporation, Japan) optical microscope.

3. Results and discussions

3.1. Effect of the injection moulding parameters

First, the samples were injection moulded in the special overmoulding mould without preforms—these were the reference samples. The tensile tests carried out on these samples yielded the reference bond strength, which was 28.8 ±0.24 MPa.

As a next step, we analysed the effect of melt temperature on bonding strength between the unfilled PP preform and the PP material (Figure 3/a). As melt temperature increased, bonding strength also increased significantly. At a low melt temperature (190 °C) the tear-off strength was very low, only 5.1 MPa, while at a melt temperature of 240 °C, tear-off strength was 14.7 MPa strength, which is still 50% lower than that of the reference sample. The reason for this could be that at higher melt temperatures, the temperature of the preform will be closer to its melt temperature, hence molecular connections are more perfect. For bonding, the surface of the preform must reach a high enough temperature for a melt layer to form. This way, the polymer molecules can diffuse across the interface (intermolecular diffusion). At a low temperature, the thickness of the melt layer is low. As melt temperature is increased, the melt layer becomes thicker and more molecular movement can be achieved.

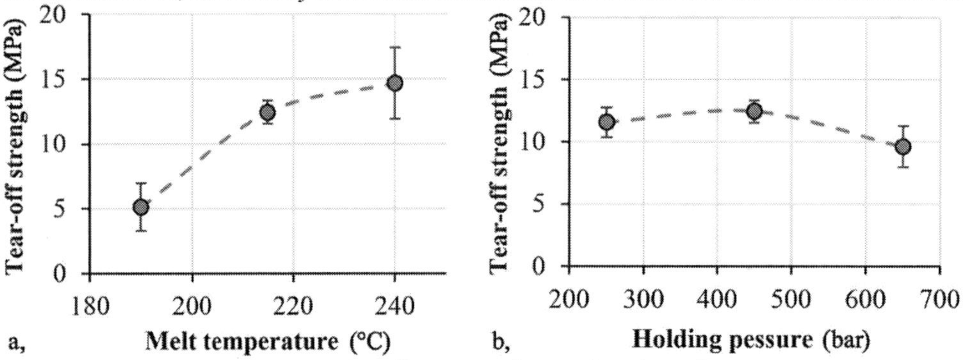

Figure 3 Bonding strength as a function of melt temperature (a) and holding pressure (b)

In the second step, we analysed the effect of holding pressure (Figure 3/b). The theory was that at higher pressures heat transfer between the surfaces is increased, hence the surface temperature of the preform and the thickness of the melt layer are also increased. This way, intermolecular diffusion is improved across the interface. Holding pressure was analysed at three levels (250, 450 and 650 bar) and was maintained for 5 seconds. Actual pressures (Figure 4.) during injection moulding were captured with cavity pressure sensors near the bonding area. Cavity pressures were about 100-150 bar lower than holding pressure set on the machine. Tear-off strength varied between 9.6 and 12.4 MPa. At low pressures (between 250 and 450 bar), the average tear-off strength increased according to the theory. Contrary to expectations, bonding strength decreased slightly at the highest pressure (650 bar); it is probably caused by residual stresses.

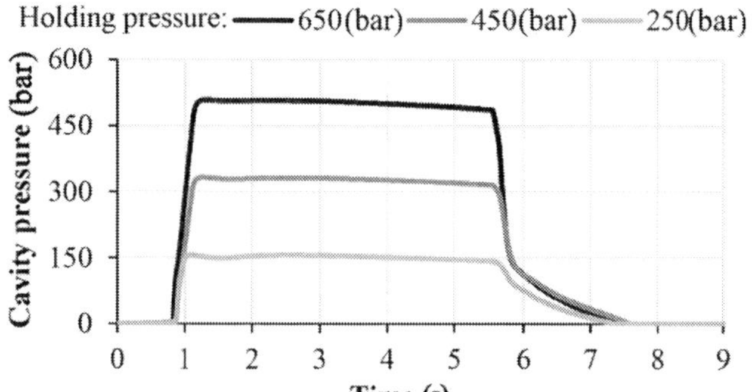

Figure 4 Cavity pressure next to the bonding surface as a function of holding pressure

3.2. The effect of the reinforcements

The effect of the glass fibre reinforcement on bonding strength was also analysed at three melt temperatures. At the lowest melt temperature (190 °C), tear-off strength was very weak; in most cases the preform and the overmoulded part separated right after clamping during the mechanical test. Average tear-off strength was about 1.2 MPa. As melt temperature increases, binding strength also increases. When melt temperature was 215 °C and 240 °C, bonding strength was 5.6 MPa and 9.8 MPa, respectively. As Figure 5 shows, the reinforcement significantly decreased bonding between the components—the difference is between 4-7 MPa. The reason for this could be that there are fibres on the surface of the preform, and connection between the fibres and the matrix is weak, hence adhesion decreases. At low melt temperature, the GF decreased strength by 75%, and at 240 °C by 35%. We found that the relative difference between the unfilled polymer and GF reinforced composites decreased when melt temperature increased. These results show the positive effect of the melt temperature.

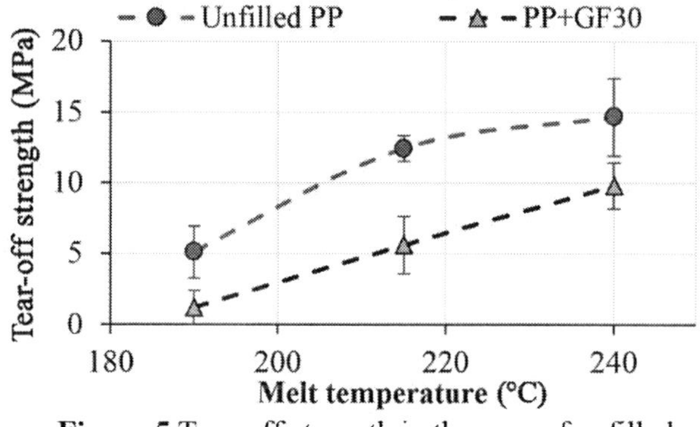

Figure 5 Tear-off strength in the case of unfilled and 30 m% glass fibre-reinforced polypropylene

3.3. The effect of glass beads

The effect of glass bead size was analysed at three different filler concentrations (10, 25 and 40 m%). With an increase in the amount of glass beads, tear-off strength decreases (Figure 6). When we overmoulded on the unfilled plates, a 12.4 MPa tear-off strength was measured. With the small glass beads 9.1 MPa, 9.15 MPa and 5.6 MPa and with the larger beads 10.4 MPa, 10.6 MPa and 8.3 MPa bonding strength was measured at GB concentrations of 10 m%, 25 m% and 40 m%. With smaller beads, average strength is also lower. The reason may be the segregation of the fillers, which was shown by microscopic analysis. The segregation of fillers means an inhomogeneous distribution of particles. This inhomogeneity can be created either along the flow path or in the cross section of the part. In our case, segregation in the cross section can cause bonding problems.

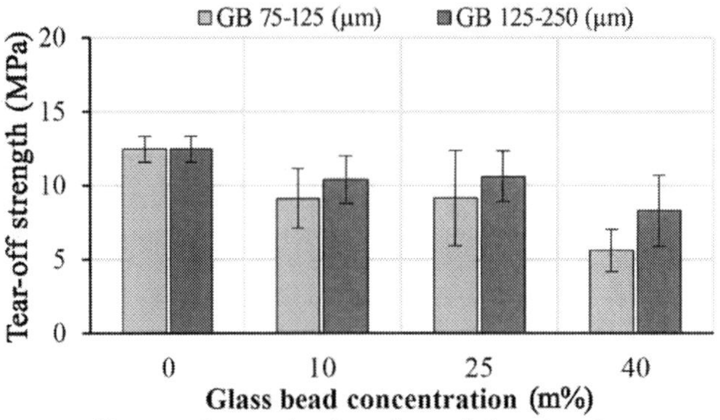

Figure 6 Tear-off strength as a the function of glass bead concentration and glass bead size

SEM images were taken about the fracture surface of the specimens (Figure 7). As the figure shows, a higher amount of glass beads can be found on the fracture surface in the case of smaller glass beads (Figure 7/a) than in the case of larger beads (Figure 7/b). When smaller particles are used, the segregation effect can be neglected. Hence the glass beads can be distributed homogeneously. When larger beads are used, segregation is more pronounced. During the filling of the cavity, a thin frozen layer develops next to the cavity wall. The flowing melt can easily tear the larger particles out of the frozen layer. In consequence, a thin layer with fewer particles evaluated near the surface.

As interfacial adhesion is weak between the filler and the matrix, the beads act as failure locations. The higher the number of glass beads close to the surface, the lower the effective bonding surface in the part. The difference between the inhomogeneous filler distributions of different glass bead sizes along the cross section of the specimen was also proved with optical microscopic images (Figure 8).

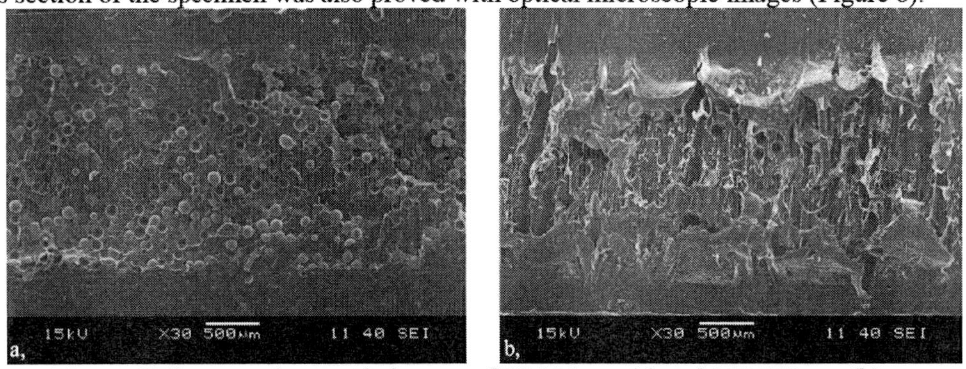

Figure 7 Electron microscopic images of 75-125 μm (a) and 125-250 μm (b) glass bead-filled polypropylene samples (PP+GB40)

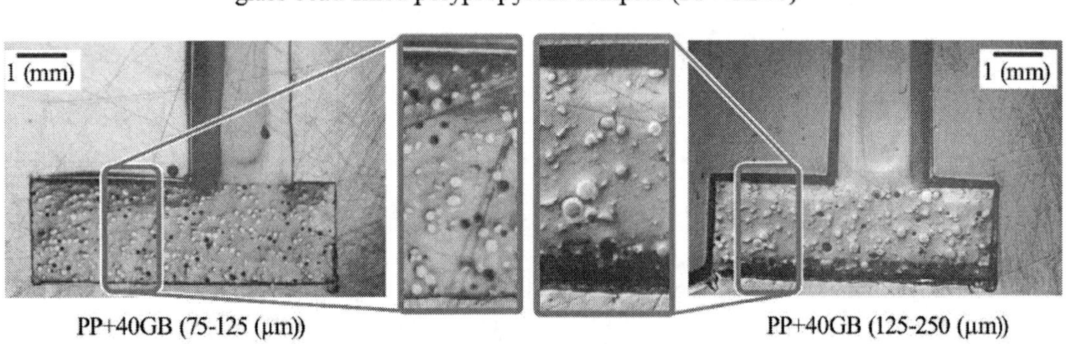

Figure 8 Microscopic images of the cross section of 75-125 μm and 125-250 μm glass bead filled polypropylene samples (PP+GB40)

4. Conclusions

We analysed the effect of glass fibre and glass beads on the bonding strength between the components of multicomponent injection moulded parts. We prepared seven different polypropylene-based glass fibre-reinforced and glass bead-filled compounds, then injection moulded 80 mm x 80 mm x 2 mm flat preforms from them. With a special mould, unfilled polypropylene was overmoulded on the preforms, and tear-off strength was measured by tensile tests. We also analysed the effect of technological parameters on unfilled specimens. We found that holding time does not have a remarkable effect on bonding strength, but with increasing melt temperature, bonding strength increased significantly. It can be explained with the thickness of the melt layer on the surface of the preform and with intermolecular diffusion. When 30 m% glass fibre was added to the matrix, bonding strength decreased. The reason is the presence of the fibres in the surface layer, which act as failure locations when interfacial adhesion is weak between the fibres and the matrix. The glass beads also have a negative effect on the bonding strength between the components of multicomponent parts. When smaller glass beads were used, strength decreased further. We explained it with the segregation of the fillers in the cross section of the preforms, which we proved with microscopic images.

Acknowledgements

The project is funded by the (NKFIH) National Research, Development and Innovation Office (NVKP_16-1-2016-0038, 2017-2.3.7-TÉT-IN-2017-00049, 2018-1.3.1-VKE-2018-00001) and by the National Research, Development and Innovation Fund (TUDFO/51757/2019-ITM, Thematic Excellence Program). It was supported by the ÚNKP-19-4 New National Excellence Program of the Ministry for Innovation and Technology. This paper was supported by the János Bolyai Research Scholarship of the Hungarian Academy of Sciences. The authors would like to thank Arburg Hungaria Ltd. for the injection moulding machine.

References

[1] Osswald T et al 2001 *Injection Molding Handbook* (Munich: Hanser)
[2] PlasticEurope 2018 *Plastics – the Facts 2018.* 1-60
[3] Xie L and Ziegmann G 2011 *J. Alloys Compd.* **509** 226-233
[4] Karsli N G et al 2014 *Compos. Part B – Eng.* **63** 154-160
[5] Tábi T et al 2015 *Express Polym. Lett.* **9** 394-400
[6] Tábi T et al 2014 *Compos. Part A – Appl. Sci.* **64** 99-106
[7] Kovacs J G and Suplicz A 2013 *J. Reinf. Plast. Comp.* **32** 1234-1240
[8] Dzulkipli A A and Azuddin M 2017 *Procedia Engineer.* **184** 663-672
[9] Ebnesajjad S 2011 *Handbook of adhesives and surface preparation* (Oxford: Elsevier)
[10] Awaja F et al 2009 *Prog. Polym. Sci.* **34** 948-968
[11] Joppich T et al 2017 *Proc. CIRP* **66** 137-142
[12] Carella A R et al 2002 *Polym. Eng. Sci.* **42** 2172-2181
[13] Boros R et al 2018 *Materials* **11** 2140
[14] Boros R et al 2019 *Express Polym. Lett.* **13** 889-897

IOP Conf. Series: Materials Science and Engineering **903** (2020) 012050 doi:10.1088/1757-899X/903/1/012050

Fatigue tests of zinc aluminium matrix syntactic foams filled with expanded perlite

A Kemény[1,2,*], B Katona[2], N Movahedi[3], T Fiedler[3]

[1] Budapest University of Technology and Economics, Faculty of Mechanical Engineering, Department of Materials Science and Technology, Budapest, Hungary
[2] MTA–BME Lendület Composite Metal Foams Research Group, Budapest, Hungary
[3] The University of Newcastle, School of Engineering, Callaghan, Australia

E-mail: alexa@eik.bme.hu

Abstract. Metal matrix syntactic foams (MMSFs) are cellular materials which have high specific energy-absorbing properties accompanied by lower density compared to the bulk material. A special type of closed-cell metal matrix foams (MMFs) is the metal matrix syntactic foam. In this research, fatigue tests were carried out of ZA27 zinc aluminium matrix syntactic foams filled with expanded perlite. The stress levels were defined from the plateau stress of the quasi-static upsetting tests (90; 70; 50 and 30%) with R=0.1 stress ratio and f=10 Hz frequency. The results were evaluated statistically.

1. Introduction

Nowadays, the mass reduction of components has a primary role in industrial applications. These components have to be lightweight with enhanced specific mechanical properties which means the improvement of the used materials and structures. Increasing the specific energy-absorbing properties of the structure to compressive loads can be achieved by using lower density and higher strength materials. One of the best solutions for this demand is the group of closed-cell metal foams.

Metal matrix syntactic foams (MMSFs) are closed-cell foams in which the second phase (or filler) consists of hollow or expanded particles. Various matrix and filler materials are used by researchers to create MMSFs. The most common matrix materials are lightweight metals such as aluminium alloys [1,2], magnesium- [3,4] or titanium alloys [5]. There are also some researches where iron- [6] or zinc alloys [7] are used as a matrix. The first filler materials in MMSFs were fly-ash microspheres [8]. However, these days the most common second phase is ceramic hollow spheres, which can be made from high purity alumina [9], silicon carbide [10] or mixed oxide ceramics [1]. There are some researches based on iron hollow spheres [11] as the filler and also some cost-efficient alternatives such as lightweight expanded clay particles [12] or expanded perlite [13].

The main load for MMSFs as energy-absorbing structures is compression. Quasi-static compressive behaviour of these materials has been widely investigated [14–22], which is a standardised test method (ISO 13314:2011 – Mechanical testing of metals – Ductility testing – Compression test for porous and cellular metals). There are also researches on dynamic properties, wear and thermal properties of these materials [23–27].

Being familiar with the fatigue properties of foam structures is crucial for the constructability and future projection of the failure method at known stress levels. There are some publications on cyclic loading of MMFs, but most of them are focused on open and closed cell foams [28–37].

2. Cyclic testing of MMFs in the literature

Yang et al. investigated the compression fatigue properties of open-cell aluminium (99.0% purity) foams that were fabricated by space-holder method with three different porosities (50; 60 and 70%). The fatigue tests were carried out with R=0.1 stress ratio and f=10 Hz frequency. The stress levels were defined from the yield strength of the foams measured with quasi-static compression tests. The fatigue strengths showed a linear decrease with the increment of the porosity. Local deformation regions were observed at strain rates ε=2% and 4% [38].

Zhou et al. also investigated the compressive fatigue properties of commercially available open-cell aluminium (6101 alloy) foams with R=0.1 stress ratio. The cyclic frequency was f=20 Hz, and the stress levels were calculated from an equation. The S-N curves were characterised by an abrupt strain jump following an incubation stage in which the forms are progressively shortened at relatively slow rates depending on the stress level [39].

Zhao et al. evaluated the tension fatigue properties of closed-cell aluminium alloy foam, specifically the damage evaluation and damage mechanism. The loading frequency was f=20 Hz with R=0.1 stress ratio. The stress levels were defined as discreet values lower than the tensile strength measured with quasi-static compression tests. They observed a large scatter in the fatigue damage of the foams due to the variability of the specimens and developed a statistical damage model [40].

Schultz et al. observed the fatigue properties of closed-cell 6061 aluminium foams produced with powder metallurgy, with an average density of 0.4 gcm^{-3} and $AlSi7Mg+15\%SiC_p$ foam produced from aluminium melt, with an average density of 0.3 gcm^{-3}. Six stress levels were tested with R=-1 stress ratio. The specific fatigue strengths were in the same range for both type foams [41].

Kim and Kim investigated the specimen aspect ratio on compressive fatigue properties of closed-cell Al-Si-Ca alloy foam. The loading frequency was f=50 Hz with R=0.1 stress ratio. The four stress levels were defined from the plateau strength obtained from quasi-static compressions. It turned out that the onset of cyclic shortening of foams with a lower aspect ratio took place earlier than the specimen with a higher ratio. Also, their fatigue strength was lower compared to the specimen with a higher aspect ratio, while the quasi-static stress-strain curves had almost the same Young's modulus, yield stress and plateau stress [42].

Katona et al. investigated the compressive fatigue properties of ceramic hollow sphere filled aluminium matrix syntactic foams with R=0.1 stress ratio. Al99.5 and AlSi12 alloys were applied as matrix material and Globocer as filler. The cyclic frequency was f=10 Hz, and the stress levels were defined from the compressive strength (σ_c) of the MMSFs measured with quasi-static compression tests. Regarding the matrix material, the softer matrix ensured higher load levels for the fatigue strength than the more rigid matrix [43].

Szlancsik et al. published on MMSFs filled with expanded perlite investigating the relation between the density (0.7 – 1.1 gcm^{-3}) and the compressive fatigue properties. The fatigue tests were carried out with R=0.1 stress ratio and f=10 Hz frequency. The stress levels were defined from the 1% compressive stress (σ_c) of the MMSFs measured with quasi-static compression tests. The fatigue criteria were at ε=2% strain rate. Both the fatigue limit and the slope of the Wöhler curve increased with the density. They also observed that the fatigue properties of MMSFs with same density but different filler materials differ [44].

Taherishargh et al. observed the fatigue properties of A356 aluminium matrix syntactic foams filled with expanded perlite focusing on the microstructure characterisation. The compressive fatigue tests were done with R=0.1 stress ratio and f=10 Hz frequency. The load levels were defined from the quasi-static compressive plateau stress of the MMSFs. Two different failure mechanisms were identified depending on the applied load level, that resulted in characteristic deformation – loading cycle curves. Traces of fatigue beachmarks and extensive plastic deformation were found on the microstructural scale [13].

There are only a few publications on the fatigue properties of metal matrix syntactic foams; therefore, it is important to produce new results in the field. Most of the research groups investigate the compressive fatigue properties with R=0.1 stress ratio factor and f=10-50 Hz frequency.

3. Materials and methods

3.1. Materials

ZA27 zinc aluminium alloy was used as matrix material, and expanded perlite (EP) particles were applied as filler obtained from Australian Perlite Pty. with a size range of 3–4 mm. The chemical composition of the used expanded perlite is 75 wt.% SiO_2, 14 wt.% Al_2O_3, 4 wt.% K_2O, 3 wt.% Na_2O, 1.3 wt.% CaO, 1 wt.% Fe_2O_3, 0.3 wt.% MgO and 0.2 wt.% TiO_2 with traces of heavy metals. ZA27 contains 25.0–28.0 wt.% Al, 2.0–2.5 wt.% Cu, 0.01–0.02 wt.% Mg, max. 0.075 wt.% Fe, max. 0.006 wt.% Pb, max. 0.006 wt.% Cd, max. 0.003 wt.% Sn and Zn as the balance.

The samples were produced with counter gravity infiltration. In this process a disk of ZA27 alloy was placed above the packed EP particles in a graphite crucible and the assembly was heated in an electric furnace to 535 °C for 30 min to melt the alloy. The EP particles were kept in place by a steel mesh. The infiltration was initiated by a 1 kg weight forcing the molten alloy between the EP particles. The solified samples were removed and machined. The samples were heat-treated at 365°C for 1 hour and cooled in water-ice mixture. A subsequent aging was at 140°C for 24 hours. The density range of the produced MMSFs was between 1.71 – 2.02 gcm^{-3} calculated from their mass divided by their volume.

3.2. Fatigue test

Samples from the produced MMSFs were tested with quasi-static compression tests to specify the density-dependent plateau stress (σ_{pl}) values. The ISO 13314:2011 standard defines the plateau stress, which is calculated between 20 and 40% strain of the measured samples in this research. The load levels of the fatigue tests were calculated as the k=30; 50; 70 and 90% of this stress value for each fatigue sample. Figure 1. shows one cycle of the fatigue tests and the most important parameters. Three samples were tested on every load level up to ~5-10-15% engineering strain.

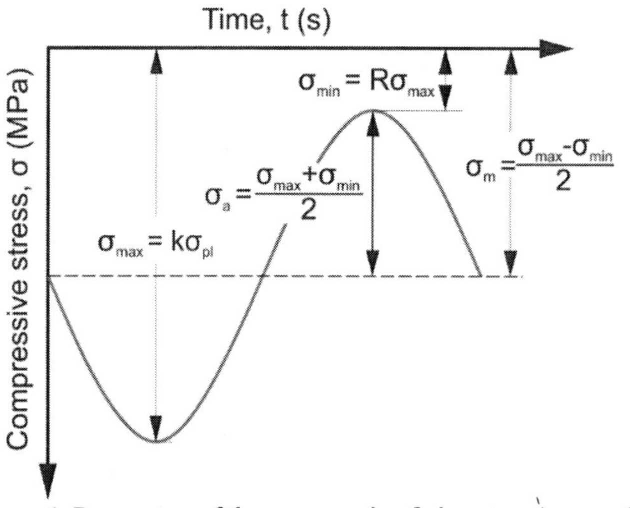

Figure 1. Parameters of the compressive fatigue test (one cycle)

The tests were carried out with f=10 Hz frequency, while the stress ratio was set to R=0.1. The fatigue limit was $2\cdot10^6$ cycles, and the failure criterion was 2% engineering strain according to other previous researches [13]. Ø20×30 mm samples were tested (the aspect ratio was H/D=1.5). For lubrication beeswax was used on the contact surfaces. The tests were performed on an Instron 8872 machine with 25 kN load cell. The displacement was measured with the crosshead displacement. The instantaneous failure of the examined MMSF with fatigue test is shown in Figure 2.

Figure 2. Failure of the fatigue tested syntactic foams

4. Results

First, the density-dependent plateau stress values were defined with quasi-static compression tests. The results are plotted in Figure 3. The equation of the fitted curve is:

$$\sigma_{pl} = 175.07\rho^2 - 582.75\rho + 525.92 \quad (1)$$

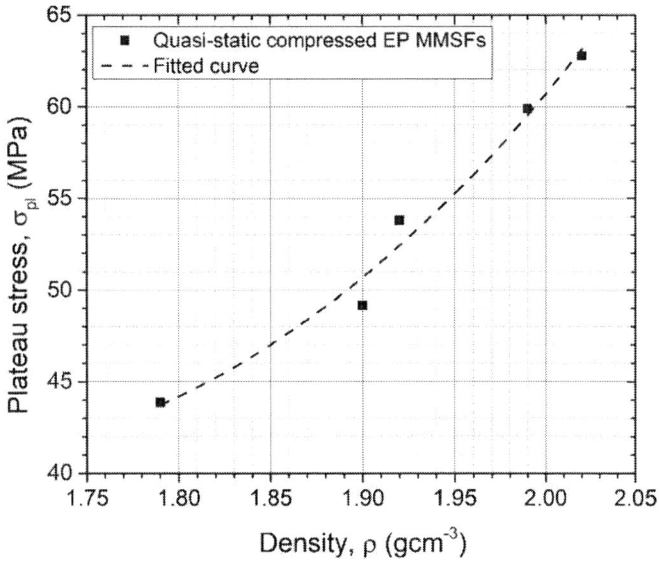

Figure 3. Plateau stress in the function of the density

Table 1. Main properties of the examined metal foams, and the applied load

Sample	Density, ρ (gcm^{-3})	Calculated plateau stress, σ_{pl} (MPa)	Load ratio, k (-)	Minimum stress, σ_{min} (MPa)	Maximum stress, σ_{max} (MPa)
EP-1	1.98	58.17	0.9	5.23	52.35
EP-2	1.71	41.34	0.9	3.72	37.20
EP-3	1.88	49.12	0.9	4.42	44.20
EP-4	1.80	44.22	0.7	3.09	30.95
EP-5	1.82	46.04	0.7	3.22	32.22
EP-6	1.86	47.81	0.7	3.34	33.46
EP-7	1.83	45.91	0.5	2.29	22.95
EP-8	1.88	49.36	0.5	2.46	24.68
EP-9	1.78	43.16	0.5	2.15	21.58
EP-10	1.87	48.49	0.3	1.45	14.54
EP-11	1.86	47.74	0.3	1.43	14.32

All samples' geometry and mass were measured before the fatigue tests. Table 1. summarises each specimen's density and the fatigue test properties.

On each load level, three specimens were examined (Figure 4.) for the appropriate statistical evaluation. In the case of each specimen, the cycle numbers at 2% engineering strain were determined. If the fatigue limit was reached before the failure criterion, the testing stopped (e.g. k=30% specimens). Based on these data, the number of cycles to failure (N_F) at a specified load level was calculated as the average of the three cycles numbers. In order to create the Wöhler-like curve of the investigated metal foam, a line was fitted on the evaluated numbers of cycles to failure – load level points by the least square method. The created Wöhler-like curve, which is only valid in the finite lifetime region, was plotted in a load ratio – number of cycles to failure graph in Figure 5.

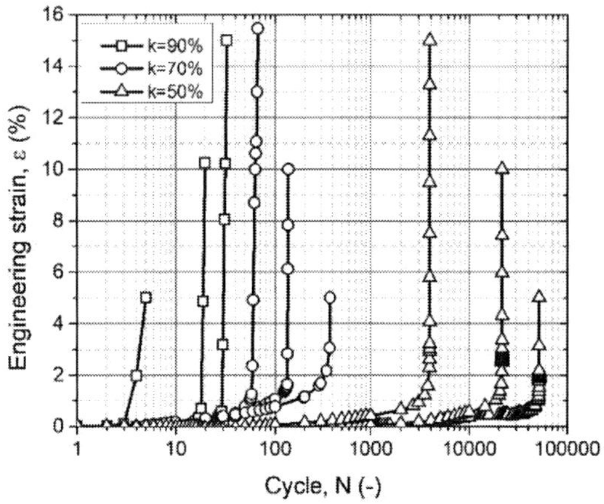

Figure 4. Engineering strain in the function of the number of cycles

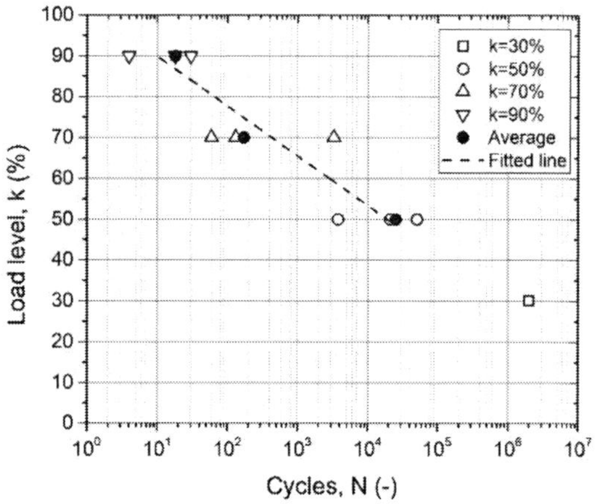

Figure 5. Wöhler-like curve

5. Conclusions

From this research, the following conclusions can be drawn:

- The relationship between the quasi-static compression plateau stress and the density of the investigated MMSFs can be described with a quadratic equation.
- The failure of the EP MMSFs is instantaneous and occurs in the upper region of the samples; however, further evaluation is needed.
- The fatigue properties of the investigated density-range EP zinc aluminium matrix foams can be determined for various load levels from the created Wöhler-like curve in the finite lifetime region.

In comparison to the results available in the literature, this research dealt with MMSFs filled with expanded perlite particles which is broaden the data on the fatigue properties of different MMFs. It is important to be aware of the stress values behind the load levels and the density of the specimens, therefore, the different MMFs can be compared properly. Another important parameter for the proper comparison is the definition of the failure criterion. With all in sight the produced ZA27-EP MMSFs have higher density and cyclic compressive stress bearing than most of the previously investigated materials, but a lot of parameters are missing for distinct conclusions to be drawn.

Acknowledgement

Supported by the ÚNKP-19-3 New National Excellence Program of the Ministry for Innovation and Technology.

References

[1] Orbulov I N 2013 *Mater. Sci. Eng. A* **583** 11–9
[2] Orbulov I N and Szlancsik A 2018 *Adv. Eng. Mater.* **20** 1–12
[3] Ding J et al 2018 *Metals (Basel).* **8** 434
[4] Anbuchezhiyan G et al 2018 *Arch. Civ. Mech. Eng.* **18** 1645–50
[5] Xue X and Zhao Y 2011 *JOM* **63** 43–7
[6] Park H et al 2019 *Mater. Sci. Eng. C* **97** 367–76
[7] Linul E et al 2019 *Compos. Part B Eng.* **167** 122–34
[8] Rohatgi P K and Guo R Q 1997 *Tribol. Lett.* **3** 339–47
[9] Santa Maria J A et al 2013 *Mater. Sci. Eng. A* **582** 415–22
[10] Luong D D et al 2013 *J. Alloys Compd.* **550** 412–22
[11] Szlancsik A et al 2015 *Materials (Basel).* **8** 7926–37
[12] Szlancsik A et al 2019 *Materials (Basel).* **12** 2023
[13] Taherishargh M et al 2017 *J. Compos. Mater.* **51** 773–81
[14] Zhang Q et al 2018 *Compos. Struct.* **183** 499–509
[15] Orbulov I N and Májlinger K 2014 *JOM* **66** 882–91
[16] Katona B and Orbulov I N 2017 *Period. Polytech. Mech. Eng.* **61** 146–52
[17] Aragon-Lezama J A et al 2015 *Mater. Sci. Eng. A* **638** 165–73
[18] Kádár C et al 2016 *Mater. Lett.* **173** 31–4
[19] Lin Y et al 2017 *Mater. Sci. Eng. A* **696** 236–47
[20] Castro G et al 2013 *Mater. Sci. Eng. A* **578** 222–9
[21] Linul E et al 2018 *J. Alloys Compd.* **740** 1172–9
[22] Taherishargh M et al 2016 *Mater. Des.* **99** 356–68
[23] Fiedler T et al 2015 *Mater. Sci. Eng. A* **626** 296–304
[24] Májlinger K 2015 *Int. J. Mater. Res.* **106** 1165–73
[25] Májlinger K et al 2016 *Tribol. Int.* **99** 211–23
[26] Rabiei A and Garcia-Avila M 2013 *Mater. Sci. Eng. A* **564** 539–47
[27] Szalóki I and Viharos Z J 2018 *Period. Polytech. Mech. Eng.* **62** 26–32
[28] Zhou J and Soboyejo W O 2004 *Mater. Sci. Eng. A* **369** 23–35
[29] Zettl B et al 2000 *Mater. Sci. Eng. A* **292** 1–7
[30] Zettl B et al 2001 *Int. J. Fatigue* **23** 565–73
[31] Vendra L J et al 2009 *Mater. Sci. Eng. A* **517** 146–53
[32] Soubielle S et al 2011 *Mater. Sci. Eng. A* **528** 2657–63
[33] Qi L et al 2017 *J. Alloys Compd.* **721** 55–63
[34] Olurin O B et al 2001 *Int. J. Fatigue* **23** 375–82
[35] Lehmhus D et al 2002 *J. Mater. Sci.* **37** 3447–51
[36] Kolluri M et al 2008 *Acta Mater.* **56** 1114–25
[37] Kashef S et al 2011 *Mater. Sci. Eng. A* **528** 1602–7
[38] Yang X et al 2019 *Int. J. Fatigue* **121** 272–80
[39] Zhou J et al 2005 *J. Eng. Mater. Technol.* **127** 40–5
[40] Zhao M et al 2016 *Int. J. Fatigue* **87** 257–65
[41] Schultz O et al 2000 *Adv. Eng. Mater.* **2** 215–8
[42] Kim A and Kim I 2008 *Acta Mech. Solida Sin.* **21** 354–8
[43] Katona B et al 2017 *Mater. Sci. Eng. A* **679** 350–7
[44] Szlancsik A et al 2018 *IOP Conf. Ser. Mater. Sci. Eng.* **426** 012045

IOP Conf. Series: Materials Science and Engineering **903** (2020) 012051 doi:10.1088/1757-899X/903/1/012051

Investigation of VVER-1200 reactor pressure vessel's material

Dorina Kovács[1], Dávid Kemény[2]

[1] Budapest University of Technology and Economics, Faculty of Mechanical Engineering, Department of Material Science and Engineering,
1111, Budapest, Műegyetem rakpart 3, Hungary
[2] Paks II. Zrt., 7030 Paks, Gagarin utca 1.

E-mail: dorina@eik.bme.hu

Abstract. At high temperature and pressure, the reactor materials are exposed to various radiation. Irradiation causes significant changes in the crystal structure and their mechanical properties. The mechanical properties of low alloy steels generally do not change so much with temperature. The aim of this research is to prepare the basic measurements to investigate 15Cr2NiMoVA steel, according to MSZ EN ISO 148-1:2017. Charpy impact test was made at different temperatures from -75 °C to 200 °C. Potentiodynamic tests were used to determine the corrosion rate. During the optical microscopy measurement, MnS inclusions were found in the examined material.

1. Introduction

The pressure reactor vessel (RPV) is one of the important equipment that ensures stable operation of the reactor by limiting variations of the reactor coolant system (RCS) pressure within allowed tolerances [1], so it limits the pressure change within permissible value during operational transient and prevents the primary coolant system from overpressure in accidental scenarios [2,3]. It follows that it is operated at high temperature and pressure, neutron radiation, thermal ageing and low cycle fatigue, which are the main environmental effects that degrade the RPV material properties during service [1,4]. According to this loading, the vessel material is degraded during the operation. In the Russian VVER type reactor [5] higher amount of Cr steel is used for pressure vessel [6]. These are the VVER-1200 water-moderated and water-cooled power reactors [7]. The protection against the irradiation, corrosion and fracture had to be guaranteed. The irradiation embrittlement, caused by long-term irradiation with high-energy neutrons, is coming with together mechanical property changes such as an increasing hardness, yield stress, and tensile strength, and a decrease of toughness [8,9]. The mechanical properties of low alloy steels generally do not change so much with temperature. The material is covered with a stainless steel layer made by cladding to protect it from the general corrosion. This layer has to be a homogenous thickness and a good sacrificial barrier against high-temperature oxidation in several microns [10]. Besides, the major parameter is the ductile to brittle transition temperature. It has to occur at the lowest temperature reachable taking into account the steel composition and mechanical properties [11].

The investigation of the reactor vessels is beloved by all. Firstly, Chernobaeva et al. [12] investigate a previous type of reactor. They used 15Cr2NiMoVA steel which received from a VVER-1000 reactor as irradiated and aged. Standard 10×10 and 'mini' 5×5 Charpy specimens were used in this research to compare the results of the different size of samples. A correlation was found between both size of samples. It makes possible to incorporate test data for mini-Charpy specimens into a database of

VVER-1000 RPV materials based on this equitation: $T_K^{10\times10} = 0.99\times T_K^{5\times5}+51\pm2\sigma$ (σ = 12.7, T is the temperature of the signed samples).

Anosov et al. [13] investigated the differences between the critical brittleness (T_C – which is the temperature of the intersection of the temperature-impact strength curve with the level of determining the transition temperature by the steel yield strength) and the brittle-viscous transition (T_T – which was characterized by the steep rise position in the curve of the temperature dependence of the absorbed energy) temperature of 15Cr2NiMoVA steel. T_C are significantly lower than the T_T. The differences is 22 °C. It can be stated, the transition temperature is better to define the brittle fracture of the VVER-vessels.

Shtrombakh et al. [14] investigated the effect of the Ni content on the thermal and radiation resistance. The previous steel was examined with modified Ni content. Lower Ni content led to the lower density of radiation-induced precipitates under irradiation.

These researches were adapted to the nuclear power plant operation of their own country. Every country has a different condition in the reactors because of its consumptions. Therefore, the aim of this paper is to get insight into the selection of the materials and prepare to investigate the material of the reactor.

2. Materials and methods

In this study 15Cr2NiMoVA steel was used with the following chemical composition: 0,14% C, 2,6% Cr, 0,31% Ni, 0,79% Mo, 0,63% V, 0,67% Mn, 0,08%P, 0,28% Si, 0,33% Cu, Fe in balanced. Ferenc Gillemot provided the Charpy results of the samples in normal and aged conditions from MTA Research Institutes. (The pre-heat treatment of the samples was unknown for us.)

Olympus optical microscope was used to investigate the cross-section of the samples. The surface of the cross-section was mechanically ground with 80 to 2500-grit SiC paper and polished with 3 μm diamond suspension. The samples were cleaned with acetone and dried with hot air. The fracture of some of the samples was observed by Olympus SZX16 stereomicroscope. The corrosion resistance of the samples was evaluated by measuring polarisation curves in boron aqueous solution at 150°C using ZAHNES IM6e electrochemical working station. The cell of the specimen was set-up as the working electrode, a Hg_2Cl_2/KCl_{tel} calomel electrode as the reference electrode and platinum were used as the counter electrode. The composition of the corroded surface was measured by Zeiss EVO MA10 scanning electron microscope (SEM) and analysed by energy-dispersive spectrometry (EDS).

3. Results and discussion

3.1. *Charpy impact test*

The results of the Charpy impact test at different temperatures and conditions were summarised in Table 1 and their transition temperature diagram is seen in Fig. 1. The transition temperature was calculated to use the hyperbolic tangent according to (1). It fits the relationship between the impact absorption energy equation (e) and test temperature (T).

$$e(T)=A+B\times \tanh\frac{T-T_0}{C} \quad (1)$$

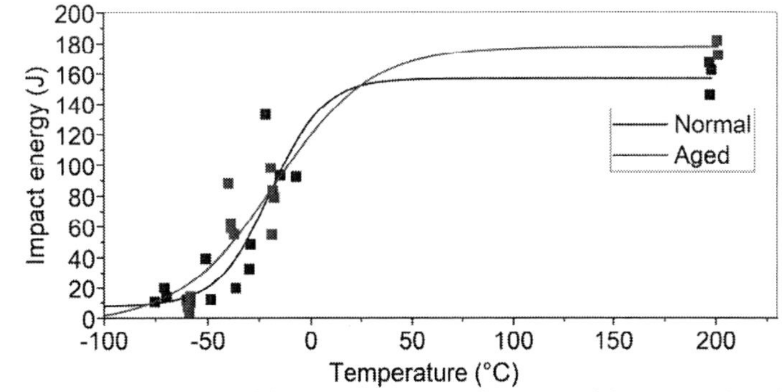

Figure 1. Transition temperature diagram of the normal and aged condition samples

where C is half of the transition temperature range from minimum to maximum value and T_0 is for the temperature when e(T) equals to the minimum value; A corresponds to the minimum energy and B to the maximum energy [15]. The transition temperature was between the brittle and ductile state is

-20 °C in normal and -7 °C in aged condition. The impact energy differences at operational temperature are 20 J, but the required value is 41 J, which lower than the results. To compare with [16] in the same condition, the results are similar, only a few J differences can be found at higher temperature.

Table 1. Results of Charpy impact test in different temperature

Normal			Aged		
Temp. (°C)	**KV (J)**	**e (mm)**	**Temp. (°C)**	**KV (J)**	**e (mm)**
-75	11	0.07	**-60**	4	0
-70	19	0.1	**-60**	14	0.05
-70	13	0.18	**-60**	9	0.09
-60	12	0	**-40**	56	1.04
-50	12	0.2	**-40**	61	0.46
-50	38	0.5	**-40**	60	0.69
-36	19	0.16	**-20**	98	0.80
-30	48	0.7	**-20**	82	0.76
-30	33	0.36	**-20**	88	0.71
-21	133	1.75	**-20**	80	0.75
-15	93	1.46	**20**	100	1.11
-10	92	1.29	**40**	116	1.60
20	86	2.15	**40**	125	1.54
40	116	2.08	**200**	174	1.78
200	163	2.15	**200**	182	1.53
200	165	2.08	**200**	182	1.52

Stereomicroscopic images are seen in Fig. 2. from the fractured surface. As it is observed, at low temperature, the material was brittle, while at high temperature it was tough. At the operational temperature, the material has to be tough because of the loadings.

Figure 2. Fractured surface after the Charpy impact test. left: at -30 °C, right: at 200 °C

3.2. Microstructure

Fig. 3 shows the optical microscopic images of the investigated material. As it is seen in Fig. 3 the microstructure was ferrite and bainite. It can be created during the pre-heat treatment. In higher magnification inclusions were found which is seen in Fig. 3. To analyse the inclusion, SEM-EDS was used to measure the composition.

The main component of the inclusion was sulphur and in large amount iron and manganese were found. The material was contained MnS inclusions which were not resulted by the irradiation because these samples have not been irradiated.

Figure 3. The microstructure of the base material in normal condition

3.3. Corrosion test

In general, the standards for the primary coolant circuit of VVER reactors water chemistry adjusted by additions of H_3BO_3, KOH, NH_4OH [17], so a 10% boron aqueous solution was used for the corrosion test. The solution was contained 10 g Boron and 1000 ml distilled water. Fig 4. shows the polarisation curves after different times. The calculated corrosion rates [18] are listed in Table 2. Based on the calculated corrosion rates, the material has bad corrosion resistance. Neither the corrosion potential nor the current density of the samples considerably changed. As it is seen, the corrosion rates were not changed over the corrosion time. In this condition, this material is not recommended for a reactor, but the corrosion rate can be decreased by cladding, which can be an X8CrNiTi18-10 stainless steel layer on the inner surface. The utilisation of a Cr content film on the inner surface of steel cladding has been considered to hinder fuel- cladding chemical interaction [19].

Table 2. Corrosion rate parameters of the measured samples

time (h)	E_{corr} (V)	i_{corr} (μA)	corr. rate (mm/year)
1	-0.57	99	0.99
5	-0.57	98	0.99
12	-0.56	98	0.99
24	-0.55	98	0.99

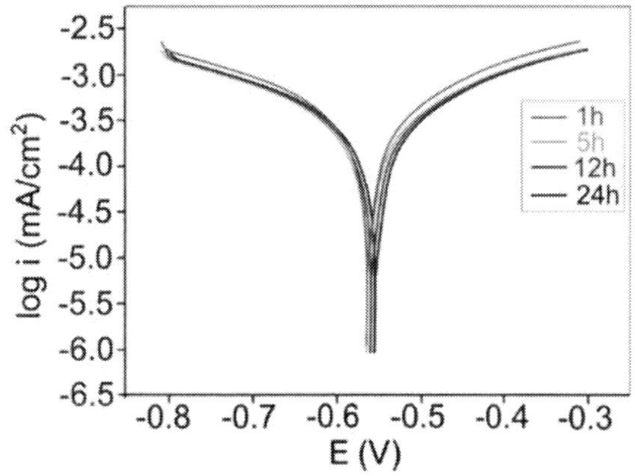

Figure 4. Polarisation curves of the normal condition sample

4. Conclusion

According to our research, the material was though based on the Charpy impact test in both of the normal and aged condition, which was examined close to the operational temperature. Based on the results, the slope of the tangent has decreased by aged specimens, which occurs the value of TTKV changing to a direction unfavourable to us. The TTKV changing should be taken into account by power plants. Next step in our research to examine the effect of neutron flux by reactor vessel how can influence the ductile to brittle transition temperature and though.

Any segregation was not found. In contrast, MnS inclusions were observed. The corrosion resistance of the pure material without the cladding layer was low.

Acknowledgement
The research reported in this paper was supported by the scholarship of the Hungarian Ministry of National Resources (National Talent Programme; NTP-NFTÖ-19-B-0036). The authors are grateful to Ferenc Gillemot for the measurements results from MTA Research Institutes and for Paks II. Zrt.

References

[1] Wang P 2019 *Appl. Math. Model.* **65** 187–206
[2] Zhong X 2019 *Ann. Nucl. Energy* **126** 133–41
[3] Szücs B 2019 *Period. Polytech. Mech. Eng.* **63** 188–94
[4] Vértesy G 2019 *Mat.***12** 963-975
[5] Dolganov K S 2019 *Nucl. Eng. Des.* **353** 110222
[6] Jeong Y H 2013 *Mat. Ag. and Deg. in Light Wat. React.: Mech. and Manag.* **3** 315-334
[7] Kursevich I P 2013 *Radiat. Mater. Sci.* **4** 575–84
[8] Kobayashi S 2012 *Philos. Mag.* **92** 3813–23
[9] Szlancsik A 2019 *Mat.* **12** 574-569
[10] Michau A 2019 *Surf. Coat. Technol.* **375** 894–902
[11] Féron D 2012 *Nucl. Corros. Sci. Eng.* 31–56
[12] Chernobaeva A A 2016 *Int. J. Press. Vessel. Pip.* **145** 23–28
[13] Anosov N 2018 *Nucl. Energy Technol.* **4** 155-161
[14] Shtrombakh Y I 2015 *J. Nucl. Mater.* **461** 292–300
[15] Cao L W 2012 *Int. J. Press. Vessel. Pip.* **93–94** 12–26
[16] Lenkey G B 2001 *Mat.-wiss. u. Werkstofftech* **32** 562–567
[17] Kritskii V G 2011 *Therm. Eng.* **58** 540–546
[18] McCafferty E 2005 *Corros. Sci.* **47** 3202–15
[19] Yeo S 2020 *J. Nucl. Mater.* **530** 151980

IOP Conf. Series: Materials Science and Engineering **903** (2020) 012052 doi:10.1088/1757-899X/903/1/012052

Neutron Irradiation of Zirconium Alloys in the Budapest Research Reactor

I Szenthe[1], F Gillemot[1], M Horvath[1], Z Hózer[1]

[1] Centre for Energy Research, Budapest, Konkoly-Thege Miklós út 29-33.

E-mail: szenthe.ildiko@energia.mta.hu

Abstract. Sustainable and increasingly safe operation is one of the constant goals of various technical and economic developments related to nuclear reactors. Reliability of the fuel elements cladding is basic for the safe and economic operation, to select the best fuel clad large number of simulations must be carried out in different operational conditions and supported by experimental results. Two types of cladding tube material had been irradiated in the a BAGIRA irradiation loop at to obtain aged clad properties. The special design of the irradiation target holder allowed the specimens to be irradiated with different fluences and at a prescribed temperature. The irradiated specimens will be tested and the measured properties will be used to simulate the aged fuel clad behaviour. This paper introduces the design, preparation and implementation of the irradiation of fuel clad materials.

1. Introduction

Ensuring structural integrity is essential for safe, reliable and economical operation of the power reactors. Knowledge of the key failure mechanisms of the fuel elements is fundamental requirement for economic production of electricity, and it requires research data.

The failure of the fuel elements clad under normal and extreme operating conditions causes large economic loss for the plant owner. To avoid the fuel elements failures, the fuel clad material is aged in the laboratory using simulated operational accidental conditions. The ageing mechanisms were also studied by computer simulations, which were compared to the experimental studies. One of the major ageing mechanism is the irradiation embrittlement caused by high fluence of fast neutron irradiation. These simulations must be carried out with validated computer programs supported by experimental results. The effect of the different ageing mechanisms should be studied, including the neutron irradiation damage.

Based on experiments performed and evaluated earlier, the fluence range where the mechanical properties of the clad suffer considerable changes is in the range of 0.5-1.5 dpa, and saturation is expected above 1-1.5 dpa. To reach this damage rate within the foreseeable time a competitive and economical neutron source has to be used.

Nowadays, there are only a few research reactors in Europe that deals with neutron irradiation. In the IAEA Research Reactor Database (RRDB) there are only 14 pieces of research reactor in Europe, except the Russians [1]. From these there are 3 TRIGA type reactors designed for education, private commercial research, non-destructive testing and isotope production, other five research reactor power have about 0-0.1 KW thermal power, they are too small for metallic material irradiation. The Belarus Giacint reactor was designed to simulate different reactor core configurations, VENUS-F reactor in

Belgium is an experimental reactor of the "zero-power critical facility" type, the SUR-100 in Germany is „zero power" reactor, suitable for teaching and research, CROCUS in Switzerland is an experimental „zero-power" reactor, mainly dedicated to teaching radiation and reactor physics, which has 200 KW thermal power, and 10^9 neutron cm^{-2} s^{-1} fast neutron scale only. At present time, IR-100 reactor in Ukraine is used to train personnel for the Ukraine's nuclear power industry. The 25 MW CABRI reactor in France is used to tests accident situations which must be considered in the safety analysis of PWRs. In addition, a pool type 30 MW thermal power MARIA in Poland is dedicated mainly for isotope production. From the remaining six reactors the 7 MW tank type WWR-M in Ukraine is available for material irradiation, but its geopolitical situation today is not good enough. The other five possibilities to perform material irradiation tests: BR-2 tank-in pool type 100 MW reactor in Belgium, HFR tank-in pool type 45 MW reactor in Netherlands, LWR-15 WWR tank type 10 MW reactor in Czech Republic, and the WWR tank type 10 MW reactor in Budapest, Hungary [2].

The Budapest Research Reactor (BRR) is operated by the Centre for Energy Research inside the Budapest Neutron Centre (BNC), which was formed in 1993. [3]. Currently 16 experimental stations are connected to the reactor: the cold neutron three-axis spectrometer (TAS), Low-Level Gamma-Spectroscopy Facility, Time of Flight Small Angle Neutron Scattering instrument, Neutron Reflectometer with Polarized Beam Option, The M(aterial) TEST neutron diffractometer, Neutron activation analysis, Neutron-Induced Prompt Gamma-ray Spectroscopy, Neutron Optics and Radiography for Material Analysis, Prompt Gamma Activation Analysis, neutron diffractometer, Small Angle Neutron Scattering, and an irradiation loop named BAGIRA [4]. Hot material testing and other laboratories are also existing within the campus providing good opportunity to carry out pre- and post-irradiation material tests.

The BRR is a VVR-type Russian designed reactor. It went critical in 1959. The initial thermal power was 2 MW. A full-scale reactor reconstruction and upgrading project began in 1986, following by 27 years of operation since initial criticality. The upgraded 10 MW reactor received the operation license in November 1993. The light-water cooled and moderated tank-type reactor with beryllium reflector has 90 fuel assemblies, the thermal neutron flux density in the flux trap is $2.5 * 10^{14}$ neutron cm^{-2} s^{-1}. The approximately maximal fast flux in the fast channel is 10^{14} neutron cm^{-2} s^{-1}.

Within the framework of a research and development contract sponsored by the National Research, Development and Innovation Fund of Hungary, two types of cladding tubes' material was selected (named E110 and E110G) for study. The project entitled „Effect of material structure changes on zirconium alloys used in nuclear power plants on fuel element integrity and environmental load". As part of the contract, Centre for Energy Research irradiated Zirconium alloys. Before the irradiation different treatment performed to simulate the environmental ageing effects during operation and accidental cases, to evaluate the combined effect of irradiation, heat treatment and corrosion.

Irradiation was performed in the BAGIRA (Budapest Advanced Gas-cooled Irradiation Rig with Aluminium structure) equipment in the BRR. The preparation and fulfill of the irradiation campaign will be presented below.

2. Planning of the irradiation campaign

To perform an irradiation, we made a detailed plan for the following:

- selection of the material to be tested and type of the samples
- determine the number of the samples
- prescribe the time of the irradiation
- prescribe the temperature of the irradiation and elaborate the design for excess gamma heat removal during irradiation
- determination of the irradiation matrix

A lot of other demand also had to be taken into consideration (e.g. fragility, encapsulation, labelling/coding, transportation, storage etc.

2.1. Selection of the samples

Because of the goal of irradiation procedure is to investigate the radiation damage of the material, to perform the post-irradiation measurement to the foreseeable time after the campaign the decay time of the relevant elements of the chemical composition has to be estimated. Based on this estimation the needed shielding for protection during the post-irradiation material testing can be assessed and planned.

The material of the tested two types of zirconium binary alloys are used as the material of the cladding tubes in the VVER type pressurized water reactors. The widely used Russian type zirconium alloy Zr1%Nb is named E110. The Russian manufacturer of the E110 cladding material is switching from an earlier electrolytic process to metal "sponge" technology. In the new E110G cladding 70% of the zirconium metal comes from this "sponge", the residual 30% is made by the iodide process. The chemical composition of E110G alloy remains the same, 99% zirconium and 1% niobium, and does not change significantly according to the manufacturer but permissible levels of certain trace element concentrations change. Independent domestic measurements are an important complement to the information provided by the manufacturer. There is a need for comparative testing of the E110 and E110G cladding materials, which have been pre-treated with different basic processes that simulate the power plant burnout sub-processes [5-11].

2.2. Determine the number of the samples

To specify the optimum number of the samples, a test matrix has been defined. The elaborated irradiation matrix of the pre-treated specimens is shown in table 1.

Table 1. Irradiation plan for as-received and pre-treated specimens

Type of pre-treatment	**For irradiation or reference**	**Planned test type**	**Material E110**	**Material E110G**	**Summa pieces of specimens**
Native	for reference	tensile test	6	6	12
		compression test	3	3	6
	irradiated	tensile test	6	6	12
		compression test	3	3	6
Heat-treated	for reference	tensile test	36	36	72
		compression test	18	18	36
	irradiated	tensile test	36	36	72
		compression test	18	18	36
Hydrogenated	for reference	tensile test	72	72	144
		compression test	36	36	72
	irradiated	tensile test	72	72	144
		compression test	36	36	72
Hydrogenated with electrolysis	for reference	compression test	3	-	3
	irradiated	compression test	3	-	3
Axial	for reference	tensile test	1	2	3
	irradiated	tensile test	1	2	3
Summa pieces for tensile test			**232**	**230**	**462**
Summa pieces for compression test			**118**	**116**	**234**

The following main factors were considered:

- the available volume for samples in the irradiation rig,
- the flux and temperature distribution in the target holder,

- the excess gamma heat removal during irradiation,
- minimum three measurements per test has to be performed to the evaluation,
- the temperature and the time of the different heat treatment simulating the operational and accidental conditions,
- the Hydrogen content in the clad material,
- the production technology (e.g. the number of the electrolysis steps)

The nominal dimensions of the Zirconium alloy tubes E110 and E110G are: outer diameter D = 9.1 mm, wall thickness = 0.685 mm. Three type of specimens are selected for material testing: 8 mm high compression test specimens, 2 mm high tensile test specimens and 50 mm high axial tensile specimens.

2.3. Calculation of the irradiation time
Irradiation time planning was based on data from irradiation performed previously in the BRR. [3-7].

Based on these experiments considerable change of the mechanical properties of the Zirconium alloys change will occur after a fast neutron irradiation with 0.5-1.5 * 10^{20} neutron cm^{-2} fluence.

More than 30% strength increase expected above the level of fast fluences 1 * 10^{20} neutron cm^{-2}.

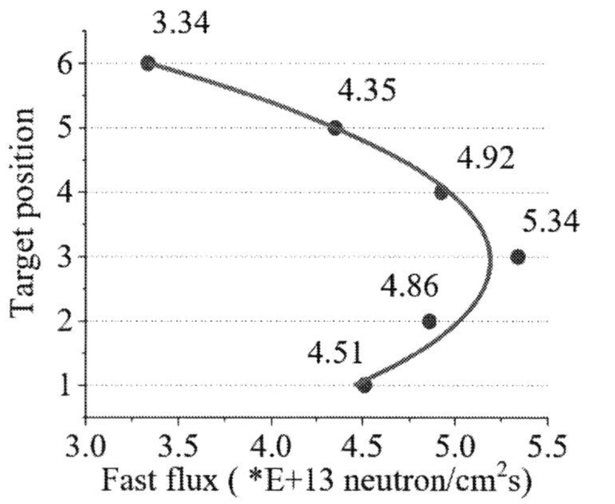

Figure 1. The fast neutron (E>1.5 MeV) flux distribution in the irradiation rig height. On the y axis the number of the targets is shown.

These data are in agreement with other literature data [12].

The fast flux distributions calculated from previous irradiations, were used to design the irradiation target. To measure the real fluence, dosimetry monitors were inserted into different positions of the target. Long decay type of the monitors used, due to the long term of the irradiation.

The fluxes calculated from the data on the dosimetry monitors in the irradiation probe, as a function of the position of the 6 sample holders, are shown on Figure 1.

The BRR operated in 10 days per cycles. To reach the required ageing minimum 5-6 reactor cycles needed. Considering the reactor schedule an optimal irradiation campaign ≈1700 hours long time selected.

2.4. Irradiation temperature control
The irradiation temperature was selected to simulate the operation conditions. To control of the irradiation temperature within the required range the heating and heat removal had to be balanced.

During neutron irradiation strong nuclear heating occurs. It depends on the position and the properties of the target. Auxiliary electric heating was applied to adjust continuously the temperature distribution. The excess nuclear heat from the target was removed through a gap between the target holders and the wall of the rig, which is cooled by the reactor cooling water. In the gap helium nitrogen gas mix is circulated. Changing the ratio of the nitrogen and helium the heat removal is controlled.

Six thermocouples measured the actual temperature. The data together with other measured parameters are stored in a computer memory. The temperature history during reactor operation can be checked remotely from the office of the research group.

3. Tool for irradiation: the BAGIRA irradiation rig
The BAGIRA rig has been operated since 1998 at BRR. Thirty-two different irradiation campaigns have been performed, testing irradiation ageing of different fission and fusion structural materials, as low alloyed and stainless steels, Al, Ti and W alloys, ceramics etc.

Aluminium sample holder can be used up to 350 °C, for higher temperature irradiation titanium sample holders can be built.

The BAGIRA device can be divided into 3 main parts:

- cooling gas mix preparation system,
- PC based temperature control and data acquisition system (gas recirculation unit, data acquisition and control unit),
- the irradiation channel, sample holder (including specimens and space filler)

The rig is operating inside of the core in place of three fuel elements.

3.1. Sample holder and targets

The rig capacity is 6 pieces of 65*21*25 mm targets. The target must be designed in such a way that it can be easily dismantled under hot-cell conditions.

To simulate the different burnout states, the samples were placed axial positions to receive different fast neutron flux.

After the pre-treatments the E110 and E110G samples were cleaned and filled into the targets. A dosimetry (Cu, Fe, Al-Co1%, Nb) monitor package for measuring flux was included in each sample packet.

Figure 2. Preparation and filling of the 65*21*25 mm targets, front- and back view of a filled target

4. The irradiation

The following data acquisition were stored by the computer at 15 second periods:

- Date,
- Time (s),
- Temperatures on 6 levels,
- Pressure of cooling gas,
- Absorbed powers of 4 electric heaters,
- Alarm messages

From the recorded data the average temperature and deviations evaluated. The average temperature of the targets summarized in table 2.

Table 2. Measured and averaged temperatures during the relevant irradiation cycles, in °C

Cycle	Time	Hours	T1	T2	T3	T4	T5	T6
1.	26.09.2017-06.10.2017	240	297	276	241	211	181	149
2.	10.10.2017-20.10.2017	240	300	282	246	206	172	148
3.	07.11.2017-21.11.2017	336	298	280	244	201	166	144
4.	28.11.2017-17.12.2017	456	300	295	265	215	174	152
5.	06.02.2018-16.02.2018	240	291	294	281	240	176	136
6.	27.02.2018-09.032018	240	285	293	286	250	188	148
Summa hours		**1752**						

During the irradiation campaign of the E110 and E110G alloys no critical alarm was detected.

After the campaign the rig was lifted from the reactor, the targets were reloaded separately and transported into the hot-cell with a special remote controlled system. In the hot cell the licensed and trained staff removed the samples from the dismantled targets. Appropriate assistance mechanisms, specially designed for this purpose, helped the work in the hot-cell. The irradiated samples are stored in labelled, covered aluminium holders.

The activity of the dosimetry monitors generally measured and evaluated according the ASTM E-261 and ASTM E-693 standards. Flux, fluence and dpa values for the actual core configuration are calculated by the Reactor Analyses Department of the Centre for Energy Research. All the received information from the irradiation campaign, the parameters of the irradiation, the measured and calculated values are documented according to quality standards and regulation of the Centre for Energy Research.

Figure 3. Dismantling of the target in the hot cell

5. Further work

The purpose of this paper was to introduce the neutron irradiation of the fuel clad samples.

Presently the samples must be stored in the hot-cell for the decay time until their activities will be reduced enough to test them. The design and production of the testing devices are going on. After testing the results will be published in a separate paper.

Acknowledgement

The study was supported by the National Research, Development and Innovation Fund of Hungary (contract number: NVKP_16-1-2016-0014).

References

[1] IAEA RR Database: https://nucleus.iaea.org/RRDB/Content/Util/MaterialIrradiation.aspx

[2] IAEA RR Database: http://nucleus.iaea.org/RRDB/RR/HeaderInfo.aspx?RId=196

[3] Horváth Á, Balaskó M, Patriskov G, Fekete T, Gillemot F, Horváth M and Uri G 2012 The contribution of Budapest Neutron Centre to the fusion materials and diagnostics development program *Hungarian Plasma Physics and Fusion Techn. Workshop (HPPW)*

[4] Gillemot F, Horváth M, Tatár L, Fekete T and Horváth Á 2008 Structural material investigations in the high temperature irradiation facility of the Budapest Research Reactor *Research Reactor Application for Materials Under High Neutron Fluence TM-34779 IAEA*

[5] Nikulin S, Rozhnov A, Belov V, Li E and Glazkina V November 2011 Influence of chemical composition of zirconium alloy E110 on embrittlement under LOCA conditions – Part 1: Oxidation kinetics and macrocharacteristics of structure and fracture *J. of Nucl. Materials* **418** Issues 1–3 p 1-7.

[6] Király M, Hózer Z, Horváth M, Novotny T, Perez-Feró E and Vér N 2019 Impact of thermal and chemical treatment on the mechanical properties of E110 and E110G cladding tubes *Nucl. Eng. and Technology* **51** Issue 2 p 518-25
[7] Budapest Database of E110 Claddings Under Accident Conditions 2007 AEKI-FRL-2007-123-01/01, NEA-1799/01 IFPE/AEKI-EDB-E110 *HAS Atomic Energy Res. Inst.*
[8] Király M, Antók D, Horváth M and Hózer Z 2018 Evaluation of axial and tangential ultimate tensile strength of zirconium cladding tubes *Nucl. Eng. and Technology* **50** no.3 425-31
[9] Ambient and high temperature mechanical properties of ZrNb1 cladding with different oxygen and hydrogen content 1999 *Enlarged Halden Programme Group Meeting*
[10] Griger Á 2004 Mechanical properties of irradiated and non-irradiated Zr1%Nb and Zircaloy claddings *International Nuclear Information System (INIS)*
[11] Kaplar A, Yegorova L, Lioutov K, Konobeyev A and Jouravkova N 2001 Mechanical Properties of Unirradiated and Irradiated Zr-1% Nb Cladding *NUREG/IA-0199*
[12] Griger Á, Horvath M, Pintérné Csordás A and Uri G 2005 Budapest Behaviour of irradiated cladding material AEKI-FRL-2005-732-01/04 Documentation, *HAS Atomic Energy Res. Inst.*

IOP Conf. Series: Materials Science and Engineering **903** (2020) 012053 doi:10.1088/1757-899X/903/1/012053

Manufacturing and investigation of aluminium matrix bimodal metal foams

B Leveles[1,2*], A Kemény[1,2], I N Orbulov[1,2]

[1] Budapest University of Technology and Economics, Faculty of Mechanical Engineering, Department of Materials Science and Technology, Budapest, Hungary
[2] MTA–BME Composite Metal Foams Research Group, Budapest, Hungary

E-mail: borbala.leveles@edu.bme.hu

Abstract. Metal matrix foams (MMFs) are mostly used in the automotive industry due to their high specific energy absorbing capacity and relatively low weight. During this research, bimodal metal matrix foams were produced, where the filler was ceramic hollow spheres (CHSs) made of high-purity alumina with the nominal diameters of $Ød_1 = 7$ mm and $Ød_2 = 2.4$ mm. The hollow spheres of different sizes were used in 1:1, 2:1 and 4:1 volume ratio, which after mixing can be considered as uniformly distributed. Al99.5 was infiltrated between the CHSs with low pressure to create a cellular material. The manufacturing parameters have a significant influence on the properties of the finished metal matrix foam. Different preheating temperatures, melt temperatures, infiltrating pressures and time was applied to achieve proper wetting and filling. Samples were evaluated based on macro images and microscopic photographs.

1. Introduction

Cellular materials in nature appeared to increase the stability of a more massive structure. Preferably, the selection was made in nature to reduce weight while maintaining or increasing strength. Metal matrix foams are mostly used in the luxury automotive, aerospace, and marine industries because of their high energy absorbing capacity and relatively low weight.

In the literature, MMFs are sorted in the class of hybrids as composites, sandwich structures and lattice materials. MMFs, in which the cells are formed by a second phase, their distribution within the material can be considered homogeneous, and their size is close to uniform, are called metal matrix syntactic foams (MMSFs). Usually, the matrix material is a lightweight metal, and the filler material is a hollow sphere or expanded material that allows the presence of air [1].

Various manufacturing technologies can produce MMFs. Three of these are widespread: powder metallurgy, stir casting and infiltration [2–4]. The most common procedure is infiltration, which may be gravity-, or pressure-assisted, depending on the wetting ratio [5,6]. In the latter technology, the matrix material is poured between the hollow spheres in a molten state, applying pressure to increase wetting. By changing the infiltration pressure, time, and temperature parameters, the quality of the given syntactic foam can be adjusted. Steel infiltration was shown in Yang and his colleagues' research, where many unsuccessful samples were created due to low preheat temperature [7–12].

Bimodal materials exhibit two distinct characteristics/distributions in one property and are nearly the same in all of the other. By bimodal metal matrix foams, the literature means MMSFs filled with CHSs of two different diameters. The proportion of CHSs of different diameters may be different from 50-50 wt.% as shown in Figure 1. [13–15].

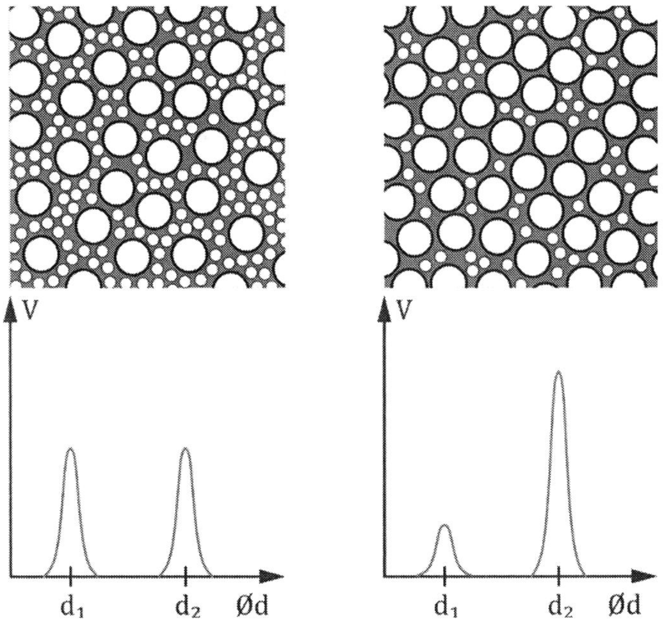

Figure 1. Bimodal distribution with two different filler ratio

Tao et al. investigated aluminium matrix syntactic foams filled with bimodal ceramic microspheres. In their research, the fabricated bimodal metal matrix foams had up to 10% higher porosity and 8% higher onset strain of densification compared to MMSFs. 6082 Al alloy was used with fine (75-125 μm) and coarse (250-500 μm) ceramic microsphere filler [14].

Orbulov et al. published a paper on the compressive characteristics of bimodal aluminium syntactic foams. AlSi12 alloy was used with smaller (150 μm) and larger (1425 μm) Globocer CHSs. The mechanical properties of the bimodal samples laid between the pure small and pure large CHS filled MMSFs, and the compressive strength could be reliably estimated by the rule of mixtures [16].

The goal of this research is the manufacture of bimodal metal foams to create a biphasic material with reduced density and higher specific compressive strength properties compared to homogeneous solid continuum material. Currently, such materials can only be manufactured under laboratory conditions without accurate knowledge of mixing methods and casting parameters. Once all condition factors become known, bimodal metal foam casting can also be industrialised with any form of shaping capability. This paper aims to cast and study the microstructure of bimodal metal foams to qualify different mixing ratios and determine their manufacturability.

2. Materials and methods

In MMSFs the matrix material absorbs and distributes the load, and through the boundary layer, the hollow spheres are kept in place and bear a part of the load. Al99.5 (1050 Al) was used as matrix material in this research to characterise the properties of bimodal metal matrix foams.

The second phase most commonly used for MMSFs are porous ceramic or hollow metal spheres. The fillers of the produced bimodal metal foams were the same in their base material and density, but two different average diameters. The CHSs are made from Al_2O_3 C795 according to EN 60672-3:1999 with nominal diameters of $Ød_1$=7 mm and $Ød_2$=2.4 mm and obtained from Hollomet GmbH. The specific properties of these CHSs are detailed in a previous research [17].

2.1. Mixing

During this research, bimodal metal foams were made with two different diameter ceramic hollow spheres as filler. The peculiarity of bimodal metal matrix foams lies in the mixing ratio of the two sized hollow spheres. First, the packing fractions of different mixing ratios of the spheres were calculated from theoretical equations [18] which has not led to total success, because the produced foams had lower packing ratios. During the producing, three mixing ratios were tested, with the ratio of large to small hollow spheres of (1:1) 50-50%, (2:1) 67-33% and (4:1) 80-20% distribution. Theoretical space fills do not differ significantly, 72.8%, 71.9% and 69.7% respectively, therefore macrostructure studies have determined the optimum and substantiality of the mixing ratios.

2.2. Manufacturing of bimodal foams

Two hollow spheres of different diameters were measured in volume ratio (and also in weight ratio due to the equal density) to the intended mixing ratio. The hollow spheres were mixed in a larger container until it was approximately uniformly distributed. After mixing, the CHSs were poured in a pre-welded and drilled, blocked hollow section into which an Al_2O_3 layer was pre-placed to prevent the flow of the molten matrix material and a steel mesh to prevent the spheres from moving, in the arrangement shown in Figure 2.

The ceramic container was preheated to the required high temperature (~100 °C below the melting point of aluminium) and then maintained to uniform heat. During this time, the matrix material was melted in an IND IF-10 induction furnace to allow rapid melting. For performing low-pressure infiltration, an insulated pipe was designed and manufactured, which was connected to a high-pressure argon bottle.

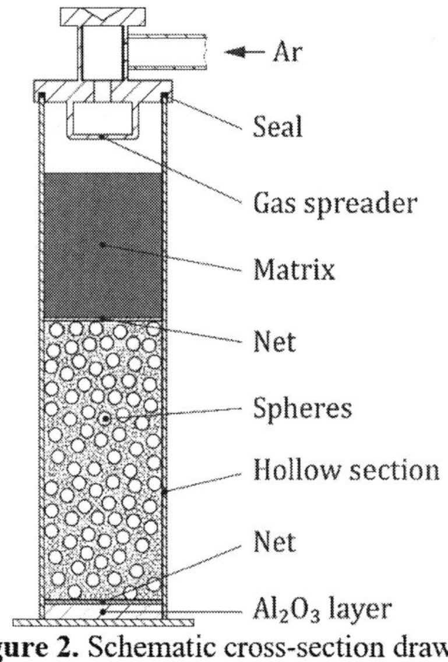

Figure 2. Schematic cross-section drawing of the infiltration layout

Figure 3. Failed MMSF production (left) and proper casting (right)

During the research, numerous bimodal metal matrix foam infiltration process was performed, some of which have failed until the proper infiltrating parameters were set. Table 1. shows the parameters to proper infiltration process with the used materials. A production was considered defective if the matrix material did not fill the space between the hollow spheres, where the aluminium did not reach the bottom of the crucible and froze too quickly, as shown in Figure 3., or that, due to too low viscosity or too high pressure, it passed completely through the hollow spheres and spilt at the bottom of the container.

The specimens were cut out using an automatic cutting machine, with a diamond particle cutting wheel, which, due to its hardness, was able to cut the block into a smooth plane cube without problems. As a result, from the 56×56×140 mm blocks (Figure 3. right) 56×56×50 mm sized cubic samples were produced.

Table 1. Infiltration parameters for proper production of bimodal MMSFs from Al99.5

Preheat (°C)	Preheat time (min)	Infiltration pressure (bar)	Infiltration time (s)	Matrix temperature (°C)
550	45	3	5	720

The surface of the resulting specimens is essential for sanding and polishing for macro and microstructural studies. Due to the different hardness of the matrix and filler and the wall porosity of the latter, this proved to be a difficult task. Unfortunately, the peeling ceramic pieces constantly caused smaller scratches on the aluminium surface, necessitating the use of a polishing technique.

3. Results and discussion

The average density of the produced bimodal metal matrix foams was 1.62±0.03 g/cm^3. It should be noted that besides the hollow spheres at the edges of the blocks more aluminium is present, and the ratio of the damaged hollow spheres also increases the value of the measured density compared to the theoretical density. This measured value is 0.2 g/cm^3 higher than the theoretical 1.4 g/cm^3, which is a reasonable rate.

3.1. Filler ratio determination

The success of producing MMSFs can first be determined by examining the actual filling rate of the CHSs and comparing it with the theoretical value. When the maximum is approached, the casting is effective. The most obvious method is to inspect the filler distribution on a flat section visually. In the first approach, the number of large and small hollow spheres were evaluated, and in the second approach, the ratio of the coverage area of all hollow spheres to aluminium (thus surface ratio). The first evaluation method has not yet yielded results. The solution for determining the filler ratio of the hollow spheres was software image analysis, for which ImageJ and Adobe Lightroom software were used with several methods. On average, Figure 4. shows a measured ratio of 55–67%, which has a significant deviation from the theoretical value of 69–73%. Identification of the proportion of small and large hollow spheres was tested with more advanced software, with the help of neural networks, but due to the inaccuracies of the scratches and grinding this method did not work flawlessly.

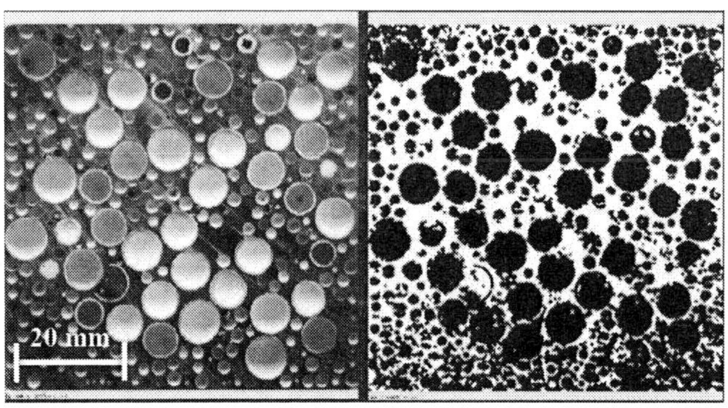

Figure 4. Image analysis result of the surface filling of CHSs

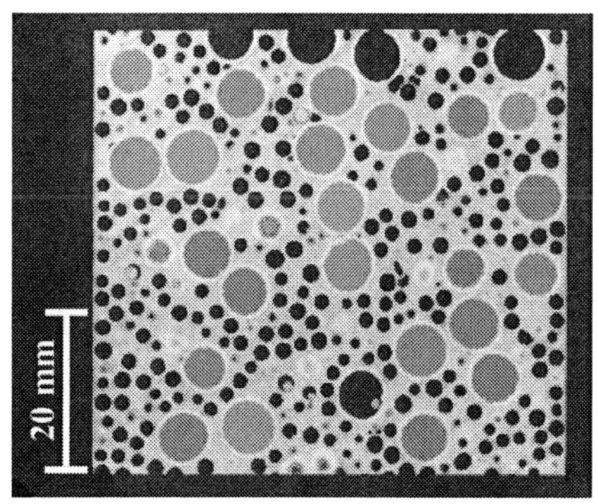

Figure 5. One 2D section of computer tomography measurement

A proper solution to measure the filler ratio is with computer tomography (CT) measurements in 3D, with which the voids can be measured with high accuracy [19]. However, the filler ratio excludes the walls of the CHSs, so their volume must be added to the cumulative void volume. Another disadvantage is that only the perfect spheres can be measured (small–blue, large–green); the ones on the borders and the spheres that are partly filled with matrix are left out (Figure 5.). From this measurement, ~50% of the bimodal metal matrix foam is made up of voids, so a calculated filler ratio is around 66%. This measurement takes quite some time to evaluate, so easier methods are needed to be developed.

3.2. Wall porosity determination

Based on microstructure studies, a relation was established between the ceramic hollow spheres and the aluminium matrix. In particular, the wall porosity of the spheres is worth investigating as it affects the ability to make cohesion with aluminium, but on the other hand, it gives a chance of failure. The porosity of the hollow spheres can be observed from polished cross sections at high magnification on optical microscopic images.

Similarly to the determination of filler distribution, wall porosity was also measured by image analysis. From the binary image, the percentage coverage of porous portions can be determined as a proportion of areas. On average, a porosity of 40-60% was observed. In Figure 6., metal microscope image made on Olympus PMG-3 illustrates the porosity of the spherical wall more closely.

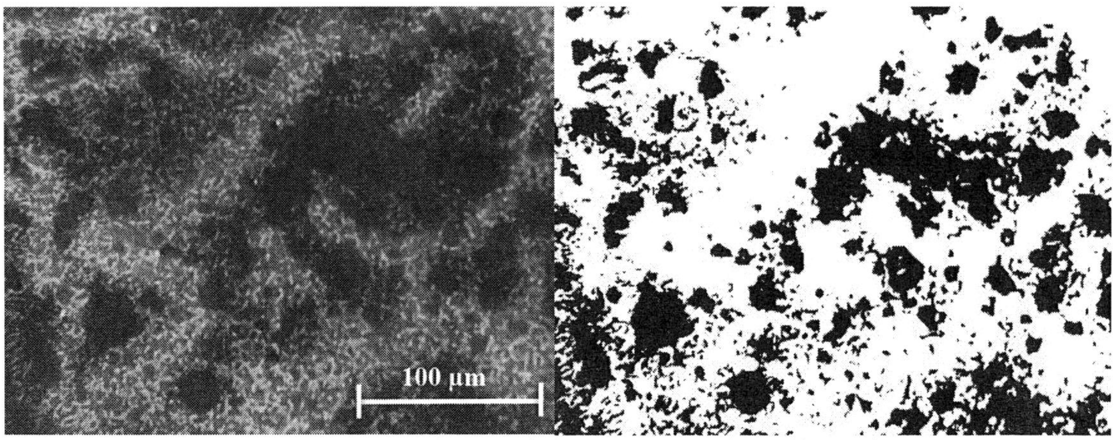

Figure 6. An approximate solution to measure the wall porosity

3.3. Boundary layer thickness

With energy dispersive X-ray spectrometry (EDS), the transition between the two materials can be examined more closely. For determining boundary layer thickness, line composition analysis was performed on a Zeiss EVO MA 10 scanning electron microscope (SEM) EDAX Z2 analyser module. The purpose of the measurement was to investigate the relation between the hollow spheres and the aluminium matrix, which is critical for structural integrity. The relation can be adhesive and cohesive. In the case of adhesion bonding, the bond depends only on the geometric properties of the surface of the CHSs. With cohesive relation, there is chemical bonding between the materials. The two phenomena often overlap, resulting in a complex relationship. To examine the boundary layer, linear EDS measurement was applied perpendicular to the transition between the hollow sphere and the matrix material. The points on the line EDS profile show the actual chemical composition in wt.%. Measurements were made on polished cross-sections starting from the wall of a hollow sphere, through the matrix material and closing on the wall of another hollow sphere (Figure 7.).

The measurement data can be used to determine the distribution of the elements in the matrix material and in the hollow spheres. The transition between the lines thus obtained determines the thickness of the boundary layer. The layer thickness was measured to be 12,50 ± 3,22 µm, which shows a strong bond between the CHSs and the Al99.5 matrix material.

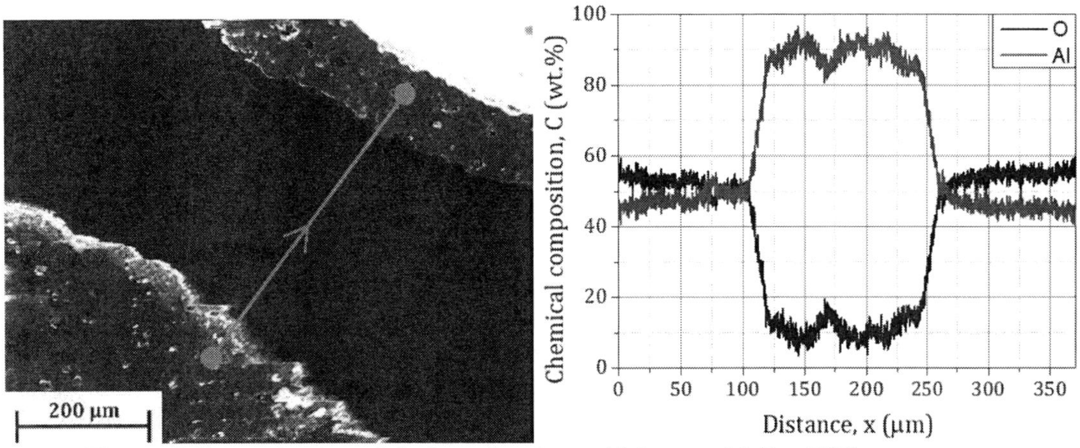

Figure 7. Determination of the boundary layer thickness with line EDS measurement

4. Conclusions

The following conclusions can be drawn from the measurements:

- CT measurements can result in determining the accurate void volume, but there is a need for simpler methods to approximate the filler ratio of bimodal MMSFs. Cross-section image analysis could be a solution to this problem if a measurement method is created, which decreases the deviation of the results.
- The measured filler ratio was ~66% so a higher ratio can be achieved with the investigated sized bimodal CHSs than with uniformly sized spheres.
- The difference from the theoretical value of the filler ratio can be explained with the high wall porosity of the CHSs, so molten aluminium can enter some spheres by cracking the thinner parts of the walls.
- There should be further investigations on the mixing methods of the different sized spheres to create an even lower density foam and approach closer to the theoretical filler ratio. The imperfect CHSs complicate this purpose.
- The line EDS measurements prove that proper infiltration parameters were defined for Al99.5/Al_2O_3 bimodal foams because a relatively thick boundary layer was formed between the CHSs and the matrix.

References

[1] Gupta N and Rohatgi P K 2018 4.15 *Compr. Compos. Mater. II* **4** 364–85
[2] Neville B P and Rabiei A 2008 *Mater. Des.* **29** 388–96
[3] Manoj et al 2018 *Mater. Sci. Eng. A* **731** 324–30
[4] Ginsztler J et al 2013 *Mater. Sci. Test. Informatics VI* **729** 68–73
[5] Castro G and Nutt S R 2012 *Mater. Sci. Eng.* A **553** 89–95
[6] Dobránszky J et al 2008 *Mater. Sci. Test. Informatics IV* **589** 137–42
[7] Eustathopoulos N et al 2001 *Mater. Sci. Eng. A* **300** 34–40
[8] Rabiei A and O'Neill A T 2005 *Mater. Sci. Eng. A* **404** 159–64
[9] Palmer R A et al 2007 *Mater. Sci. Eng. A* **464** 85–92
[10] Orbulov I N 2013 *Mater. Sci. Eng. A* **583** 11–9
[11] Bárczy T and Kaptay G 2009 *Mater. Sci. Forum* **473–474** 297–302
[12] Castro G and Nutt S R 2012 *Mater. Sci. Eng. A* **535** 274–80
[13] Tao X F and Zhao Y Y 2012 *Mater. Sci. Eng. A* **549** 228–32
[14] Tao X F et al 2009 *Mater. Des.* **30** 2732–6

[15] Brouwers H J H 2006 *Phys. Rev. E - Stat. Nonlinear, Soft Matter Phys.* **74** 1–15
[16] Orbulov I N et al 2019 *Compos. Part A Appl. Sci. Manuf.* **124** 105479
[17] Kemény A and Károly D 2019 *Acta Mater. Transylvanica* **2** 27–31
[18] Brouwers H J H 2013 *Phys. Rev. E - Stat. Nonlinear, Soft Matter Phys.* **87** 1–8
[19] Kozma I et al 2014 *Period. Polytech. Mech. Eng.* **58** 87–91

Effect of Beracha-I type color etchant on the ferrite phase in different type Fe-C alloys

József Bálint Renkó [1], Attila Bonyár [2], Péter János Szabó [1]

[1] Budapest University of Technology and Economics, Department of Materials Science and Engineering, H-1111 Budapest, Bertalan L. street 7, Hungary
[2] Budapest University of Technology and Economics, Department of Electronics Technology, H-1111 Budapest, Egry J. street 18, Hungary

Email: renko.jozsef@edu.bme.hu

Abstract. In material science, and particularly in metallography, various etching processes are frequently used to visualize grain structure. For some etching processes, it is sufficient to immerse the specimen into the etchant only for a few seconds, but for color etching, samples must be kept in the etchant for several minutes to achieve a colored surface layer. In our research, a measurement environment was created, in which the process of color etching can be observed in real-time. A closed microfluidic system ensured the controlled laminar flow of the etchant. The surface of the observed spherical graphite cast iron and DC01 low carbon steel was monitored from the beginning to the end of etching. Comparing the etching of the spherical graphite cast iron and the low carbon steel samples give a more comprehensive picture of the effect of Beraha-I type color etchant on the ferritic structure. Though the etchant reacts with the same ferritic phase in both cases, the etching process develops in different ways. The layer formation was much faster in the ferritic phase of DC01 steel, compared to ferrite in cast iron specimens.

1. Introduction

Etching is an essential step in the sample preparation of metallic materials. The purpose of this process is to reveal the microstructure of the materials for further microscopic examination. The advantage of etching compared to other methods is that it is cheaper and requires less preparation, such as other technologies, like electron microscopy [1-2].

One of the typical etching processes is color etching [3]. The test sample is immersed in a container filled with the etchant. The etching agent reacts with the surface. As a result, a thin, plastic layer begins to form on the surface of the sample. This layer is a chemical mixture (usually compound) of the examined metal and the etchant. Depending on the type of etchant, the film may also be oxide, sulfide, selenium, chromate, or complex molybdate [4]. The formed layer is transparent, and its thickness increases with the etching time. In practice, a layer can reach a thickness of 40-500 nm during a measurement. Depending on its thickness, the film interferes with light waves reflected from the metal surface, which causes colors specified by the interference criterion [5]. If the thickness of the layer satisfies the conditions of destructive interference, then the wavelength components will fall out of the reflected light, causing the observed color. Therefore:

$$d = m * \frac{\lambda}{4n} \quad (1)$$

In Eq. (1), m is an odd integer, λ is the incident light wavelength, and n is the refractive index of the film. If a given light component falls out of the visible light range due to interference, the surface below the layer will appear as a complement to the light that is lost [6]. As a result, the layer absorbs light of different wavelengths periodically, depending on the thickness of the layer. Thus, the color of the layer and the grain beneath it changes cyclically. It is also proved that the layer thickness depends on the etching time and the etching speed of the grain [7-8].

Depending on the investigated materials, different types of etchants are used [9-14]. Since our tests were carried out with Beraha-I type etchant, the specimens used were chosen accordingly. Beraha-I is an anisotropic etchant for ferritic materials, so the etching rate is highly dependent on the orientation of the individual grains. As a result, particles with different orientations are etched at different speeds, which causes a definite difference in their color after a given etching time. Although the ferrite grains differ in color due to the different etching speed, the process allows the individual particles to be separated.

Our current work aims to investigate how the Beraha-I type etchant reacts with different ferritic phases under the same conditions. By using the recorded etching processes of the investigated spherical graphite cast iron and low carbon steel samples, it was possible to compare the reaction of materials with different chemical compositions.

2. Experimental

2.1. Instrumentation

To carry out the planned investigations, it was necessary to create a measurement setup capable of keeping the etching under controlled conditions. Besides, the surface of the sample must be monitored by microscopy throughout the whole process. Thus, a microfluidic cell was designed and created to follow the etching of the samples. In order to design a microfluidics cell, however, the geometric dimensions and setup of the tools available must be known. For optical microscopy, an Olympus BX51 type LMPlan FI 50 × / 0.50 long-range objective microscope was used. The microscope was connected to a DP72 type digital camera, so microscopic images were recorded directly. The continuous laminar flow of the etchant was provided by a syringe pump.

2.2. Design and fabrication of the microfluidic cell

To create the microfluidic cell, the technical specifications of the instruments already available had to be considered, such as the focal length of the microscope objective and the orientation of the surface to be examined (Fig. 1). The focal length of the used objective was 2 mm; therefore, the microfluidic cell must be designed so that the surface of the sample should be accessible to 2 mm. In the designed system, this is accomplished with a 0.4 mm thick flow channel and a 0.7 mm thick, cell sealing glass sheet. This leaves an additional 0.9 mm for the objective to approach the microfluidic cell.

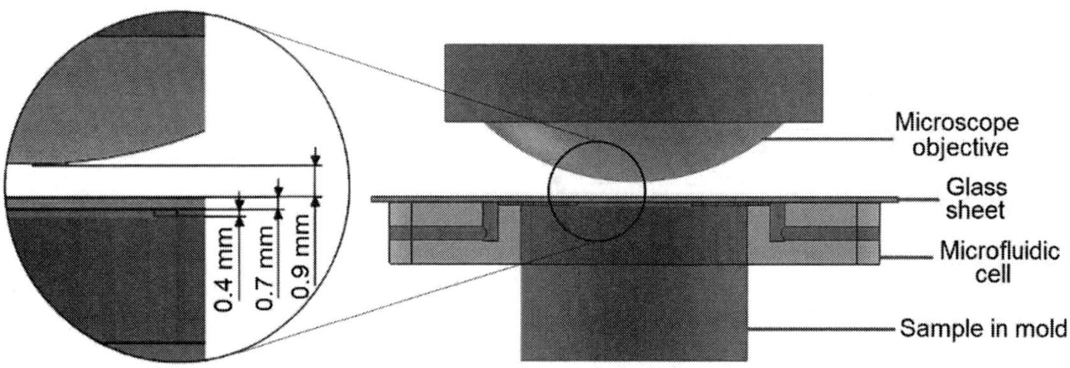

Figure 1. Design considerations for the microfluidic cell. The flow channel is 0.4 mm thick, while the glass sheet is 0.7 mm. Thus, the objective has an additional 0.9 mm to approach the examined surface.

Since the etchant would also react with the equipment, materials inert to the etchants must be used. The microfluidic system had to be adequately sealed in order to isolate the etchant from other equipment. Besides, it must provide laminar flow within the whole cell.

The designed microfluidic cell is presented in Fig. 2. The base of the cell is poly-dimethyl-siloxane, which is a chemically inert, flexible material. It is frequently used in the field of microfluidics [15]. The curing agent and the silicon pre-polymer were mixed in the prescribed ratio of 1:10. The mixture was filled into a molding form, created by an Objet Eden 250 type 3D printer. The filled molding form was set into a pre-heated oven at 100°C for 60 min to cure the polymer. After polymerization, the cell was bonded to a 1 mm thick glass sheet with corona discharge for 1 min. Further details of the technology were described elsewhere [15]. The upper, still open part of the cell is closed by the sample itself. The volume of the flow channel in the closed cell is 205 mm^3.

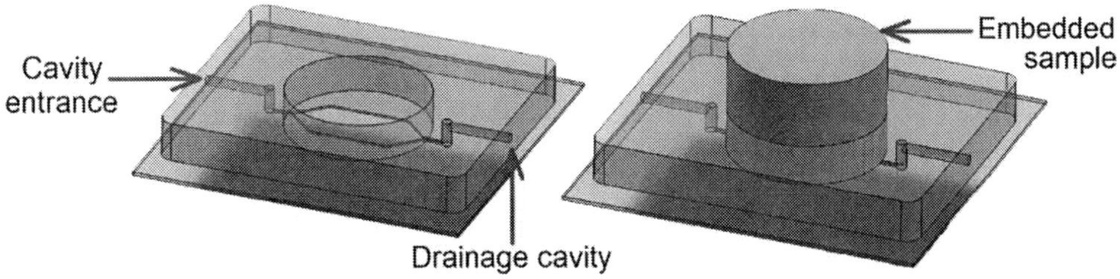

Figure 2. 3D CAD model of the microfluidic cell and microfluidic cell with an embedded sample

2.3. Materials and preparations

Spherical graphite cast iron and DC01 low carbon steel were used for the tests. The chemical composition of the examined samples is shown in Table 1.

Table 1. Chemical composition of the examined materials

Chemical composition (wt%)											
Spherical cast iron						**DC01 Low carbon steel**					
C	**Si**	**Mn**	**Mg**	**S**	**P**	**C**	**Mn**	**Cr**	**Mo**	**Ni**	**S**
3.59	2.43	0.36	0.063	0.012	0.046	0.05	0.524	0.082	0.025	0.062	0.013

After 4% nital pre-etching, it was assessed that the grain structure of the low carbon steel was too fine. At the intended magnification, the grains could not be distinguished. Therefore, the DC01 steel samples were heat-treated before testing [16]. After heating up to 850 °C for 3 hours, the samples were annealed at a cooling rate of 1 °C / min. After the approx.14 hours of annealing, the steel samples were embedded, then ground and polished. After polishing, the resulting particle diameter was checked by a repeated 4% nital etching.

The Beraha-I etchant used for the tests was freshly mixed by dissolving 3g $K_2S_2O_5$ and 10g $Na_2S_2O_3$ in 100 ml distilled water. In all cases, etching was performed in an air-conditioned laboratory at 22 °C.

2.4. Measurement setup

After the sample and microfluidic cell preparations, the measurement setup was prepared, as shown in Fig. 3. The figure shows how the syringe pump is connected to the microfluidic cell. Changes of the sample surface were tracked through the Olympus BX51 microscope's large focal length lens. During the etching of both samples, the flow rate was set on the syringe pump at 110 μl / min.

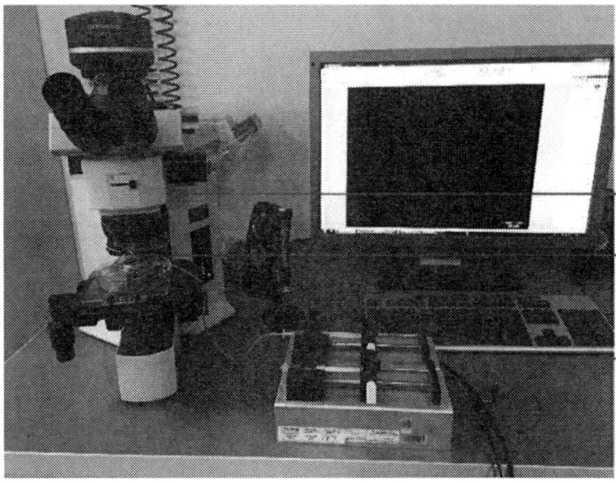

Figure 3. Measuring setup for optical microscopy

3. Results and discussion

Screenshots were made from the video recorded during etchings. Fig. 4. shows the surface of the spherical graphite cast iron sample and Fig. 5. shows the surface of the DC01 low carbon steel sample.

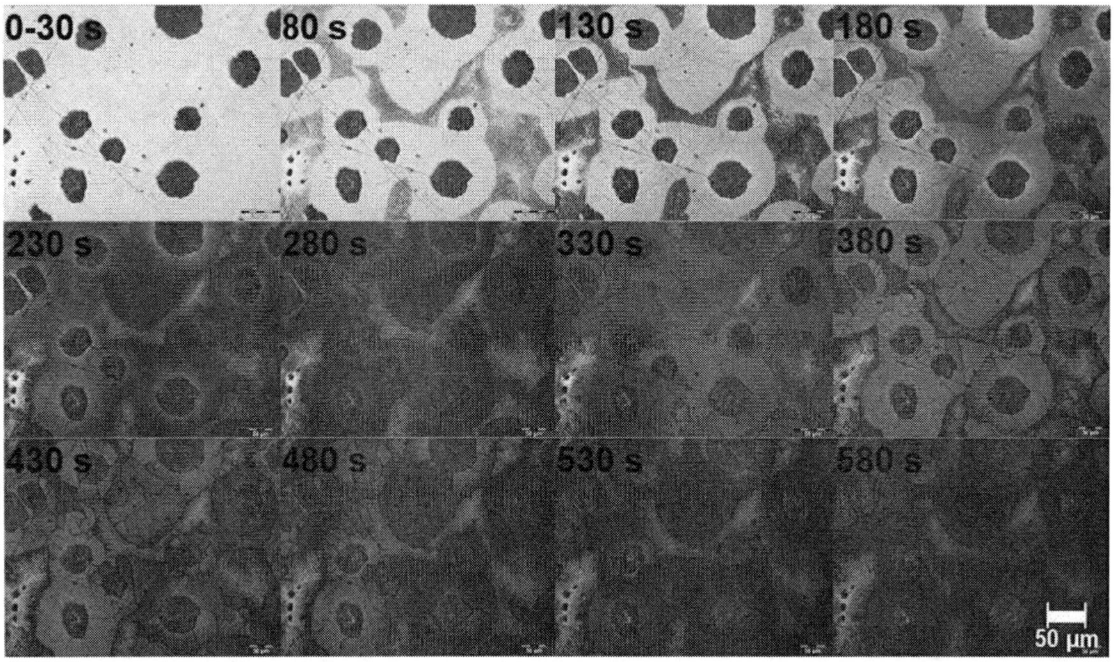

Figure 4. Spherical graphite cast iron etched with Beraha-I. Pictures are made at every 50 seconds

The observed area in cast iron could be easily identified several times due to the constant visibility of the graphite particles. During the etching of the cast iron sample, ferritic phases were attacked by Beraha-I. In the first 30 seconds, there were no visible changes. As we have shown in our previous article, the etchant was supposed to take so long to break the oxide layer on the surface [17]. From 30 seconds on, the perlite in cast iron sample darkens, as the etchant attacked the ferritic phase of the less corrosion-resistant fabric element. After approximately 180 seconds, the etching of the pure ferritic phases also begins. From this point on, the cyclic color change of the ferrite can be observed. The ferrite phase in the perlite is over etched after approximately 300 seconds, from which time no significant color change occurs in the perlite phase. The ferritic phases around the graphite particles changed color much longer. They needed approximately 580 seconds to reach the same state as perlite.

IOP Conf. Series: Materials Science and Engineering **903** (2020) 012054 doi:10.1088/1757-899X/903/1/012054

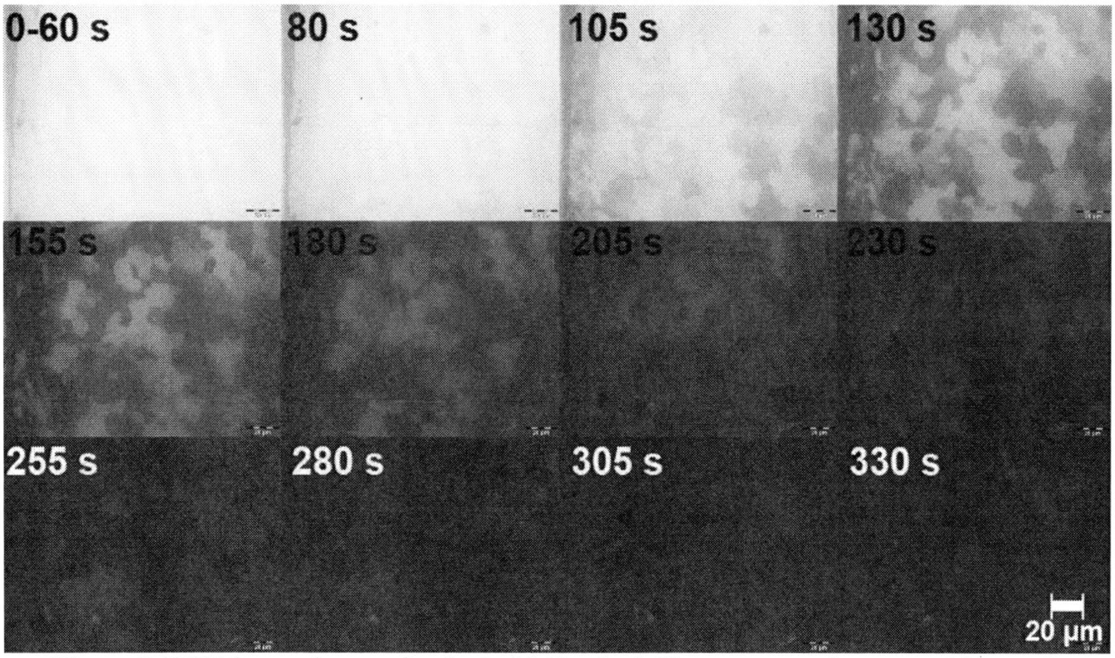

Figure 5. DC01 low carbon steel etched with Beraha-I. Pictures are made at every 25 seconds

The DC01 low carbon steel shows a significant difference compared to the cast iron specimen, though it has a similar ferritic structure. After 95 seconds, the particles became visible on the surface of the steel sample. It means that breaking through the surface oxide layer was much easier compared to cast iron's ferritic phases. All this suggests that the whole etching process took place at a much faster rate. While in the cast iron sample, the grains were over etched at approximately 580 seconds and darkened completely, the same condition of the steel sample was reached approx. 240 seconds earlier.

4. Conclusions

The ferrite phase of perlite in the cast iron sample was similar to the etching of low carbon steel's ferritic phase. In both of the cases breaking through the oxide layer took approximately 90 seconds, followed by a cyclic color change of approximately 270 seconds. The purely ferritic grains of cast iron were much more resistant to Beraha-I, requiring approximately 160 sec to begin layering. The etching and deposition rates were significantly slower here. The periodic color change was observed almost twice as long time, approximately 420 seconds long. As the test conditions were the same, the difference in etching can only be attributed to the difference in the chemical composition of the examined samples.

In order to gain a deeper understanding of the etching processes, the experiment should be extended later by a quantitative approach, such as the one presented in our previous work [17]. Comparing the results obtained would make it possible to quantify the visual difference.

Acknowledgment
This paper was supported by the ÚNKP-19-3-I-BME-266 New National Excellence Program of the Ministry for Innovation and Technology.

References

[1] Vander Voort G F 1984 Metallography, Principles and Practice, McGraw-Hill Book Co.
[2] Kovács D et al 2019, *Period. Polytech. Mech. Eng.* **63(3)** 214-219
[3] Schaberger E et al 2000 *Prakt. Metallogr.* **37** 419-434
[4] Kardos I et al 2007 *Mater. Sci. Forum* 2007 **389** 537-538
[5] Jenkins F A et al 1981 Fundamentals of Optics 4th ed. *McGraw-Hill Inc*

[6] Szabo P J et al 2010 *Mater. Char.* **61** 814-817
[7] Bonyár A et al 2012, *Microsc. Microanal.* **18(6)** 1389-1392
[8] Szabó P J et al 2011 *Micron* **43** 349-351
[9] Nádai L et al 2013 *Period. Polytech. Mech. Eng.* **57(2)** 53-57
[10] Varbai B et al 2018 *Period. Polytech. Mech. Eng.* **62(3)** 247-253
[11] Vander Voort G F 2004 ASM Handbook **9** 493–512
[12] Beraha E 1970 *Prakt. Metallogr.* **7(5)** 242–248
[13] Beraha E 1971 *Prakt. Metallogr.* **8(9)** 547–550
[14] Beraha E, 1974 *Prakt. Metallogr.* **11** 271–275
[15] Bonyár A et al 2010 *Procedia Eng.* **5** 291–294
[16] Vander Voort G F 2001 *Heat Treating Progress* **1(2)** 25-32
[17] Bonyár A et al 2019, Mater. Char. **156** 109844

IOP Conf. Series: Materials Science and Engineering **903** (2020) 012055 doi:10.1088/1757-899X/903/1/012055

Numerical modelling of the thinning behaviour of sheet metal parts

V Gál[1] and Zs Lukács[2]

[1] PhD Student, University of Miskolc, Hungary
[2] Associate Professor, University of Miskolc, Hungary

E-mail: metgv@uni-miskolc.hu

Abstract. In the automotive industry there are several software for numerical simulations to predict the behaviour of the material, which can reduce the expensive try-out experiments and dies cost. These software uses different mesh types. The goal of this work is to compare the different mesh element types in sheet metal forming simulations from the viewpoint of excess thinning. The benchmark of the 2018 NUMISHEET conference was the base of this comparison. This benchmark investigates the formability of a 2.8 mm thick, hot rolled sheet with 440 MPa tensile strength. The FE modelling of the process was made with thin shell-, thick shell-, and brick element types, thereafter; it was validated by the minimum thickness of the part based on physical data measurement. To better identify the differences due to the mesh elements, the thickness distribution of the part was investigated, too.

1. Introduction

Most parts in the automotive industry – particularly in the Body in White production – are produced from sheet metal. To achieve the RFT (Right for the First Time) production, the behavior of the material needs to be predicted. The more complex the geometry of the part is, the more complex stamping tool required. To increase the productivity and to reduce the expenses of tryout dies, the finite element simulation of the deformation of the sheet metal has a key role [1].

In sheet metal forming simulations commonly three set of parameters are used, to determine the plastic deformation. The Yield surface, which determines the onset of the plastic deformation. The Flow curves that show the material answer for the plastic deformation. Nowadays, in sheet metal forming, the Forming Limit Diagram (FLD) with the Forming Limit Curve (FLC) is regarded as the most important materials characteristics, which belongs to the limit strains resulting in the fracture of the material. The procedure to determine these constitutive parameters can be very difficult both with measurement, and theoretical methods.

The difficulty of determining these parameters are caused for example by the high strength materials (e.g. ultra high strength steel-s or high strength aluminium alloys) or the elevated temperatures measurement that required for hot forming simulations. All of these needs more difficult equipment. Today the problems with the measuring equipment for flow-curves (and of course the Yield-surfaces) considered as solved. However, the FLD's depends much more things and has several drawbacks, like the strain path or sheet thickness dependence, or even the differences between the calculated and the measured limit strain thanks to the friction dependence [2].

The accurate description of these parameters is the main target of the numerical simulations, because only in this way the behavior of the material can be predicted. Despite the FLDs are the most widely used material parameter in automotive industry, in most cases the Flow curve and Yield-surface can be enough for FEM due to the general prescription for the maximum thinning 30%, which is determined in the automotive industry. Based on this, the cracking, or the excess thinning of a material is not acceptable.

To achieve a successful simulation, by the accurate material card (which include the Flow curve, the Yield-surface and the FLD), the real manufacturing environment (die geometries and kinematics) and the accurate boundary conditions need to be generated. Based on this, the importance of the geometry types of the applied mesh element unequivocal. Otherwise -as you can see in the Figure 1. and Figure 2.- there are some cases when the results of the simulation is far from the reality. (The initial sheet thickness of the Figures below was 2 mm.)

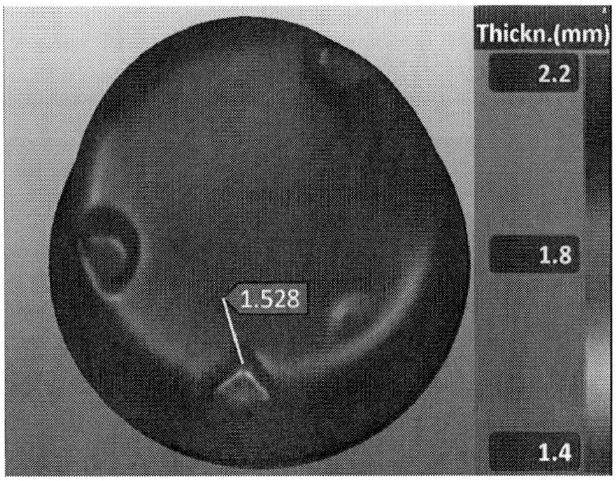

Figure 1. Results of a complex drawing operation of a 'Cone' part with shell element

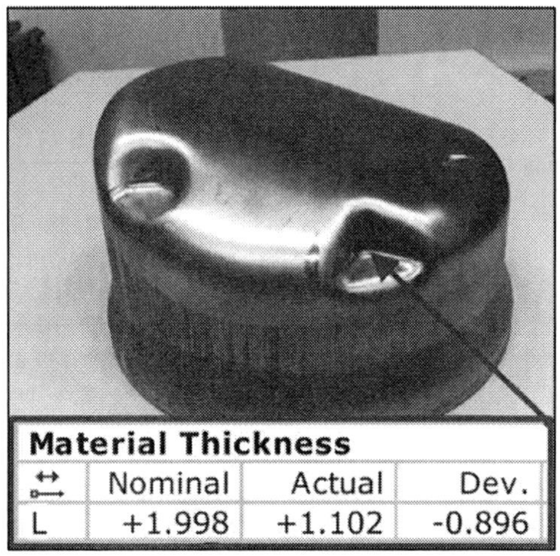

Figure 2. The physical result of the operation

For sheet metal forming simulations, the thin shell is the commonly used element type. For simple deformation cases, this element is appropriate. Nevertheless, in the cases of complex forming, especially when there is a forming operation in the opposite of the mean direction, inaccuracy may occur. This is one of the reasons why the dedicated FEM codes give the opportunity to use different element types.

Since different types of mesh elements or meshing methods may cause different results [3][4], the main goal of this work to examine the effect of these for local thinning. The base of the comparison was the benchmark of the 2018 NUMISHEET conference. In this work, the thinning behavior of an industrial part, was investigated with different mesh element types. The thickness dependence on die height, and the distribution of a sheet thickness along the cross section was examined.

2. Experimental setup

To have a better knowledge of the huge difference between the simulation and the real part that has shown above, the critical phase of the deformation needed to be examined. The excess thinning and the greatest difference occurred at one embossment of the four. The first thing that can be seen from the Figure 1. and Figure 2., that to create these embossments opposite operation to the drawing direction needed. To examine the current operation cross-section of the dies and the sheet was investigated (Figure 3.). In accordance with the industrial experience, the excess thinning was caused, because in the second step of the forming, while the material has no possibility to flow-in because the die of the first drawing step holding it back.

The huge calculating time comes with the complex geometry justified the decision to find an easier geometry, which create the same deformation condition, but make the calculating time shorter, and in the same time make the evaluation simpler too. The NUMISHEET conference, which is one of the largest conference in the field of numerical simulations of sheet metal forming, every second year has a benchmark project. To achieve this project the geometry of a part and tools, the appropriate constitutive equations, and other important simulation parameters used to be given.

One of the main tasks of the benchmark of 2018 NUMISHEET was the examination of an industrial part thinning behaviour in case of different die heights. Figure 4. shows the geometries of the benchmark tools, and it can be seen, that they can cause comparable deformation condition of the sheet, that has been shown in Figure3.

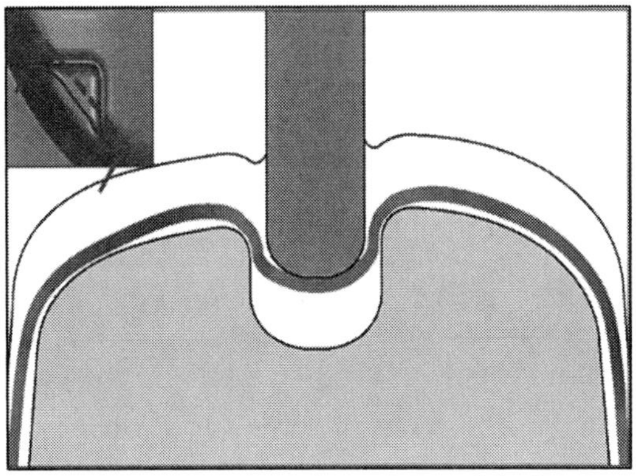

Figure 3. Cross-section of the critical part

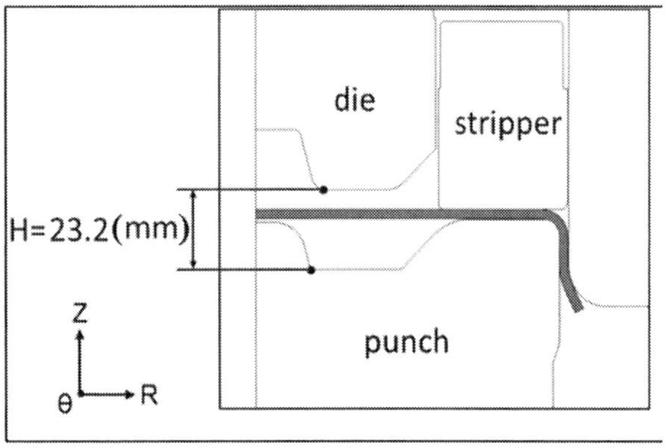

Figure 4. The setup of the NUMISHEET part

The blank had a diameter 246 mm and a thickness 2,8 mm, which is suitable for the measurement of thickness changing. The first step of the forming operation, the application of a 50 kN holding force with the stripper. After that, the top die starts moving downward with constant velocity of 10 mm/s. The deformation caused by the die can be separated into two stages. In the first stage, the outer flange of the part has been made, thereafter in the second stage, when no material flow occurs thanks to the cup wall and the blank holder force, the centre boss has been formed. Thanks to the first stage and the stripper, the thinning will be advanced, and a fractur will occur by the centre box.

The material data was given by the description of the benchmark. For the flow curve uniaxial tensile tests were performed. For the contours of plastic work biaxial tensile test with cruciform specimen in seven different linear stress paths were performed. Also, the results were given of the physical measurement the drawing operation show the context between the die height (H) and the minimum thickness of the sheet in the second stage when the drawing operation occurs till cracking.

3. Finite element simulation

The numerical simulations of the drawing process were carried to predict the necking behaviour using to different FE codes. First, the AutoForm FE code was used, because nowadays this program package is the most widely applied one for automotive sheet metal parts [1]. Secondly, the DEFORM, which is a process simulation system for forming simulation with solid elements.

The need of the two software can be explained with the mesh element type we planned to compare. For triangular thin shell element AutoForm R8 were used. For thick shell element also the AutoForm R8 was used since the software provides the opportunity to choose the element type before the simulation. The meshing was made automatically by the AutoForm, which use adaptive remeshing during the deformation with both element type in case of inadequate element size. The DEFORM FE code usually use tetrahedral element, but provide the brick element as an option too. In this case the brick element is better than the tetrahedral because of the number of elements in the thickness direction.

Three different element types in two different software were used. The number of element layers was 11 with both shell type and 9 for the brick mesh (because of the rule of the software). The generating of the simulation was almost the same, in the DEFORM and AutoForm^R8. The tool surfaces were imported from the benchmark and because its deformation will not playing a role, rigid type of material was choose for those. Constant static friction coefficient of 0.15 was used between the blank and the dies. The same material properties (flow curve, and yield surface), and the same boundary condition were used for both software.

The main goal of the measurements was to compare the increase of the excess thinning both in virtual and the physical measured ways. Due to this, the occurrence of the cracking was eliminated, and the last step (the die height) of the examination was the same as the physical measurement in the benchmark.

4. Results and discussion

The goal of the forming simulations was to examine the thinning behaviour of a hot rolled sheet metal in case of complex drawing operation. The simulation results for the minimum sheet thickness with different die heights ('H' in the Figure 4.) can be seen in the Figure 5.

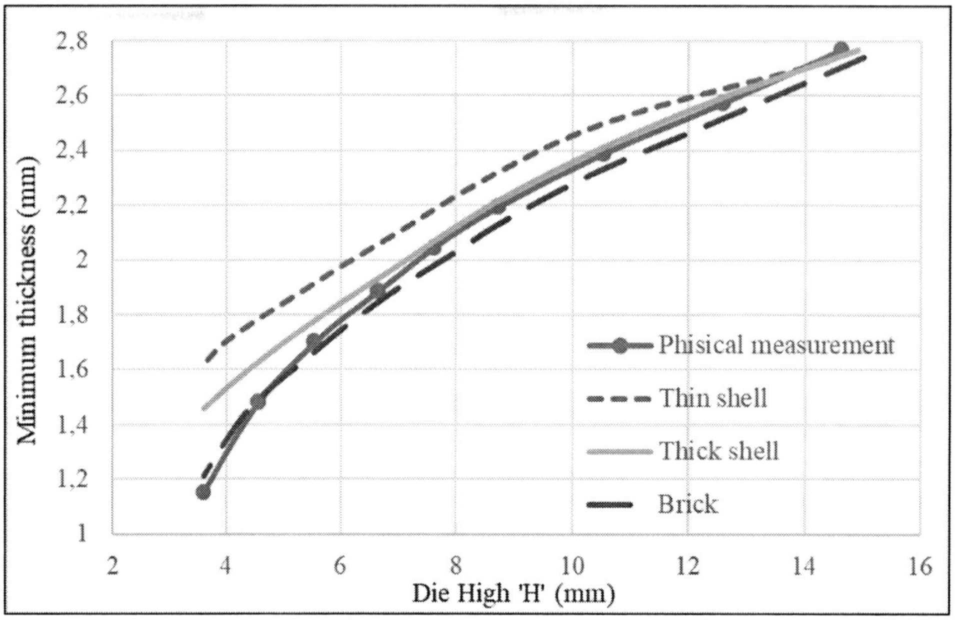

Figure 5. The minimum thickness as a function of die height

The decrease of the thickness at the beginning of the deformation is almost the same, but with the increasing deformation is different for each element type. It can be observed that with the developed deformation, the thin shell element show increasing difference from the physical measured points. At the end of the measurement (before the fracture occurs) the difference of thin shell from the physical measurement is more than 0,4 mm, so the maximum thinning is only 40% instead of almost 60%.

The thick shell compared to the thin has thickness values much closer to the measured ones. In the uniform strain path of the deformation, the thickness of the specimen change as in the physical measurement. It has also been observed, that the difference starts increase when the necking occurs. The difference between the physical and FE measurement was only 11% of the original sheet thickness at the end of the simulation.

The best results of all was given by the brick type element. The thickness values are moving with the physical measurement. Based on this it can be observed, that the simulation of the excess thinning shows a strong dependence of the used element type in case when the deformation is reserve direction to the main direction.

To have a better knowledge of the differences examination of the cross-section in the last measured point before the cracking occur has been made in case of the different mesh elements. Since there was no exact information of the thickness distribution of the current physical part, the best fit element type from the Figure 5. has been choose as a base for comparison. We have investigated the sheet thickness as the function of arc length measured from the axis of the symmetry.

The thickness comparison can be seen on the Figure 6. and it shows huge differences between the thickness distributions. It is important to notice, that the highest changes in the thickness occurs at the simulations with brick mesh in the *A* and the *B* phases.

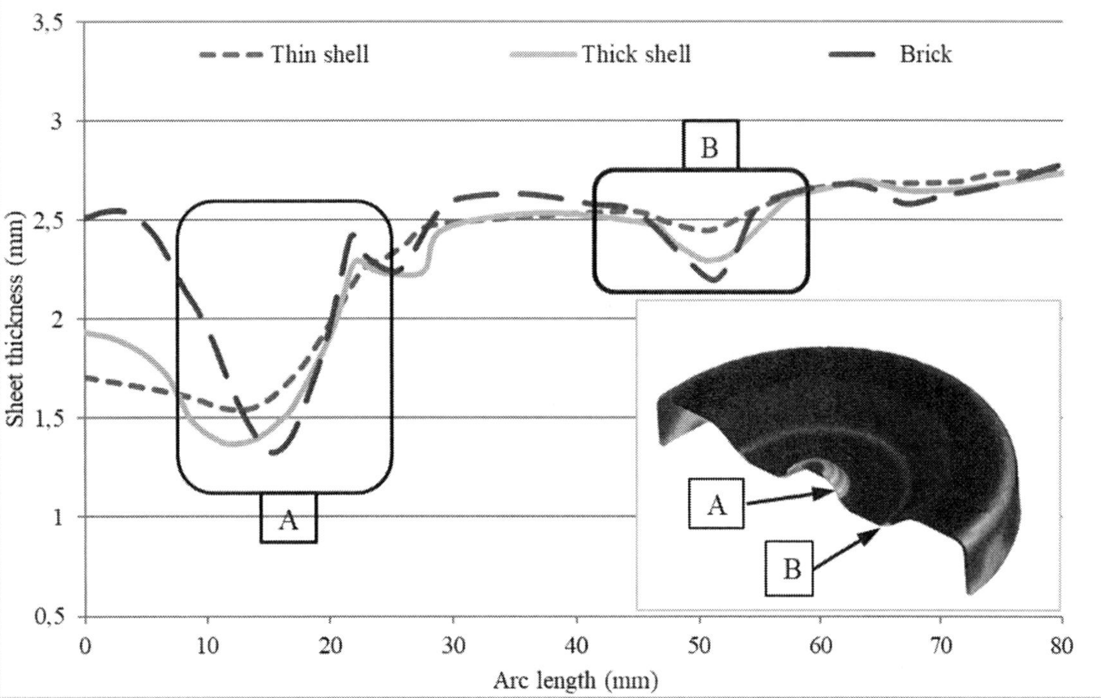

Figure 6. Thickness distribution as a function of the arc length

The thick shell element approaches the distribution of the thickness, at the critical points, the local minimum can be seen as by the brick element, but the values of the thinning are smaller than it need to be. As it have been expected on the basis of measurement above, the thin shell element has the worst effect for the thinning in this case. The thickness distribution actually has no critical minimum, so the excess thinning has not developed at the illustrated *A* and *B* points. The almost permanent minimum thickness surround the axis suggest that, in case of complex forming operation the mesh from thin shell makes a homogeneous distribution of the elongations of the elements. The phenomenon can be explained by the fact that the thin shell mesh does not consider the stresses in the thickness direction.

Taking the elements suitability into account in case of automotive application, the comparison of the calculating times also needed. The simulation with the thin shell need 8 minute, the thick shell 12 minute, the brick mesh 4 hour computing time, on the same computer setup.

5. Conclusion

In the present research work FE simulations of a benchmark of the 2018 NUMISHEET conference have been performed to study the effect of the different element types for the excess thinning. It has been observed in the FE simulations, that the thick shell element in the viewpoint of the change of the minimum thickness with the die height, and also presumably the thickness change along the cross-section, show better results, than the thin shell element. The inadequate efficiency of the thin shell element can be explained with that these element does not count the stress in the direction of the thickness. High reliability was achieved in the simulation of complex thinning with brick elements.

Based on the results it can be declared, that the applied elements have significant effect on the modelling of thinning behaviour depending on the complexities of the forming operations in case when there is a forming part of the operation which direction is reverse to the main direction. The best results for complex forming operation with the brick element, the least acceptable result with shell element can be achieved in case of complex deformation.

To convince the above described results, simulation was carried out with the part geometry that can be seen in the Figure 1. and Figure 2. The results were given by thin shell elements were only compared with thick shell elements in the software AutoForm^R8. There was no possibility to check the results with brick mesh element, due to the system requirements and calculating time caused by the dimensions of the part. It can be clearly seen in the Figure 7., that the simulations with thick shell elements has been approached better the reality than the shell element. The difference between the real part and the results of the thick shell also can be noted, but the thick shell can give better warning (the excess thinning was more visible than for thin shell).

Figure 7. The difference of the thickness results between thin and thick elements with a real part

The simulations with thin shell elements has the shortest calculating time, so it is perfectly suitable in case of the generally used stretching kind operations. In case of complex operation, the brick element can provide the best fitting with the reality, but in the viewpoint of the calculating time of the simulation too, the best option for the automotive industry can be the thick shell element.

Acknowledgement
The described article/presentation/study was carried out as part of the EFOP-3.6.1-16-2016-00011 “Younger and Renewing University – Innovative Knowledge City – institutional development of the University of Miskolc aiming at intelligent specialisation” project implemented in the framework of the Szechenyi 2020 program. The realization of this project is supported by the European Union, co-financed by the European Social Fund.

References

[1] Tisza, M. (2013); Recent development trends in sheet metal forming Int. J. Microstructure and Materials Properties, Vol. 8, Nos. 1/2, 2013
[2] Béres. G, Weltsch. Z, Lukács. Zs, Tisza. M; Prediction of stress- and strain-based forming limits of automotive thin sheets by numerical, theoretical and experimental methods AIP Conference Proceedings 1960, 160002 (2018); doi: 10.1063/1.5035028
[3] Li Kaiping, A. M. Habraken, H. Bruneel (1995); Simulation of square-cup deep-drawing with different finite elements; Journal of Materials Processing Technology 50 (1995) 81-91
[4] Alessandro R, Roberto D’A, Saverio A; Comparison of Meshing Strategies in THR Finite Element Modelling. Materials 2019, 12, 2332; doi:10.3390/ma12132332

IOP Conf. Series: Materials Science and Engineering **903** (2020) 012056 doi:10.1088/1757-899X/903/1/012056

A comparative analysis of characteristics of cutting forces at micro-milling of hardened steels

B Z Balázs[1] and M Takács[2]

[1,2] Budapest University of Technology and Economics, Faculty of Mechanical Engineering, Department of Manufacturing Science and Engineering, Budapest, Hungary

E-mail: balazs@manuf.bme.hu

Abstract. Micro-milling is one of the most important technologies to produce miniature components. Mainly due to the size reduction, the process has many special characteristics, such as the relatively large tool run-out, the relatively large deformation and the cutting edge radius of the micro-milling tool. The cutting force is often correlated to the phenomena mentioned above, its analysis is therefore often required. The main objective of the present paper is to analyse and compare the characteristics of cutting forces in micro-milling of different hardened steels. A systematic series of experiments were conducted on a hardened AISI H13 hot-work tool steel and a Böhler M303 martensitic corrosion resistance steel. The hardness of both materials is 50 HRC. A 500 µm diameter two fluted carbide micro end mill and a five-axis micromachining centre were applied. The effects of the cutting parameters are analysed by the analysis of variance (ANOVA). Experimental results show that the workpiece material has a significant influence on the cutting forces.

1. Introduction

Recently, the demand from many areas of the industry for miniature components is highly growing, which requires the development of micromachining technologies. Therefore, these processes become strongly researched [1-3]. One of the most essential machining technologies to produce micro components is micro-milling, which is capable of machining complex 3D geometries with high precision in a wide range of the materials [4, 5]. The tool geometries and the kinematics of micro-milling are similar to those used at macro dimensions. However, the knowledge cannot be directly adapted from macro dimensions due to the size reduction (to several orders of magnitude) [6]. Furthermore, the anisotropy and the grain size of the material become more significant in micro-dimensions [7], and the further problems are the relatively large run-out and the relatively large cutting edge radius (r_β) of the tool. The latter causes the problem of the minimum chip thickness (h_{min}). In the case of cutting below the h_{min} value, the ploughing is the main material removal mechanism [8], which can characterise by the high frictional forces [9]. In this case, the material is deformed elastically and plastically, and there is no chip formation [10]. The chip thickness changes continuously due to the kinematics of micro-milling, the ploughing effect is reduced, and the chip starts to form when the minimum chip thickness value is reached [10]. The value of h_{min} is considered by many researchers as a function of the cutting edge radius. The ratio of h_{min}/r_β is uniformly distributed between 0.2-0.4 [11, 12]. The cutting below the minimum chip thickness also reduces the surface quality due to the stronger ploughing effect [13].

The investigation of cutting forces and the effect of tool wear on them are also the subject of numerous scientific studies. Kuram and Ozcelik [14] investigated milling strategies on Ti6Al4V. They observed that the increasement of the rotational speed (12 000 – 28 000 rpm) reduced the tool wear at down milling and zig-zag strategy in the case of low feed rates (75 mm/min). However, it affected the tool wear negatively at the up milling strategy. Under these conditions, the surface quality improved, and the range of the F_y and F_x forces decreased (except the down milling). The cutting forces also depend on the milling strategy: the forces are smaller in the case of up milling than in the case of down milling, when machining AISI 1045 [15]. According to Mian et al. [16], the material of the workpiece significantly influences the chip formation, the deformation mechanisms of the workpiece and the flow, if the theoretical chip thickness is smaller than the cutting edge radius or the material grain size. Based on the experimental study by Gao et al. [9], a greater cutting edge radius may result in a greater negative effective-rake angle, which affects the shear and ploughing forces. These ploughing forces start to increase significantly below the minimum chip thickness, which entails a significant increase in the specific cutting force. Thus, the forces may increase due to the excessive reduction of the feed due to the ploughing phenomenon. Biró and Szalay [17] analysed the specific cutting force and identified a new boundary chip thickness in the range of micro chip formation. It was found by Aramcharoen and Mativenga [18] that the size effect is significant when the ratio of h_{min}/r_{β} is smaller than 1, in the case of micro-milling of a very fine-grain AISI H13.

2. Aims and experimental conditions

The main aim of this research is the detailed experimental and theoretical investigation of micro-milling of such hardened steels, which can be used for micro-injection moulding tools. The industry requires not only the hardness of the materials; in many cases, the corrosion resistance is also important. Present paper deals with a detailed comparison of micro-milling of AISI H13 hot-work tool steel and a Böhler M303 martensitic stainless steel in terms of the cutting forces. Based on the certificate of the heat treatment, the hardness of both materials is 50±1 HRC. The chemical composition of the materials is shown in Table 1.

Table 1. Chemical composition of the materials used [19, 20]

	C (wt%)	Si (wt%)	Mn (wt%)	Cr (wt%)	Mo (wt%)	Ni (wt%)	V (wt%)
AISI H13	0.39	1.10	0.40	5.20	1.40	-	0.95
Böhler M303	0.27	0.30	0.65	14.50	1.00	0.85	-

The experiments were performed on a VHTC 130M five-axis micromachining centre with a maximum spindle rotational speed of 60,000 rpm. The tools were commercially available, 500 μm diameter, two fluted, coated carbide micro end mills. The main geometric characteristics of the tools are as follows: γ=-10°, α=13.7°, and λ=25°. The micro-milling tools have a 50 μm corner radius, which increases the stability of the cutting edge. Pre-worn tools were used during the experiments to minimise the effect of the tool wear on the results.

A Kistler 9257A three-component dynamometer was combined with a Kistler 5080A charge amplifier to measure the cutting force components. The data was recorded by a NI USB-4431 sound and vibration device and a measurement program developed in LabVIEW environment. A Dino-Lite AM413T5 (magnification of 500x) digital microscope was installed inside the machine to check the condition of the tool.

Table 2. Design of experiments

Factors	**f_z (μm)**	**a_p (μm)**	**Milling strategy**	**Material**
Levels	*7*	*3*	**3**	**2**
Values	1, 2, 4, 6, 8, 10, 12	25, 50, 100	Up-, Down-, Full groove milling	AISI H13 Böhler M303

Full factorial experiments were designed and carried out (according to Table 2) to investigate the effect of the depth of cut, the feed per tooth and the milling strategies on the cutting forces of the micro-milling process. During the experiments, a cutting speed of 90 m/min was used, which correlates with the maximum available spindle speed. An a_e=500 μm (=tool diameter) radial depth of cut was set at the groove milling, while an a_e=250 μm was applied in up milling and down milling.

The recorded force signals were evaluated using a program developed in LabVIEW environment. The values of cutting forces were calculated as follows: (i) first, the collected force data were cut to a 100 consecutive-rotations-long signal when the impact of the entry and the exit phenomenon was negligible; then (ii) the maximum values ($F_{max,i}$) were calculated in each tool rotation; finally (iii) the average of them was calculated based on Equation 1.

$$F = 100^{-1} \cdot \sum_{i=1}^{100} F_{max,i} \tag{1}$$

3. Results and discussion

3.1. Cutting forces at groove milling

The characteristics of cutting forces provide indirect information on chip removal processes, and on material removal mechanisms, therefore, its analysis is often required [21–23]. Furthermore, cutting force is also in a close correlation to the deformation of the cutting tool [24], to the characteristics of surface roughness [25] and are essential for the foundation of process monitoring applications [26]. The influence of feed per tooth, depth of cut, and type of the material on the cutting forces can be seen in Figure 1.

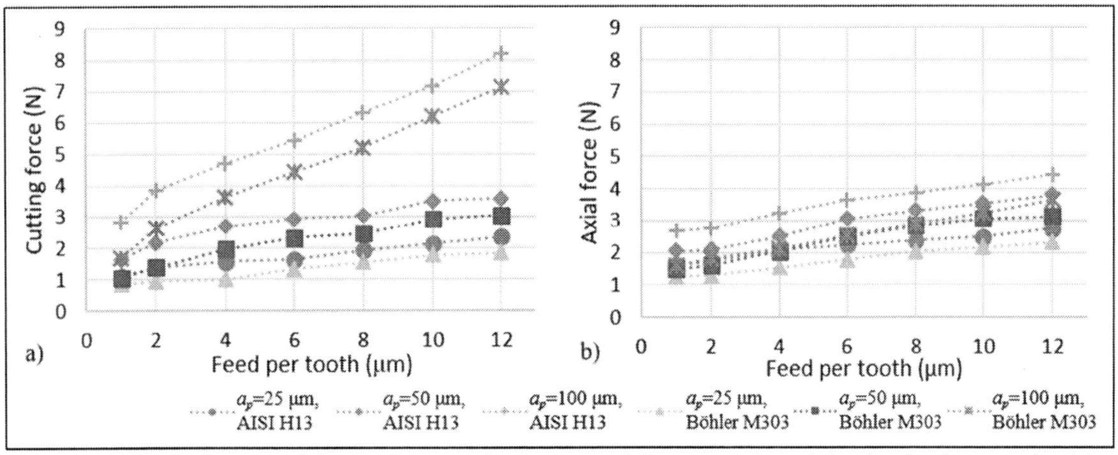

Figure 1. The influence of the feed per tooth, the type of the material, and the depth of cut on the (a) cutting force and on the (b) axial force component

The influence of the analysed input parameters is clear in the investigated parameter range. The response values increase with the increasing of the feed per tooth. In addition, the effect of the depth of cut was to be found similar. These results were expected due to the increasing chip cross-section. In the case of the Böhler M303 material, the difference in the values of axial forces (F_z) is less significant in the case of a_p=50 μm or a_p=100 μm depth of cut parameters was set. Therefore, in this case, the value of a_p is recommended to be doubled in order to increase productivity, which results in only a slight increase in the axial force.

The ratio of the axial force to the cutting force (F_z/F_c) was also analysed in order to understand the changing of the specific axial force component of the process. This parameter describes whether the axial cutting force is large compared to the main cutting force, or not. It was found that the ratio decreases with the increasing of the depth of cut, and the increasing of the feed per tooth also reduces it slightly. For example, in the case of the AISI H13, a_p=25 μm, and f_z=1 μm factors were applied, the ratio is F_z/F_c=1.487, while, if the f_z is increased to 12 μm, the ratio decreased to F_z/F_c=1.175. In the case of a_p=100 μm, the values of the ratio are significantly smaller: F_z/F_c=0.946 (at f_z=1 μm), and F_z/F_c=0.544 (at f_z=12 μm). In contrast, when the a_p value is equal to the corner radius of the tool, there is a reduction in the values up to 4 μm feed per tooth value, however, the further increase of the feed results in a smaller increase in the values. The possible reason for the relatively high F_z component can be found in the relatively large corner radius of the cutting tool. The effect of the relatively small radius becomes more significant in micro sizes. The higher ratio of F_z forces can results in a more stable cutting condition because the tool is loaded in the most rigid direction (axial). In contrast, the cutting force causes deflection of the micro-milling tool in a radial direction, which is one of the main problems in the case of micro tools with low stiffness.

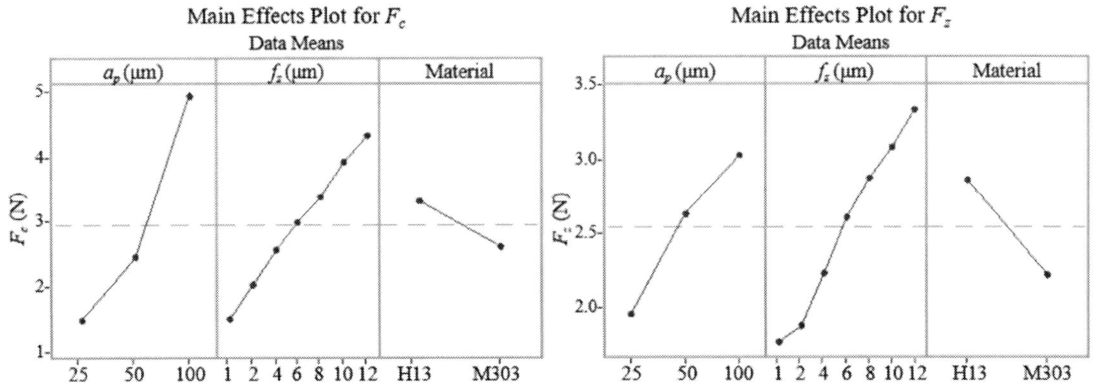

Figure 2. Main effects plot for (a) F_c and (b) F_z

The results were also analyzed by analysis of variance (ANOVA) using the Minitab program. The results obtained are shown in Figures 2. and 3. The $F_{crit}=f(p, df_1, df_2)=4.0847$ at a $p=0.05$ significance level. Based on the ANOVA results, the effect of all of the three examined factors is significant. For F_c, the effect of a_p is the most significant (F-value: 2171.24, P-value: 0.000), followed by the feed per tooth (F-value: 966.38, P-value: 0.000), and by the workpiece material (F-value: 127.90, P-value: 0.000). The two hardened materials used have a uniform hardness of 50±1 HRC; in contrast, there are significant differences in cutting forces.

Table 3. ANOVA table: F_c versus a_p; f_z; Material

Source	DF	Adj SS	Adj MS	F-Value	P-Value
Model	7	144.686	20.6694	477.15	0.000
Linear	3	139.920	46.6399	1076.68	0.000
a_p	1	94.054	94.0537	2171.24	0.000
f_z	1	41.862	41.8618	966.38	0.000
Material	1	5.540	5.5405	127.90	0.000
2-Way Interactions	3	15.166	5.0555	116.71	0.000
a_p*f_z	1	14.317	14.3170	330.51	0.000
a_p*Material	1	0.839	0.8388	19.36	0.000
f_z*Material	1	0.008	0.0084	0.19	0.662
3-Way Interactions	1	0.007	0.0072	0.17	0.685
a_p*f_z*Material	1	0.007	0.072	0.17	0.685
Error	34	1.473	0.0433		
Total	41	146.159			

On average, 30% higher forces were measured in the case of the AISI H13 material was applied than the Böhler M303 one. The possible reason could be found in the different structures of the different workpiece materials. The main effect of the material is 0.7 N for the F_c and 0.65 N for the F_z in the investigated parameter range. So, the demand in the industry of the corrosion resistance of the materials could result in even better conditions in terms of the cutting forces. The interaction plots are shown in Figure 3. The values shown in Table 3 show that the interaction between the depth of cut and the feed per tooth, and between the a_p and the material is statistically significant.

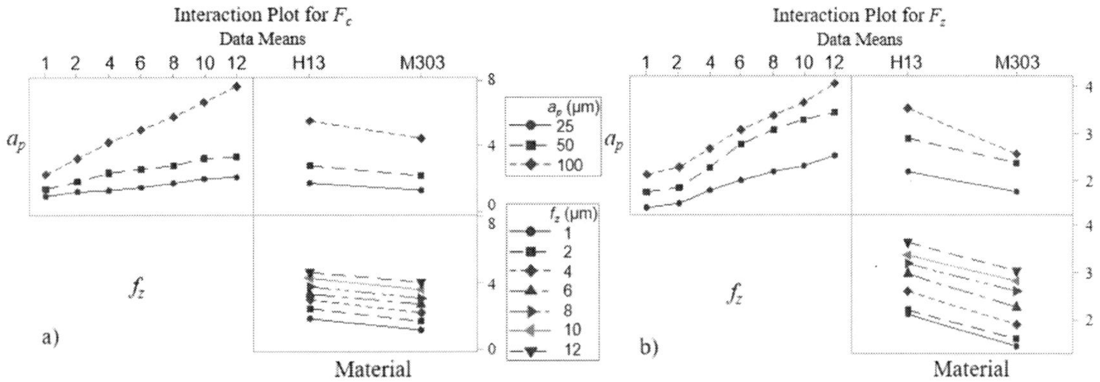

Figure 3. Interaction plot for (a) F_c and (b) F_z

The results of ANOVA are similar in the case of the F_z force component to the results of F_c forces. The main effects plot for F_z is shown in Figure 3b. The most significant factor is the feed per tooth (F-value: 331.22, P-value: 0.000), followed by the depth of cut (F-value: 187.31, P-value: 0.000), and by the workpiece material (F-value: 115.22, P-value: 0.000).

3.2. Cutting forces at different milling strategies

This chapter deals with the comparison of the milling strategies (up milling and down milling) with special regards to the forces measured during the machining. The influence of the feed per tooth, depth of cut, and type of the material on the cutting forces can be seen in Figure 4.

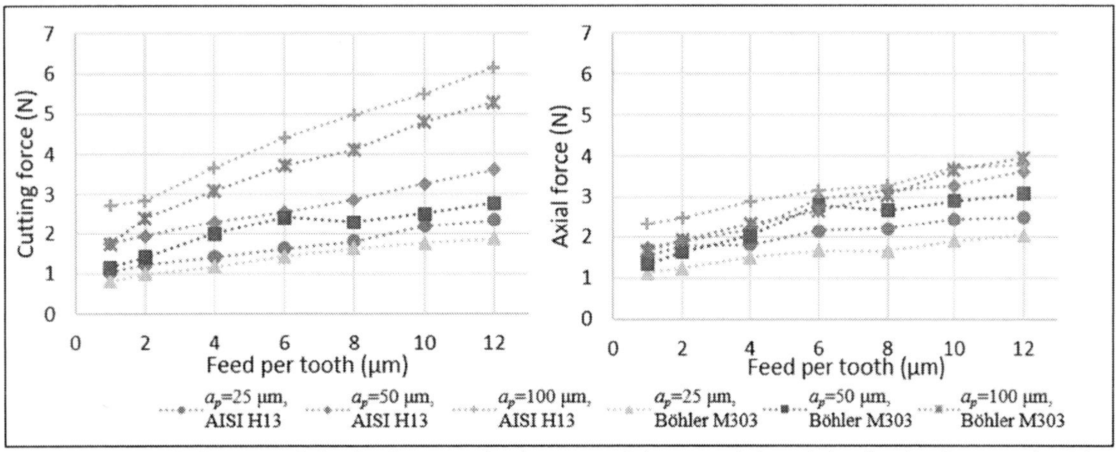

Figure 4. The influence of the feed per tooth, the type of the material, and the depth of cut on the (a) cutting force and on the (b) axial force component

The trends of the cutting forces and the axial forces are similar to the results of the full groove milling process, as was expected before, the forces increase with the increasing of the chip cross-section. In the case of F_z component is concerned, a significant increase was observed at both materials and both milling strategies, at a_p=50 µm and after f_z=4 µm. This change occurs when the ratio of chip thickness to the cutting edge radius is $h/r_\beta=1\pm\delta$ (where δ is an experimental error). The possible reason for these phenomena could be found in the different dominant material removal mechanisms (shear and ploughing) appearance.

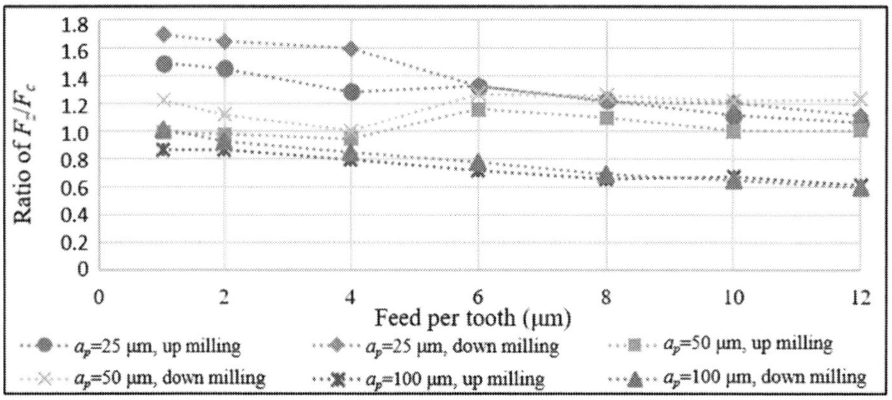

Figure 5. The influence of feed per tooth, depth of cut and strategy on the F_z/F_c

The ratio of F_z/F_c was also analyzed in the case of up milling and down milling. A decrease of the ratio was also observed with the increases of the feed per tooth and with the increases of the depth of cut (Figure 5.). However, there is an exception at a_p=50 μm (this depth of cut is equal to the corner radius of the tool), which proper reason needs additional experiments in the future. There is a minimum point at both strategies when the feed per tooth value is 4 μm (which is equal to the assumed cutting edge radius of the tool). In the case of the hot-work tool steel, the ratio is on average 11% lower at up milling than at down milling. In contrast, it is on average, 5.5% lower at down milling in the case of Böhler M303.

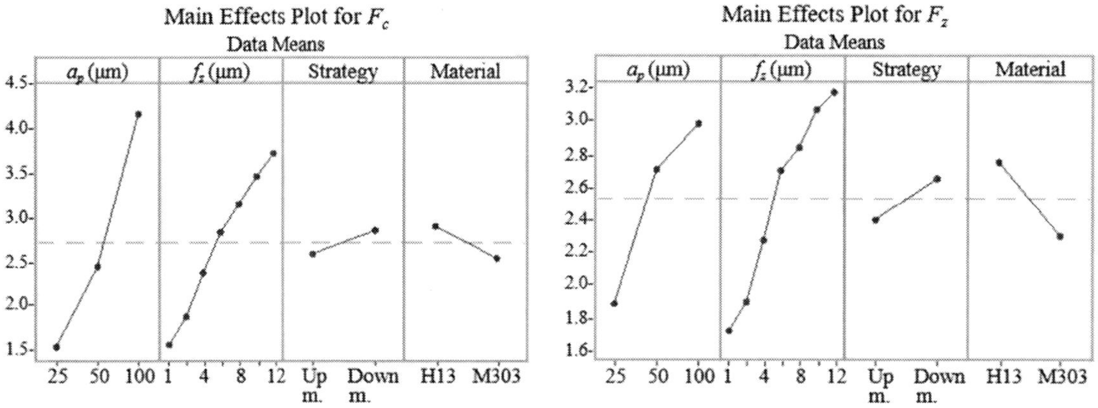

Figure 6. Main effect plots for (a) F_c and (b) F_z

Based on the results of the analysis of variance, the effect of all of the four analysed factors is statistically significant. The related F-values and P-values are shown in Table 4, ordered by the size of their effect. The interactions in terms of the analysed factors are also statistically significant in the case of F_c. Nevertheless, the effect of the interactions are statistically not significant in the case of the F_z component is concerned.

Table 4. F-values and P-values for F_c and F_z

F_c			F_z		
Factor	**F-value**	**P-value**	**Factor**	**F-value**	**P-value**
a_p	2913.09	0.000	f_z	196.01	0.000
f_z	1421.38	0.000	a_p	127.79	0.000
Material	89.45	0.000	**Material**	37.94	0.000
Strategy	46.55	0.000	**Strategy**	10.29	0.002

Based on the statistical analysis, the use of up milling strategy results in smaller forces. This result is similar to that found in our previous study when AISI 1045 steel was micro-milled [7]. Smaller forces result in smaller deflection of the tool, which could provide more accurate geometry. Moreover, it could be a positive effect on tool life, too. The burr formation is not part of the present study, but according to many researches, the top burr is smaller on the up milling side than on the down milling side [7, 27, 28]. Considering the results as well as other considerations, the use of up milling strategy is recommended when milling with a half-diameter of the tool.

The effect of the materials is also noteworthy in the half-diameter milling strategies. In the case of the AISI H13 was machined, the F_c forces were 21% higher at the up milling strategy and 8% higher at the down milling than in the case of the Böhler M303 was cut while the F_z forces were 21% and 24% higher, respectively.

4. Conclusions

Micromachining experiments were conducted on a VHTC 130M 5-axis machining centre using commercially available, 500 μm diameter, two fluted, coated carbide micro-milling tools. A Böhler M303 martensitic stainless steel and AISI H13 hot-work tool steel with 50±1 HRC hardness were micro-milled. Based on the present research work, the following conclusions can be drawn:

- It was observed that the feed per tooth and the depth of cut increase the cutting forces, due to the increasing chip cross-section.
- ANOVA results show that the effect of all of the examined factor is statistically significant at a confidence level of p=0.05.
- Experimental results show that the material type has a significant influence on the cutting forces. The Böhler M303 martensitic stainless steel can be micro-milled on average 30% smaller forces than the AISI H13. Thus, the demand from the industry for the corrosion resistance may result in more favourable cutting conditions in terms of cutting forces during micro-milling.
- It was found that the effect of the corner radius is significant in micro sizes due to the size reduction. It increases the F_z force component at a small depth of cut values.
- The F_z/F_c ratio decreases with the increase of the feed per tooth and the depth of cut. An exception was observed where the depth of cut is equal to the corner radius when the half-diameter strategies were applied. The ratio is on average 11% higher at down milling than at up milling in the case of AISI H13. In contrast, the ratio of F_z/F_c is up to 5.5% smaller at down milling than at up milling in the case of Böhler M303. In terms of the force ratio, differences can be seen in the trends when the depth of cut is the same as the corner radius.
- The milling strategies were compared based on the characteristics of cutting forces. It is recommended the use of the up milling strategy in the case of milling with a radial depth of cut of the half-diameter of the cutting tool.

Acknowledgement
The research reported in this paper was partly supported by the National Research, Development and Innovation Fund (TUDFO/51757/2019-ITM, Thematic Excellence Program), and by the European Commission through the H2020 project EPIC under grant No. 739592. The research work introduced herein was partly supported by the research project K 132430 (Transient deformation, thermal and tribological processes at fine machining of hard metal surfaces) provided by the National Research, Development and Innovation Office. The authors gratefully thank the support of companies Fraisa and Böhler. We especially thank the help of our colleague, Ádám Jacsó in this research.

References

[1] Csala V et al 2015 *Acta Polytech. Hung.* **12:2** 53–69
[2] Kuzin V V et al 2016 *Refract. Ind. Ceram.* **57** 283–287
[3] Anand R S and Patra K 2014 *Mach. Sci. Technol.* **18** 323-347
[4] Alting L et al 2003 *CIRP Ann.* **52** 635–657

[5] Zhu K and Yu X 2017 Mech. Syst. Signal Process. **93** 80–91
[6] Câmara M A et al 2012 J. Mater. Sci. Technol. **28** 673–685
[7] Balázs B Z et al 2017 *Proc. Int. Conf. Innov. Technol.* 161–164
[8] Sun Q et al 2017 *Procedia Eng.* **174** 760–766
[9] Gao S et al 2017 *Int. J. Adv. Manuf. Technol.* **89** 191–205
[10] Bissacco G et al 2005 *J. Mater. Process. Technol.* **167** 201–207
[11] Lai X et al 2008 *Int. J. Mach. Tools Manuf.* **48** 1–14
[12] Vipindas K et al 2018 *Int. J. Adv. Manuf. Technol.* **97** 711–722
[13] Dib M H M et al 2018 *Precis. Eng.* **51** 244–262
[14] Kuram E and Ozcelik B 2016 *Int. J. Adv. Manuf. Technol.* **84** 691–703
[15] Takács et al 2017 *Proc. Int. Conf. Innov. Technol.* 181–184
[16] Mian A J et al 2011 *Int. J. Mach. Tools Manuf.* **51** 383–394
[17] Biró I and Szalay T *2017 Int. J. Adv. Manuf. Technol.* **88** 2735–2743
[18] Aramcharoen A and Mativenga P T 2009 *Precis. Eng.* **33** 402–407
[19] http://www.boehler.hu/hu/W302ISOBLOC.php Accessed: 21-Jun-2018
[20] https://www.boehler.hu/hu/M303.php Accessed: 08-May-2019
[21] Geier N et al 2019 *Compos. Part Appl. Sci. Manuf.* **125** 105552
[22] Pereszlai C and Geier N 2020 *Int. J. Adv. Manuf. Technol.*
[23] Biró I et al 2018 *Procedia CIRP* **77** 279–282
[24] Takács M 2006 Sokkristályos ötvözetek mikroforgácsolása keményfém szármaróval
[25] Geier N and Pereszlai C 2020 *Period. Polytech. Mech. Eng.* **64** 67–80
[26] Jáuregui J C et al 2018 *IEEE Access* **6** 6400-6410
[27] Saptaji K and Subbiah S 2017 *Procedia Eng.* **184** 137–144
[28] Biermann D and Steiner M 2012 *Procedia CIRP* **3** 97–102

IOP Conf. Series: Materials Science and Engineering **903** (2020) 012057 doi:10.1088/1757-899X/903/1/012057

Preparation and characterization of poly(lactic acid)/boehmite alumina composites for additive manufacturing

L Lendvai[1] and I Fekete[1]

[1] Széchenyi István University, Audi Hungaria Faculty of Vehicle Engineering, Department of Materials Science and Engineering, Győr, Hungary

E-mail: lendvai.laszlo@sze.hu

Abstract. The purpose of this work is to investigate the suitability of boehmite alumina (BA) reinforced poly(lactic acid) (PLA) for additive manufacturing with the fused filament fabrication method. PLA filaments with 0-10 wt.% BA content were produced *via* melt compounding using a twin-screw extruder. Dumbbell-shaped and prismatic specimens were manufactured then by feeding the prepared filaments into a desktop 3D printer. Mechanical, morphological and melt flow properties of the developed samples were determined. The addition of BA decreased the melt flow rate of PLA, however not so much that it would have hindered its suitability for 3D printing. With increasing BA content both the strength and the stiffness of the samples increased slightly at constant deformability and toughness. Scanning electron microscopic images revealed a homogenous dispersion of BA particles within the PLA matrix, although remaining agglomerates were also observed.

1. Introduction

Additive manufacturing (AM) – generally referred to as 3D printing – is a rapidly developing technology that received a significant amount of scientific interest over the last two decades. AM technologies are based on the principle of building products through a layer-by-layer process and offer numerous advantages, including reduced production time and cost and the opportunity to fabricate complex objects that would be otherwise impossible. Numerous AM techniques have been developed up until now, each of them operating on different principles [1]. Additive manufacturing can be used either to fabricate parts directly or to prepare molds for common polymer processing technologies, such as injection molding. This latter is a cost efficient way to produce low series of 3D objects [2].

The most widespread methods of AM are the fused deposition modeling (FDM) and the fused filament fabrication (FFF). These similar technologies use filaments of polymers (or metals with low melting temperature) for the printing process. The only difference lies within the heating of the building chamber: while FDM machines are generally applied with temperature control in the build chamber, the FFF ones are mostly non-industrial equipment without this feature. Due to their low cost in use and maintenance both are widely used for biomedical applications, in the aerospace and automotive industry and recently in the everyday life as well [3]. Various commodity and engineering plastics (e.g. polystyrene, polyamide 6, polyethylene terephthalate) are suitable for FDM and FFF printing, however the most common ones are acrylonitrile butadiene styrene and poly(lactic acid) (PLA).

In recent years there has been an increasing interest in the development of bio-based and biodegradable polymers due to the persistently escalating petroleum consumption [4, 5]. PLA is considered as one of the most promising candidates. It is a semi-crystalline thermoplastic polyester that can be prepared from renewable resources and it is also biodegradable. Considering its mechanical properties, PLA could substitute petrochemical polymers in the packaging, agricultural and automotive industry and it is excellent for biomedical applications as well due to its biocompatibility. Generally, PLA exhibits a high enough strength and stiffness to substitute common petrol-based plastics such as polystyrene or polyethylene terephthalate. However, because of its low rate of crystallization from melt it can become quite amorphous when exposed to rapid cooling [6].

A considerable amount of literature has been published on the development of PLA-based polymer blends and composites with enhanced properties in order to increase its range of applications [7]. A widely used method to improve the mechanical properties of PLA is to reinforce it with suitable micro- or nanofillers (graphene, cellulose, layered silicates, etc.). Previous studies have reported that the incorporation of such reinforcing particles does not only improve the mechanical properties of PLA, but also beneficially affects its processability when printing it with FDM and FFF [8].

Among these micro- and nanofillers used as reinforcement, an aluminum oxyhydroxide, namely boehmite alumina (BA) is one of the most promising candidate due to its relatively low price and availability in large industrial scales. In the past decade a number of researchers have sought to determine the effects of BA on the properties of various polymers [9], including PP [10] and PLA [11]. The incorporation of BA particles is reported to improve the mechanical properties, the thermal stability and the flame resistance of the polymer matrix as well.

So far, no previous study has investigated the possibility of using BA reinforced PLA composite filaments for FDM or FFF printing. The aim of this present work is to produce PLA/BA composites with a BA content up to 10 wt.% and to validate their suitability for FFF printing applications. The effect of BA as reinforcement on the melt flow rate, the mechanical and morphological properties of 3D printed objects are presented and evaluated in this paper.

2. Experimental

2.1. *Materials*

An extrusion grade PLA (2003D) of high molecular weight (M_n = ~100 500 g/mol, M_w = ~180 500 g/mol [12]) obtained from NatureWorks LLC (Minnetonka, MN, USA) was used as polymer matrix. It has a melting temperature of ~170 °C and a D-isomer content of ~4%. As reinforcement Disperal® 40 grade BA of Sasol GmbH (Hamburg, Germany) was applied. This specific type has an Al_2O_3 content of 80 m%, an average particle agglomerate size of 35 μm, and a surface area of 100 m^2/g. The list of abbreviations and formulations of the prepared samples are listed in Table 1.

Table 1. The composition and abbreviation of the prepared samples

Sample code	PLA (wt.%)	BA (wt.%)
PLA	100	-
PLA_2.5BA	97.5	2.5
PLA_5BA	95	5
PLA_10BA	90	10

2.2. *Preparation of the filaments*

The PLA and the BA powder were dried at 80 °C for 4 hours prior to processing. Melt compounding of the components was performed using a co-rotating twin-screw extruder (LTE 20-44, Labtech Engineering Co., Ltd., Samutprakan, Thailand) with intermeshing screws (diameter of 20 mm, L/D ratio of 40). The temperature of the heating zones from feed hopper to die end was set to 155 °C, 160 °C, 160 °C, 165 °C, 165 °C, 170 °C, 170 °C, 175 °C, 180 °C, 185 °C and 185 °C, respectively. The screw speed was 30 rpm, the diameter of the die was 2 mm. In this way, neat PLA and BA-reinforced PLA composite filaments with a diameter of 1.75 ± 0.08 mm were produced.

2.3. Printing of the filaments by fused filament fabrication

The filaments prepared by extrusion were used to manufacture dumbbell-shaped specimens (type: 3 according to EN ISO 8256) for tensile tests and prismatic specimens (80 mm x 10 mm x 4 mm) for impact tests (Figure 1). Samples were prepared with a Creality CR-10 FFF machine. A linear infill with horizontal processing direction and raster angle (±45°) between the layers was used. The printing temperature was 210 °C, the nozzle diameter was 0.4 mm and the temperature of the building platform was set to 40 °C. The layer height was 0.2 mm and the filling rate was 100% for all samples.

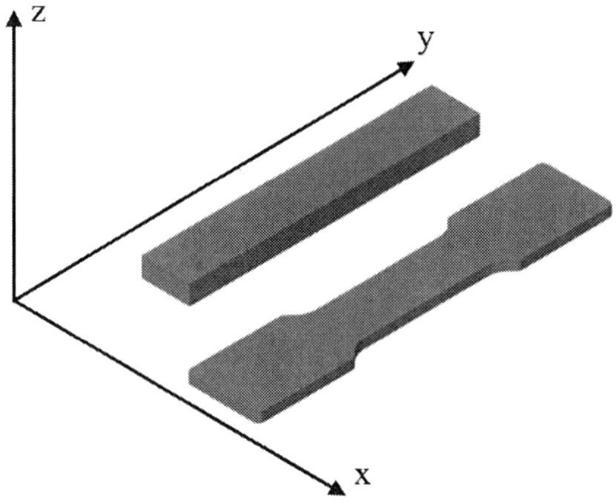

Figure 1. The dumbbell-shaped and the prismatic specimen's orientation on the building platform

Computed tomographic (CT) analysis was performed on the 3D printed samples in order to exclude the possible defects presence. It was also determined, that the average porosity of the 3D printed samples was approximately 0.18%.

2.4. Measurements and characterization

The measure of melt flow rate (MFR) was conducted using an apparatus Ceast 7026.000 (Ceast S.p.A., Pianezza, Italy). The applied load was 2.16 kg, the temperature of the cylinder was set to 210 °C.

Tensile mechanical properties, i.e., yield strength, elongation at break and Young's modulus were determined with an Instron® 3344 (Instron Ltd., Norwood, USA) universal tensile testing machine equipped with a 2 kN force sensor. The tests were carried out at a crosshead speed of 5 mm/min. The results reported in here are the averages of five parallel measurements.

The Charpy impact toughness was measured on a Ceast 6545 (Ceast S.p.A., Pianezza, Italy) pendulum-type testing machine equipped with an impact hammer of 2 J. The specimens were unnotched, rectangular bars. The results reported are the averages of five parallel measurements.

A scanning electron microscope (SEM) Hitachi S-3400N (Hitachi Ltd.,Tokyo, Japan) with an acceleration voltage of 10 kV was used to investigate the fracture surfaces of the 3D printed samples at various magnifications. The surfaces of the samples were coated by a gold layer in a Quorum Technologies SC7620 high-vacuum sputter coater (Quorum Technologies Ltd, Laughton, UK).

3. Results and Discussion

3.1. Melt flow rate

The MFR of the neat PLA and its composites at various concentration of BA are represented in Figure 2. During FFF-based 3D printing the suitable materials are highly limited depending on their melt viscosity. If the viscosity is lower than the optimal, then the material is not able to provide structural support. However, when it is too high, then the extrusion may fail because of the molten polymer getting stuck within the nozzle [13]. Measuring the MFR of a polymer can be helpful to predict its processing behaviour as it is inversely proportional to the viscosity. Based on the literature [14] the printing of a material with lower than 1 g/10 min MFR might fail due to its high viscosity. According to the results listed in Figure 1 the addition of BA particles definitely influences the melt flow rate of PLA. The MFR value of neat PLA (8 g/10 min) dropped when 2.5 wt.% BA was added (to 5.8 g/10 min). This is most probably due to the rigid BA particles hampering the motion of polymer chains. With further BA incorporated, the MFR decreased steadily, bottoming at 5 g /10 min at 10 wt.% BA content. These results refer to the suitability of the developed PLA/BA composites for FFF-based printing even at maximum (10 wt.%) BA loading.

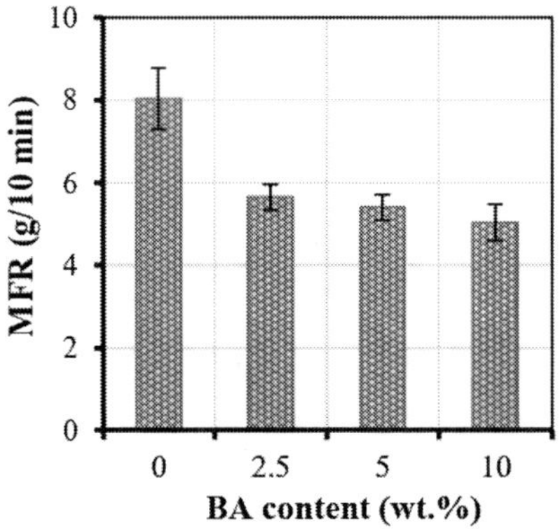

Figure 2. The melt flow rate (MFR) of neat PLA and its composites with various (0-10 wt.%) BA content

3.2. *Tensile mechanical properties*

The characteristic stress-strain curves of the 3D printed PLA specimens containing 0-10 wt.% BA are shown in Figure 3. It can be clearly seen, that the tensile properties of PLA and the PLA/BA composites are close. The neat PLA sample without BA reinforcement exhibited a brittle behavior. Fracture occurred without a necking phenomenon. Even though the presence of rigid fillers – such as BA – tend to cause polymers to show a more brittle behavior, in here, mainly due to the poor initial toughness of PLA only a slight embrittlement can be observed.

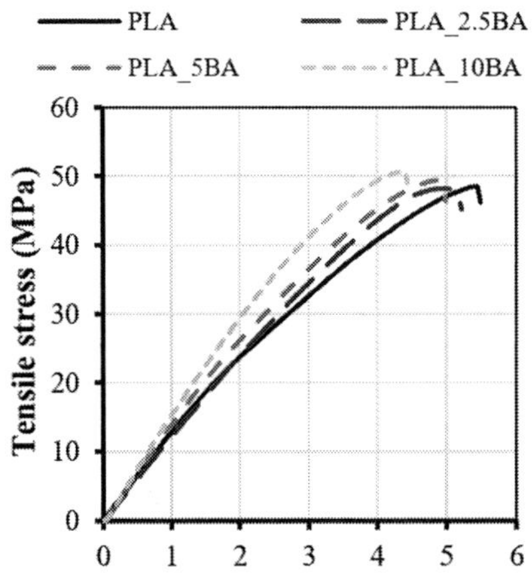

Figure 3. Stress-strain curves of the 3D printed PLA specimens containing 0-10 wt.% BA

The tensile mechanical properties of the samples are listed in Table 2. According to the results, there was a slight improvement in yield strength with increasing BA content (from ~48 MPa to ~50 MPa), albeit only within the deviation range. The rise of Young's modulus as a function of BA concentration was more remarkable. The value measured for the sample PLA_10BA (~1.45 GPa) is relatively 20% higher than that of parent PLA (~1.23 GPa). Interestingly, the modulus values described in the literature for this specific PLA grade are significantly higher (2-3 GPa). A potential explanation for this discrepancy can be the different processing techniques applied for the sample preparation (extrusion, compression moulding, injection moulding, etc.). One must also consider that prior to processing with those conventional techniques, PLA is usually dried. However, due to the relatively long period of sample manufacturing with FFF printing, the filament could regain its moisture content during the process easily, therefore the drying of PLA is mostly neglected. As PLA is highly sensitive to moisture (even a small amount of water can hydrolyze PLA in its molten state), the lack of drying may be the reason behind this drop measured in the Young's modulus. Regarding the deformability it can be concluded, that the presence of BA does not affect it. All the samples showed an elongation at break of about 5.3%.

Table 2. Yield strength, Young's modulus and elongation at break of the 3D printed neat PLA and the PLA/BA composite specimens containing various amount (0-10 wt.%) of BA

Sample code	Yield strength (MPa)	Young's modulus (GPa)	Elongation at break (%)
PLA	48.0 ± 2.1	1.23 ± 0.06	5.3 ± 0.7
PLA_2.5BA	48.7 ± 2.0	1.35 ± 0.05	5.4 ± 0.7
PLA_5BA	50.0 ± 1.3	1.37 ± 0.05	5.2 ± 0.2
PLA_10BA	50.2 ± 1.7	1.45 ± 0.05	5.2 ± 0.7

Mean ± standard deviation

3.3. Impact test results
Figure 4 displays the impact strength of the prepared samples as a function of BA content. Based on the results it can be concluded that the incorporation of BA particles does not affect the toughness of PLA. All the samples exhibited an impact strength of about 12 kJ/m^2. This is in good agreement with the tensile test results. Recall, that PLA is a polymer of low toughness in itself. Thus, the presence of rigid BA particles does not cause the PLA to embrittle even more.

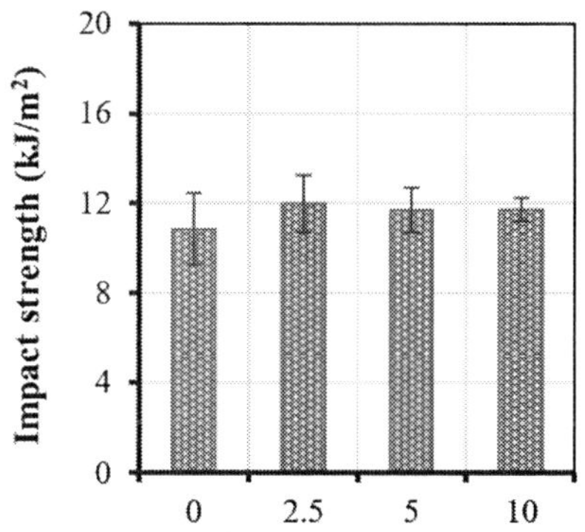

Figure 4. Charpy impact strength of the 3D printed neat PLA and the PLA/BA composite specimens containing various amount (0-10 wt.%) of BA

3.4. Morphology
Figure 5 reveals SEM images of the impact fracture surfaces of the 3D printed PLA and PLA/BA specimens. According to Figure 5/a the neat PLA has a rigid, homogenous fracture surface. This is in good accordance with the results of the mechanical tests. Those samples containing BA show a rigid fracture surface as well, however, with BA agglomerates of 10-30 μm also included. The presence of agglomerates refers to the fact that the decomposition of BA particle bundles during melt compounding was imperfect. Even though these agglomerates on the micrographs do not seem to be the starting points of the crack propagations during the failure of the sample, they still can act as failure sites. This might be the reason for the marginal enhancement observed in the strength.

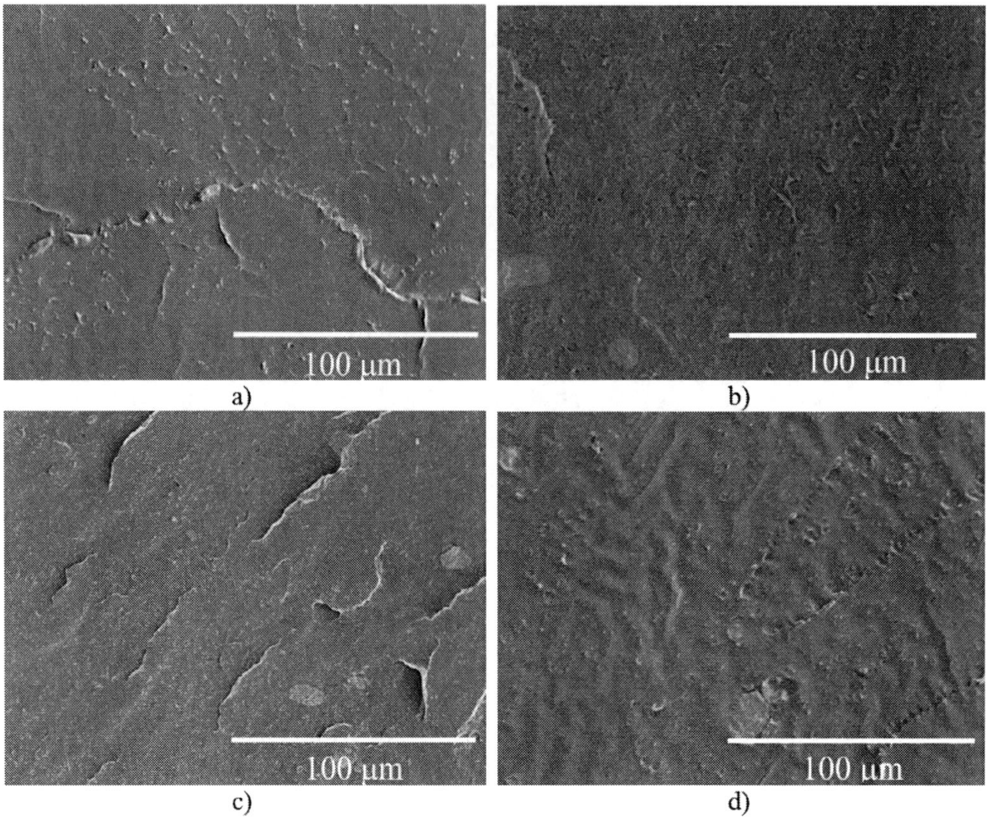

Figure 5. SEM micrographs of neat PLA (a), PLA_2.5BA (b) PLA_5BA (c) and PLA_10BA at lower magnification

At higher magnification (Figure 6) it can be clearly seen that besides the agglomerates, the PLA/BA composites also contain BA particles dispersed individually (or as much smaller bundles). The dispersion of these small particles seems homogenous within the polymer matrix. This refers to a partial disintegration of the boehmite alumina agglomerates. One should consider, however, that the morphology of the fracture surface does not necessarily correspond with that of the bulk, still this is generally accepted in the literature.

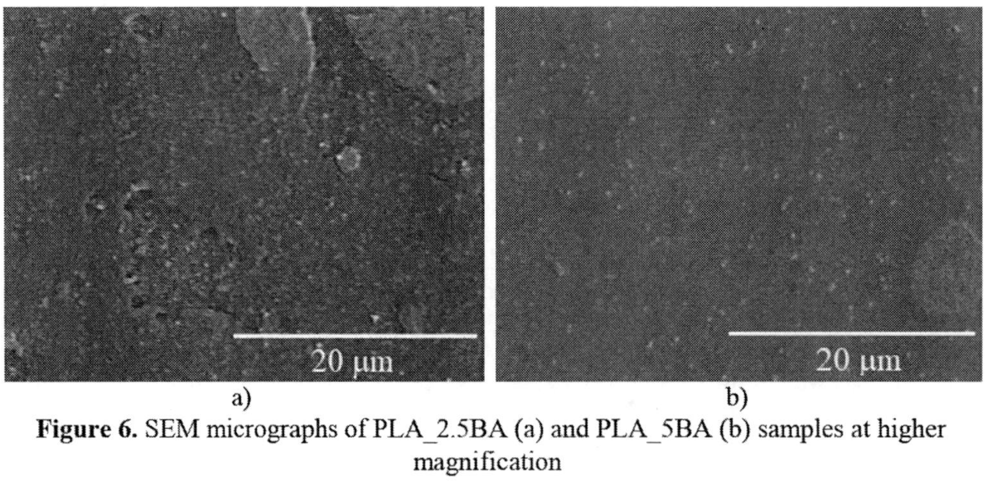

Figure 6. SEM micrographs of PLA_2.5BA (a) and PLA_5BA (b) samples at higher magnification

4. Conclusions

This study has shown that PLA/BA composites are generally suitable for 3D printing with an FFF equipment. In order to demonstrate this, filaments with 0-10 wt.% BA reinforcement were produced by melt compounding. The melt flow rate analyses substantiated the assumption that these newly developed materials can be processed with FFF technology. Specimens for the mechanical and morphological tests were 3D printed out of these filaments.

According to the mechanical test results the PLA/BA composites outperformed neat PLA in Young's modulus and there was also a slight improvement in yield strength. The elongation and impact strength was, however, not affected, all the results were within deviation range. The SEM images revealed a fine dispersion of BA particles in the PLA matrix, although remaining agglomerates were observed as well.

Acknowledgements

The research was carried out as part of the EFOP-3.6.2-16-2017-00016 project in the framework of the New Széchenyi Plan. The project was also supported by the Ministry of Human Capacities through the project NTP-NFTÖ-19.

References

[1] Ngo T D, Kashani A, Imbalzano G, Nguyen K T Q and Hui D 2018 Additive manufacturing (3D printing): A review of materials, methods, applications and challenges *Composites Part* B **143** 172-96
[2] Tábi T, Kovács N K, Sajó I E, Czigány T, Hajba S and Kovács J G 2016 Comparison of thermal, mechanical and thermomechanical properties of poly(lactic acid) injection-molded into epoxy-based rapid prototyped (PolyJet) and conventional steel mold *J. Therm. Anal. Calorim.* **123** 349-61
[3] Liu Z, Wang Y, Wu B, Cui C, Guo Y and Yan C 2019 A critical review of fused deposition modeling 3D printing technology in manufacturing polylactic acid parts *Int. J. Adv. Manuf. Tech.* **102** 2877-89
[4] Singh T, Gangil B, Patnaik A, Biswas D and Fekete G 2019 Agriculture waste reinforced corn starch-based biocomposites: effect of rice husk/walnut shell on physicomechanical, biodegradable and thermal properties *Mater. Res. Express* **6** 045702
[5] Singh T, Pruncu C I, Gangil B, Singh V and Fekete G 2020 Comparative performance assessment of pineapple and Kevlar fibers based friction composites *J. Mater. Res. Technol.* **9** 1491-9
[6] Garlotta D 2001 A literature review of poly(lactic acid) *J. Polym. Environ.* **9** 63-84
[7] Lendvai L and Brenn D 2020 Mechanical, morphological and thermal characterization of compatibilized poly(lactic acid)/thermoplastic starch blends *Acta Technica Jaurinensis* **13** 1-13
[8] Coppola B, Cappetti N, Di Maio L, Scarfato P and Incarnato L 2018 3D Printing of PLA/clay nanocomposites: influence of printing temperature on printed samples properties *Materials* **11** 1947
[9] Karger-Kocsis J and Lendvai L 2018 Polymer/boehmite nanocomposites: A review *J. Appl. Polym. Sci.* **135** 45573
[10] Lendvai L 2020 Water-assisted production of polypropylene/boehmite composites *Period. Polytech., Mech. Eng.* **64** 128-35
[11] Das K, Ray S S, Chapple S and Wesley-Smith J 2013 Mechanical, thermal, and fire properties of biodegradable polylactide/boehmite alumina composites *Ind. Eng. Chem. Res.* **52** 6083-91
[12] Kmetty A, Litauszki K and Reti D 2018 Characterization of different chemical blowing agents and their applicability to produce poly(lactic acid) foams by extrusion *Appl. Sci.* **8** 1960
[13] Wang X, Jiang M, Zhou Z, Gou J and Hui D 2017 3D printing of polymer matrix composites: A review and prospective *Composites, Part* B **110** 442-58
[14] Cataldi A, Rigotti D, Nguyen V D H and Pegoretti A 2018 Polyvinyl alcohol reinforced with crystalline nanocellulose for 3D printing application *Mater. Today Commun.* **15** 236-44

IOP Conf. Series: Materials Science and Engineering **903** (2020) 012058 doi:10.1088/1757-899X/903/1/012058

The influence of the base material - filler metal pairing on the fatigue crack propagation of 700 MPa strength category welded joints

J Lukács[1] and HFH Mobark[2]

[1] Professor, University of Miskolc, Miskolc, Hungary
[2] PhD Student, University of Miskolc, Miskolc, Hungary

E-mail: janos.lukacs@uni-miskolc.hu

Abstract. In the field of welding of high-strength structural steels, the *700 MPa* strength category represents a kind of boundary. Overmatched (OM) consumables are also available, but the behavior of various mismatch types under cyclic loading condition is not yet clear. In order to know the fatigue crack propagation resistance of *700 MPa* strength category steels and their gas metal arc welded joints fatigue crack growth tests were performed on statistically representative samples. Matching/overmatching (M/OM) mismatch type was tested, the tests results were analyzed and fatigue crack propagation limit curves were determined. The different mismatch types, our and previous results were compared. It can be concluded that M type is the optimal and M/OM type is more beneficial than simple OM type.

1. Introduction

Nowadays, the reliability of a structural element or a structure is one of the most important, if not the most important, feature of the structural element or the structure itself. This is a requirement on the one hand; on the other hand this is a task for the structural and technological designers, manufacturers, operators and maintainers of the structure. The regrettable accidents always draw attention to this fact; and summary studies about these form global messages from which all stakeholders can learn [1].

Figure 1 shows based on [2] a non-exhaustive overview of different international organizations and groups working on prevention, preparedness and response to chemical and industrial accidents. The figure provides a summary of key tools and methodologies developed by the different organizations, which can be seen in accordance with the timeline of the accidents.

The purpose of the eMARS (Major Accident Reporting System, later renamed eMARS after going online) is to facilitate exchange of lessons learned from accidents and near misses involving dangerous substances in order to improve chemical accident prevention and mitigation of potential consequences [3]. There are other accident report sites in order to share the experiences and observations worldwide (e.g. The Central Reporting and Evaluation Office for Major Accidents and Incidents in Process Engineering Facilities (ZEMA) [4], The Japanese Failure Knowledge Database [5]).

The reliability of a structural element or structure is closely related to its integrity [6], and different structural types (e.g. bridges [7], transporting pipelines [8], ships [9], and vehicles [10]) exhibit characteristics. Material quality, construction, applied manufacturing technology and the inhomogeneities affect both the reliability and the integrity. This may be particularly the case for welded structures made of high-strength steels, where the manufacturing without material

IOP Conf. Series: Materials Science and Engineering **903** (2020) 012058 doi:10.1088/1757-899X/903/1/012058

discontinuities can be guaranteed only in principle. In addition, material discontinuities can occur during the operation of the structures, and the most dangerous of them are the cracks.

Organisation	Prevention	Preparedness	Response	Post-accident	Learning
OECD	Guiding Principles for Chemical Accidents, Preparedness and response				eMARS
UNECE	Transboundary Effects of Industrial Accidents Convention				eMARS
EU	Seveso-III-Directive, Civil Protection Mechanism			Environment Liability Directive	eMARS
JEU		UN Disaster Assessment and Coordination Mechanism, Flash Environmental Assessment Tool			
UN Environment	Flexible Framework, APELL, Responsible production toolkit				
UNISDR	Sendal Framework for Disaster Risk Reduction 2015-2030				
WHO		International Health Regulation			Event Management System (EMS)
	Public health management of chemical incidents				
EPSC	Member network				Member network

Policy, no intervention — Intervention based — Regulation/Legislation/Convention

Figure 1. Summary of key tools and methodologies according with the timeline of the accidents (based on [2])

In the field of welding of first generation high-strength structural steels [11], the *700 MPa* strength category represents a kind of boundary. Overmatched consumables are also available for gas metal arc welding of such steels, but the behavior of various mismatch types under cyclic loading condition is not yet clear. This is especially true for fatigue crack propagation resistance of such high-strength steels and their welded joints.

The microstructural and strength inhomogeneities of the welded joint raise the question of how the local properties influence the global behavior. The question, though not with complete accuracy, can be answered using statistical approach.

A *0.08%C-1.5%Mn (weight%)* microalloyed steel, designated as RD480, was tested under different microstructural conditions, both tensile and fatigue crack growth behavior were investigated. The various microstructural conditions were produced by means of heat treatments followed by water quench, in which the material samples were kept at the temperatures of *800°C*, *950°C* and *1200°C* [12]. The different microstructures result characteristic differences on the exponent of the Paris equation [13].

A soft buffer layer (BL) between the weld metal (WM) and the high strength base material (BM) was prepared, and BM, welded BM without a BL and welded BM with a BL were investigated. Notches were cut both in the BM and the WM; therefore four different cases of the crack propagation were studied and compared with the BM and each other. Both the BL and the notch location have influenced the fatigue crack growth behavior of the configurations [14].

Seeing that high strength steels are more sensitive to notches or initial cracks, the study of fatigue crack growth behavior in the different microstructural zones of the heat-affected zone (HAZ) is essential. Single edge notched bend (SENB) specimens were used for the characterization of fracture mechanical behavior of thermally simulated heat affected zones [15]. Gleeble physical simulator was applied reproducing of different microstructures, based on previously measured heating curves on real welded joints. The effect of the stress ratio (R) on the threshold stress intensity factor range and the crack growth curves was analyzed and significant influences were observed.

Fracture toughness (K_{Ic}) and fatigue crack growth rate data are the two key fracture mechanical values which are used for safe-life analysis of different engineering structures or structural elements. Testing on numerous specimens is costly, time dependent and time costuming way for the determination of these necessary parameters. A numerical approach, called virtual testing technic has been developed which enables engineers to generate fracture analytically by eliminating unnecessary tests [16].

Based on the previously mentioned facts, the aims of this work are as follows:

- characterization the fatigue crack growth (FCG) resistance of different high strength steels and their gas metal arc welded (GMAW) joints in *700 MPa* strength category;
- investigation of the mismatch effect on the fatigue crack growth characteristics of the GMAW joints;
- determination of fatigue crack propagation limit curves for the investigated steels and their GMAW joints, based on simple crack growth relationship [13], [17].

2. Investigations

The chemical composition and the basic mechanical properties of the investigated BMs and filler metals (FM) are summarized in Tables 1-2, respectively; the sources of the data are quality certificates. The thicknesses of the used RUUKKI Optim 700QL and SSAB Weldox 700E plates were 30 mm and 15 mm, respectively; the diameters of the applied welding consumables were 1.2 mm.

Table 1. Chemical composition of the investigated base materials and used filler metals (weight%)

Material designation	C	Si	Mn	P	S	Cr	Ni	Mo
RUUKKI Optim 700QL	0.16	0.31	1.01	0.010	0.001	0.61	0.21	0.205
INEFIL NiMoCr	0.80	0.50	1.60	0.007	0.007	0.30	1.50	0.250
SSAB Weldox 700E	0.14	0.30	1.13	0.007	0.001	0.30	0.04	0.167
Böhler Union X85	0.07	0.68	1.62	0.010	0.010	0.29	1.73	0.61
Böhler Union X90	0.1	0.8	1.8	N/A	N/A	0.35	2.3	0.6
Material designation	**V**	**Ti**	**Cu**	**Al**	**Nb**	**B**	**N**	**Zr**
RUUKKI Optim 700QL	0.010	0.016	0.015	0.041	0.001	0.0015	0.003	N/A
INEFIL NiMoCr	0.090	N/A	0.120	N/A	N/A	N/A	N/A	N/A
SSAB Weldox 700E	0.011	0.009	0.01	0.34	0.001	0.002	0.003	N/A
Böhler Union X85	<0.01	0.08	0.06	<0.01	N/A	N/A	N/A	<0.01
Böhler Union X90	N/A	N/A	N/A	N/A	N/A	N/A	N/A	N/A

Table 2. Basic mechanical properties of the investigated base materials and used filler metals

Material designation	$R_{p0,2}$ (MPa)	R_m (MPa)	A (%)	CVN impact energy (J)
RUUKKI Optim 700QL	809	850	17.0	-40°C: 106
INEFIL NiMoCr	750	820	19.0	-40°C: 60; -20°C: 90; 20°C: 120
SSAB Weldox 700E	791	836	17.0	-40°C: 165
Böhler Union X85	$\geq$ 790	$\geq$ 880	$\geq$ 16.0	-50°C: $\geq$ 47; 20°C: $\geq$ 90
Böhler Union X90	890	950	15.0	-60°C: 47; 20°C: 90

Matching (M), overmatching (OM) and matching/overmatching (M/OM) mismatch conditions were used, where the M/OM condition means matched (M) root layers and overmatched (OM) filler layers. The different base material – filler metal pairings were as follows:

- RUUKKI Optim 700QL – INEFIL NiMoCr (O7-INE-M): matching (M);
- SSAB Weldox 700E – Böhler UNION X85 (W7-X85-M): matching (M);
- SSAB Weldox 700E – Böhler UNION X90 (W7-X90-OM): overmatching (OM);
- SSAB Weldox 700E – Böhler UNION X85/Böhler UNION X90 (W7-X89/X90-M/OM): matching/overmatching (M/OM).

The welding equipment was a DAIHEN VARSTROJ WELBEE P500L power source. The dimensions of the welded plates were *300 mm* long and *125 mm* width. In order to attain approximately equal stress distribution (residual stress field) X-grooved (double V-grooved) welding joints were used, with *80°* groove angle, *2 mm* root opening, and *1 mm* land thickness. During the welding, the plates were rotated, regularly, after each layer. Based on the industrial practice, solid wires and *18%* CO_2 *+ 82% Ar* gas mixture (M21) were applied in all cases. The root layers (2 layers for both thicknesses) were made by a qualified welder; while the other layers (16 layers for *30 mm* and 6 layers for *15 mm* thicknesses) were made by automated welding car. The welding technological parameters were selected based on both theoretical considerations and real industrial applications, as follows (root layers / filler layers):

- preheating temperature (T_{pe}) and interpass temperature (T_{ip}): *150 / 180°C*;
- welding current (I): *130-140 / 280-300 A*;
- voltage (U): *19.0-20.5 / 28.5-29.5 V*;
- welding speed (v_w): *200 / 400 mm/min*;
- linear energy (E_v): *700-750 / 1000-1000 J/mm*;
- calculated critical cooling time ($t_{8.5/5}$): *7.0-8.0 / 9.0-11.0 s*.

The FCG tests were executed on single edge notch bend (SENB) specimens. The position of the notches in the BM specimens correlated with the rolling direction (marked: T-L, L-T, T-S, and L-S), and in the welded joints (WJ) with the 21 and 23 joint directions (marked: 21W and 23W). According to these designations the nominal width (W) values of the specimens were *13 / 18 mm* and *26 / 28 mm* for both BMs and WJs. The positions of the notches, the notch distances from the centerline of the WJs, were different; therefore, the positions of the notches and the crack paths represent the most important and most typical crack directions in a real WJ (either part of our statistical approach). Figure 2 shows the macro-structure of the M/OM welded joint with a notch location and hardness indentations (see later), as example. Tests were executed in as-welded conditions; post-weld heat treating was not applied after welding. The FCG investigations were performed with tensile stress, the stress ratio was $R = 0.1$. The test equipment was MTS type electro-hydraulic universal testing equipment (MTS 312); sinusoidal loading wave form was applied, at room temperature, and on laboratory air. The loading frequency was not constant during the FCG tests, it was higher at the two-thirds of crack growth ($f = 20$ *Hz*), and it was lower at the last third ($f = 5$ *Hz*). The length of the fatigue cracks during their propagation was registered with optical method, hundredfold magnification ($N = 100x$) was applied using industrial video camera.

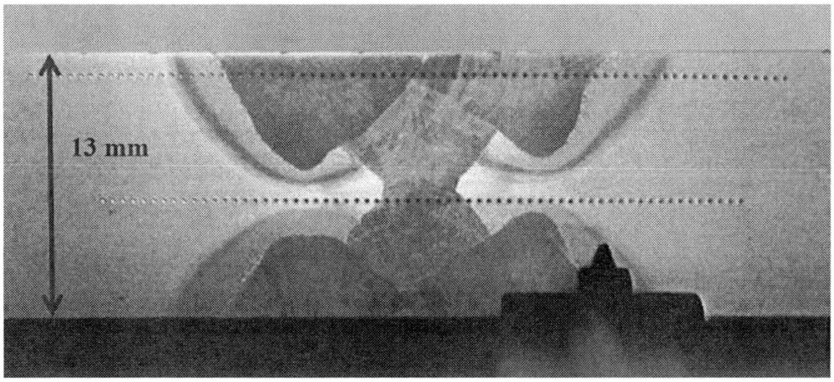

Figure 2. Macro-structure of the M/OM welded joint with a notch location and hardness indentations (polished and etched, 2% HNO_3)

Reicherter UH250 hardness tester was used for measuring of Vickers hardness. *HV10* values were measured along two lines, along sub-surface and root lines, in both T-L/21W and T-S/23W specimens.

3. Results

The hardness distributions along the sub-surface and root lines in both types of specimens can be seen in Figure 3. The curves reflect the influence of the applied matched and overmatched filler metals.

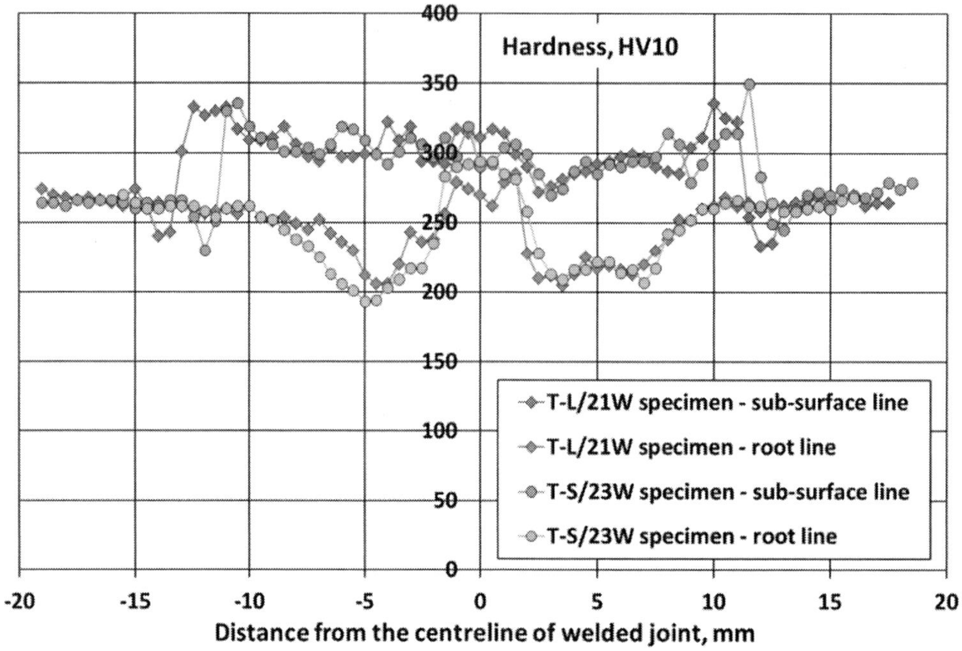

Figure 3. Hardness distributions of matched/overmatched (M/OM) welded joint

The crack length vs. number of cycles curves for SSAB Weldox 700E – Böhler UNION X85/Böhler UNION X90, matching/overmatching (M/OM) pairing in T-L/21W orientation shows Figure 4, and the calculated stress intensity factor range vs. fatigue crack growth rate values can be seen in Figure 5, in both orientations (T-L/21W and T-S/23W).

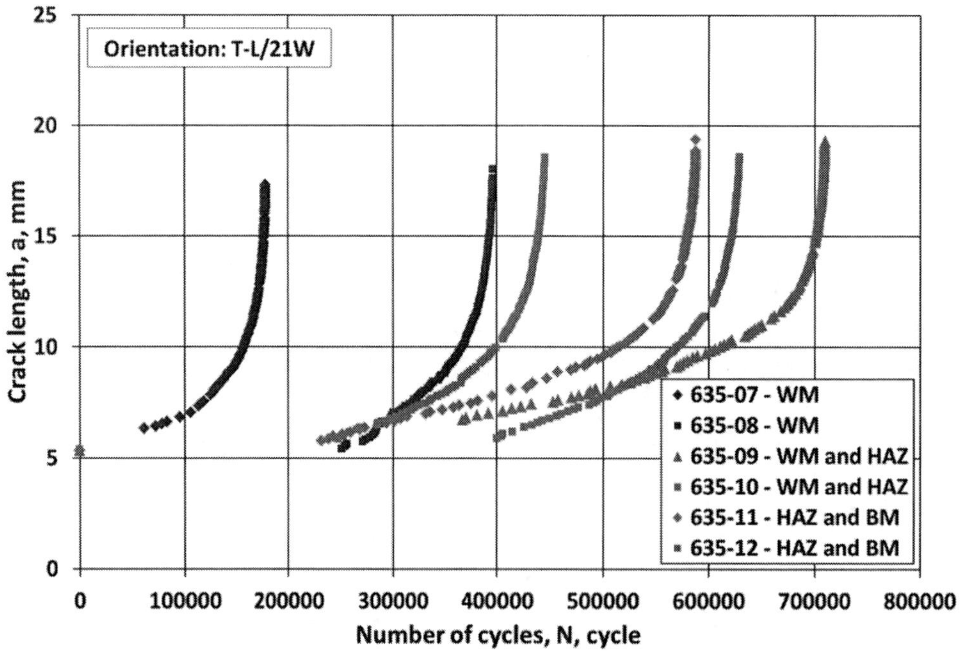

Figure 4. Crack length vs. number of cycles curves for SSAB Weldox 700E – Böhler UNION X85/Böhler UNION X90 pairing

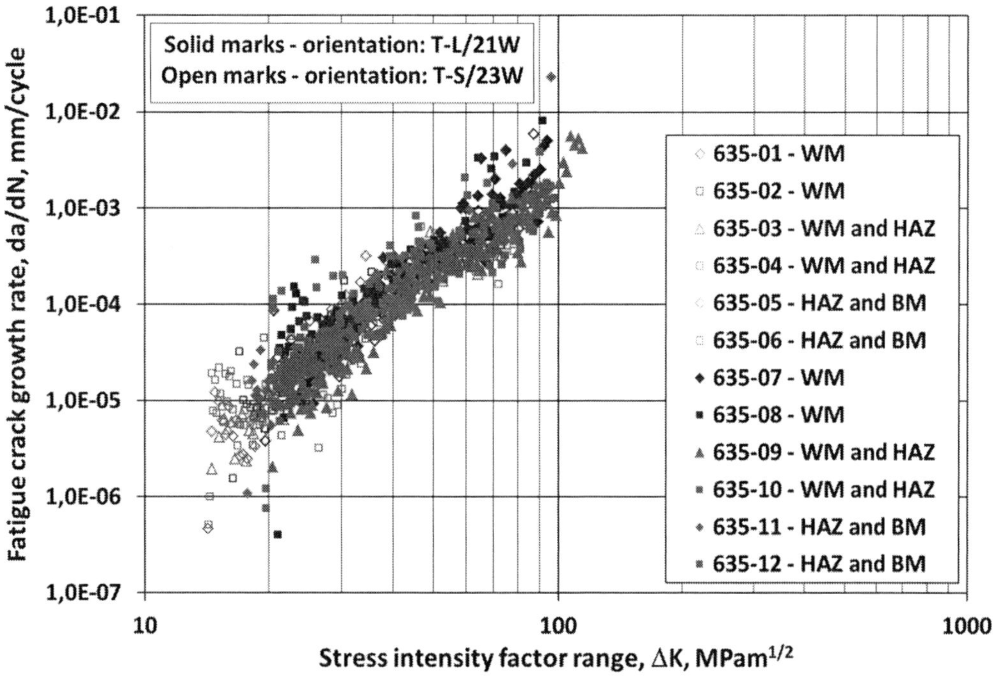

Figure 5. Kinetic diagrams of fatigue crack propagation for SSAB Weldox 700E – Böhler UNION X85/Böhler UNION X90 pairing

In Figures 4-5 the specimen ID consists of two parts: the first part (635) means the number of welded joint, the second part (01-12) the serial number of the tested specimen cut from the joint. The "cloud" in Figure 5 demonstrates both the expanse of the calculated results and the partial overlapping of the results between the two investigated orientations.

Based on the experimental data and results, fatigue crack propagation limit curves were determined using a previously developed six steps method [18] (other part of our statistical approach). Table 3. summarizes the average values of Paris-Erdogan exponents ($n_{average}$) for all statistical samples and the parameters of the determined fatigue crack propagation limit curves. Based on these data Figure 6 demonstrates the limit curves for all cases.

Table 3. Average values of Paris-Erdogan exponents and the parameters of the determined fatigue crack propagation limit curves

Material designation and mismatch condition	Orientation	$n_{average}$	n	C (mm/cycle, $MPa^{1/2}$)	ΔK_{fc} ($MPa^{1/2}$)	Source
O7-INE-M	T-S/23W	1.81	1.20	6.52E-06	93	[19]
W7-BM	T-L, L-T	2.43	1.70	8.09E-07	101	[20]
	T-S	2.19	1.50	2.06E-06	75	[20]
W7-X85-M	T-L/21W	5.10	4.10	1.12E-11	105	[20]
	T-S/23W	4.15	2.30	4.93E-08	80	[20]
W7-X90-OM	T-L/21W	3.68	1.85	4.02E-07	96	[20]
	T-S/23W	3.43	1.90	3.19E-07	61	[20]
W7-X89/X90-M/OM	T-L/21W	3.29	2.67	8.88E-09	90	this study
	T-S/23W	3.11	2.85	3.87E-09	67	this study

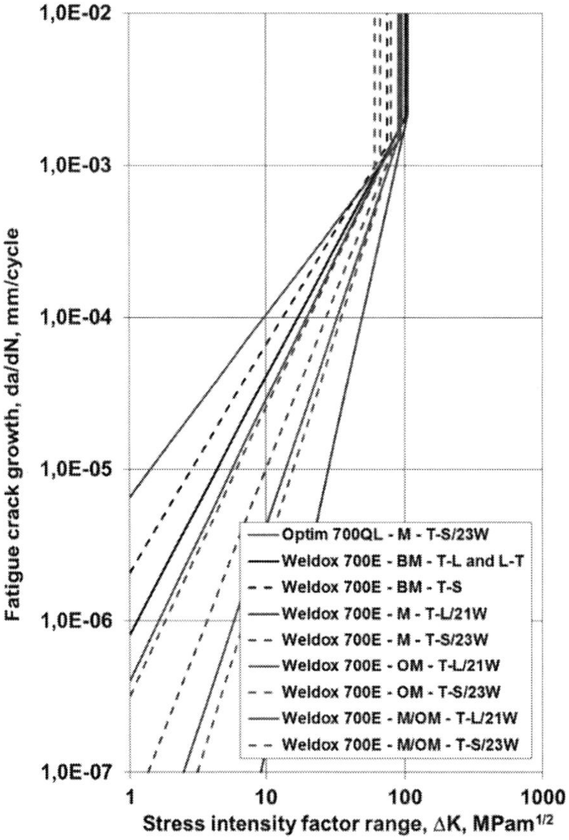

Figure 6. Determined fatigue crack propagation limit curves

4. Conclusions

Based on our investigations and their results, the following conclusions can be drawn.

The results of the achieved fatigue crack growth (FCG) investigations justified the accuracy and necessity of statistical approaches. This means on the one hand the selection of notch directions and locations in the base material and welded joints (the number of the tested specimens), on the other hand the processing and evaluation of the measured data.

The applied gas metal arc welding (GMAW) technology, the different technological parameters during the welding of root and filler layers are appropriate for fabrication welded joints with adequate and repeatable quality. The adequate resistance to fatigue crack propagation is an important part of this quality. Neither welding discontinuities nor abnormal FCG behavior were not found.

The FCG resistance of the investigated Weldox 700E base material is in the longitudinal and transversal directions the same, however different in the thickness direction.

The welding causes disadvantageous influence on the FCG resistance of the investigated Weldox 700E high strength steel. The average values of the Paris-Erdogan exponents (n) of the matching (M), overmatching (OM) and matching/overmatching (M/OM) conditions of the investigated Weldox 700E welded joints were statistically higher than the exponent of the base material. The mismatch phenomenon has significant effect on the FCG characteristics on the Weldox 700E welded joints; M type is the optimal and M/OM type are more beneficial than simple OM type (see Figure 6). The FCG resistance of the investigated Optim 700QL and Weldox 700E matched welded joints was significantly different (see Figure 6 in that case, too).

Applying the previously developed method [18] and using the determined and evaluated results, fatigue crack propagation limit curves can be defined for both the base material and the GMA welded joints. The calculated limit curves adequately reflect the FCG characteristics of the investigated materials (see Figure 6), and can be applied for safe-life or structural integrity calculations.

References

[1] EEA Technical report No 13/2010 2010 *Mapping the impacts of natural hazards and technological accidents in Europe. An overview of the last decade* (Copenhagen: European Environment Agency) p 146

[2] Anon. 2017 *International efforts for industrial and chemical accidents prevention, preparedness and response* (Inter-Agency Coordination Group for Industrial and Chemical Accidents) https://www.unisdr.org/we/inform/publications/62198 p 34 Last accessed 3 May 2020

[3] https://emars.jrc.ec.europa.eu/EN/emars/content Last accessed 3 May 2020

[4] https://www.infosis.uba.de/index.php/en/site/13947/zema/index.html Last accessed 3 May 2020

[5] http://www.sozogaku.com/fkd/en/ Last accessed 3 May 2020

[6] Lukács J, Nagy Gy, Harmati I, Koritárné FR and Kuzsella Lné KZs: *Selected chapters from structural integrity of engineering structures* 2012 (Miskolc: Miskolci Egyetem) ISBN 978-963-358-000-4 (in Hungarian) p 334

[7] Helmerich R 2005 *Alte Stähle und Stahlkonstruktionen – Materialuntersuchungen, Ermüdungsversuche an originalen Brückenträgern und Messungen von 1990 bis 2003* Forschungsbericht 271 (Berlin: Bundesanstalt für Materialforschung und -prüfung) p 86

[8] Anon. 2016 *The Pipeline Defect Assessment Manual. Edition 2. A Report to the PDAM Joint Industry Project October 2016* (Penspen International Ltd.) p 456

[9] Fricke W First published online 06 March 2017 *Fatigue and Fracture of Ship Structures. Encyclopedia of Maritime and Offshore Engineering*, https://doi.org/10.1002/9781118476406.emoe007 p 12 Last accessed 3 May 2020

[10] Eds Keeler S, Kimchi M and Mooney P J April 2017 *Advanced High-Strength Steels Application Guidelines. Technical Version 6.0* https://www.worldautosteel.org/projects/advanced-high-strength-steel-application-guidelines/ p 314 Last accessed 3 May 2020

[11] Keeler S May 2012 *The Science of Forming Automotive Steels – Online Technical Resources* MetalForming p 54-55

[12] Laurito DF, Baptista CARP, Torres MAS and Abdalla AJ 2010 *Microstructural effects on fatigue crack growth behavior of a microalloyed steel* Procedia Engineering pp 1915-1925

[13] Paris P and Erdogan F Dec 1963 *A Critical Analysis of Crack Propagation Laws* Journal of Basic Engineering **85(4)** pp 528-533

[14] Zhang C, Lu P, Hu X and Song X 2012 *Effect of buffer layer and notch location on fatigue behavior in welded high-strength low-alloy* Journal of Materials Processing Technology **212** pp 2091-2101

[15] Maier B, Guster Ch, Tichy R and Ecker W 2013 *Influence of different microstructures of the welding zone on the fatigue crack growth behaviour of HSLA steels* (13th International Conference on Fracture, June 16–21, 2013, Beijing, China) p 7

[16] Farahmand B 2008 *Application of Virtual Testing for Obtaining Fracture Allowable of Aerospace and Aircraft Materials* (Multiscale Fatigue Crack Initiation and Propagation of Engineering Materials: Structural Integrity and Microstructural Worthiness) ed GC Sih (Springer Science+Business Media B.V.) pp 1-22

[17] BS 7910:2013+A1:2015 2015 *Guide to methods for assessing the acceptability of flaws in metallic structures* (BSI Standards Limited)

[18] Lukács J 2003 *Fatigue crack propagation limit curves for different metallic and non-metallic materials* Materials Science Forum **414-415** pp 31-36.

[19] Dobosy Á 2017 *Design limit curves for cyclic loaded structural elements made of high strength steels* (Miskolc: Miskolci Egyetem) PhD Thesis (in Hungarian)

[20] Lukács J and Dobosy Á 2019 *Matching effect on fatigue crack growth behaviour of high-strength steels GMA welded joints* Welding in the World **63** pp 1315-1327

Investigation of the energy absorption properties of cross-linked polyethylene foams

M Tomin[1] and Á Kmetty[1,2]

[1]Department of Polymer Engineering, Faculty of Mechanical Engineering, Budapest University of Technology and Economics, Műegyetem rkp. 3., H-1111 Budapest, Hungary
[2]MTA–BME Research Group for Composite Science and Technology, Műegyetem rkp. 3., H-1111 Budapest, Hungary

E-mail: kmetty@pt.bme.hu

Abstract. We performed the dynamic mechanical investigation of cross-linked polyethylene (XL-PE) foams, which are the most commonly used polymer foams for damping purposes. Our experiments were primarily focused on analyzing the energy-absorbing capability of foams with different densities and studying the relationship between cell structure and shock absorption. The cell structure and energy absorption properties of the foams were determined by mechanical tests and microscopic examination. The samples were subjected to falling dart and falling weight impact tests using different weight geometries and impact energy. Our experiments showed that the impact damping properties of foams are significantly influenced by the deformation mechanisms in the cellular structure during dynamic loading since excessive deformation of the cell walls leads to the compaction of the foam, and therefore a significant reduction in impact damping capability.

1. Introduction

Today, the industrial use of polymeric foams is expanding due to their low density and weight, as well as their outstanding thermal and acoustic insulation, impact damping and energy absorption properties [1-4]. The leading industries are increasingly using foams, and the global polymer foam market is continuously growing. The most significant segments are the construction industry, where outstanding heat insulation is required, the car industry, which specially focuses on weight reduction, the packaging industry with high shock-absorption requirements, and the sports industry [5, 6]. Processing technologies are rapidly improving, and now foaming can be used even for upcycling to reduce the load on the environment [7].

The most important feature of polymer foam products used in sports is their outstanding impact damping and energy absorption capability. In many sports, several different mats are used for safety purposes or for providing an appropriate surface for sports activities. In these cases, the primary task is to protect the health of the athlete and prevent sports injuries [8]. Foams are suitable for this function because of their unique cellular structure. Under load, the cells undergo different types of deformation (bending of cell edges, cell wall buckling), so these materials can absorb a huge amount of energy while keeping the maximum stress below a certain value [9].

Current sports mats are produced from cross-linked polyethylene foams, rebonded polyurethane foams, ethylene-vinyl-acetate foams, and elastomeric foams, of which XL-PE foams are the most commonly used [6]. Compared to conventional foams, they have better thermal stability and greater form stability due to the cross-links between the molecular chains. From the point of view of manufacturing technology, there are physical cross-linking by radiation or chemical cross-linking. In this case, peroxide compounds are used most commonly, which remove hydrogen atoms from the polymer chain and form free radicals. These radicals start combining, resulting in cross-linked bonds [10-12].

In recent years, increasing attention has been paid to the study of the energy-absorbing properties of polymer foams. In the case of sports mats, dynamic mechanical tests are the most popular, but the difference of falling weight geometries and impact energy in sports-specific standards make comparison with previous experimental results difficult [13, 14]. Dynamic mechanical tests can be low impact velocity tests, such as the falling dart [15, 16] and falling weight impact tests [17], while for high impact velocities the Hopkinson bar can be used [18]. However, the relationship between cellular structure, foam density, and energy absorption is not fully understood, and the previously mentioned testing methods are not compared, so the aim of our research is to investigate cross-linked polyethylene foams dynamically, analyze the effect of density on energy absorption properties and compare the two most commonly used testing methods for low-velocity impacts.

2. Materials and methods

This chapter presents the investigated materials, the investigating methods, and measuring devices used.

2.1. Materials

We performed the tests on two physically cross-linked polyethylene foams with different densities, supplied by UFM Bt. (Hungary, Mosonmagyaróvár). The main properties of these commercially available foams are summarized in Table 1 [19].

Table 1 Main properties of the cross-linked polyethylene foams used [19]

Property	Unit	Measurement standard	XL-PE 45	XL-PE 60
Density	(kg/m^3)	ISO 845	45 ± 6	60 ± 8
Tensile strength	(kPa)	ISO 1798	> 215	> 380
Elongation at break	(%)	ISO 1798	> 210	> 300
Compressive strength	(kPa)	ISO 3386/1 (at 10 (%) comp.)	> 17	> 40
Compression set	(%)	ISO 1856 (22 (h), 23 (°C), 25 (%))	≤ 6	≤ 4
Thermal conductivity	(W/mK)	DIN 52612 (at 10 (°C))	0.039	0.051
Shore hardness	(-)	ISO 868	> 30	> 35

2.2. Methods

2.2.1. Density measurement. The densities given in the manufacturer's datasheet were verified by hydrostatic density tests with a Radwag analytical balance according to formula (1):

$$\rho_f = \frac{m_a}{m_a - m_l} \cdot \rho_l \tag{1}$$

where ρ_f (kg/m^3) is the calculated density of the foam, m_a (kg) is the specimen's weight in air, m_l (kg) is the specimen's weight in liquid, and ρ_l (kg/m^3) is the density of the measuring liquid at room temperature. We used distilled water as measuring liquid for the calculations. Prior to the mechanical tests, the thickness of the samples was also determined with a Mitutoyo digital micrometer.

2.2.2. Scanning electron microscopy. The microscopic images were taken with a JEOL JSM 6380LA scanning electron microscope. The samples were coated with a gold-palladium alloy for appropriate conductivity.

2.2.3. Falling dart impact tests. The falling dart impact tests were performed with a Ceast 9350 impact tester machine. The parameters for the tests are summarized in Table 2, while the layout of the test is shown in Figure 1. For the tests, samples with the size of 100 (mm) x 100 (mm) x 18 (mm) were cut out from the extruded foam sheets.

Table 2 Main parameters of the falling dart impact tests

Property	Unit	Value
Impactor geometry	(-)	cylinder, hemispherical end
Impactor diameter	(mm)	20
Applied mass	(kg)	5.41
Falling height	(mm)	942
Impact energy	(J)	50
Temperature	(°C)	23
Support type	(-)	hollow

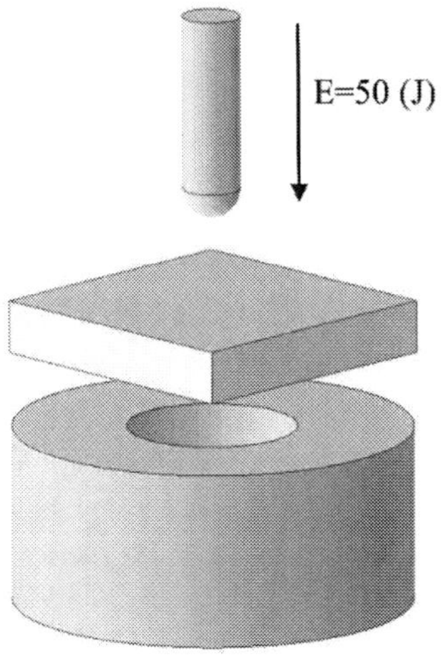

Figure 1 The layout of falling dart impact tests

The force-time diagrams were recorded during the tests, from which perforation energy (2) and ductility index (3) were calculated according to the following:

$$E_{perf} = \frac{E_{total}}{v} \tag{2}$$

where E_{perf} (J/mm) is the perforation energy, E_{total} (J) is the total absorbed energy during penetration, while v (mm) is the thickness of the sample.

$$DI = \frac{E_{total} - E_{Fmax}}{E_{total}} \tag{3}$$

where DI (-) is the ductility index, E_{total} (J) is the total absorbed energy during penetration, while E_{Fmax} (J) is the energy absorbed until maximum load.

2.2.4. Falling weight impact tests. The falling weight impact tests were performed on a custom-designed falling weight impact tester, which was built according to the wrestling mat regulation set up by United World Wrestling [20]. The applied parameters for the tests are summarized in Table 3, while the layout of the test is shown in Figure 2. For these tests, samples with the size of 200 x 200 x 18 (mm) were cut from the extruded foam sheets.

IOP Conf. Series: Materials Science and Engineering **903** (2020) 012059 doi:10.1088/1757-899X/903/1/012059

Table 3 Main parameters of the falling weight impact tests

Property	Unit	Value
Impactor geometry	(-)	cylinder, flat end
Impactor diameter	(mm)	100
Applied mass	(kg)	10.025
Falling height	(mm)	414 ± 0.4
Impact energy	(J)	40.7
Temperature	(°C)	23
Support type	(-)	solid

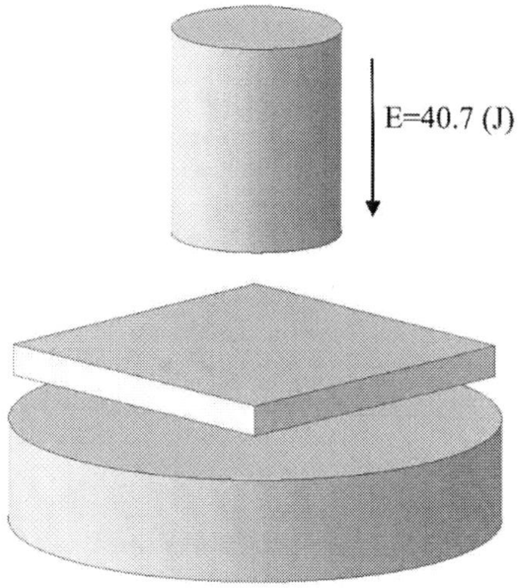

Figure 2 The layout of falling weight impact tests

The falling weight impact tests consisted of eight drops on each sample to show the effect of repetitive impacts on damping properties. The time between impacts was one minute. In the evaluation of the results, we focused on four different parameters:

- The maximum deceleration during the impact: a (g=9,81 m/s^2) – which refers to the gravitational acceleration.
- The amount of absorbed energy during the impact: E (%)
- The maximum deformation of the samples: p (mm)
- The duration of the collision: t (ms)

3. Results

This chapter contains the evaluation of the results of density, scanning electron microscope, falling dart and falling weight tests.

3.1. Density results

The results of the hydrostatic density tests and the measured thickness of the samples are shown in Table 4. The densities of the tested foams are almost the same as the density specified by the manufacturer. The XL-PE 45 and XL-PE 60 foams have the same thickness, so the different densities can be compared.

Table 4 Measured density and thickness

Property	Unit	XL-PE 45	XL-PE 60
Measured density	(kg/m^3)	43.82 ± 0.68	58.30 ± 0.27
Measured thickness	(mm)	17.87 ± 0.22	17.86 ± 0.27

3.2. Results of scanning electron microscopy

The scanning electron microscopic images (Figure 3) show that the tested foams have a closed-cell structure, that is, the foam cells are isolated from each other. The XL-PE 45 sample has an average cell diameter of 108 ± 42 (μm), whereas the XL-PE 60 foam has an average cell size of 87 ± 30 (μm), so the higher density foam has smaller cells. Another important difference is that the higher density foam has thicker cell walls, which effects a more rigid structure. It is likely that this type of difference has a significant effect on the shock absorption properties since the cells are more resistant to deformation.

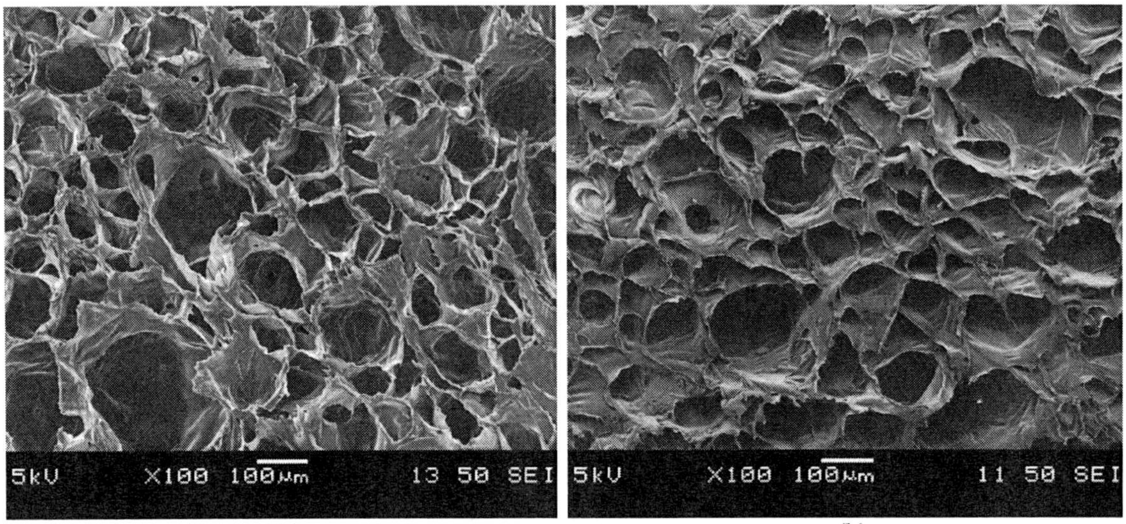

(a) (b)

Figure 3 SEM images of XL-PE 45 (a) and XL-PE-60 (b) samples

3.3. Falling dart impact tests

During the falling dart impact tests, the samples were completely punctured, hence we compared the dynamic mechanical properties of the two foams by calculating perforation energy and the ductility index. The recorded force-time diagrams (Figure 4) show that during the penetration of the dart, small vibrations and harmonics appeared, but they did not significantly influence the test, so the results are valid. The noise during the test was caused by the measurement layout, which did not consist of any clamping; if a clamping ring had been used, the cellular structure of the foams would have been deformed, and it would have affected the results.

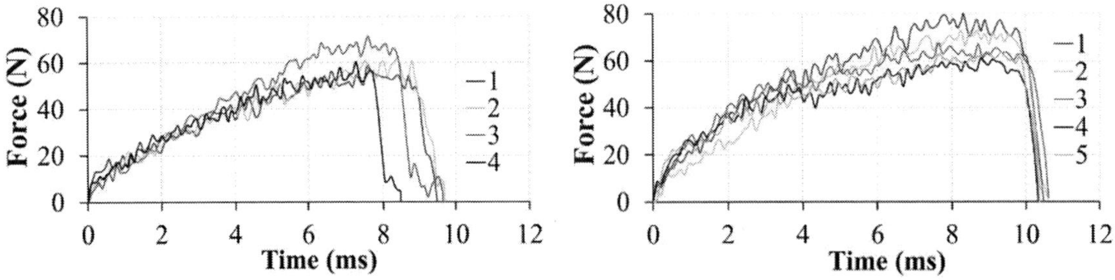

(a) (b)

Figure 4 Force-time curves of the falling dart tests in case of XL-PE 45 (a) and XL-PE 60 (b)

In the case of XL-PE 60 samples, total rupture time exceeded 10 (ms), while in the case of XL-PE 45 type foam it was 9 (ms). The reason for this difference is that the foam with thicker cell walls was more resistant, therefore the maximum force during the test was also higher. These properties also

IOP Conf. Series: Materials Science and Engineering **903** (2020) 012059 doi:10.1088/1757-899X/903/1/012059

influenced the ductility of the fractured material and the amount of energy absorbed. The stiffer cell structure resulted in higher perforation energy and a higher ductility index (Table 5).

Table 5 Results of the falling dart impact tests

Property	Unit	XL-PE 45	XL-PE 60
E_{perf}	(J/mm)	0.09 ± 0.01	0.12 ± 0.01
DI	(-)	0.13 ± 0.08	0.22 ± 0.09

The fact that the specimens were fully ruptured suggests that the impact damping properties of foams cannot be tested with this method. Using solid support can resolve this issue.

3.4. Falling weight impact tests

The results of the falling weight impact tests show that the higher density foam effected a lower maximum deceleration on the falling weight and was also able to absorb more energy (Table 6). This can be explained by the stiffer cell structure.

The XL-PE 45 sample showed approximately 1 (mm) larger deformation, which explains the higher maximum deceleration, since due to the higher compression, the cells became more compact and were likely to approach the behavior of the solid material.

Table 6 Results of the falling weight impact tests

Property	Unit	XL-PE 45	XL-PE 60
p	(mm)	15.88 ± 0.16	14.81 ± 0.47
a	(g)	130.03 ± 12.96	113.24 ± 26.87
E	(%)	45.09 ± 3.50	55.76 ± 7.41
t	(ms)	16.85 ± 0.72	16.76 ± 1.23

We also examined the effect of repetitive impacts. As the tests progressed, the maximum deceleration in both samples during impact increased continuously while the absorbed energy decreased continuously (Figure 5). This indicates that the impact damping property of the foams decreases in the case of repetitive impacts, as the material gradually hardens.

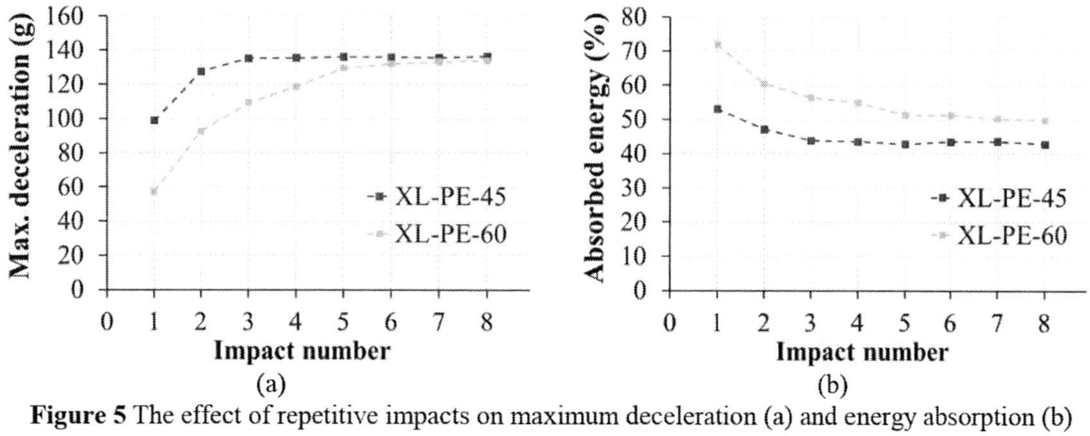

Figure 5 The effect of repetitive impacts on maximum deceleration (a) and energy absorption (b)

The reason for this phenomenon is that during the collision, the cells were deformed and compressed, as a result of which opposite cell walls probably met, and the one minute between the impacts is probably not enough for the cells to return to their original condition. This results in a reduction of the impact damping and energy-absorbing properties of the foams during repetitive impacts.

This method was more suitable for investigating the impact damping properties of foams. The solid support represents a model closer to the real-life use of sports mats, where they are laid down on the floor to reduce the force on the athletes.

4. Conclusions

We performed the dynamic mechanical investigation of two cross-linked closed-cell polyethylene foams with different densities, by falling dart and falling weight impact tests. Based on the results, the following conclusions can be drawn:

- The cross-linked polyethylene foam with a density of 60 (kg/m^3) has higher perforation energy, ductility index, and better impact damping and energy-absorbing properties compared to the foam with the lower, 45 (kg/m^3) density.
- The impact damping properties of polymer foams are significantly influenced by the deformation mechanisms in the cellular structure during dynamic loads, since excessive deformation of the cell walls leads to the compaction of the foam structure. This significantly reduces impact damping capability.
- In falling dart impact tests with a hollow support, the foams are fully ruptured, therefore this method is inappropriate for investigating the impact damping properties of foams. The use of a solid support similar to the support in the falling weight impact tester can resolve this issue.

As the use of cross-linked polyethylene foams in the sports industry is increasing and falling weight impact tests are also used in the packaging industry, our research results can be utilized in many industries to select the appropriate density and cell structure of the foam.

Acknowledgments

This paper was supported by the National Research, Development and Innovation Office, Hungary (K 132462). This research was supported by the ÚNKP-19-4 New National Excellence Program of the Ministry for Innovation and Technology; János Bolyai Research Scholarship of the Hungarian Academy of Sciences; and United World Wrestling.

References

[1] Liu S, Duvigneau J and Vancso G J 2015 *Eur. Polym. J.* **65** 33-45
[2] Verdejo R, Stämpfli R, Alvarez-Lainez M, Mourad S, Rodriguez-Perez M A, Brühwiler P A and Shaffer M 2009 *Compos. Sci. Technol.* **69** 1564-9
[3] Brückner K, Odenwald S, Schwanitz S, Heidenfelder J and Milani T 2010 *Procedia Eng.* **2** 2789-93
[4] Heidari A and Fasihi M 2019 *Express Polym. Lett.* **13** 429-42
[5] Zhang Z X, Dai X R, Zou L, Wen S B, Sinha T K and Li H 2019 *Express Polym. Lett.* **13** 948-58
[6] Mills N 2007 *Polymer Foams Handbook: Engineering and Biomechanics Applications and Design Guide* (Oxford, Elsevier) pp 235-50
[7] Dogossy G and Szabó V 2020 *Period. Polytech. Mech. Eng.* **64** 81-7
[8] Lyn G and Mills N 2001 *Sports Eng.* **4** 153-63
[9] Avalle M, Belingardi G and Montanini R 2001 *Int. J. Impact Eng.* **25** 455-72
[10] Dias D B and Silva L G A 2007 *Radiat. Phys. Chem.* **76** 1696-7
[11] Sipaut C S, Mansa R F, Yugis A R, Ibrahim M N M, Ariff Z M and Abdullah A A 2012 *Cell. Polym.* **31** 145-64
[12] Danaei M, Sheikh N and Taromi F A 2005 *J. Cell. Plast.* **41** 551-62
[13] EN 12503 2013 *Sports mats*
[14] EN 1621 2014 *Motorcyclists' Protective Clothing against mechanical impact*
[15] Avalle M and Scattina A 2014 *Lat. Am. J. Solids Stru.* **11** 200-22
[16] Velasco J I, Martínez A B, Arencón D, Rodríguez- Pérez M A and de Saja J A 1999 *J. Mater. Sci.* **34** 431-8
[17] Ruiz-Herrero J L, Rodríguez-Pérez M A and de Saja J A 2005 *Polym. Test.* **24** 641-7
[18] Liu J, Saletti D, Pattofatto S and Zhao H 2014 *Polym. Test.* **36** 101-9
[19] https://ufm.hu/hu/ 2020.01.11.

IOP Conf. Series: Materials Science and Engineering **903** (2020) 012059 doi:10.1088/1757-899X/903/1/012059

[20] United World Wrestling 2020 *Regulations For The Licensing Of Mats (https://unitedworldwrestling.org)*

IOP Conf. Series: Materials Science and Engineering **903** (2020) 012060 doi:10.1088/1757-899X/903/1/012060

Examination of correlation between the granulometric properties of molding and core sand mixtures and their production parameters

Henrietta Hudák[1], Dr. László Varga[2]

[1,2] University of Miskolc, Institute of Foundry, Miskolc-Egyetemváros 3515, Hungary

E-mail:[1] ontheni@uni-miskolc.hu;

Abstract. Foundry technology uses a lot of several natural materials. The chemically bonded sand mixtures main component is the foundry sand approx. 90-98%. Sand properties depend on it has chemical and mineralogical composition; mainly particle-size distribution and shape of grains and its size and sand surface texture. A comparative measurement of 3 foundry sand with different surface quality was carried out. Chemically bonded sand mixtures were prepared to measure their gas permeability and 3-point bending strength. A new qualifier number, CQ_i was used to compare our investigations.

1. Introduction

Numerous types of natural or nowadays widely-used synthetic (artificially produced) sand are applied for the preparation of molds and cores in foundry technology. The sand mixture is made from base sand and added bonding agents from which molds and core sand can be made. The molding and core sand mixtures are quite complex systems. Many factors affect the stability of molds and cores. The strength of cores depends on the sand's granulometric features, the bonding agents (quantity, quality, different curing conditions) and the bulk density of core sand. [1-4]

The aim of the research is to create a complex qualification for the important casting features of cores. The quality of molding and core sand mixtures is mainly determined by the parameters of foundry sand, that is, its granulometric parameters (sand grain's form and surface, size, structure and specific surface). Beyond the granulometric features of the refractory matrix, the parameters of core sand also depend on the type and quantity of the bonding agent and the bulk density. Since the current qualification methods of granulometric features are imprecise and not clearly quantified, the core sand parameters and the different setting parameters cannot be compared.

The relevance of the topic is also confirmed by the project being carried out by the Chair of Metal Forming and Casting (UTG) at the Technical University of Munich. From input parameters such as the manufacturing process and the materials used, they calculate effective physical properties like strength, gas permeability and thermal conductivity. They can accelerate development by modelling and simulating the process.

They state we need to consider the following properties since they influence the physical properties of the composites: grain shape of the sand, size of the grains, size distribution of the grains in the sand, volume content of the binder, chemical composition of the binder. [5-6]

IOP Conf. Series: Materials Science and Engineering **903** (2020) 012060 doi:10.1088/1757-899X/903/1/012060

2. Material and measurements

2.1. Investigated materials

Quartz sand of two and one of synthetic foundry sand was used for our tests (polish quartz sand - SREDNI (from Grudzeń-Las mine), GBM 45 (from Badger mine USA) and Bauxite sand W55). The investigated foundry sand types were all different in shape, but it was important their average grain size needed to be nearly the same. Figure 1-3. shows the grain morphology of different sands while a grain size distribution of them can be seen on Figure 4.

Figure 1. Stereo and scanning electron microscopy (SEM) image, EDS analisys of SREDNI quartz sand.

Figure 2. Stereo and scanning electron microscopy (SEM) image, EDS analisys of GBM 45 quartz sand.

Figure 3. Stereo and scanning electron microscopy (SEM) image, EDS analisys of Bauxite sand W55 sand.

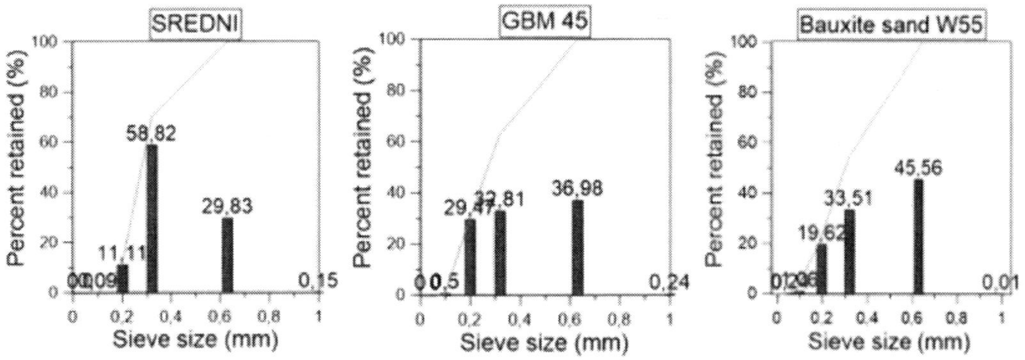

Figure 4. Grain size distribution of the used sands.

Results of granulometry analyses show that the properties of the various types of foundry sands are quite different.

Table 1. Measurement data of the investigated sands.

		SREDNI	GBM 45	Bauxite sand W55
Average granularity (d_{50})	**(mm)**	0.319	0.303	0.325
AFS granularity number	-	52.66	61.7	57.44
Homogeneity degree	**(%)**	64	44	55
Angularity coefficient of sand	-	1.38	1.14	1.08
Specific surface (A_{BLAINE})	**(cm^2/g)**	125	105	100
pH	-	5.73	6.8	6.3
Density	**(g/cm^3)**	2.6	2.65	3.1
Bulk density	**(g/cm^3)**	1.45	1.6	1.91
Grain shape	-	angular	sub-rounded	rounded

The results of the granulometric analysis show summarized in Table 1. The difference between the different sands is not significant considering that the average grain size. The Bauxite sand W55 has the lowest surface area due to its rounded grain shape and high sphericity. SRDENI quartz sand has the highest surface area since it is the most angular sand.

2.2. Measurements

No-bake binding sand mixtures were made from a different type of sands. Binder quantity and quality were the same. We used phenolic no-bake resin 1 % (instead of the mass of sand refer in respect of sand surface) (Furtolit 4003) and hardener 0.4% (in respect of resin mass) (Härter RS 20). For different types of sand equal binder layer thickness is achieved.

Specimens (Figure 5.) of standard size made from laboratory sand rammer (compressed with 3, 5, 7 and 9 rams) of sand mixture with the Multiserw-Morek laboratory mixer, within 2 minutes of mixing time (resin - 1 minute and catalyst 1 minute).

Figure 5. Specimens and their core box.

I carried out a 3-point bending flexural test and a air permeability test on samples made from the sand mixture. Air permeability test was carried out on standard cylinder samples (50x50 mm). The 3-point bending flexural ($\sigma_{bending}$) test was carried out on standard prism samples; like by the cylinder samples I used a rammer machine with rams of 3, 5, 7 and 9. The bending strength test was made after 30 minutes, 1 hour and 24 hours. The parameters of granulometric and bulk density were examined in a correlation system of a new qualification method.

3. Rating method

The comparison based on bulk density is incorrect, include with an error when comparing these sand types. The cause of the error is the diverging densities of the different materials which affect bulk density. In the case of sand cores with the same bulk density, different results can be obtained due to the different material densities. I would like to correct this fault, eliminate it with a new qualification parameter, which characterizes the total granulometric and bulk density conditions of sand cores at the same time.

One single index includes the core sand's volume, surface and bulk density not dependent on material density; therefore, it is eligible for the complex qualification of core parameters.

$$CQ_i = Air\ quantity\ \%/(SM)$$

where:

CQ_i = Core Quality Index
Air quantity % = Percentage of air in the sand core
SM = Sand Module (cm)

$$SM = V_{sand}/SA_{sand}\ (cm)$$

where:

V_{sand} = Volume of sand in sand core (cm^3)
SA_{sand} = Total surface area of sand in sand core (cm^2)
The following data are required to determine the value of SM (Sand Module):

- Volume of sand core = V_{core} (cm^3)
- Mass of sand core (sand) = m_{core} (g)
- Density of sand (material density) = ρ (g / cm^3)
- Specific surface area of sand = A_{BLAINE} (cm^2 / g)

For each specimen, the mass was measured to thereby determine the Air quantity %, the SM, and the CQ_i values.

4. Results

Figure 6. shows the air permeability test results and the Figure 7. shows the bending strength test results of sand mixtures. The air permeability values shown in the diagram are the result of one measurement each. The illustrated bending strength measurement results come from the average of the 3 measurements. The curves put on the chart were placed for better visibility and easier comprehension of the results. In case of bending strength values of curves were inscribed only for 3 rams and 9 rams values.

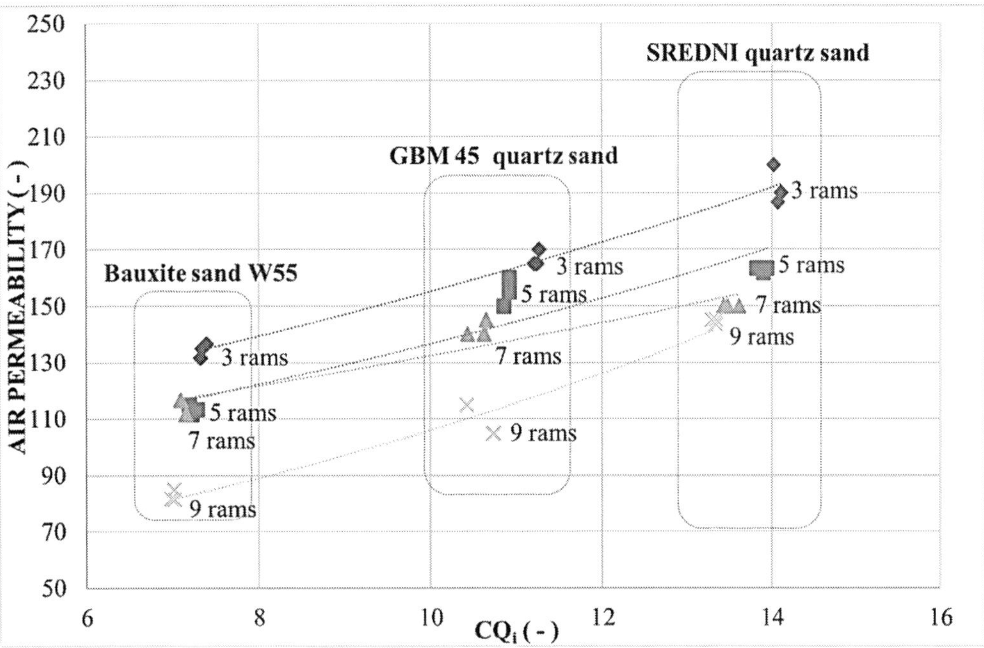

Figure 6. The air permeability test results.

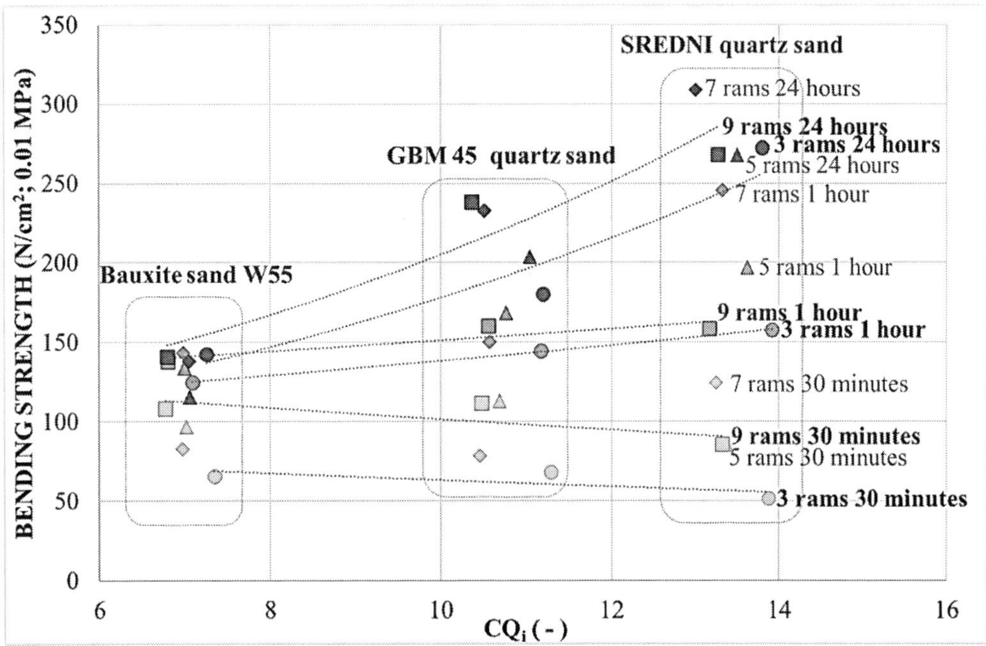

Figure 7. The bending strength test results.

In each used foundry sand case, the number of rams increase in raise gives a lower permeability. By the connection of CQ_i and air permeability, it is visible that higher CQ_i values have higher air permeability values. The sand core samples with the lowest CQ_i were made of bauxite sands (Bauxite sand W55), this sand type was the most rounded of the three, it has the smallest specific surface area. The highest CQ_i value is given to the sand core samples made of Polish quartz sand (SREDNI), and this sand type is the most angular compared to the other sands and it has the largest specific surface area.

The rounded grains have better compaction because it has less porosity between the grains and thus it has lower air permeability. Due to the lower porosity, the proportion of air in the specimen is lowered, so the CQ_i value is also low.

In addition, angular shape grains reduce the bulk density of the sand thereby increasing the air permeability of core sand.

In case of values of bending tests after 30 minutes and 1 hour there are no significant strength differences; the reason for this is that the bonding agent has not reached its final strength. In case of the 24-hours bending tests, the higher CQ_i values have higher bending strength values. While there was no significant change over time in the strength values of Bauxite sand, until then the 24-hours strength values increased nearly tripled in the case of Polish quartz sand compared to 30 minutes strength values. Angular grains (SREDNI) show higher bending strength than rounded grains (Bauxite sand). Because a rounded shape grain has a less contiguous surface like a subangular or angular grain.

5. Conclusion

The air rate (%) or air quantity (%) of samples made from different sand is similar; as a consequence, the sand volume within the samples is nearly the same. Furthermore, as the average grain size of the sand is similar, that is, the sand types are similar from granulometric (sieve analysis) viewpoint, the difference of bending strength values can be put down to variable specific surface areas.

As a result, the bonding agent's thickness and the characteristic of the bonding bridge cause diverse bending values. The strength of cores has two parts: the adhesive and cohesive force among the bonding agent and the sand grains. During my examinations, I assumed that due to similar granulometry the cohesive force of strength is similar in case of the application of diverse sand types. The growth derives from the increase of the adhesive force. This statement is underlined by the results of the bending test. It is visible from the results that there is no significant difference between the strength values after 30 minutes values because the bonding agent has not reached its full cohesive force; thus, the increase of adhesive force cannot be detected within the strength values. After 1 and 24 hours cohesive force develops within the strength values, so the effect of the adhesive force can be detected. The biggest difference among values can be seen during the bending test made after 24 hours.

References

[1] Dipl.-Ing. Bernhard Johannes Stauder 2018 *Dissertation*
[2] John R. Brown, 1999 *Foseco Non-Ferrous Foundryman's Handbook*
[3] Eckhart Flemming, Werner Tilch 1993 *Formstoffe und Formverfahren*
[4] Bakó Károly 1976 *Öntödei formázóanyagok*
[5] M. Schneider at all 2018 *International Journal of Solids and Structures*
[6] Florian Ettemeyer at all 2019 *International Journal of Solids and Structures*

AUTHOR INDEX

Baki, R. 291
Bakonyi, Eszter Szatmáriné 269
Balázs, B Z 399
Barkóczy, P 229
Barkóczy, Péter 277
Baumli, P 7
Benke, M 129
Berczeli, Miklós 177, 185
Bonyár, Attila 387
Boros, R 355
Bubenko, M. 38
Bubonyi, Tamás 269, 277
Budavari, Imre 30
Budavári, I 243
Buza, Gábor 193
Chaouadi, R 75
Chinh, Nguyen Quang 298
Choe, Heeman 340
Csenge, Tóth 221
Csincsi 291
Csordás, A Pintér 22
Czigany, T 334
Czinege, I 161
Delville, R 22
Dogossy, G 347
Dorina, Kovács 215
Dávid, Pammer 215
Fegyverneki, Gy. 38
Fegyverneki, Gy 14
Fehér, K 283
Fekete, I 407
Felde, I 137
Ferenczi, Tibor 116
Ficzere, P 53
Fiedler, T 362
Filep, Ádám 193
Forgács, György 269
Fried, Z 137
Fris, Dorottya Réka 153
Gacsi, Zoltan 81, 116
Gasparics, A 75
Gillemot, F 75, 373
Gonda, V 283
Gubicza, Jeno 340
Gubicza, Jenö 298
Gyarmati, G 14
Gácsi, Zoltán 277
Gál, V 393
Han, Gigap 340
Harangozó, D 161
Hargitai, H 110
Hegedus, Balazs 81, 116
Hegyes, T 229
Herbath, B. 207
Hlavács, A 129
Horvath, M 373
Horváth, M 22
Hudák, Henrietta 430
Hung, Pham Tran 340
Hózer, Z 22, 373
Ibriksz, T 235
Illés, L 22
Janovszky, D 7, 129
Juhász, Laura 269
Karpati, Viktor 30, 81, 116
Katona, B 362
Kazup, Agota 81, 116
Kemény, A 362, 380
Kemény, Dávid 368
Kim, D 22
Kincses, D 328
Kmetty, Á 122, 422
Knaup, J. 102
Kocsis, Bence 306
Koncsik, Zs 89
Kovacs, K 207
Kovács, Dorina 368
Kovács, Krisztián Norbert 221
Kovács, Péter Zoltán 46
Kovári, A 1, 312
Kristály, F 199
Krállics, Gy 129
Kulcsár, K 67, 110
Kárpáti, V 129
Kókai, E 97
Kónya, J 67, 110
Lendvai, L 407
Leveles, B 380
Litauszki, K 122
Lukács, J 414
Lukács, N L 53
Lukács, Zs 89, 393
Maloveczky, Anna 193
Mende, T 14
Mertinger, V 129, 199
Miklós, Kemény Dávid 215

Mikó, T 129
Mobark, Hfh 414
Molnar, D 38
Molnar, Daniel 30, 116
Molnar, J. 291
Monika, T. 38
Movahedi, N 362
Mádi, L 243
Mészáros, L 145
Nagy, E 199
Nagy, Melinda 269
Novotny, T 22
Orbulov, I N 328, 380
Palotás, A B 129
Perez-Feró, E 22
Petrény, R 145
Por, G. 291
Pámer, Á. 291
Pázmán, J 283
Péter, Barkóczy 269
Péter, Jenei 340
Renkó, József Bálint 387
Santillana, I Aviles 129
Schyns, M 22
Semperger, O V 355
Semperger, Orsolya Viktória 59
Somlyai-Sipos, L 7
Suplicz, András 59
Suplicz, A 355
Sycheva, A 7
Szabados, O. 291
Szabó, Ferenc 153, 169
Szabó, P. 291
Szabó, P 22
Szabó, Péter Jánoš 387
Szabó, V A 347
Szalai, Sz 161
Szenthe, I 75, 373
Szilvia, Szilvia 269
Szlancsik, A 328
Szénási, S 137
Tajti, Ferenc 177, 185
Takacs, Laszlo 169
Takács, M 399
Tancsics, F 235
Temesi, T 334
Terjék, D V 97
Terricabras, A Gallifa 129
Tisza, M 258
Tokár, M 14
Tomin, M 422
Trampus, Peter 320
Tóka, Bence 46
Tóth, Á. D. 102
Uytdenhouwen, I 75
Varga, Dr. László 430
Varga, Lajos Károly 306
Varga, L 243
Verö, B 283
Vida, Ádám 193
Vértesy, G 75
Wadas, T 258
Weltsch, Z 251
Wizner, K 1, 312
Zsoldos, Ibolya 306
Zsoldos, I 67